INVERSE SEMIGROUPS

INVERSE SEMIGROUPS

The Theory of Partial Symmetries

Mark V Lawson
University College of North Wales

World Scientific
Singapore • New Jersey • London • Hong Kong

Published by

World Scientific Publishing Co. Pte. Ltd.

P O Box 128, Farrer Road, Singapore 912805

USA office: Suite 1B, 1060 Main Street, River Edge, NJ 07661

UK office: 57 Shelton Street, Covent Garden, London WC2H 9HE

British Library Cataloguing-in-Publication Data
A catalogue record for this book is available from the British Library.

INVERSE SEMIGROUPS
The Theory of Partial Symmetries

ISBN 981-02-3316-7

Printed in Singapore by Uto-Print

This book is dedicated to the memory of my father
Alfred Mark Lawson (1930–1993)

'We are all things that make and pass, striving upon a hidden mission, out to the open sea.' H. G. Wells

Preface

Introduction

An appreciation of symmetry appears to be a feature of the human mind: even the earliest human artifacts were shaped in ways which transcended the purely utilitarian. The mathematical theory of symmetry grew out of investigations into the solutions of polynomial equations. Developing ideas of Lagrange and Abel, Galois discovered that the nature of these solutions could be determined by studying certain symmetries of the equation. It was from this work that the concept of a group as a mathematical device for measuring symmetry gradually evolved.

The theory of groups is one of the most successful branches of algebra with applications which range from error–correction in communication systems to the existence of elementary particles. But although group theory is certainly concerned with symmetry, it is by no means the case that the converse is true. This was highlighted in spectacular fashion in 1984, when Shechtman et al announced the discovery of a metallic phase whose diffraction pattern showed icosahedral symmetry, something which had long been assumed impossible on the basis of a result from group theory known as the 'crystallographic restriction'. The fault lay not in group theory, but in the assumptions underlying the translation of the intuitive idea of symmetry into a mathematical one. As in all translations, something was lost in the process.

In this book, we concentrate on just one aspect of our intuitive idea of symmetry which fails to be captured by groups: namely, the relationship between the parts and the whole. To see that group theory does not capture this relationship in a satisfactory way, one need only consider those natural forms, modelled by fractals, which are not highly symmetrical in the classical sense, but do possess self-similarity properties in which the whole is repeated at smaller scales. We instinctively feel that such forms possess symmetry but, due to their global irregularity, not of a kind which can be detected by groups. Self-similarities are examples of what we term partial symmetries, by which we mean symmetries between the parts of a structure. Partial symmetries are common in mathematics not only on their own terms, but also in situations where the principal interest may be centred on symmetries; this is because the first step in constructing a symmetry is frequently to construct a partial symmetry and then to extend it to a symmetry. Galois' theory of equations makes essential use of this idea.

The mathematical structures which encode information about partial symmetries are certain generalisations of groups called inverse semigroups. Like groups, inverse semigroups first arose in questions concerned with the solutions of equations, but this time in Lie's attempt to find the analogue of Galois theory for differential equations. The symmetries of such equations form what are now termed Lie pseudogroups, and inverse semigroups are, several times removed, the corresponding abstract structures. In addition to their early appearance in differential geometry, inverse semigroups have found a number of other applications in recent years including: C^*-algebras; tilings, quasicrystals and solid-state physics; combinatorial group theory; model theory; and linear logic.

One of my aims in writing this book was to show that inverse semigroups had the potential to act as a unifying concept for applications of partial symmetries, in much the same way that groups have acted as a unifying concept for applications of symmetries.

Structure of the book

The book is notionally divided into three parts:

(I) The first two chapters present the two main themes of the book: the relationship between inverse semigroups and partial symmetries; and the relationship between partial symmetries and global symmetries. For the *cognoscenti*, Chapter 1 revolves around the Wagner–Preston representation theorem, and Chapter 2 around the McAlister covering theorem.

(II) Chapters 3, 5, 6, 7 and 9 present the essentials of inverse semigroup theory, concentrating particularly on the theory of E-unitary and 0-E-unitary inverse semigroups. Chapter 9 is rich in applications of inverse semigroups, most notably to the theory of tilings and quasi-crystals.

(III) Chapters 4, 8 and 10 contain detailed applications of category theory to inverse semigroups. Chapters 4 and 8 incorporate Ehresmann's theory of ordered groupoids into mainstream inverse semigroup theory by applying it to the theory of E-unitary covers and idempotent pure extensions. Chapter 10 is, in some ways, a continuation of Chapter 1, in that it shows in what sense every inverse semigroup can be regarded as a semigroup of partial symmetries of some structure. This work has its origins in

Girard's use of inverse semigroup theory in his geometry of interaction programme for linear logic.

In order to read this book, I have assumed that the reader has followed a basic undergraduate mathematics course. A familiarity with category theory, sufficient to understand what is meant by an equivalence of categories, would be useful at a number of points in the text, but only essential for the theory developed in Chapters 4, 8 and 10.

Acknowledgements

This book would not have been completed without the goodwill and assistance of many friends and colleagues. Although I had been toying with the idea of writing a book for some time, the impetus to start work came from Douglas Munn. Douglas's own contributions to inverse semigroup theory, together with those of Don McAlister, form the backbone of this subject as a glance at the bibliography will show.

I have been lucky in obtaining support for my research from a number of institutions. I began my study of Ehresmann's *Oeuvres* when I was a Junior Research Fellow at Lincoln College, Oxford. My thanks to the Fellows of Lincoln, in particular Vivian Green and David Edwards, for electing me to the Fellowship, and to Peter Neumann and Hilary Priestley who supported my first painful efforts to formalise what I knew about semigroup theory. Subsequently, I spent a year in Darmstadt, Germany supported by the Royal Society in Prof. Karl Hofmann's research group where I was able to continue the work initiated in Oxford interspersed with *Kaffee und Kuchen.* The British Council supported a visit to Australia, and the Hungarian Academy of Sciences and the Royal Society supported a visit to Hungary. In Australia, I discussed with David Easdown, Tom Hall, Gordon Preston, Bob Sullivan, and Peter Trotter many aspects of inverse and general semigroup theory. In Hungary, early drafts of the first few chapters were tried out by the Budapest–Szeged Semigroup Seminar: namely, Pham Ngoc Ánh, Laszlo Márki, György Pollák and Mária Szendrei. Special thanks are due to Laci and Mária, whose comments caused me to rethink my approach, and whose corrections saved me from many embarrassing errors.

The contents of Chapter 9 owe a special debt to Johannes Kellendonk, who patiently explained his work to me in both Bangor and Berlin; I am grateful to Prof. Dr. U. Pinkall, Sprecher of Sonderforschungsbereich 288, at the Technische Universität, Berlin for the invitation to visit Berlin and for the financial support provided by the DFG. Paul Goldstein first alerted me to Johannes's work during a Gregynog Pure Mathematics Colloquium here at the University of Wales.

Des FitzGerald read a number of chapters and his comments helped to sharpen up the presentation considerably. Conversations with John Fountain, Vicky Gould, Peter Jones and Don McAlister helped clarify my thinking about the main arguments of this book. Thanks are also due to my research students: Peter Hines and Helen James; Peter first alerted me to the work of Girard and subsequently took the first steps in trying to understand self-similarity from an inverse semigroup-theoretic perspective, and Helen's eagle-eye spotted many typos. John Hickey helped out with numerous references and background information. Many others supported this project by providing offprints of their work or by making encouraging noises from the side-lines.

A number of colleagues at Yr Ysgol Fathemateg, Prifysgol Cymru, Bangor assisted with the production of this book: Julie Thistlethwaite and Siân Jones typed some of the chapters, and Garreth Roberts and Chris Wensley provided computer expertise; *diolch yn fawr iawn i chi.*

Lakshmi Narayan of World Scientific waited patiently for this book to see the light of day.

Finally, I would like to express my indebtedness to two outstanding books: Howie's 'Introduction to semigroups' is the canonical text for anyone interested in semigroup theory, whereas Petrich's 'Inverse semigroups' has been an invaluable reference throughout the preparation of this book.

The cover design was supplied by Michael Roesgen and Johannes Kellendonk and shows part of an aperiodic tiling derived by projection from $\mathbb{Z}^7$.

Contents

Preface **vii**

1 Introduction to inverse semigroups **1**
1.1 The origins of inverse semigroups 1
1.2 Pseudogroups and local structures 10
1.3 Symmetry 15
1.4 Abstract inverse semigroups 17
1.5 Representation theorems 36
1.6 Notes on Chapter 1 39

2 Extending partial symmetries **47**
2.1 Partial symmetries 48
2.2 E-unitary covers 56
2.3 Congruences 58
2.4 The minimum group congruence 62
2.5 Notes on Chapter 2 69

3 The natural partial order **75**
3.1 The associated groupoid 76
3.2 Green's relations 82
3.3 Primitive inverse semigroups 93
3.4 The bicyclic monoid 97
3.5 Notes on Chapter 3 103

4 Ordered groupoids **107**
4.1 Inductive groupoids 108
4.2 Ordered groupoids from $*$-semigroups 116
4.3 Ordered congruences 124
4.4 Notes on Chapter 4 130

5 Extensions of inverse semigroups **133**
5.1 The Kernel–trace description of a congruence . . . 134
5.2 Idempotent-separating extensions . . . 138
5.3 λ-semidirect products . . . 147
5.4 Inverse ω-semigroups . . . 160
5.5 Notes on Chapter 5 . . . 167

6 Free inverse semigroups **171**
6.1 The existence of free inverse semigroups . . . 171
6.2 Solving the word problem . . . 174
6.3 The structure of free inverse semigroups . . . 185
6.4 Munn normal form . . . 187
6.5 Notes on Chapter 6 . . . 194

7 E-unitary inverse semigroups **197**
7.1 Division theorems . . . 198
7.2 The P-theorem . . . 208
7.3 The Möbius inverse monoid . . . 216
7.4 F-inverse semigroups . . . 221
7.5 F-inverse covers . . . 226
7.6 Notes on Chapter 7 . . . 230

8 Enlargements **233**
8.1 A coordinate-free P-theorem . . . 233
8.2 E-unitary covers . . . 239
8.3 The maximum enlargement theorem . . . 248
8.4 Applications . . . 259
8.5 Notes on Chapter 8 . . . 270

9 0-E-unitary inverse semigroups **273**
9.1 The structure of a class of 0-E-unitary inverse semigroups . . . 274
9.2 Kellendonk's topological groupoid . . . 278
9.3 Polycyclic monoids . . . 285
9.4 McAlister semigroups . . . 304
9.5 Tiling semigroups . . . 318
9.6 Notes on Chapter 9 . . . 327

10 Category actions and inverse semigroups **331**
10.1 Inverse semigroups from category actions . . . 332
10.2 A coordinatisation theorem . . . 335
10.3 Category actions from inverse semigroups . . . 341
10.4 Functors between systems and semigroups . . . 345
10.5 Equivalent systems . . . 347

10.6 Composing the functors 353
10.7 Special cases 358
10.8 Notes on Chapter 10 369

Bibliography **373**

Index **405**

Chapter 1

Introduction to inverse semigroups

In this chapter, we trace the development of inverse semigroup theory from its origins in the theory of pseudogroups of transformations to the Wagner–Preston representation theorem. We also examine the rôle of inverse semigroups in mathematics, arguing that they capture aspects of our intuitive understanding of symmetry in a more complete way than groups.

1.1 The origins of inverse semigroups

Inverse semigroups were first explicitly defined by V. V. Wagner in a short note published in 1952 [419], and independently by Gordon Preston in 1954 [325], [326] and [327]. Earlier work by Clifford [45], [46] and Rees [335], [336] can be viewed as contributions to inverse semigroup theory, as can the contemporaneous paper of Clifford [47]. Much of Charles Ehresmann's work in the 1940s and 1950s was also concerned with inverse semigroup theory, although written from a different mathematical perspective.

Inverse semigroups were introduced as part of the legacy of two nineteenth-century mathematical enterprises: Klein's Erlanger Programm and Lie's theory of infinite continuous groups.

The Erlanger Programm

In the course of the nineteenth century, understanding of the nature of geometry underwent a profound seachange. Euclidean geometry, which had previously occupied a privileged position, came to be seen as just one amongst a great variety of different geometries. But with this diversity came the problem of how these geometries should be classified and how the relationships between them should be established. The central idea of Klein's Erlanger Programm [88] was that this could be achieved using the group of symmetries of the geometry.

The symmetry group of a geometry consists of all its structure-preserving bijections; thus from every geometry we can construct a group. However, Klein argued that this group could in turn be used to reconstruct the geometry. This is because the symmetry group induces an equivalence relation between geometric figures: two figures are equivalent if one can be mapped to the other by means of a symmetry. Geometric properties are properties of equivalence classes of figures rather than individuals, and so the geometry can be viewed as the invariant theory of the group.

This group-theoretic classification of geometries can also be used to establish relationships between geometries. Let $\mathcal{G}$ be a geometry with symmetry group G, and suppose that G' is a subgroup of G. Figures equivalent with respect to G' are equivalent with respect to G, but not conversely. Thus the subgroup G' can be used to construct a geometry $\mathcal{G}'$ which makes finer distinctions between figures. We say that the geometry $\mathcal{G}'$ is *richer* than the geometry $\mathcal{G}$. In this way, the classical non-Euclidean geometries and the circle, sphere and line geometries of Möbius, Lie and Plücker can all be constructed from projective geometry, as can the Euclidean, equiform and affine geometries. It is not surprising that Cayley was moved to write that 'projective geometry is all geometry' [21] (see also [436]); but, in fact, there were geometries which simply could not be accommodated within the group-theoretic framework. The following passage from the important 1932 book by Veblen and Whitehead [418] indicates one source of the difficulty:

> But long before the Erlanger Programm had been formulated there were geometries in existence which did not properly fall within its categories, namely the Riemannian geometries. [...] Riemannian spaces [...] are metric spaces [...] whose groups of automorphisms reduce to the identity. Such a geometry obviously cannot be characterised by a group.

J. H. C. Whitehead [433] in his obituary of Elie Cartan makes the point that, between the formulation of the Erlanger Programm and the formulation of General Relativity, the foundations of geometry were dominated by Klein's approach, but that

> After the discovery of general relativity, which was based on Riemannian geometry, it was realized that the Erlanger Program (sic) was no longer adequate as a general description of geometry.

Nevertheless, the fact that the Erlanger-Programm did not work for all geometries simply spurred mathematicians into trying to generalise it. This point was explicitly made by Veblen and Whitehead:

> There is, therefore, a strong tendency among contemporary geometers to seek a generalization of the Erlanger Programm which can replace it as a definition of geometry by means of the group concept.

In other words, if the group of symmetries is not sufficient to classify the geometry perhaps there is a more general structure which can. One particular way of generalising the Erlanger Programm was based on the concept of a pseudogroup introduced by Sophus Lie.

Pseudogroups

In the 1880s, Sophus Lie began to study what he termed continuous groups, partly as an attempt to develop a Galois theory for differential equations [299]. In modern terminology, the objects Lie studied consisted of families of partial diffeomorphisms of a manifold defined by a system of partial differential equations which were closed under composition and inverses[1]. Lie distinguished two classes of such groups: the finite and the infinite. The finite continuous groups are essentially representations by diffeomorphisms of an abstract, finite-dimensional Lie group. The infinite continuous groups are not groups at all but merely group-like. These structures have come to be called *Lie pseudogroups* (see [2], [3], [324], and [349]).

A Lie pseudogroup without its differential equations is just a collection of partial homeomorphisms between the open subsets of a topological space which is closed under composition and inverses. Veblen and Whitehead [418] recognised in such structures, which they called simply *pseudogroups*, suitable algebraic vehicles for generalising the Erlanger Programm to the developing theory of differential manifolds:

> The geometric objects and the spaces which they determine are classified by means of the pseudo-group of regular point transformations, two objects belonging to the same class if, and only if, they are equivalent. It is this pseudo-group, rather than the group [...], which is relevant, because a geometric object is not necessarily defined over the whole of a simple manifold [...]. This classification of geometric objects is in the spirit of the Erlanger Programm, equivalence under the pseudo-group being an inevitable generalization of equivalence in the narrower sense.

[1] A *diffeomorphism* is a homeomorphism which is infinitely differentiable and whose inverse is also infinitely differentiable.

Thus the Erlanger Programm is generalised by replacing groups of symmetries by pseudogroups. Such pseudogroups can be used to define many of the basic structures of differential geometry; we explain precisely how this is done in Section 1.2. Focussing on pseudogroups in this way raises a new question. Groups of transformations are representations of abstract groups, so it is natural to ask what the corresponding abstract structures are for pseudogroups. It is precisely this question which served as the impetus for defining inverse semigroups. Before describing how this question was answered, we formalise some of the terms we have been using.

Partial bijections and their properties

Let X and Y be any two sets. A *partial function* f from X to Y is a function from a subset of X to a subset of Y. The subset of X consisting of all those elements $x \in X$ for which $f(x)$ is defined is called the *domain (of definition) of* f, which we denote by $\operatorname{dom} f$. The *image* of f is the subset $\operatorname{im} f = f(\operatorname{dom} f)$ of Y.

Two special classes of partial functions are particularly important. For any two sets X and Y there is a unique *empty partial function from X to Y* which we denote by 0_{YX}. We use the term *empty function* to refer to any partial function of this type and allow the context to determine which one we mean. For every subset A of X the identity function on A, denoted 1_A, is a partial function from X to itself. Such partial functions are termed *partial identities.* The partial identity function on $\operatorname{dom} f$ is denoted by $\mathbf{d}(f)$ and the partial identity function on $\operatorname{im} f$ is denoted by $\mathbf{r}(f)$. The identity function 1_X on X and the identity function $1_\emptyset$ on the empty subset of X, which is just the empty function from X to itself, are especially significant. When the underlying set X is clear, we denote these functions by 1 and 0 respectively.

Let g be a partial function from X to Y and f a partial function from Y to Z. Then their *composite* is a partial function $f \circ g$ from X to Z, where the domain of $f \circ g$ is given by

$$\operatorname{dom}(f \circ g) = g^{-1}(\operatorname{dom} f \cap \operatorname{im} g)$$

and if $x \in \operatorname{dom}(f \circ g)$ then $(f \circ g)(x) = f(g(x))$. The image of $f \circ g$ is $f(\operatorname{dom} f \cap \operatorname{im} g)$. The case where $\operatorname{dom} f$ and $\operatorname{im} g$ have empty intersection causes no problems: $f \circ g$ is just the empty function. We usually write fg rather than $f \circ g$.

We shall concentrate on those partial functions which induce bijections between their domains and images; we call such partial functions *partial bijections.* All partial identities and empty functions are partial bijections. If f is a partial bijection from X to Y then we denote by f^{-1} the partial bijection from Y to X which is the *inverse* of f. Thus the domain of f^{-1} is $\operatorname{im} f$ and its image is $\operatorname{dom} f$. The composition of partial bijections is again a partial bijection.

Some important properties of partial bijections are described below.

Proposition 1 *Let X, Y and Z be sets, and let $f: X \to Y$ be a partial bijection.*

(1) $f^{-1}f = 1_{\operatorname{dom} f}$, *a partial identity on X, and* $ff^{-1} = 1_{\operatorname{im} f}$, *a partial identity on Y.*

(2) *For a partial bijection $g: Y \to X$, the equations $f = fgf$ and $g = gfg$ hold if, and only if, $g = f^{-1}$.*

(3) $(f^{-1})^{-1} = f$.

(4) $1_A 1_B = 1_{A \cap B} = 1_B 1_A$ *for all partial identities 1_A and 1_B where $A, B \subseteq X$.*

(5) $(gf)^{-1} = f^{-1}g^{-1}$ *for any partial bijection $g: Y \to Z$.*

Proof (1) Observe that $x \in \operatorname{dom}(f^{-1}f)$ precisely when $x \in \operatorname{dom} f$. Thus the domains of f and $f^{-1}f$ are the same. But $f^{-1}f$ is the identity function on its domain and so $f^{-1}f = 1_{\operatorname{dom} f}$, as required. The other result is proved similarly.

(2) Suppose that $f = fgf$ and $g = gfg$. Let $y \in \operatorname{dom} g$ and put $x = g(y)$. Then $x = (gfg)(y)$, so that $x = g(f(x))$. But g is a partial bijection, and so $y = f(x)$. Hence $x = f^{-1}(y)$, which gives $g \subseteq f^{-1}$. Now let $y \in \operatorname{dom}(f^{-1})$ and put $x = f^{-1}(y)$, so that $f(x) = y$. Then $(fgf)(x) = y$ so that $f(g(y)) = y$. But f is a partial bijection and so $g(y) = x$, which gives $f^{-1} \subseteq g$. Consequently, we have shown that $f^{-1} = g$. The converse is clear.

(3) This follows from (2).

(4) We show that $1_A 1_B = 1_{A \cap B}$. Let $x \in \operatorname{dom}(1_A 1_B)$. Then $(1_A 1_B)(x) = 1_A(1_B(x)) = x$. Thus $1_A 1_B$ is a partial identity. Since $\operatorname{dom}(1_A 1_B) = A \cap B$, it follows that $1_A 1_B = 1_{A \cap B}$.

(5) We have that

$$gf(f^{-1}g^{-1})gf = g(ff^{-1})(g^{-1}g)f = g(g^{-1}g)(ff^{-1})f = gf$$

since by (4) partial identities commute. We may similarly show that

$$(f^{-1}g^{-1})gf(f^{-1}g^{-1}) = f^{-1}g^{-1}.$$

Thus $(gf)^{-1} = f^{-1}g^{-1}$ by (2). ■

The collection of all partial bijections between sets forms a category, but we shall mainly be interested in the set of all partial bijections from a set X

to itself. This forms a monoid[2], denoted $I(X)$, called the *symmetric inverse monoid on X*.

Recall that an *idempotent* in a semigroup is any element equal to its square. An important property of symmetric inverse monoids is the following; its significance will become apparent in Theorem 3.

Proposition 2 *Let $I(X)$ be the symmetric inverse monoid on the set X. Then the idempotents of $I(X)$ are precisely the partial identities on X. In particular, the idempotents form a commutative, idempotent subsemigroup.*

Proof By Proposition 1(4), every element of the form 1_A is an idempotent of $I(X)$. Suppose now that f is an idempotent. Then $f = fff$, so that $f^{-1} = f$ by Proposition 1(2). Thus $f = ff = f^{-1}f = 1_{\mathrm{dom}\, f}$ by Proposition 1(1). ■

We can now return to the problem we posed earlier: what are the abstract structures which are represented by pseudogroups? This received two, distinct solutions.

Inverse semigroups

The first solution was provided independently by Wagner and Preston with their introduction of the class of *inverse semigroups*; these were defined to be semigroups S satisfying the following two conditions:

(1) S is *regular*. This means that for every element $a \in S$ there is an element b, called an *inverse* of a, satisfying $a = aba$ and $b = bab$.

(2) The idempotents of S commute.

From Propositions 1 and 2 symmetric inverse monoids really are inverse semigroups. Wagner and Preston then proved that every inverse semigroup could be embedded in some symmetric inverse monoid. This result is proved in Section 1.5.

Soon after inverse semigroups were introduced, the following theorem was proved.

Theorem 3 *Let S be a regular semigroup. Then the idempotents of S commute if, and only if, every element of S has a unique inverse.*

Proof Let S be a regular semigroup in which the idempotents commute and let u and v be inverses of x. Then

$$u = uxu = u(xvx)u = (ux)(vx)u,$$

[2]It may be worth pointing out at this point that a *semigroup* is just a set with an associative binary operation, whereas a *monoid* is a semigroup with an identity.

where both ux and vx are idempotents. Thus, since idempotents commute, we have that

$$u = (vx)(ux)u = vxu = (vxv)xu = v(xv)(xu).$$

Again, xv and xu are idempotents and so

$$u = v(xu)(xv) = v(xux)v = vxv = v.$$

Hence $u = v$.

To prove the converse, we begin by proving that in a regular semigroup the product of two idempotents e and f has an idempotent inverse. Let $x = (ef)'$ be any inverse of ef. Consider the element fxe. It is idempotent because

$$(fxe)^2 = f(xefx)e = fxe,$$

and it is an inverse of ef because

$$(fxe)ef(fxe) = (fxe)^2 = fxe \text{ and } ef(fxe)ef = (ef)x(ef) = ef.$$

Now let S be a semigroup in which every element has a unique inverse. We shall show that $ef = fe$ for any idempotents e and f. By the result above, $f(ef)'e$ is an idempotent inverse of ef. Thus $(ef)' = f(ef)'e$ by uniqueness of inverses, and so $(ef)'$ is an idempotent. Every idempotent is self-inverse, but on the other hand, the inverse of $(ef)'$ is ef. Thus $ef = (ef)'$ by uniqueness of inverses. Hence ef is an idempotent. We have shown that the set of idempotents is closed under multiplication. It follows that fe is also an idempotent. But $ef(fe)ef = (ef)(ef) = ef$, and $fe(ef)fe = fe$ since ef and fe are idempotents. Thus fe and ef are inverses of ef. Hence $ef = fe$. ∎

It follows that inverse semigroups are precisely the semigroups S in which for each element $s \in S$ there exists a unique element $s^{-1} \in S$ such that $s = ss^{-1}s$ and $s^{-1} = s^{-1}ss^{-1}$. The element s^{-1} is called the *inverse* of s in S. An *inverse subsemigroup* of an inverse semigroup is a subsemigroup closed under inverses. A *Pseudogroup* is an inverse semigroup of partial homeomorphisms between open subsets of a topological space[3].

Ordered groupoids

We have already indicated that inverse semigroups were only one solution to the problem of abstractly characterising pseudogroups. The second solution, due to Charles Ehresmann, took as its starting point a different definition of the composition of partial bijections.

[3] There is little consensus in the literature on the definition of pseudogroups; this is about the most generous.

Let g be a partial function from X to Y and f a partial function from Y to Z. The *restricted product* of f and g, denoted by $f \cdot g$, is defined only when $\operatorname{dom} f = \operatorname{im} g$, in which case $f \cdot g = fg$. In terms of our notation, the restricted product $f \cdot g$ is defined precisely when $\mathbf{d}(f) = \mathbf{r}(g)$. We shall call the product we defined earlier the *full product* when we wish to distinguish it from the restricted product.

The two ways of composing partial bijections stem from two different ways of viewing partial bijections. When they are considered with respect to the restricted product, we are actually regarding them as *total* functions on their domains of definition, whereas the full product is the product of partial bijections regarded as *partial* functions.

Partial bijections under the restricted product do not form a semigroup, instead they form a *groupoid* in the sense of category theory; that is, a category in which every morphism is invertible. The theories of groupoids and inverse semigroups are not equivalent. However, in addition to the restricted product of partial functions, a partial order can be defined on the collection of partial functions. If f and g are two partial functions from X to Y such that

$$\operatorname{dom} f \subseteq \operatorname{dom} g \text{ and } f(x) = g(x)$$

for all $x \in \operatorname{dom} f$ then we write $f \subseteq g$ and say that f is a *restriction* of g. Some important properties of this partial ordering in terms of the full product are described below.

Proposition 4 *Let $f, g \colon X \to Y$ be partial bijections between the sets X and Y.*

(1) *If $f \subseteq g$ then $f^{-1} \subseteq g^{-1}$.*

(2) *Let $h, k \colon Y \to Z$ be partial bijections between the sets Y and Z. If $f \subseteq g$ and $h \subseteq k$ then $hf \subseteq kg$.*

(3) *$f \subseteq g$ precisely when there is a partial identity $1_A \in I(X)$ such that $f = g1_A$.*

Proof The proofs of (1) and (2) are straightforward.
(3) Suppose that $f \subseteq g$. Let $1_A = f^{-1}f$ where $A = \operatorname{dom} f$ by Proposition 1(1). By (2) above, $f \subseteq g1_A$. In particular, $\operatorname{dom} f \subseteq \operatorname{dom}(g1_A)$. Let $x \in \operatorname{dom}(g1_A)$. Then $1_A(x)$ is defined so that $x \in \operatorname{dom} f$. Thus $\operatorname{dom} f = \operatorname{dom}(g1_A)$. Hence $f = g1_A$.

Conversely, suppose that $f = g1_A$ for some partial identity 1_A. Let $x \in \operatorname{dom}(f)$. Then $f(x)$ is defined and so $(g1_A)(x)$ is defined. It follows that $x \in \operatorname{dom} g$. Thus $\operatorname{dom} f \subseteq \operatorname{dom} g$. But observe that with the above x, we have that $(g1_A)(x) = g(x)$. Thus $f \subseteq g$. ■

Whereas Wagner and Preston found axioms satisfied by the semigroup $(I(X), \circ)$, Ehresmann found axioms satisfied by the structure $(I(X), \cdot, \subseteq)$ within the theory of ordered groupoids. These two approaches are in fact equivalent; this is proved properly in Chapter 4, but we can easily provide some insight into the nature of this equivalence here.

If f is a partial function from X to Y and X' is a subset of $\operatorname{dom} f$ then we may define a new partial function $(f \mid X')$, the *restriction of f to X'*, to be the partial function from X to Y with $\operatorname{dom}(f \mid X') = X'$ such that $(f \mid X')(x) = f(x)$ for all $x \in X'$. Likewise, if Y' is a subset of $\operatorname{im} f$ then we define $(Y' \mid f)$, the *corestriction of f to Y'*, to be the partial function from X to Y with $\operatorname{dom}(Y' \mid f) = f^{-1}(Y')$ such that $(Y' \mid f)(x) = f(x)$ for all $x \in \operatorname{dom}(Y' \mid f)$.

The essence of the relationship between inverse semigroups and ordered groupoids is contained in the following result.

Proposition 5 *Let $f, g \in I(X)$, and put $A = \operatorname{dom} f \cap \operatorname{im} g$. Then*

$$fg = (f \mid A) \cdot (A \mid g).$$

Proof From the definition of the restricted product

$$\operatorname{dom}((f \mid A) \cdot (A \mid g)) = \operatorname{dom}((A \mid g)),$$

and from the definition of the corestriction

$$\operatorname{dom}((A \mid g)) = g^{-1}(A) = g^{-1}(\operatorname{dom} f \cap \operatorname{im} g).$$

Thus

$$\operatorname{dom}((f \mid A) \cdot (A \mid g)) = \operatorname{dom}(fg).$$

The result is now clear. ■

Concluding remarks

In this section, we have tried to explain how inverse semigroups arose from an attempt to generalise the Erlanger Programm to differential geometry. However, it is important to stress that pseudogroups are just one of the ways in which the group concept is generalised in differential geometry. Kirill Mackenzie in the introduction to his book [209] provides a broader perspective for the ideas of this section:

> The modern, rigorous concept of a group is far too restrictive for the range of geometrical applications envisaged in the work of Lie.

> There have thus arisen the concepts of Lie pseudogroup, of differentiable and of Lie groupoid, and of principal bundle—as well as various related infinitesimal concepts such as Lie equation, graded Lie algebra and Lie algebroid—by which mathematics seeks to acquire a precise and rigorous language in which to study the symmetry phenomena associated with geometrical transformations which are only locally defined.

1.2 Pseudogroups and local structures

In this section, we describe how pseudogroups can be used to construct geometric structures in differential geometry. In the 1950s, Ehresmann analysed this construction in terms of category theory. This work and its applications to inverse semigroup theory are described in Section 8.3.

Charts and atlases

At its simplest, differential geometry concerns spaces which look locally like pieces of $\mathbb{R}^n$ and pseudogroups provide the glue to hold these pieces together. We now introduce the mathematics which will enable us to make this idea precise.

Let X and Y be topological spaces. In what follows, the space X will be the *model space* and our aim will be to construct geometric structures on Y which are locally like pieces of X. A *chart* from X to Y is a homeomorphism ϕ: $U \to V$ between open subsets of X and Y respectively. Thus a chart is nothing other than a special kind of partial bijection. An *atlas* from X to Y is a collection of charts from X to Y such that the union of the images of the charts is Y; thus $1_Y = \bigcup_{\phi \in \mathcal{A}} \phi\phi^{-1}$. If this latter condition does not hold we shall say that $\mathcal{A}$ is a *partial atlas*.

If $\mathcal{A}$ is a partial atlas from X to Y, and $\mathcal{B}$ is a partial atlas from Y to Z, then $\mathcal{BA}$ is the partial atlas from X to Z defined by

$$\mathcal{BA} = \{\psi\phi\colon \psi \in \mathcal{B} \text{ and } \phi \in \mathcal{A}\}.$$

If $\mathcal{A}$ is a partial atlas from X to Y then

$$\mathcal{A}^{-1} = \{\phi^{-1}\colon \phi \in \mathcal{A}\}$$

is a partial atlas from Y to X.

Intuitively, the existence of an atlas from X to Y means that Y can be described by a family of overlapping sets each of which looks like a piece of X. An atlas, in the geographical sense, provides a good example of an atlas in our sense from $\mathbb{R}^2$ to a sphere.

The problem now arises of dealing with the overlaps between different charts, and it is here that pseudogroups get into the picture.

Let $\phi_i: U_i \to V_i$ and $\phi_j: U_j \to V_j$ be any charts in the (partial) atlas $\mathcal{A}$ from X to Y. Then we can form the partial homeomorphism

$$\phi_j^{-1}\phi_i: \phi_i^{-1}(V_i \cap V_j) \to \phi_j^{-1}(V_i \cap V_j)$$

between the open subsets $\phi_i^{-1}(V_i \cap V_j)$ and $\phi_j^{-1}(V_i \cap V_j)$ of X. Partial homeomorphisms such as these are called *transition functions* of the atlas. Thus the transition functions of a (partial) atlas $\mathcal{A}$ from X to Y belong to the pseudogroup $\Gamma(X)$ of all partial homeomorphisms between the open subsets of X. This property can be succinctly expressed by the equation $\mathcal{A}^{-1}\mathcal{A} \subseteq \Gamma(X)$.

In order to construct geometric structures, we shall need to know under what circumstances partial bijections can be glued together.

Proposition 1 *Let $f, g: X \to Y$ be partial bijections between the sets X and Y.*

(1) *$f \cup g$ is a partial function precisely when fg^{-1} is an idempotent.*

(2) *$f \cup g$ is a partial bijection precisely when $f^{-1}g$ and fg^{-1} are idempotents.*

(3) *Let $\{\phi_i: i \in I\}$ be a family of partial bijections from X to Y. Then $\bigcup \phi_i$ is a partial bijection if, and only if, $\phi_i \cup \phi_j$ is a partial bijection for all $i, j \in I$.*

(4) *Let $\{\phi_i: i \in I\}$ be a family of partial bijections from X to Y such that $\bigcup \phi_i$ is a partial bijection. Then $\bigcup \phi_i^{-1}$ is a partial bijection from Y to X and*

$$(\bigcup \phi_i)^{-1} = \bigcup \phi_i^{-1}.$$

(5) *Let $\{\phi_i: i \in I\}$ be a family of partial bijections from X to Y such that $\bigcup \phi_i$ is a partial bijection. Then for any partial bijection $\phi: W \to X$ the union $\bigcup \phi_i\phi$ is a partial bijection from W to Y, such that*

$$(\bigcup \phi_i)\phi = \bigcup \phi_i\phi.$$

Similarly, for any partial bijection $\psi: Y \to Z$ the union $\bigcup \psi\phi_i$ is a partial bijection from X to Z such that

$$\psi(\bigcup \phi_i) = \bigcup \psi\phi_i.$$

Proof (1) Suppose that $f \cup g$ is a partial function. Then f and g agree on $\operatorname{dom} f \cap \operatorname{dom} g$. By Proposition 1.1.1, the domain of the idempotent $e = f^{-1}fg^{-1}g$ is just $\operatorname{dom} f \cap \operatorname{dom} g$. It is easy to check that $fe = ge$. But $fe = fg^{-1}g$ and $ge = gf^{-1}f$. Thus $fg^{-1}g = gf^{-1}f$. Composing on the right by g^{-1} we obtain $fg^{-1} = g(f^{-1}f)g^{-1}$, which is an idempotent.

Conversely, suppose that fg^{-1} is an idempotent. Then $fg^{-1} = (fg^{-1})^{-1} = gf^{-1}$, and so $(fg^{-1})(gf^{-1}) = gf^{-1}$. Hence $fg^{-1}gf^{-1} = gf^{-1}$. Multiplying by f on the right, we obtain $f(g^{-1}g) = g(f^{-1}f)$. It is now easy to check that f and g agree on $\operatorname{dom} f \cap \operatorname{dom} g$, and so $f \cup g$ is a partial bijection.
(2) This follows from (1), together with the observation that $f \cup g$ is a partial bijection precisely when both $f \cup g$ and $f^{-1} \cup g^{-1}$ are functions.
(3) Suppose that $\phi = \bigcup \phi_i$ is a partial bijection. For any ϕ_i and ϕ_j, we have that $\phi_i^{-1}\phi_j \subseteq \phi^{-1}\phi$ and $\phi_i\phi_j^{-1} \subseteq \phi\phi^{-1}$. Thus by (2) the union $\phi_i \cup \phi_j$ is a partial bijection.

Conversely, suppose that $\phi_i \cup \phi_j$ is a partial bijection for all $i, j \in I$. Let $\phi = \bigcup \phi_i$. Firstly, ϕ is a partial function, for suppose that $\phi(x) = y$ and $\phi(x) = z$. Then $\phi_i(x) = y$ and $\phi_j(x) = z$ for some $i, j \in I$. But $\phi_i \cup \phi_j$ is a partial function and so $y = z$. Now suppose that $\phi(x) = \phi(y)$ for some $x, y \in \operatorname{dom} \phi$. Then $\phi_i(x) = \phi_j(y)$ for some $i, j \in I$. But $\phi_i \cup \phi_j$ is a partial bijection. Thus $x = y$. Hence ϕ is a partial bijection from X to Y.
(4) Straightforward.
(5) Suppose that $\bigcup \phi_i$ is a partial bijection. Then by (3), $\phi_i \cup \phi_j$ is a partial bijection for any $i, j \in I$. Consider now $\phi_i\phi$ and $\phi_j\phi$. Then

$$(\phi_i\phi)^{-1}(\phi_j\phi) = \phi^{-1}(\phi_i^{-1}\phi_j)\phi.$$

But by (2), the product $\phi_i^{-1}\phi_j$ is an idempotent and so $(\phi_i\phi)^{-1}(\phi_j\phi)$ is an idempotent. Similarly, $(\phi_i\phi)(\phi_j\phi)^{-1}$ is an idempotent. Thus $\phi_i\phi \cup \phi_j\phi$ is a partial bijection for all $i, j \in I$. Hence $\bigcup \phi_i\phi$ is a partial bijection by (3). The fact that $(\bigcup \phi_i)\phi = \bigcup \phi_i\phi$ follows easily from the observation that the domains of both functions are equal. The other assertions are proved similarly. ∎

On the basis of (2) above, partial bijections f and g are said to be *compatible*, written $f \sim g$, if $f \cup g$ is a partial bijection. An inverse subsemigroup S of $I(X)$ is said to be *complete* if the union of every non-empty compatible subset of S belongs to S. Clearly, $I(X)$ is complete.

Let X be a topological space and $\Gamma = \Gamma(X)$ the pseudogroup of all partial homeomorphisms between open subsets of X. The union of every compatible subset of $\Gamma(X)$ is not only a partial bijection but also a partial homeomorphism between open subsets of of X. A *complete pseudogroup* on X is an inverse subsemigroup Γ' of Γ with the property that the union of every non-empty compatible subset of Γ' belongs to Γ'.

Constructing local structures

We have now assembled enough definitions to explain how to define geometric structures by means of pseudogroups.

Let X and Y be topological spaces, where X is the model space, and let Γ' be a complete pseudogroup on X. An atlas $\mathcal{A}$ from X to Y is said to be *compatible* with Γ' if $\mathcal{A}^{-1}\mathcal{A} \subseteq \Gamma'$. We are therefore requiring that all the transition functions belong to a specified inverse subsemigroup of $\Gamma(X)$.

Let $\mathcal{A}$ and $\mathcal{B}$ be atlases from X to Y compatible with Γ'. If $\mathcal{A}^{-1}\mathcal{B} \subseteq \Gamma'$, then we say that $\mathcal{A}$ and $\mathcal{B}$ are *compatible modulo* Γ', denoted $\mathcal{A} \sim \mathcal{B} \pmod{\Gamma'}$.

Proposition 2 *Let $\Gamma'(X, Y)$ be the set of all atlases from X to Y compatible with the complete pseudogroup Γ'.*

(1) *The relation $\mathcal{A} \sim \mathcal{B} \pmod{\Gamma'}$ is an equivalence relation on $\Gamma'(X, Y)$.*

(2) *Each equivalence class contains a maximum element.*

Proof (1) Reflexivity of the relation follows from the assumption that each atlas in $\Gamma'(X, Y)$ is compatible with Γ'. The relation is symmetric, for if $\mathcal{A}^{-1}\mathcal{B} \subseteq \Gamma'$ then $\mathcal{B}^{-1}\mathcal{A} \subseteq \Gamma'$, since Γ' is closed under inverses. We now prove transitivity. Let $\mathcal{A} \sim \mathcal{B}$ and $\mathcal{B} \sim \mathcal{C} \pmod{\Gamma'}$, where

$$\mathcal{A} = \{\theta_i \colon i \in I\}, \; \mathcal{B} = \{\phi_j \colon j \in J\} \text{ and } \mathcal{C} = \{\psi_k \colon k \in K\}.$$

Let $\theta_i \in \mathcal{A}$ and $\psi_k \in \mathcal{C}$. We shall show that $\theta_i^{-1}\psi_k \in \Gamma'$ which will prove that $\mathcal{A}^{-1}\mathcal{C} \subseteq \Gamma'$. Since $\mathcal{B}$ is an atlas from X to Y, we have that $\bigcup \phi_j\phi_j^{-1} = 1_Y$. Thus

$$\theta_i^{-1}\psi_k = \theta_i^{-1}1_Y\psi_k = \theta_i^{-1}(\bigcup \phi_j\phi_j^{-1})\psi_k$$

and so

$$\theta_i^{-1}\psi_k = \bigcup(\theta_i^{-1}\phi_j)(\phi_j^{-1}\psi_k),$$

by Proposition 1. Now $\theta_i^{-1}\phi_j \in \mathcal{A}^{-1}\mathcal{B}$ and $\phi_j^{-1}\psi_k \in \mathcal{B}^{-1}\mathcal{C}$ so that $\theta_i^{-1}\phi_j$ and $\phi_j^{-1}\psi_k$ are both elements of Γ'. Thus $\theta_i^{-1}\phi_j\phi_j^{-1}\psi_k \in \Gamma'$, since Γ' is closed under composition. By assumption, the union of all the elements $\theta_i^{-1}\phi_j\phi_j^{-1}\psi_k$ is an element of Γ. Thus $\theta_i^{-1}\psi_k \in \Gamma'$ since Γ' is a complete pseudogroup. Hence $\mathcal{A}^{-1}\mathcal{C} \subseteq \Gamma'$.

(2) Let $\{\mathcal{A}_i \colon i \in I\}$ be an equivalence class mod Γ'. Put $\mathcal{A} = \bigcup \mathcal{A}_i$. Clearly $\mathcal{A}$ is an atlas from X to Y. Observe that

$$\mathcal{A}^{-1}\mathcal{A} = (\bigcup \mathcal{A}_i)^{-1}(\bigcup \mathcal{A}_j) = \bigcup \mathcal{A}_i^{-1}\mathcal{A}_j.$$

But $\mathcal{A}_i^{-1}\mathcal{A}_j \subseteq \Gamma'$ for all $i, j \in I$ since $\{\mathcal{A}_i \colon i \in I\}$ is an equivalence class mod Γ'. Thus $\mathcal{A}^{-1}\mathcal{A} \subseteq \Gamma'$. Similarly, $\mathcal{A} \sim \mathcal{A}_i \pmod{\Gamma'}$ for every $\mathcal{A}_i$. Thus $\mathcal{A}$ is the maximum element of the equivalence class. ■

An atlas $\mathcal{A}$ from X to Y which is the maximum element in its equivalence class mod Γ' is called a *complete atlas compatible with* Γ', and is said to define a Γ'*-structure* on Y. Such structures are often called *local structures*. Any atlas $\mathcal{A}$ from X to Y determines a Γ'-structure on Y, namely the one determined by the maximum element of the equivalence class containing $\mathcal{A}$. Local structures are the analogues of the geometries in the Erlanger Programm and the pseudogroup Γ' replaces the group.

Two examples of local structures

There are many important examples of local structures in differential geometry, such as foliations and fibre bundles, which can be constructed by means of atlases. We shall content ourselves with describing just two.

The first example is just the familiar definition of a differential manifold. Let Y be a hausdorff space with a countable basis of open sets, let $X = \mathbb{R}^n$ and let Γ' be the inverse semigroup of all diffeomorphisms between open subsets of $\mathbb{R}^n$. This is a complete pseudogroup. Then an atlas from $\mathbb{R}^n$ to Y compatible with Γ' defines a *smooth differential manifold* on Y.

Our second example is very different and illustrates how the nature of the pseudogroup is crucial in determining the nature of the resulting structure. Let X and Y be topological spaces. A *local homeomorphism* θ: $X \to Y$ is a continuous function such that for each $x \in X$ there is an open subset U_x of X containing x such that $V_x = \theta(U_x)$ is open in Y and the restriction of θ to U_x is a homeomorphism onto V_x.

Given such a local homeomorphism, an atlas can be constructed from Y to X as follows. For each $x \in X$ put

$$\phi_x = (\theta \,|\, U_x)\colon U_x \to V_x$$

and let $\mathcal{A} = \{\phi_x^{-1}\colon x \in X\}$. By assumption, each element of $\mathcal{A}$ is a chart from Y to X and by construction the union of the images of the elements of $\mathcal{A}$ is X. Thus $\mathcal{A}$ is an atlas from Y to X. Let $\phi_x^{-1}, \phi_y^{-1} \in \mathcal{A}$. Since $\phi_x, \phi_y \subseteq \theta$ the product $(\phi_x^{-1})^{-1}\phi_y^{-1} = \phi_x\phi_y^{-1}$ is an identity function on an open subset of Y. Let Γ' be the inverse semigroup of all identities on open subsets of Y. This is a complete pseudogroup. Then $\mathcal{A}$ is an atlas from Y to X compatible with Γ'. Thus starting with a local homeomorphism from X to Y, we have constructed an atlas from Y to X compatible with Γ'.

We now show that the converse is true. Let $\mathcal{A} = \{\phi_i^{-1}\colon V_i \to U_i\}$ be an atlas from Y to X compatible with Γ'. Put $\theta = \bigcup \mathcal{A}^{-1}$. To see that θ is a well-defined function let $\phi_i^{-1}, \phi_j^{-1} \in \mathcal{A}$ where $\phi_i^{-1}\colon V_i \to U_i$ and $\phi_j^{-1}\colon V_j \to U_j$. By assumption $(\phi_i^{-1})^{-1}\phi_j^{-1} = \phi_i\phi_j^{-1}$ is an identity function on an open subset of Y. Consequently by Proposition 1, $\phi_i \cup \phi_j$ is a well-defined function from $U_i \cup U_j$ to $V_i \cup V_j$. Thus θ is a well-defined function. It is continuous because it is a union of continuous functions. Finally, θ is a local homeomorphism for if $x \in X$ then $x \in U_i$ for some i. But then $(\theta \,|\, U_i) = \phi_i$ which is a homeomorphism.

1.3 Symmetry

The Erlanger Programm was based on the group of symmetries of a geometry, but the more general work on local structures was based on pseudogroups. This suggests that groups may not be the most general way of representing symmetry algebraically. For this reason, we now examine the nature of the relationship between symmetry and groups.

Not just groups?

Symmetry is one of the most important organising principles in the natural sciences, and the way that symmetries are treated mathematically is by means of groups. Artin, writing in [6], says:

> In modern mathematics the investigation of the symmetries of a given mathematical structure has always yielded the most powerful results. Symmetries are maps which preserve certain properties.

Rose [361], citing Artin himself, adds:

> And this is why groups play a fundamental role in mathematics today. The symmetries of a structure *are* the permutations of the underlying set X which are structure-preserving. (My emphasis)

Thus the structure-preserving bijections of a structure are precisely the symmetries of that structure, and the totality of such symmetries forms a group.

But is it true that *all* symmetry phenomena can be accommodated by group theory? In fact, a number of authors have questioned the relationship between group theory and symmetry.

Mackay has advocated what he terms *generalised* or *post-classical crystallography* [207], [208]. Although his ideas go beyond the consideration of symmetry alone, being influenced by computer science, he does consider notions of local symmetry, hierarchic structures and self-similarity which, whilst relating to symmetry, are hard to accommodate within the group-theoretic framework. The advent of quasicrystals, in particular, has called into question a number of basic tenets of crystallography. As Marjorie Senechal [392] writes:

> It suggested that we may not know what 'long-range order' means, nor what a 'crystal' is, nor how 'symmetry' should be defined.

Stewart and Golubitsky in their book on symmetry [399] are prompted by the existence of quasi-crystals to explicitly challenge the 'group-theoretic paradigm of symmetry' (pages 266–268). This paradigm is further challenged by Alan Weinstein [431]:

> Mathematicians tend to think of symmetry as being virtually synonomous with the theory of groups ... In fact, though groups are indeed sufficient to characterize homogeneous structures, there are plenty of objects which exhibit what we clearly recognize as symmetry, but which admit few or no non-trivial automorphisms.

It is time, therefore, to reconsider the basis of our assumption that groups capture what we mean by symmetry. To do this, it would be as well to look again at what we mean by the word 'symmetry'.

What is symmetry?

In his book 'Symmetry' [432], Weyl begins by isolating two meanings of the word symmetry in everyday language. In the first, symmetry:

> ... denotes that sort of concordance of several parts by which they integrate into a whole. (page 3)

The second meaning is that of bilateral symmetry,

> ... [this] is a strictly geometric and, in contrast to the vague notion of symmetry defined before, an absolutely precise concept. (page 4)

It is from the second definition that the mathematically precise notion of a group is derived.

The process of transforming an intuitive idea into a mathematically precise one is familiar from other areas of mathematics. For example, the intuitive notion of a continuous function is one whose graph can be drawn without breaks. The mathematically precise definition requires the full panoply of ϵs and δs. However, the same procedure is not usually adopted for the notion of symmetry, despite there being no *a priori* grounds for treating symmetry differently from continuity. This is almost certainly due to the great success group theory has enjoyed. Yet it is clear that the intuitive meaning of symmetry which Weyl describes is richer than its group-theoretic formalisation. In particular, it contains the idea of the relationship between the parts and the whole, which is not evident in the group-theoretic interpretation. This tends to support the claim that groups do not fully capture our intuitions about symmetry.

Partial symmetries

The above analysis motivates the following definition. A *partial symmetry* of a structure is an isomorphism between two substructures of the structure; we shall also use the term *partial automorphism*. A partial symmetry in this sense is in fact a good candidate for a symmetry in Weyl's intuitive sense, but the adjective 'partial' serves to distinguish this particular aspect of symmetry from the classical, group-theoretic notion; sometimes we shall use the term *global symmetry* to emphasise the fact that we are dealing with an everywhere defined symmetry. The partial symmetries of a structure will, in many cases, form an inverse semigroup. In this way, the relationship between inverse semigroups and partial symmetries is a generalisation of the relationship between groups and symmetries. An example from [399] provides an effective dramatisation of the distinction between global symmetries and partial symmetries.

Consider two plane figures: the equilateral triangle and the Sierpiński gasket, a two-dimensional figure analogous to the Cantor set. Classically, the symmetry groups of these structures are the same. But intuitively, the Sierpiński gasket contains many partial symmetries which are not described by the group of symmetries alone. In particular, the gasket can be regarded as consisting of three identical gaskets, each half the size of the original. So it has three partial symmetries which shrink it towards each corner and halve its size. Each of these partial symmetries has an inverse which is defined on each of the smaller gaskets. By composing these partial symmetries and their inverses we can construct other partial symmetries between the copies of the gasket at the different levels. In Section 9.3, the partial symmetries of the Cantor set are explicitly described in inverse semigroup-theoretic terms.

1.4 Abstract inverse semigroups

We have seen that pseudogroups and, more generally, semigroups of partial bijections closed under inversion are examples of inverse semigroups, but it remains to be proved that every inverse semigroup can be faithfully represented by partial bijections; this is precisely the Wagner–Preston representation theorem which is the subject of the next section. In this section, we set out the basic theory of inverse semigroups: this includes the algebraic properties of inverse semigroups connected with inverses, ideals and homomorphisms, and the order-theoretic properties which arise from the natural partial order and the compatibility relations.

Properties of inverses

Inverse semigroups come equipped with an operation of inversion which behaves very much like the one defined for groups.

Proposition 1 *Let S be an inverse semigroup.*

(1) *For any $s \in S$, both $s^{-1}s$ and ss^{-1} are idempotents and $s(s^{-1}s) = s$ and $(ss^{-1})s = s$.*

(2) $(s^{-1})^{-1} = s$ *for every $s \in S$.*

(3) *For any idempotent e in S and any $s \in S$ the element $s^{-1}es$ is an idempotent.*

(4) *If e is an idempotent in S then $e^{-1} = e$.*

(5) $(s_1 \dots s_n)^{-1} = s_n^{-1} \dots s_1^{-1}$ *for all $s_1, \dots, s_n \in S$ where $n \geq 2$.*

Proof (1) For the first, observe that $(s^{-1}s)^2 = s^{-1}(ss^{-1}s) = s^{-1}s$. A similar proof shows that ss^{-1} is an idempotent. The remaining assertions are immediate.
(2) Clearly s is a solution of the equations $s^{-1} = s^{-1}xs^{-1}$ and $x = xs^{-1}x$. The result now follows from the uniqueness of inverses.
(3) We have that $(s^{-1}es)^2 = s^{-1}e(ss^{-1})es = s^{-1}e^2(ss^{-1}s) = s^{-1}es$, using the fact that idempotents commute.
(4) Immediate.
(5) The case $n = 2$ is immediate from the definition and the fact that idempotents commute. The general case can now be proved by induction. ■

The idempotents ss^{-1} and $s^{-1}s$ behave as left and right identities respectively for the element s by property (1) above. We shall often write $\mathbf{d}(s) = s^{-1}s$ and $\mathbf{r}(s) = ss^{-1}$. The set of all idempotents of a semigroup S is denoted by $E(S)$, and if A is any subset of S, then $E(A) = A \cap E(S)$.

Properties (2) and (5) of Proposition 1 are shared by a wider class of semigroups. A *semigroup with involution* S is a semigroup equipped with a unary operation $s \mapsto s^{-1}$ satisfying the following two axioms:

(IS1) $(s^{-1})^{-1} = s$ for all $s \in S$.

(IS2) $(st)^{-1} = t^{-1}s^{-1}$ for all $s, t \in S$.

Thus all inverse semigroups are semigroups with involution.

It is natural to call the operation $s \mapsto a^{-1}sa$ *conjugation* of s by a; conjugation of idempotents is again an idempotent. This observation leads to the following useful result.

Lemma 2 *Let S be an inverse semigroup.*

(1) *For every idempotent e and element s there exists an idempotent f such that $es = sf$.*

(2) *For every idempotent e and element s there exists an idempotent f such that $se = fs$.*

Proof We prove (1); the proof of (2) is similar. Put $f = s^{-1}es$, an idempotent by Proposition 1. Then

$$sf = s(s^{-1}es) = (ss^{-1})es = e(ss^{-1})s = es.$$

■

There is a suggestive, but at the same time, somewhat mysterious, example from Girard [97] which is related to the preceding result; it is only the most elementary part of an application of inverse semigroups to linear logic. Think of the idempotents of an inverse semigroup as *messages* and the (non-idempotent) elements of the inverse semigroup as *wires*. If e is a message and s a wire, then the process of transmitting e through s is represented by es. By Lemma 2, the result is sf; thus, as Girard writes, 'wires propagate messages'.

We have already noted that an inverse subsemigroup of an inverse semigroup is a subsemigroup closed under inverses. The proof of the following is straightforward.

Proposition 3 *Let X be a non-empty subset of an inverse semigroup S. Then the intersection of all inverse subsemigroups of S which contain X is an inverse subsemigroup of S. It consists precisely of all products of elements drawn from the set $X \cup X^{-1}$.* ■

The subsemigroup defined above is called the *inverse subsemigroup generated by X* and is denoted by $\langle X \rangle$.

Clearly, the idempotents of an inverse semigroup form an inverse subsemigroup. More generally, any inverse subsemigroup containing all the idempotents is said to be *full*.

The exact relationship between groups and inverse semigroups is spelt out by the following.

Proposition 4 *Groups are precisely the inverse semigroups with exactly one idempotent.*

Proof Clearly, groups are inverse semigroups having exactly one idempotent. Conversely, let S be an inverse semigroup with exactly one idempotent, e say.

Then $s^{-1}s = e = ss^{-1}$ for each $s \in S$. But $es = (ss^{-1})s = s = s(s^{-1}s) = se$, and so e is the identity of S. Hence S is a group. ■

Let S be an inverse semigroup. If S has an identity which we wish to distinguish then we say that S is an *inverse monoid*; if it has a zero element which we wish to distinguish then we say that S is an *inverse semigroup with zero*. Every (inverse) semigroup may be converted into an (inverse) monoid or (inverse) semigroup with zero by *adjoining an identity* or *adjoining a zero* in the following way. Let S be a semigroup. We define a monoid S^1 as follows. If S is a monoid then $S^1 = S$. If S is not a monoid then $S^1 = S \cup \{1\}$ together with the multiplication in S extended to S^1 by defining $1s = s = s1$ for all $s \in S$ and $11 = 1$. Then S^1 is a monoid and if S is inverse then S^1 is an inverse monoid. In a completely analogous way, we may define the semigroup with zero S^0 which is inverse if S is inverse.

Ideals

Let A and B be subsets of a semigroup S. Then their product, denoted AB, is just the set of all possible products of elements of A with elements of B. If $A = \{a\}$ or $B = \{b\}$, singleton sets, then we write aB instead of $\{a\}B$ and Ab instead of $A\{b\}$.

Let S be a semigroup. A subset I of S is a *left* (resp. *right*) *ideal* if $sa \in I$ (resp. $as \in I$) for each $a \in I$ and $s \in S$. *Ideals* are the subsets which are both left and right ideals. The intersection of any non-empty set of left (resp. right, two-sided) ideals is again a left (resp. right, two-sided) ideal. In particular, for each element $s \in S$ there is a smallest left (resp. right, two-sided) ideal containing s called the *principal left* (resp. *right, two-sided*) *ideal* containing s. We can describe these ideals more explicitly as follows. Let S be an arbitrary semigroup. The principal left ideal containing an element s is S^1s, and s is said to be a *generator* of this ideal. Likewise, the principal right ideal containing s is sS^1, and the principal ideal containing s is S^1sS^1. In an inverse semigroup, $s = (ss^{-1})s = s(s^{-1}s)$ and so $s \in Ss$ and $s \in sS$. Thus the principal left ideal containing s is Ss, the principal right ideal containing s is sS, and the principal two-sided ideal containing s is SsS.

If G is a group and $g \in G$ is any element then $gG = G = Gg$. For this reason, ideals play no part in group theory. However, the situation is quite different in inverse semigroup theory. For example, the structure of the principal right ideals turns out to be important in proving the Wagner–Preston theorem.

Lemma 5 *Let S be an inverse semigroup.*

(1) $aS = aa^{-1}S$ *for all* $a \in S$, *and* aa^{-1} *is the unique idempotent generator of* aS

(2) $Sa = Sa^{-1}a$ *for all* $a \in S$, *and* $a^{-1}a$ *is the unique idempotent generator of* Sa.

(3) $eS \cap fS = efS$ *where* e *and* f *are idempotents.*

(4) $Se \cap Sf = Sef$ *where* e *and* f *are idempotents.*

Proof (1) We have that $aS = (aa^{-1})aS \subseteq aa^{-1}S \subseteq aS$. Thus $aS = aa^{-1}S$. Now let e be any idempotent such that $aS = eS$. Then $aa^{-1}S = eS$. Thus $aa^{-1} = es$ and $e = aa^{-1}t$ for some $s, t \in S$. But then $eaa^{-1} = aa^{-1}$ and $aa^{-1}e = e$. However, idempotents commute and so $aa^{-1} = e$.
(2) Similar to (1).
(3) Let $a \in eS \cap fS$. Then $ea = a$ and $fa = a$. Thus $(ef)a = e(fa) = ea = a$, and so $a \in efS$. Conversely, if $a \in efS$ then $ea = a$ and $fa = a$, by commutativity of idempotents. Thus $a \in eS \cap fS$.
(4) Similar to (3). ■

The natural partial order

In addition to its algebraic structure, the symmetric inverse monoid is ordered via the restriction order. Using the characterisation of the restriction order given in Proposition 1.1.4(3), we define the relation $\leq$ on any inverse semigroup S as follows:

$$s \leq t \Leftrightarrow s = te$$

for some idempotent e. Before showing that this really does define a partial order, we show, amongst other things, that the side on which the idempotent appears in the definition is irrelevant.

Lemma 6 *Let S be an inverse semigroup. Then the following are equivalent:*

(1) $s \leq t$.

(2) $s = ft$ *for some idempotent* f.

(3) $s^{-1} \leq t^{-1}$.

(4) $s = ss^{-1}t$.

(5) $s = ts^{-1}s$.

Proof (1) $\Rightarrow$ (2). Let $s = te$. Then $s = ft$ for some idempotent f by Lemma 2.
(2) $\Rightarrow$ (3). Let $s = ft$ for some idempotent f. Then $s^{-1} = t^{-1}f$. Thus by definition $s^{-1} \leq t^{-1}$.
(3) $\Rightarrow$ (4). Let $s^{-1} \leq t^{-1}$. Then $s^{-1} = t^{-1}e$ for some idempotent e. Taking inverses we obtain $s = et$. But $es = s$ and so $ess^{-1} = ss^{-1}$. Thus $s = ss^{-1}t$.
(4) $\Rightarrow$ (5). Let $s = ss^{-1}t$. Then $s = te$ for some idempotent e by Lemma 2. But $se = s$ and so $s^{-1}se = s^{-1}s$. Hence $s = ts^{-1}s$.
(5) $\Rightarrow$ (1). Immediate. ■

Property (3) above should be noted: inversion does not reverse the relation.

The following concept is needed below. Let $(P, \leq)$ be a partially ordered set (or *poset*). A subset Q of P is said to be an *order ideal* if $x \leq y \in Q$ implies that $x \in Q$. The smallest or *principal* order ideal of P containing an element x is the set $[x] = \{y \in P\colon y \leq x\}$. More generally, if A is any subset of P then

$$[A] = \{y \in P\colon y \leq a \text{ for some } a \in A\}$$

is the *order ideal generated by A*.

Proposition 7 *Let S be an inverse semigroup.*

(1) *The relation $\leq$ is a partial order on S.*

(2) *For idempotents $e, f \in S$ we have that $e \leq f$ if, and only if, $e = ef = fe$.*

(3) *If $s \leq t$ and $u \leq v$ then $su \leq tv$.*

(4) *If $s \leq t$ then $s^{-1}s \leq t^{-1}t$ and $ss^{-1} \leq tt^{-1}$.*

(5) *$E(S)$ is an order ideal of S.*

Proof (1) Since $s = s(s^{-1}s)$, the relation is reflexive. Now let $s \leq t$ and $t \leq s$. Then $s = ts^{-1}s$ and $t = st^{-1}t$, so that

$$s = ts^{-1}s = st^{-1}ts^{-1}s = st^{-1}t = t.$$

Thus the relation is antisymmetric. Finally, suppose that $s \leq t$ and $t \leq u$. Then $s = te$ and $t = uf$ for some idempotents e and f. Hence $s = te = (uf)e = u(fe)$. Thus $s \leq u$.
(2) Suppose that $e \leq f$. Then $e = fi$ for some idempotent i. But then $fe = e$ and so $e = fe = ef$. The converse is clear.
(3) Let $s \leq t$ and $u \leq v$. Then there are idempotents e and f such that $s = te$ and $u = vf$. Thus $su = tevf$. By Lemma 2, $ev = vi$ for some idempotent i. Thus $su = tv(if)$, and so $su \leq tv$.
(4) Immediate from Lemma 6(3) and (3) above.

(5) Immediate from the definition of the natural partial order and the fact that $E(S)$ is closed under multiplication. ■

The order defined above is called the *natural partial order* on S.

Property (3) of Proposition 7 is shared by a broader class of semigroups. Let S be a semigroup and let $\leq$ be a partial order defined on S. Then $\leq$ is said to be *compatible* with the multiplication if for all $a, b, c, d \in S$ we have that $a \leq b$ and $c \leq d$ implies that $ac \leq bd$. In this case, the semigroup S is said to be *partially ordered* by $\leq$.

Observe that if T is an inverse subsemigroup of S, then the natural partial order defined in T agrees with the restriction to T of the natural partial order on S.

Let $(P, \leq)$ be a poset. If $z \leq x, y$ then z is said to be a *lower bound* of x and y. If z is the largest of the lower bounds it is called the *greatest lower bound* and denoted by $x \wedge y$. A *meet semilattice* is a poset in which every pair of elements has a greatest lower bound.

Proposition 8 *Let S be any semigroup. Define a relation $\leq$ on $E(S)$ by*

$$e \leq f \Leftrightarrow e = ef = fe.$$

Then $\leq$ is a partial order on $E(S)$. If S is an inverse semigroup then $(E(S), \leq)$ is a meet semilattice.

Proof Let $e \in E(S)$, then $e \leq e$ since e is an idempotent. Thus $\leq$ is reflexive. Suppose that $e \leq f$ and $f \leq e$. Then $e = ef = fe$ and $f = fe = ef$. Thus $e = f$, and so $\leq$ is antisymmetric. Finally, suppose that $e \leq f$ and $f \leq g$. Then $e = ef = fe$ and $f = fg = gf$. Thus $e = ef = e(fg) = (ef)g = eg$. Similarly, $e = ge$, and so $e \leq g$. Thus $\leq$ is transitive.

Now suppose that S is an inverse semigroup. Let $e, f \in E(S)$. Then $(ef)e = (fe)e = fe = ef$, by commutativity. Thus $ef \leq e$. Similarly, $ef \leq f$. Now let $i \leq e, f$. Then $i(ef) = (ie)f = if$ since $i \leq e$, and $if = i$ since $i \leq f$. Thus $i \leq ef$. Hence $e \wedge f = ef$. It follows that $(E(S), \leq)$ is a meet semilattice. ■

Because of Proposition 8, the set of idempotents of an inverse semigroup is often referred to as its *semilattice of idempotents*. In the case of the symmetric inverse monoid on the set X, the semilattice of idempotents is isomorphic to the set of all subsets of X ordered by set-inclusion.

In Proposition 4 we saw that groups are precisely the inverse semigroups with a unique idempotent. We may now characterise the inverse semigroups in which every element is idempotent.

Proposition 9 *Meet semilattices are precisely the inverse semigroups in which every element is an idempotent.*

Proof Let $(P, \leq)$ be a meet semilattice. It is routine to check that P is a commutative semigroup with respect to the operation $\wedge$, the greatest lower bound. Clearly, $e = e \wedge e$ for each element $e \in P$. Thus $(P, \wedge)$ is an inverse semigroup in which every element is idempotent. The converse is immediate by Proposition 8. ■

The properties of the natural partial order lead to an alternative characterisation of groups.

Proposition 10 *Let S be an inverse semigroup. Then the natural partial order is the equality relation if, and only if, S is a group.*

Proof Suppose that the natural partial order is the equality relation. If e and f are two idempotents then $ef \leq e, f$. But the natural partial order is equality and so $e = f$. Thus S has exactly one idempotent. The result now follows from Proposition 4. The converse is clear. ■

The compatibility relations

In Proposition 1.2.1, we proved that if $f, g \in I(X)$ then $f \cup g$ is a partial function precisely when fg^{-1} is an idempotent, and $f \cup g$ is a partial bijection precisely when fg^{-1} and $f^{-1}g$ are idempotents. We use these results to motivate the following definitions, just as we used the restriction ordering on partial bijections to motivate the definition of the natural partial order.

For all $s, t \in S$, the *left compatibility relation* is defined by

$$s \sim_l t \Leftrightarrow st^{-1} \in E(S),$$

the *right compatibility relation* is defined by

$$s \sim_r t \Leftrightarrow s^{-1}t \in E(S),$$

and the *compatibility relation*, the intersection of the above two relations, is defined by

$$s \sim t \Leftrightarrow st^{-1}, s^{-1}t \in E(S).$$

It is clear that all three relations are reflexive and symmetric, but none of them need be transitive (see Theorem 2.4.4 for a characterisation of the inverse semigroups having a transitive compatiblity relation). Furthermore the left and right compatibility relations are usually distinct (see Theorem 3.2.7 for a characterisation of the precise class of inverse semigroups in which the left and right compatibility relations are distinct). The next three lemmas describe some of the basic properties of these relations.

Lemma 11 *Let S be an inverse semigroup and let $s, t \in S$.*

(1) *$s \sim_l t$ if, and only if, the greatest lower bound $s \wedge t$ of s and t exists and $(s \wedge t)^{-1}(s \wedge t) = s^{-1}st^{-1}t$.*

(2) *$s \sim_r t$ if, and only if, the greatest lower bound $s \wedge t$ of s and t exists and $(s \wedge t)(s \wedge t)^{-1} = ss^{-1}tt^{-1}$.*

(3) *$s \sim t$ if, and only if, the greatest lower bound $s \wedge t$ of s and t exists and*

$$(s \wedge t)(s \wedge t)^{-1} = ss^{-1}tt^{-1} \text{ and } (s \wedge t)^{-1}(s \wedge t) = s^{-1}st^{-1}t.$$

Proof We prove (1); the proof of (2) is similar and (3) is a consequence of (1) and (2). Suppose that $s \sim_l t$. Put $z = st^{-1}t$. Then $z \leq s$ and $z \leq t$, since st^{-1} is an idempotent. Let $w \leq s, t$. Then $w^{-1}w \leq t^{-1}t$ and so $w \leq st^{-1}t = z$. Hence $z = s \wedge t$. Also

$$z^{-1}z = (st^{-1}t)^{-1}(st^{-1}t) = t^{-1}ts^{-1}st^{-1}t = s^{-1}st^{-1}t.$$

Conversely, suppose that $s \wedge t$ exists and

$$(s \wedge t)^{-1}(s \wedge t) = s^{-1}st^{-1}t.$$

Put $z = s \wedge t$. Then $z = sz^{-1}z$ and $z = tz^{-1}z$. Thus $sz^{-1}z = tz^{-1}z$, and so $st^{-1}t = ts^{-1}s$. Hence $st^{-1} = ts^{-1}st^{-1}$, which is an idempotent. Thus $s \sim_l t$. ■

Lemma 12 *Let S be an inverse semigroup.*

(1) *If $s \sim_l t$ then*
$$s \wedge t = st^{-1}t = ts^{-1}t = ts^{-1}s = st^{-1}s.$$

(2) *If $s \sim_r t$ then*
$$s \wedge t = ss^{-1}t = st^{-1}s = tt^{-1}s = ts^{-1}t.$$

(3) *If $s \sim t$ then*
$$s \wedge t = st^{-1}t = ts^{-1}t = ts^{-1}s = st^{-1}s = ss^{-1}t = tt^{-1}s.$$

Proof We prove (1); the proof of (2) is similar and (3) is immediate from (1) and (2). We have already proved that $s \wedge t = st^{-1}t$. Since st^{-1} is an idempotent, we have that $st^{-1} = (st^{-1})^{-1} = ts^{-1}$. Thus $st^{-1}t = ts^{-1}t$. By symmetry we also have that $s \wedge t = ts^{-1}s$ and $s \wedge t = st^{-1}s$. ■

Lemma 13 *Let ρ be any one of the three relations $\sim_l, \sim_r$, and $\sim$. Then the following two properties hold.*

(1) *$s \,\rho\, t$ and $u \,\rho\, v$ imply that $su \,\rho\, tv$.*

(2) *$s \leq t$, $u \leq v$ and $t \,\rho\, v$ imply that $s \,\rho\, u$.*

Proof We shall prove the results for $\rho = \sim_l$. The other compatibility relations can be handled similarly.
(1) Let $s \sim_l t$ and $u \sim_l v$. Then $st^{-1}, uv^{-1} \in E(S)$. But

$$su(tv)^{-1} = s(uv^{-1})t^{-1} \leq st^{-1},$$

since uv^{-1} is an idempotent. Hence $su(tv)^{-1}$ is an idempotent, and so $su \sim_l tv$.
(2) Let $s \leq t$ and $u \leq v$ and $t \sim_l v$. Then $su^{-1} \leq tv^{-1} \in E(S)$. Thus $s \sim_l u$. ■

A subset A of an inverse semigroup is said to be *compatible* if any pair of elements in A are compatible.

Lemma 14 *Let S be an inverse semigroup and let $s, t \in S$.*

(1) *If $s \sim_l t$ and $s^{-1}s \leq t^{-1}t$ then $s \leq t$.*

(2) *If $s \sim_r t$ and $ss^{-1} \leq tt^{-1}$ then $s \leq t$.*

(3) *$[s]$ is a compatible subset of S.*

Proof We prove (1); the proof of (2) is similar. We have immediately that $s \leq st^{-1}t$. But st^{-1} is an idempotent and so $(st^{-1})t \leq t$. Hence $s \leq t$.
(3) Straightforward by Lemma 13(2). ■

For inverse semigroups S with zero there is a refinement of the compatibility relation which is also important. Elements $s, t \in S$ are said to be *orthogonal*, denoted by $s \perp t$, if $s^{-1}t = 0 = st^{-1}$. Clearly, orthogonal elements are compatible. In the symmetric inverse monoid $I(X)$, a pair of elements $f, g \in I(X)$ satisfy $f \perp g$ precisely when f and g have disjoint domains and disjoint images.

Properties of meets and joins

The operation of inversion is an order automorphism of the natural partial order and so a non-empty subset has a join (resp. meet) precisely when the set of inverses of all elements of the subset has a join (resp. meet). As we now show, the semilattice of idempotents is closed under joins and meets whenever they exist.

Lemma 15 *Let S be an inverse semigroup and A a non-empty set of idempotents.*

(1) *If $\bigwedge A$ exists, then it is an idempotent.*

(2) *If $\bigvee A$ exists, then it is an idempotent.*

Proof (1) This is immediate from the fact that the idempotents form an order ideal.
(2) Let $a = \bigvee A$. Then $e \leq a$ for each $e \in A$. Thus $e \leq a^{-1}a$ for each $e \in A$. Hence $a \leq a^{-1}a$, so that a is an idempotent. ■

Potentially any non-empty subset of an inverse semigroup may have a meet, but the same is not true of joins.

Lemma 16 *Let S be an inverse semigroup and let A be a non-empty subset of S such that $\bigvee A$ exists. Then any two elements of A are compatible.*

Proof Let $a, b \in A$. By definition $a, b \leq \bigvee A$. Thus $a \sim b$ by Lemma 14(3). ■

We say that an inverse semigroup is *complete* if every non-empty compatible subset has a join. Thus symmetric inverse monoids are complete, and complete pseudogroups, in the sense of Section 1.2, are complete. An inverse semigroup is said to be *meet complete* if every non-empty subset has a meet.

Meets and joins display complementary behaviour with respect to the semigroup operations. We begin by considering how to calculate $\mathbf{d}$ and $\mathbf{r}$ of a meet or join of elements. For meets, we proved in Lemma 11 that $s \wedge t$ exists and satisfies $\mathbf{d}(s \wedge t) = \mathbf{d}(s) \wedge \mathbf{d}(t)$ only when $s \sim_l t$. For joins, however, there are no such restrictions.

Proposition 17 *Let S be an inverse semigroup and let $A = \{a_i\colon i \in I\}$ be any non-empty subset of S.*

(1) *If $\bigvee a_i$ exists then $\bigvee a_i^{-1}a_i$ exists and $(\bigvee a_i)^{-1}(\bigvee a_i) = \bigvee a_i^{-1}a_i$.*

(2) *If $\bigvee a_i$ exists then $\bigvee a_i a_i^{-1}$ exists and $(\bigvee a_i)(\bigvee a_i)^{-1} = \bigvee a_i a_i^{-1}$.*

Proof We prove (1); the proof of (2) is similar. Let $a = \bigvee a_i$. Then $a_i \leq a$ implies $a_i^{-1}a_i \leq a^{-1}a$. Thus the set $\{a_i^{-1}a_i\colon i \in I\}$ is bounded above by $a^{-1}a$. Now suppose that $a_i^{-1}a_i \leq b$ for some $b \in S$ and for all $i \in I$. Then

$$a_i \leq a_i b \leq ab \text{ for all } i \in I.$$

Thus $a = \bigvee a_i \leq ab$. But then $a = (aa^{-1})ab = ab$, so that $a^{-1}a = a^{-1}ab$. Hence $a^{-1}a \leq b$. It follows that $\bigvee a_i^{-1}a_i = a^{-1}a$. ■

We now consider how meets and joins behave with respect to multiplication. For joins, the strongest result we can prove is the following.

Proposition 18 *Let S be an inverse semigroup, $A = \{a_i : i \in I\}$ a non-empty subset of S and $s \in S$.*

(1) *If $a = \bigvee a_i$ and $a_i a_i^{-1} \leq s^{-1}s$ for all $i \in I$ then $\bigvee sa_i$ exists and $sa = \bigvee sa_i$.*

(2) *If $a = \bigvee a_i$ and $a_i^{-1}a_i \leq ss^{-1}$ for all $i \in I$ then $\bigvee a_i s$ exists and $as = \bigvee a_i s$.*

Proof We prove (1); the proof of (2) is similar. Since $a_i \leq a$ for all $i \in I$, we have that $sa_i \leq sa$ for all $i \in I$. Thus the set $\{sa_i : i \in I\}$ is bounded above by sa. Now suppose that $sa_i \leq b$ for some $b \in S$ and for all $i \in I$. Then $s^{-1}sa_i \leq s^{-1}b$ and so $a_i \leq s^{-1}b$ since $a_i a_i^{-1} \leq s^{-1}s$. Thus $a \leq s^{-1}b$ and so $sa \leq ss^{-1}b \leq b$. It follows that $\bigvee sa_i = sa$. ∎

However, meets behave well with respect to multiplication.

Proposition 19 *Let S be an inverse semigroup, $A = \{a_i : i \in I\}$ a non-empty subset of S and $s \in S$.*

(1) *If $a = \bigwedge a_i$ exists, then $\bigwedge sa_i$ exists and $\bigwedge sa_i = sa$.*

(2) *If $a = \bigwedge a_i$ exists, then $\bigwedge a_i s$ exists and $\bigwedge a_i s = as$.*

Proof We prove (1); the proof of (2) is similar. By definition, $a \leq a_i$ for all $i \in I$, and so $sa \leq sa_i$ for all $i \in I$. Thus the set $\{sa_i : i \in I\}$ is bounded below by sa. Now let $b \leq sa_i$ for some $b \in S$ and for all $i \in I$. Then

$$s^{-1}b \leq s^{-1}sa_i \leq a_i,$$

so that $s^{-1}b \leq a$. Hence $ss^{-1}b \leq sa$. Now $b \leq sa_i$ and so

$$bb^{-1} \leq (sa_i)(sa_i)^{-1} = sa_i a_i^{-1} s^{-1} \leq ss^{-1}.$$

Thus $ss^{-1}b = b$, and so $b \leq sa$. It follows that $\bigwedge sa_i = sa$. ∎

The behaviour of joins under multiplication leads to the following definition. An inverse semigroup is said to be *left infinitely distributive* if, whenever A is a non-empty subset of S for which $\bigvee A$ exists, then $\bigvee sA$ exists for any element $s \in S$ and $s(\bigvee A) = \bigvee sA$. *Right infinitely distributive* is defined analogously. Using inversion, it is clear that an inverse semigroup is left infinitely distributive precisely when it is right infinitely distributive. A semigroup which is both left and right infinitely distributive is called *infinitely distributive.*

Proposition 20 *Let S be an inverse semigroup. Then the following are equivalent:*

(1) *S is infinitely distributive.*

(2) *The semilattice $E(S)$ is infinitely distributive.*

(3) *For all non-empty subsets A and B of S, if $\bigvee A$ and $\bigvee B$ exist, then $\bigvee AB$ exists and $\bigvee AB = (\bigvee A)(\bigvee B)$.*

Proof The fact that (1) implies (2) follows from Lemma 15, and it is immediate that (3) implies (1), so we need only prove that (2) implies (3). Let

$$A = \{a_i \colon i \in I\} \text{ and } B = \{b_j \colon j \in J\}$$

be non-empty subsets of S such that $a = \bigvee A$ and $b = \bigvee B$. Clearly, the set AB is bounded above by ab. We prove that $\bigvee AB = ab$.

Let g be any element which bounds the set AB. Then for all $a_i \in A$ and $b_j \in B$ we have that $a_i b_j \leq g$. But $a_i \leq a$ and $b_j \leq b$ so that

$$a_i^{-1} a_i b_j b_j^{-1} \leq a^{-1} g b^{-1}.$$

Now

$$a^{-1} a b_j b_j^{-1} = (\bigvee A)^{-1} (\bigvee A) b_j b_j^{-1} = (\bigvee a_i^{-1} a_i) b_j b_j^{-1}$$

by Proposition 17. However,

$$(\bigvee a_i^{-1} a_i) b_j b_j^{-1} = \bigvee (a_i^{-1} a_i b_j b_j^{-1})$$

since $E(S)$ is infinitely distributive, so that

$$a^{-1} a b_j b_j^{-1} = \bigvee a_i^{-1} a_i b_j b_j^{-1};$$

but

$$a_i^{-1} a_i b_j b_j^{-1} \leq a^{-1} g b^{-1},$$

so that

$$a^{-1} a b_j b_j^{-1} \leq a^{-1} g b^{-1}.$$

Now

$$a^{-1} a b b^{-1} = a^{-1} a (\bigvee B)(\bigvee B)^{-1} = a^{-1} a (\bigvee b_j b_j^{-1})$$

by Proposition 17. However

$$a^{-1} a (\bigvee b_j b_j^{-1}) = \bigvee a^{-1} a b_j b_j^{-1}$$

since $E(S)$ is infinitely distributive, so that

$$a^{-1} a b b^{-1} = \bigvee a^{-1} a b_j b_j^{-1}.$$

Consequently $a^{-1} a b b^{-1} \leq a^{-1} g b^{-1}$. Thus

$$ab = a(a^{-1} a b b^{-1}) b \leq a a^{-1} g b^{-1} b \leq g.$$

Hence $ab = \bigvee AB$. ■

Homomorphisms

Homomorphisms between inverse semigroups are just semigroup homomorphisms; the following results show that we do not need to require anything stronger. We shall need the following definition. If $(P, \leq)$ and $(P', \leq)$ are posets, then a function θ: $P \to P'$ is said to be *order-preserving* if $x \leq y$ implies that $\theta(x) \leq \theta(y)$. It is worth adding at this point that an order-preserving function θ is an *order isomorphism* when it is bijective and its inverse is also order-preserving.

Proposition 21 *Let θ: $S \to T$ be a homomorphism between inverse semigroups.*

(1) $\theta(s^{-1}) = \theta(s)^{-1}$ *for all* $s \in S$.

(2) *If e is an idempotent then $\theta(e)$ is an idempotent.*

(3) *If $\theta(s)$ is an idempotent then there is an idempotent e in S such that $\theta(s) = \theta(e)$.*

(4) $\mathrm{Im}\,\theta$ *is an inverse subsemigroup of T.*

(5) *If U is an inverse subsemigroup of T then $\theta^{-1}(U)$ is an inverse subsemigroup of S.*

(6) *The function θ is order-preserving.*

(7) *Let $a, b \in S$ such that $\theta(a) \leq \theta(b)$. Then there exists an element $a' \in S$ such that $a' \leq b$ and $\theta(a') = \theta(a)$.*

Proof (1) Clearly, $\theta(s)\theta(s^{-1})\theta(s) = \theta(s)$ and $\theta(s^{-1})\theta(s)\theta(s^{-1}) = \theta(s^{-1})$. Thus by uniqueness of inverses we have that $\theta(s^{-1}) = \theta(s)^{-1}$.
(2) $\theta(e)^2 = \theta(e)\theta(e) = \theta(e)$.
(3) If $\theta(s)^2 = \theta(s)$, then $\theta(s^{-1}s) = \theta(s^{-1})\theta(s) = \theta(s)^{-1}\theta(s) = \theta(s)^2 = \theta(s)$.
(4) Since θ is a semigroup homomorphism $\mathrm{im}\,\theta$ is a subsemigroup of T. By (1), $\mathrm{im}\,\theta$ is closed under inverses.
(5) Straightforward.
(6) Let $a \leq b$. Then $a = be$ for some idempotent e. Then $\theta(a) = \theta(b)\theta(e)$ and $\theta(e)$ is an idempotent. Thus $\theta(a) \leq \theta(b)$.
(7) Put $a' = ba^{-1}a$. Then $a' \leq b$, and $\theta(a') = \theta(b)\theta(a^{-1}a) = \theta(a)$. ■

Property (1) shows that homomorphisms between inverse semigroups are automatically *involution homomorphisms.*

Property (3) is a special case of a result which also holds for regular semigroups, where it is known as *Lallement's lemma.*

Because of property (7), we say that inverse semigroup homomorphisms *reflect* the natural partial order.

If $\theta\colon S \to T$ is a surjective homomorphism then we say that S is a *cover* of T. More generally, if U is an inverse subsemigroup of S and $\theta\colon U \to T$ is a surjective homomorphism then we say that T *divides* S.

A homomorphism between monoids is required to preserve the identities for it to be a *monoid homomorphism*. Likewise, a homomorphism between semigroups with zero is required to preserve the zeros if it is to be a *homomorphism of semigroups with zero*.

A homomorphism from an inverse semigroup S to a symmetric inverse monoid is called a *representation of S by partial bijections*. If the homomorphism is injective then the representation is said to be *faithful*.

A homomorphism $\phi\colon S \to T$ between inverse semigroups is said to be *join-preserving* if for every subset $A \subseteq S$ such that $\bigvee A$ exists, then $\bigvee \phi(A)$ exists in T and $\phi(\bigvee A) = \bigvee(\phi(A))$. The homomorphism is said to be *meet-preserving* if for every subset $A \subseteq S$ such that $\bigwedge A$ exists, then $\bigwedge \phi(A)$ exists in T and $\phi(\bigwedge A) = \bigwedge(\phi(A))$.

Join completions

We will show that every inverse semigroup can be embedded in a complete, infinitely distributive inverse semigroup. This result is used to embed E-unitary inverse semigroups into semidirect products of semilattices by groups in Theorem 7.1.5.

A subset A of an inverse semigroup S is said to be *permissible* if it is a compatible order ideal. The set of all permissible subsets of S is denoted $C(S)$.

Lemma 22 *Let S be an inverse semigroup, and let A be a permissible subset of S.*

(1) *If $s, t \in A$ and $s^{-1}s = t^{-1}t$ then $s = t$.*

(2) *If $s, t \in A$ and $ss^{-1} = tt^{-1}$ then $s = t$.*

(3) *Both*

$$A^{-1}A = \{a^{-1}a\colon a \in A\} \text{ and } AA^{-1} = \{aa^{-1}\colon a \in A\},$$

are order ideals of S.

Proof (1) Let $s, t \in A$. Then s and t are compatible. But $s \sim t$ and $s^{-1}s = t^{-1}t$ implies that $s = t$ by Lemma 14.
(2) Proof similar to that of (1).
(3) We prove the results stated for $A^{-1}A$. Let $a^{-1}b \in A^{-1}A$, where $a, b \in A$.

By assumption $a \sim b$ and so $z = aa^{-1}b = a \wedge b$ exists by Lemma 12. But A is an order ideal and so $z \in A$. Also

$$z^{-1}z = (aa^{-1}b)^{-1}aa^{-1}b = b^{-1}aa^{-1}b = (a^{-1}b)^{-1}a^{-1}b = a^{-1}b.$$

Thus $A^{-1}A \subseteq \{a^{-1}a \colon a \in A\}$. The reverse inclusion is immediate. Finally, we show that $A^{-1}A$ is an order ideal. Let $e \leq a^{-1}a$. Put $b = ae$. Then $b \leq a$ so that $b \in A$ since A is an order ideal. But $b^{-1}b = ea^{-1}ae = e$. ∎

Define $\iota \colon S \to C(S)$ by $\iota(s) = [s]$. We now have the following result.

Theorem 23 *Let S be an inverse semigroup. Then $C(S)$ is a complete, infinitely distributive inverse semigroup. The function $\iota \colon S \to C(S)$ is an injective homomorphism, and every element of $C(S)$ is a join of a non-empty subset of $\iota(S)$.*

Proof We begin by proving that $C(S)$ is an inverse semigroup under multiplication of subsets.

To prove closure, we have to show that if $A, B \in C(S)$ then AB is a compatible order ideal. First, AB is an order ideal, for if $z \leq ab$ where $a \in A$ and $b \in B$, then $z = a(bz^{-1}z)$. But $bz^{-1}z \leq b$, and B is an order ideal, and so $bz^{-1}z \in B$. Thus $z \in AB$. To show that AB is a compatible subset of S, let $ab, cd \in AB$ where $a, c \in A$ and $b, d \in B$. Now $(ab)^{-1}cd = b^{-1}(a^{-1}c)d$. By assumption, $a \sim b$, so that $a^{-1}c$ is an idempotent. Thus $b^{-1}(a^{-1}c)d \leq b^{-1}d$. But again, $b^{-1}d$ is an idempotent. Hence $(ab)^{-1}cd$ is an idempotent. We may similarly show that $ab(cd)^{-1}$ is an idempotent. Thus $AB \in C(S)$.

Closure under inverses is straightforward to prove.

To prove that $C(S)$ is regular, we show that $AA^{-1}A = A$. Clearly, $A \subseteq AA^{-1}A$. To prove the reverse inclusion let $ab^{-1}c \in AA^{-1}A$ where $a, b, c \in A$. By Lemma 22, $ab^{-1} = uu^{-1}$ for some $u \in A$. Thus $ab^{-1}c = uu^{-1}c$. By Lemma 22, $u^{-1}c = v^{-1}v$ for some $v \in A$. Now

$$v^{-1}v = u^{-1}c(u^{-1}c)^{-1} = u^{-1}cc^{-1}u \leq u^{-1}u,$$

and $uv^{-1} = vu^{-1}$ since uv^{-1} is an idempotent. Thus

$$ab^{-1}c = uu^{-1}c = (uv^{-1})v = vu^{-1}u = v.$$

Finally, to prove that $C(S)$ is inverse, we show that idempotents commute. First we identify the idempotents. Suppose that $A^2 = A$. Let $a \in A$. Then $a = bc$ for some $b, c \in A$. Thus $a = aa^{-1}a = a(bc)^{-1}a = (ac^{-1})(b^{-1}a)$. But $a \sim c$ and $a \sim b$ so that ac^{-1} and $b^{-1}a$ are both idempotents. Thus a is an idempotent. It follows that $A \subseteq E(S)$. Conversely, it is clear that any order ideal of $E(S)$ is an idempotent of $C(S)$. That the idempotents commute is now immediate.

We now characterise the natural partial order in $C(S)$. By Lemma 6, $A \leq B$ when $A = AA^{-1}B$. We have already seen that AA^{-1} consists of idempotents. Thus $A = AA^{-1}B$ implies that $A \subseteq B$, since B is an order ideal. Thus $A \leq B$ implies $A \subseteq B$. Conversely, suppose that $A \subseteq B$. Then $A \subseteq AA^{-1}B$ is immediate. To prove the reverse inclusion, let $aa^{-1}b \in AA^{-1}B$ where $a \in A$ and $b \in B$. Then $a, b \in B$ implies $a^{-1}b$ is an idempotent, since $a \sim b$. Thus $a(a^{-1}b) \leq a$ and so $aa^{-1}b \in A$.

Next, we show that $C(S)$ is complete. Observe first, that $A \sim B$ precisely when $A \cup B \in C(S)$. Given our description of the natural partial order, completeness is now immediate: the join of a compatible subset is just the union of its elements. Infinite distributivity is a straightforward consequence of the description of the join.

The function $\iota(s) = [s]$ is well-defined because $[s]$ is a permissible subset by Lemma 14. It is easy to check that it is an injective homomorphism. If $A \in C(S)$ then $A = \bigcup\{[a]\colon a \in A\}$. Thus every element of $C(S)$ is a join of a non-empty compatible subset of $\iota(S)$. ■

The universal property of the embedding $\iota\colon S \to C(S)$ is described below.

Theorem 24 *If $\theta\colon S \to T$ is any homomorphism to a complete, infinitely distributive inverse semigroup, then there is a unique join-preserving homomorphism $\theta^*\colon C(S) \to T$ such that $\theta^*\iota = \theta$*

Proof Clearly any homomorphism preserves the compatibility relation. Thus if $A \in C(S)$, then $\{\theta(a)\colon a \in A\}$ is a compatible subset of T. We may therefore define

$$\theta^*(A) = \bigvee\{\theta(a)\colon a \in A\}.$$

Observe that $\theta^*([s]) = \theta(s)$. Thus $\theta^*\iota = \theta$.

First we show that θ^* is a homomorphism. Let $A, B \in C(S)$. Then

$$\theta^*(A) = \bigvee\{\theta(a)\colon a \in A\} \text{ and } \theta^*(B) = \bigvee\{\theta(b)\colon b \in B\}.$$

Thus

$$\theta^*(A)\theta^*(B) = \bigvee\{\theta(ab)\colon ab \in AB\} = \theta^*(AB),$$

by Proposition 20, since T is infinitely distributive. Thus θ^* is a homomorphism.

To show that θ^* is join-preserving, let $\{A_i\colon i \in I\}$ be a family of elements in $C(S)$ such that $\bigvee_{i \in I} A_i$ exists. Since homomorphisms preserve the compatibility relation, $\{\theta^*(A_i)\colon i \in I\}$ is a compatible family in T and so $\bigvee_{i \in I} \theta^*(A_i)$

exists in T. We need to show that $\theta^*(\bigcup_{i\in I} A_i) = \bigvee_{i\in I} \theta^*(A_i)$. But by definition

$$\theta^*(\bigcup_{i\in I} A_i) = \bigvee\{\theta(a)\colon a \in \bigcup_{i\in I} A_i\}$$

which is equal to $\bigvee_{i\in I} \theta^*(A_i)$.

The function θ^* is unique with the stated properties since by Theorem 23 every element $A \in C(S)$ is a join of elements from $\iota(S)$. Thus two join-preserving homomorphisms from $C(S)$ which agree on all elements of the form $[s]$ must agree on all elements of $C(S)$. ■

The category of complete, infinitely distributive inverse semigroups together with join-preserving homomorphisms forms a subcatgeory of the category of inverse semigroups and homomorphisms; indeed, the former is a *reflective subcategory* of the latter. The proof of the following is now immediate from standard category theory (see [210]).

Theorem 25 *The function $S \mapsto C(S)$ is the object part of a functor from the category of inverse semigroups and semigroup homomorphisms to the category of complete, infinitely distributive inverse semigroups and join-preserving homomorphisms.* ■

Meet completions

We have seen that each inverse semigroup may be embedded in a complete inverse semigroup. Likewise, each inverse semigroup may be embedded in a meet complete inverse semigroup. Meet completions will not figure prominently in this book, so we shall mainly content ourselves with a statement of results; references to proofs may be found in the notes to this section at the end of the chapter.

For every non-empty subset X of an inverse semigroup S define

$$[X]^\uparrow = \{s \in S\colon x \leq s \text{ for some } x \in X\},$$

the *closure* of X in S. A subset is said to be *closed* if it equals its closure. A non-empty, closed subset A of an inverse semigroup S is called a *coset* if $A = AA^{-1}A$. The set of cosets of S is denoted by $K(S)$. We first justify the use of the term coset.

Proposition 26 *Let A be a non-empty subset of a group G. Then $A = AA^{-1}A$ if, and only if, A is a coset of a subgroup of G.*

Proof Suppose that $A = AA^{-1}A$. Put $H = AA^{-1}$. Then

$$HH = AA^{-1}AA^{-1} = AA^{-1} = H$$

and $H^{-1} = H$. Thus H is a subgroup of G. Let $a \in A$ be any element. Then clearly $Ha \subseteq A$. To prove the reverse inclusion holds let $b \in A$. Then $ba^{-1} \in H$ and $b = (ba^{-1})a$. Thus $b \in Ha$. Hence $A = Ha$. The converse is straightforward. ■

A further justification for the term coset comes from the theory of representations of inverse semigroups.

Proposition 27 *Let $\theta: S \to I(X)$ be a representation of the inverse semigroup S. Let $x, y \in X$. Put $A = \{s \in S: \theta(s)(x) = y\}$. Then if A is non-empty, it is a coset.*

Proof Let $a, b, c \in A$. Then

$$\theta(ab^{-1}c)(x) = \theta(a)\theta(b)^{-1}\theta(c)(x) = x.$$

Thus $ab^{-1}c \in A$. Let $a \in A$ and $a \leq b$ for some $b \in S$. Then $\theta(a)(x) = y$ and so $\theta(b)(x)$ is defined and equals y. Thus $b \in A$. Hence A is a coset. ■

The above result is the first step in the theory of inverse semigroup representations. Let θ: $S \to I(X)$ be a representation of the inverse semigroup S by partial bijections. The relation τ_θ is defined on the set X by $(x, y) \in \tau_\theta$ if there exists $s \in S$ such that $\theta(s)(x) = y$. The relation τ_θ is symmetric and transitive but not necessarily reflexive (it is a *partial equivalence relation*). If τ_θ is reflexive then the representation θ is said to be *effective*; this simply means that every element of X appears in the domain of some element of the image of S under θ. A representation which is not effective can clearly be 'cut down' to one that is by replacing X with the subset of all those $x \in X$ for which $(x, x) \in \tau_\theta$. A representation is said to be *transitive* if for all $x, y \in X$ there exists $s \in S$ such that $\theta(s)(x) = y$. The theory of effective, transitive representations can be developed as a generalisation of transitive group representations using the general notion of coset.

We now return to meet completions. The intersection of any non-empty set of cosets is either empty or a coset. For every non-empty subset X of the inverse semigroup S we define $j(X)$ to be the intersection of all cosets containing X. Define a binary operation on $K(S)$ as follows:

$$A \otimes B = j(AB),$$

the smallest coset containing AB.

Define ι: $S \to K(S)$ by $\iota(s) = [s]^\uparrow$.

Theorem 28 *Let S be an inverse semigroup. Then $(K(S), \otimes)$ is a meet complete inverse semigroup. The function ι: $S \to K(S)$ is an injective homomorphism, and each element of $K(S)$ is the meet of a non-empty subset of $\iota(S)$.*

■

The universal property of the embedding ι: $S \to K(S)$ is described below.

Theorem 29 *Let θ: $S \to T$ be a homomorphism from an inverse semigroup S to a meet complete inverse semigroup T. Then there is a unique meet-preserving homomorphism θ^*: $K(S) \to T$ such that $\theta^* \iota = \theta$.* ■

The category of meet complete inverse semigroups together with meet-preserving homomorphisms forms a subcatgeory of the category of inverse semigroups and homomorphisms; indeed, the former is a *reflective subcategory* of the latter. The proof of the following is now immediate from standard category theory (see [210]).

Theorem 30 *The function $S \mapsto K(S)$ is the object part of a functor from the category of inverse semigroups and semigroup homomorphisms to the category of meet complete inverse semigroups and meet-preserving homomorphisms.* ■

The construction of $K(S)$ may, of course, be carried out when S is a group. The inverse semigroup which arises in this case is called a *group coset semigroup*; such semigroups are discussed further in Section 4.2.

1.5 Representation theorems

In Section 1.1, we introduced inverse semigroups and indicated that every such semigroup could be represented by a semigroup of partial bijections. We now have more than enough theory to prove this result.

The Wagner–Preston representation theorem

Theorem 1 *Let S be an inverse semigroup. Then there is a set X and an injective homomorphism θ: $S \to I(X)$ such that*

$$a \leq b \Leftrightarrow \theta(a) \subseteq \theta(b).$$

Proof For each element $a \in S$, define θ_a: $a^{-1}aS \to aa^{-1}S$ by $\theta_a(x) = ax$. This is clearly well-defined, since $aS = aa^{-1}S$ by Lemma 1.4.5. Observe that $\theta_{a^{-1}}$: $aa^{-1}S \to a^{-1}aS$ and that $\theta_{a^{-1}}\theta_a$ is the identity on $a^{-1}aS$ and $\theta_a\theta_{a^{-1}}$ is the identity on $aa^{-1}S$. Thus θ_a is a bijection and $\theta_a^{-1} = \theta_{a^{-1}}$. Define θ: $S \to I(S)$ by $\theta(a) = \theta_a$. This is well-defined by the above.

Next we show that $\theta_a\theta_b = \theta_{ab}$. By definition and Lemma 1.4.5,

$$\operatorname{dom}\theta_a \cap \operatorname{im}\theta_b = a^{-1}aS \cap bb^{-1}S = a^{-1}abb^{-1}S.$$

Hence by definition and Lemma 1.4.5,

$$\mathrm{dom}(\theta_a\theta_b) = \theta_b^{-1}(a^{-1}abb^{-1}S) = b^{-1}a^{-1}aS = b^{-1}a^{-1}abS.$$

Thus $\mathrm{dom}(\theta_a\theta_b) = \mathrm{dom}(\theta_{ab})$. It is immediate from the definitions that $\theta_a\theta_b$ and θ_{ab} have the same effect on elements, and so θ is a homomorphism.

Suppose that $a \leq b$. Then, in particular, $a^{-1}a \leq b^{-1}b$ so that $a^{-1}aS \subseteq b^{-1}bS$. Let $x \in a^{-1}aS$. Then

$$\theta_b(x) = bx = b(a^{-1}ax) = ax = \theta_a(x).$$

Hence $\theta_a \subseteq \theta_b$.

Conversely, suppose that $\theta_a \subseteq \theta_b$. By definition $a^{-1}aS \subseteq b^{-1}bS$. Now $a^{-1} \in a^{-1}aS$. By assumption, $\theta_a(a^{-1}) = \theta_b(a^{-1})$. Thus $aa^{-1} = ba^{-1}$ and so $a = ba^{-1}a$. Hence $a \leq b$. It is immediate from this result that θ is injective.

■

We have now come full circle. Starting with partial bijections, whose introduction was motivated by pseudogroups and an enquiry into the meaning of symmetry, we have defined an abstract class of algebraic structures, inverse semigroups, which can be faithfully represented by means of partial bijections.

A refinement of the Wagner–Preston theorem

The Wagner–Preston theorem can be somewhat refined; the resulting theorem (Theorem 4) is an important ingredient in the theory of inverse semigroups described in Chapter 10. First we need some definitions; these will be given in their 'left-handed' forms. In all cases, we may make the obvious 'right-handed' definitions.

Let S be a semigroup and X a set. Then S is said to *act on X on the left* if there is a function $S \times X \to X$, denoted by $(a, x) \mapsto a \cdot x$, such that $(ab) \cdot x = a \cdot (b \cdot x)$. A set X equipped with a left S-action is called a *left S-system*. If S is a monoid with identity 1 and satisfies the condition that $1 \cdot x = x$ for all $x \in X$ then the action is called a *monoid action*.

Let X be a left S-system, $X' \subseteq X$ and $S' \subseteq S$. Then

$$S' \cdot X' = \{a \cdot x \colon a \in S' \text{and } x \in X'\}.$$

If either S' or X' is a singleton set then we omit brackets as usual.

Let X and Y be two left S-systems. A function $\theta\colon X \to Y$ is called a *left S-homomorphism* if

$$\theta(a \cdot x) = a \cdot \theta(x).$$

A bijective left S-homomorphism is called a *left S-isomorphism.*

Let X be a left S-system. A subset Y of X such that $S \cdot Y \subseteq Y$ is said to be *left S-invariant.* Such a subset is a left S-system in its own right called a *left S-subsystem.* If a left S-subsystem Y of X is of the form $S \cdot x$ for some $x \in X$ then we say that Y is *cyclic.*

Every monoid S acts on itself on the left by left multiplication. The left S-subsystems under this action are precisely the left ideals of S. A left S-homomorphism between two left ideals I and J of S is a function θ: $I \to J$ such that $\theta(ax) = a\theta(x)$ where $x \in I$ and $a \in S$. An important class of left S-homomorphisms may be constructed as follows. Let $a \in S$ and define the function ρ_a: $S \to S$ by $\rho_a(x) = xa$. Observe that $\rho_a(sx) = sxa = s\rho_a(x)$. Thus in fact ρ_a: $S1 \to Sa$ and ρ_a is a left S-homomorphism, called a *right translation.* For this reason, left S-homomorphisms for a semigroup S acting on itself on the left are often called *partial right translations.*

For any semigroup S let $\hat{S}$ denote the set of all *right* S-isomorphisms between right ideals of S. There is no mathematical significance in our choice of handedness: it is simply convenient given previous results. Clearly, $\hat{S}$ is a subset of $I(S)$.

Proposition 2 *$\hat{S}$ is an inverse subsemigroup of $I(S)$ for any semigroup S.*

Proof First of all, the inverse of a right S-isomorphism is again a right S-isomorphism. To see this, let α be a right S-isomorphism and let $t \in \operatorname{im}\alpha$ and $r \in S$. Now $\alpha^{-1}(t) \in \operatorname{dom}\alpha$, so that $\alpha(\alpha^{-1}(t)r) = \alpha(\alpha^{-1}(t))r = tr$. Hence $\alpha^{-1}(tr) = \alpha^{-1}(t)r$.

Next, we observe that if I is a right ideal contained in $\operatorname{dom}\alpha$ then $\alpha(I)$ is a right ideal contained in $\operatorname{im}\alpha$, for let $r' \in \alpha(I)$ and $s \in S$. Then $r' = \alpha(r)$ for some $r \in I$. But then $rs \in I$ so that $\alpha(rs) \in \alpha(I)$. However, $\alpha(rs) = \alpha(r)s = r's$.

Now let α, β be two right S-isomorphisms. Suppose that the intersection of $\operatorname{dom}\alpha$ and $\operatorname{im}\beta$ is non-empty. From above, $\beta^{-1}(\operatorname{dom}\alpha \cap \operatorname{im}\beta)$ is a right ideal contained in $\operatorname{dom}\beta$ and $\alpha(\operatorname{dom}\alpha \cap \operatorname{im}\beta)$ is a right ideal contained in $\operatorname{im}\alpha$. As usual, $\alpha\beta$ is a bijection between these two right ideals. It only remains to check that $\alpha\beta$ is a right S-homomorphism. But

$$(\alpha\beta)(sr) = \alpha(\beta(sr)) = \alpha(\beta(s)r) = \alpha(\beta(s))r = (\alpha\beta)(s)r.$$

■

Denote by $\check{S}$ the set of all right S-isomorphisms between *principal* right ideals of S.

Lemma 3 *Let S be an inverse semigroup S. Then $\check{S}$ is an inverse subsemigroup of $\hat{S}$.*

Proof The result follows from two observations: the intersection of any two principal right ideals in an inverse semigroup is again a principal right ideal by Lemma 1.4.5; and the image of a principal right ideal under a right S-isomorphism is again a principal right ideal. ■

We may now prove our refinement of the Wagner–Preston theorem.

Theorem 4 *Every inverse semigroup S is isomorphic to the inverse semigroup $\check{S}$ of all right S-isomorphisms between principal right ideals of S.*

Proof Define the function $\theta: S \to \check{S}$ by $\theta(a) = \theta_a$ where $\theta_a: a^{-1}aS \to aa^{-1}S$ is given by $\theta_a(x) = ax$. It is immediate from the proof of Theorem 1, that θ is a well-defined injective homomorphism. It remains to show that θ is surjective.

Let $\phi: xS \to yS$ be a right S-isomorphism. Put $a = \phi(xx^{-1})$. We shall show that $\phi = \theta_a$. Firstly, $yy^{-1}a = a$ since $a \in yS$. Also

$$a = \phi(xx^{-1}) = \phi(xx^{-1}xx^{-1}) = \phi(xx^{-1})xx^{-1} = axx^{-1}.$$

Thus $a = axx^{-1}$. By assumption, ϕ is surjective and so there exists $a' \in xS$ such that $\phi(a') = yy^{-1}$. However

$$yy^{-1} = \phi(a') = \phi(xx^{-1}a') = \phi(xx^{-1})a' = aa'.$$

Thus $aa'a = yy^{-1}a = a$. Since $a' \in xS$ it follows that $a'aa' \in xS$ thus

$$\phi(a'aa') = \phi(a')aa' = \phi(a')yy^{-1} = yy^{-1} = \phi(a').$$

But ϕ is injective and so $a' = a'aa'$. Hence $a' = a^{-1}$. In particular, $aa^{-1} = yy^{-1}$. Clearly, $a^{-1}, a^{-1}a \in xS$ and

$$\phi(a^{-1}a) = \phi(a^{-1})a = yy^{-1}a = a.$$

But $\phi(xx^{-1}) = a$. Thus by injectivity, we have that $a^{-1}a = xx^{-1}$. Finally, if $xs \in xS$ then

$$\phi(xs) = \phi(xx^{-1}xs) = \phi(xx^{-1})xs = axs = \lambda_a(xs),$$

where λ_a is left multiplication by a. We have therefore proved that $\phi = \theta_a$, and so θ is a surjection and therefore an isomorphism. ■

1.6 Notes on Chapter 1

Section 1.1

It may be worth expanding slightly on my references to the work of Clifford and Rees. Clifford's 1941 paper [45] deals with the construction of what are now termed Clifford semigroups; these are the inverse semigroups with central idempotents (see Section 5.2). His 1942 paper [46] introduces Brandt semigroups; these are the inverse completely 0-simple semigroups (see Section 3.3). Finally, Clifford's 1953 paper [47] describes bisimple inverse monoids. This work is generalised to arbitrary inverse semigroups in Chapter 10. Turning now to Rees' work: his 1947 paper [335] is described in Section 2.4, whereas his 1948 paper [336] is the forerunner of work on inverse ω-semigroups discussed in Section 5.4.

Another paper which prefigures inverse semigroup theory was discussed by Schein [384]; published by Gołąb [101] in 1939 it contains ideas which anticipate the theory of E-unitary semigroups (see Chapter 7).

The Erlanger Programm is reprinted in Klein [88], but I have relied heavily on Chapter 12 of Behnke [21] for the account given here, together with Yaglom [436]. A critical appraisal of the Erlanger Programm may be found in [122].

Lie's work on infinite continuous groups is discussed in [435]. A discussion of Lie's original motivation may be found in [299]. Some of the earliest work on the structure of Lie pseudogroups was carried out by Cartan, who classified the primitive Lie pseudogroups in $\mathbb{R}^n$ into 6 families; Thom takes these pseudogroups as the starting point for some speculations on the nature of matter in [410] (see also [411] for the rôle of pseudogroups in the definition of form). An overview of Cartan's work on Lie pseudogroups may be found in Whitehead's obituary in [433].

To the best of my knowledge, a proper history of the origins of pseudogroups has yet to be written. My account was based on Ehresmann's *Oeuvres complètes* [71], Reinhart [349] and Schein [382], [384]. I also consulted Boothby [33] and Olver [299] for background on differential manifolds and Lie groups; and Albert and Molino [2], [3], Appendix D of [256], and Pommaret [324] for background on Lie pseudogroups.

The relationship between symmetric inverse monoids and abstract inverse semigroups is a special case of more general results connected with categories of partial functions. Such categories have become increasingly important in recent years, principally because of their applications in theoretical computer science. A useful survey of some of this work may be found in [359]. See also [391] for the semigroup analogue of some of these ideas.

Wagner and Preston both introduce inverse semigroups as regular semigroups with commuting idempotents. Theorem 3 is due to Liber [201] and independently Munn and Penrose [278].

Groupoids were introduced by Brandt [34]; groupoid theory predates the full-blown theory of categories developed by Eilenberg and MacLane. A fuller account of Brandt's work may be found in the paper by Fritzsche and Hoehnke [89]. Ehresmann's work on ordered groupoids emerged from his earlier work in geometry and Lie pseudogroups. This may be found reprinted in Volume I of

[71], and is discussed in the same volume by Paulette Libermann, particularly on pages 518–519.

It is also interesting to note that both Wagner and Ehresmann were partly motivated by Veblen and Whitehead's book. See [382] for Wagner's connection with this book, and for Ehresmann's connection see page 337 of Volume II-1 of [71]. Although Preston's paper does not mention it explicitly, Preston was also motivated by the theory of pseudogroups, as he told me in a conversation at the Hobart Semigroup Conference in 1994. Preston's supervisor was J. H. C. Whitehead. An account of Wagner's work together with a complete bibliography may be found in [382].

Section 1.2

The notion of structures compatible with a pseudogroup of transformations may be found in a paper of Ehresmann dated 1947 (this is work number 20 in the list of papers to be found in [71]). The source appears to be [418]. Nice descriptions of the pseudogroup theory of local structures may be found in [60] and [65], and formed the basis for this section. An interesting summary of Ehresmann's ideas on the rôle of pseudogroups in defining local structures may be found on page 471 of Volume I of [71]. What led him to local structures is described on page 475.

The pivotal paper for Ehresmann's work on local structures is [66]; this is the first of his papers to rely heavily on category theory, and summarises, in categorical terms, his work on local structures. All of his subsequent work in this area, such as [67], [68], [69], and parts of the book [70], has its roots in this paper. But the motivation for the paper comes from his earlier, non-categorical, work on local structures. The approach is very close to the one adopted by Eilenberg in [72]. Some comments on Ehresmann's work on local structures together with a more modern, topos-theoretic, approach may be found in [173]. Local structures and category theory are also the subject of [105] and [196].

The existence of local structures compatible with a given pseudogroup is of great interest. The paper by Jekel [152] provides a very good introduction to this area. The case of foliations is particularly important [123], [124] and [349]. The properties of foliations are themselves closely connected with properties of associated pseudogroups: see [284], [323], and [362] as well as the references to foliation theory.

The idea of defining structures by means of pseudogroups was also an important motivation in Wagner's work. He attempted to describe atlases algebraically, since atlases are just as important in defining local structures as pseudogroups. However, atlases link different structures and so their elements cannot be composed pairwise. It is necessary instead to define a ternary operation on an atlas. The axiomatisation of this operation is one of Wagner's other achievements. The approach Wagner adopted in describing atlases in terms of ternary operations has been developed in contemporary mathematical terms in [172] and [430].

Pseudogroups are very closely related to topological groupoids: a topological groupoid can be constructed from a pseudogroup, and the invertible local sections of a topological groupoid form a pseudogroup. A number of early papers describe this relationship: [106], [416], and [417]. See also [152] and [349]. Modern versions of this idea in contemporary language (the theory of locales) may be found in [174], [173]. A construction of topological groupoids from arbitrary inverse semigroups is described in Section 9.2. This connection is important because C^*-algebras may be constructed from topological groupoids [175], [350]. In this way, inverse semigroups are becoming increasingly important in C^*-algebra theory [302].

Section 1.3

I should perhaps begin by saying that my remarks on group theory are a critique not a criticism. The books by Artin [6] and Rose [361] are excellent expositions of their respective areas. It is simply that my aims are different.

In [208], Mackay states that whereas classical crystallography is concerned with generating patterns by applying groups to motifs, he is concerned in generating structure from programs applied to motifs. It is worth noting that general semigroup theory is intimately connected with theoretical computer science (via automata theory and formal languages). Thus inverse semigroup theory combines elements of classical algebra and of computer science. This theme is taken up in Chapter 2 where I discuss the type II conjecture. I am grateful to Ian Stewart for supplying the references to Mackay's work.

Some attempts at alternative algebraic descriptions within crystallography were also carried out by Fichtner in relation to polytypic structures (layered structures, such as silicon carbide) [77], [78], [79] and [80]. I am unable to judge the significance of Fichtner's work, but it contains explicit references to partial symmetries in the context of crystallography.

Weinstein's article [431] is the most recent one in which the identification of group theory and symmetry is questioned. He mentions inverse semigroups in passing.

The idea of using semigroups, rather than groups, to study the symmetry of fractals is mentioned in [333], again with reference to the Sierpiński gasket. Of course, iterated function systems generate semigroups, but my concern is with the symmetry of a figure rather than the way it is generated.

Section 1.4

The natural partial order and the compatibility relation were introduced by Wagner in [419] and [420] respectively. The orthogonality relation arose from work of myself and Peter Hines on Girard's linear logic [135].

Properties of the natural partial order with respect to arbitrary joins and meets have been considered by Domanov [61], Rinow [357], and Schein [376]. All of Ehresmann's work on ordered groupoids shows a great concern for order-theoretic properties which stems from the origin of his work in the theory of pseudogroups and local structures. Proposition 17 was proved by Rinow [357]; Proposition 18 by Domanov [61]; and Proposition 20 by Schein [376]. The surprising property of meets in Proposition 19 was noted by Schein [376], and is important in Leech's work on inverse monoids with arbitrary meets [198].

Completions of inverse semigroups were first considered by Ehresmann, who was motivated by his work on the algebraic foundation of differential geometry. A subset A of an inverse semigroup is termed an *atlas* by Ehresmann if it satisfies $A = AA^{-1}A$; such sets arise in constructing differential manifolds by taking collections of charts: see Sections 1.2 and 8.3. The paper by Rinow [357] follows in the Ehresmann tradition. It essentially constructs $C(S)$, although Rinow is more interested in obtaining completions where the embedding map preserves certain order-theoretic properties. The later, but independent, paper by Schein [376] is in the Wagner tradition. Schein goes on to show that the translational hull of S is an inverse subsemigroup of $C(S)$. It consists of those permissible subsets H where $H^{-1}H$ and HH^{-1} are retract ideals of $E(S)$ (where a *retract ideal* I of $E(S)$ is an ideal with the property that for every $e \in E(S)$ the intersection $E(S) \cap [e]$ is also a principal ideal). Schein also proves that the translational hull is the idealiser of S in $C(S)$.

The semigroup $C(S)$ plays a crucial rôle in the theory of S-systems. Schein [377] proves that an inverse semigroup is self-injective precisely when it is complete, infinitely distributive and E-reflexive. Shoji [393] constructs the injective hull of an E-reflexive semigroup S as a quotient of $C(S)$. The class of E-reflexive inverse semigroups is discussed in Section 3.2.

The work on meet completions is by Leech [198] with some independent contributions by Lawson [186]. But there are precursors to this work in a paper of Schein [374]. The notion of cosets in semigroups is discussed by Schein [387].

That cosets of subgroups can be characterised as those non-empty subsets such that $A = AA^{-1}A$ is due to Dubreil [62]. The construction of group coset semigroups is due to Schein [374] and was developed by McAlister [239], and McAlister and Reilly [246]. They were also studied, independently, by Joubert [162], who uses them in his work on topological foliations.

Section 1.5

The Wagner–Preston representation theorem was proved independently by Wagner [419] and Preston [327].

A different kind of refinement of the representation theorem arises from the theory of algebras. Bredikhin [35] and Szabó [403] proved that an inverse subsemigroup S of the symmetric inverse monoid $I(X)$ is the semigroup of all partial automorphisms of an algebraic structure on X if, and only if, the following three conditions hold:

(B1) $\bigcap A \in S$, for all non-empty $A \subseteq S$.

(B2) If A is a non-empty chain in S then $\bigcup A \in S$.

(B3) If $\{(x,x)\}, \{(y,y)\} \in S$ then $\{(x,y)\} \in S$.

An abstract characterisation of such inverse subsemigroups was obtained by Bredikhin [35] and Domanov [61] as follows. An *algebraic lattice* is a lattice isomorphic to the substructure lattice of an algebra. An inverse monoid S is isomorphic to the inverse monoid of all partial automorphisms of an algebra if, and only if, the following two conditions hold:

(A1) $(E(S), \leq)$ is an algebraic lattice.

(A2) Every non-empty subset of $(S, \leq)$ has a meet, and every non-empty chain has a join.

The above result can be used to provide an alternative proof using only inverse semigroup theory of a result due to Schmidt [389]: if G is a group and E an algebraic lattice, then there exists an algebra whose automorphism group is isomorphic to G and whose lattice of substructures is isomorphic to E [254].

At least two other lines of enquiry have been pursued in this area. The first is to examine the completeness properties of partial automorphism semigroups: in particular, to characterise those algebras enjoying nice completeness properties; an example of this is [390]. The second is to discover the extent to which algebras within a class can be characterised by their partial automorphism semigroups; an example of this may be found in [303]

Goberstein's survey article [99] is a useful reference for work on partial automorphism semigroups upto 1984. Deeper questions on inverse semigroups of partial automorphisms are discussed in Chapter 2. The notion of dual partial automorphisms is discussed in Section 4.2.

The topological analogue of the semigroup of partial automorphisms of an algebraic structure is the inverse monoid of all partial homeomorphisms between the open subsets of a topological space. The lattice of idempotents in such a semigroup is of a type known as a *frame*. This is a complete lattice in which meets distribute over arbitrary joins. The relationship between frames and topological spaces is more complex than the relationship between algebraic lattices and substructure lattices (see Johnstone [153]). However, inverse semigroups having a frame of idempotents are natural algebraic models of pseudogroups. It was just such inverse semigroups (or rather, their ordered groupoid analogues) which were the basis of Ehresmann's work. In Section 5.2, we prove one algebraic result on inverse semigroups of partial homeomorphisms.

Chapter 2

Extending partial symmetries

In Chapter 1, we argued that inverse semigroups capture our intuitive ideas about symmetry in a more complete way than groups. Our argument was based on examples drawn from geometry together with an examination of the meaning of the word symmetry. The Wagner–Preston representation theorem assured us that inverse semigroups were the abstract counterparts of semigroups of partial symmetries. In this chapter, we shall reinforce this argument by taking as our point of departure a number of applications of partial symmetries in algebra.

All of these examples, both geometric and algebraic, serve to underline one of the main themes of this book: that inverse semigroups are intimately concerned with the theory of partial symmetries. However, the algebraic applications we consider in Section 2.1 introduce our second main theme: the relationship between partial and global symmetries. Just as partial symmetries can be abstractly described by means of inverse semigroups, so the relationship between partial and global symmetries can be abstractly described in inverse semigroup-theoretic terms. This relationship finds its algebraic expression in the classes of E-unitary inverse semigroups and factorisable inverse monoids, and in the notion of an E-unitary cover. These ideas are examined in Sections 2.2 and 2.4.

Examples of E-unitary inverse semigroups permeate this book: in Section 2.4 we show how left reversible cancellative monoids can be used to construct E-unitary inverse semigroups; in Section 3.4 we study the bicyclic monoid, an important example of an E-unitary semigroup; the free inverse semigroups described in Chapter 6 are E-unitary; finally the analogous class of 0-E-unitary inverse semigroups forms the basis of Chapter 9, where numerous examples and applications are discussed. The structure theory of E-unitary inverse semigroups is the subject of Chapter 7 and Section 8.1.

Factorisable inverse monoids are briefly discussed in Section 4.2 (Theorems 4.2.6 and 4.2.7) and their relationship with E-unitary inverse semigroups is examined in Section 7.1.

In Section 8.2, we obtain a proper abstract understanding of the relationships between factorisable inverse monoids, E-unitary semigroups and E-unitary covers.

2.1 Partial symmetries

In this section, we describe some of the ways in which partial symmetries are used in model theory and combinatorial group theory. We conclude by summarising the key ideas which arise, and which form the basis of the abstract theory introduced in Section 2.2.

Model theory

Superficially, model theory appears to be concerned with the manipulation of formulae, but Roland Fraïssé formulated an approach to the subject based on partial bijections [85], [86], [87], and consequently partial bijections permeate the field. A good example of this is provided by the important concept of homogeneity. Structures in which every partial automorphism between finite substructures can be extended to an automorphism are said to be *homogeneous*[1]. Hodges [137] observes that homogeneous structures are those possessing 'many symmetries'. This accords well with the motivation for studying inverse semigroups outlined in Section 1.3. Homogeneity is also closely connected with quantifier-elimination: a first-order theory Σ has quantifier-elimination if, and only if, every model of Σ has an elementary extension which is a homogeneous model of Σ (see [39] for a proof).

Structures in which partial automorphisms can be extended to automorphisms are important more generally in algebra. Here are two examples. Let $L : K$ be a field extension. Then the set of all isomorphisms between subfields of L which contain K and fix the elements of K pointwise form an inverse submonoid of $I(L)$, which we term the *Galois inverse monoid* of the extension. The group of units of this inverse monoid is just the usual Galois group of the field extension. In the case of finite, normal field extensions every element of the Galois inverse monoid can be extended to an element of the Galois group; this result is instrumental in setting up a correspondence between the structure of a finite, separable, normal field extension and the structure of its Galois

[1]There are, in fact, a number of different notions of homogeneity.

group.

Another example of the same phenomenon occurs in the theory of inner product spaces. Let V be a finite-dimensional vector space over a field K equipped with a non-degenerate, symmetric bilinear form; that is, with an inner-product. A linear isomorphism between two subspaces of V which preserves the inner product is called a *partial isometry.* The collection of all partial isometries forms an inverse submonoid of $I(V)$. The group of units of this monoid is just the usual group of isometries. In this case, the link between the inverse monoid and its group of units is provided by Witt's lemma [178], which states that, when the characteristic of the field is not equal to 2, every partial isometry of V can be extended to an isometry. This result is important in understanding the structure of such vector spaces.

More generally, Aschbacher [7] uses the term *Witt property* to describe structures having the property that every partial automorphism is the restriction of an automorphism. It is interesting to note that Clark, Krauss and Hill [44] show that the classical theory of countable abelian torsion groups of finite Ulm type can be completely developed in terms of the extension properties of classes of partial automorphisms.

If a structure is not homogeneous then we can try to embed it in a structure of the same type in such a way that every partial automorphism of the smaller structure extends to an automorphism of the larger structure; we shall call this the *HNN-property* after the following example. It is a theorem of Higman, Neumann and Neumann [134], that any group G can be embedded in a group H in such a way that every partial automorphism of G is induced by conjugation by an element of H, and thus by an automorphism of H.

The HNN-property is important in model theory. If the structure in question is finite, it is natural to require that the extension also be finite. Thus Hrushovski [150] proved that every finite graph X can be embedded in a finite graph Z in such a way that every partial automorphism of X can be extended to an automorphism of Z. This result was extended to arbitrary relational structures by Herwig [128], and to some special classes of graphs in [129]. Results of this kind are important in establishing the small index property for the generic structures in the class being considered, and this in turn has applications to reconstructing a structure from its automorphism group; consult [138] and [129] for more information.

We conclude this section with a simple example of these ideas: we establish the Witt-property and the HNN-property for sets.

Proposition 1 *Let X be a set.*

(1) *$I(X)$ has the property that every element is beneath a bijection if, and only if, X is finite.*

(2) *If X is infinite then X can be embedded in a set Z of the same cardinality with the property that every element of $I(X)$ is beneath a bijection of Z.*

Proof (1) Let X be finite. If $f \in I(X)$ then

$$|X \setminus \operatorname{dom} f| = |X \setminus \operatorname{im} f|.$$

Let g be any bijection from $X \setminus \operatorname{dom} f$ to $X \setminus \operatorname{im} f$. The function $f \cup g$ is a permutation of X which extends f.

Conversely, suppose that X is infinite. Pick any $x \in X$. Then $|X| = |X \setminus \{x\}|$ and so there is a bijection $f \colon X \to X \setminus \{x\}$. Clearly f is not the restriction of any bijection on X. Thus if X is infinite it is not the case that every element of $I(X)$ is beneath a bijection.

(2) Let Y be any set disjoint from X and having the same cardinality as X. Put $Z = X \cup Y$. Clearly $I(X)$ is a subsemigroup of $I(Z)$. Let $f \in I(X)$ regarded as an element of $I(Z)$. Then we have that

$$|Z \setminus \operatorname{dom} f| = |Y| + |X \setminus \operatorname{dom} f| = |Y| = |Z|$$

and

$$|Z \setminus \operatorname{im} f| = |Y| + |X \setminus \operatorname{im} f| = |Y| = |Z|.$$

Thus f can be extended to a permutation of Z. ■

Combinatorial group theory

The theory of finitely-generated free groups is a particularly fertile area for applications of partial symmetries. It is also connected with some of the deepest results in contemporary semigroup theory. These arose out of attempts to prove Rhodes' type II conjecture. We shall mention the type II work only in passing here, and discuss it in more detail in the notes to this section.

Let Σ be any non-empty set called, in this context, an *alphabet.* The set Σ^* consists of all finite sequences of elements of Σ; the empty sequence is denoted by 1. Elements of Σ^* are called *strings* (or *words*) over Σ. A typical non-empty element of Σ^* is thus of the form $(x_1, \ldots, x_n)$ where $x_i \in X$. The *length* of this string is n. The set $\Sigma^\dagger$ is just the set Σ^* with the empty string removed. A binary operation called *concatenation* can be defined on Σ^* as follows: if $u = (x_1, \ldots, x_m)$ and $v = (y_1, \ldots, y_n)$ are strings then their concatenation uv is the string $(x_1, \ldots, x_m, y_1, \ldots, y_n)$. With respect to this operation, Σ^* is a monoid and $\Sigma^\dagger$ is a semigroup. The set Σ may be embedded in Σ^* or $\Sigma^\dagger$ by defining $\iota(x) = (x)$. It is usual to write strings without brackets and so we write $x_1 \ldots x_n$ rather than $(x_1, \ldots, x_n)$. Let x be a string in Σ^*. If $x = yz$, we say that y is a *prefix* of x and z is a *suffix*. If $x = uyv$ then y is called a *factor* of x.

To understand the importance of the semigroup Σ^* and the monoid $\Sigma^\dagger$, we need the following definition. Let $\mathcal{S}$ be a category of semigroups. The objects of this category are semigroups of a particular kind and the morphisms between them are a class of semigroup homomorphisms; we call these morphisms $\mathcal{S}$-homomorphisms. Let X be a non-empty set, S a semigroup in $\mathcal{S}$, and $\iota: X \to S$ a function. We say that (S, ι) is a *free $\mathcal{S}$-semigroup on X* if for every semigroup $T \in \mathcal{S}$ and function $\kappa: X \to T$ there is a unique $\mathcal{S}$-homomorphism $\theta: S \to T$ such that $\theta\iota = \kappa$. The category of semigroups just consists of semigroups and their homomorphisms, whereas the category of monoids consists of monoid homomorphisms. The proof of the following is straightforward.

Theorem 2 *For every non-empty set X, $(X^\dagger, \iota)$ is the free semigroup on X and (X^*, ι) is the free monoid on X.* ∎

Let X be a set and denote by X^{-1} a set disjoint from and bijective with X; the element of X^{-1} in bijective correspondence with $x \in X$ is denoted by x^{-1}. Form the free monoid $FMI(X) = (X \cup X^{-1})^*$, and define a unary operation on this semigroup as follows: $1^{-1} = 1$ and for each $y \in X \cup X^{-1}$ put $y^{-1} = x^{-1}$ if $y = x \in X$ and put $y^{-1} = x$ if $y = x^{-1} \in X^{-1}$. If $y_1 \dots y_n \in (X \cup X)^\dagger$ then define $(y_1 \dots y_n)^{-1} = y_n^{-1} \dots y_1^{-1}$. This unary operation turns $FMI(X)$ into a monoid with involution. In fact, $FMI(X)$ is the free monoid with involution on X; this simple result is proved in Theorem 6.1.1.

Free groups can easily be described in terms of $FMI(X)$. A string in $FMI(X)$ which contains no factor of the form xx^{-1} or $x^{-1}x$ for any $x \in X$ is said to be *reduced.* Any string in $FMI(X)$ can be transformed into a reduced string by deleting all occurrences of factors of the form $x^{-1}x$ and xx^{-1} until a reduced string is reached. It is a result of combinatorial group theory that, however this is done, the resulting reduced string is unique [212]. If u is a string then $\text{red}(u)$ denotes the corresponding reduced string. The free group, $FG(X)$, on X can be regarded as consisting of all reduced strings together with a product defined by $u \cdot v = \text{red}(uv)$.

Any subset of a free semigroup, monoid or group is called a *language.* Such languages arise naturally in combinatorial (semi)group theory. We are particularly interested in those languages which can be described in effective ways. The notion of an automaton provides one way in which this can sometimes be achieved.

An *automaton* is a 5-tuple $\mathbf{A} = (S, \Sigma, s_0, \delta, F)$ consisting of a set of *states* S, an alphabet Σ, an *initial state* $s_0 \in S$, a *transition function* δ which is a partial function $\delta: S \times \Sigma \to S$, and a set of *final states* $F \subseteq S$. Each letter $a \in \Sigma$ induces a partial function of S given by $s \mapsto \delta(s, a)$. Define a function γ from Σ to the monoid of partial functions of S (composed from left-to-right) by $(s)\gamma(a) = \delta(s, a)$. Then γ extends to a homomorphism, also denoted by γ, from Σ^* to the monoid of partial functions of S. The image of this homomorphism is called the *transition monoid* of **A**. The function $\delta^*: S \times \Sigma^* \to S$ is defined by $\delta^*(s, w) = (s)\gamma(w)$. The *language recognised by* **A** is the set $L(\mathbf{A})$ of all strings $x \in \Sigma^*$ such that $\delta^*(s_0, x) \in F$.

An automaton is said to be *injective* if each letter $a \in \Sigma$ induces a partial bijection on the set of states. An *inverse automaton* is a special kind of injective automaton: the alphabet Σ is of the form

$$\{a\colon a \in X\} \cup \{a^{-1}\colon a \in X\}$$

and the partial bijections induced by a and a^{-1} are mutually inverse in the symmetric inverse monoid $I(S)$. In describing an inverse automaton, it is enough to specify the partial bijection corresponding to either a or a^{-1}. An inverse automaton is called a *group automaton* if each of the partial bijections induced by a and a^{-1} is actually a bijection. The transition monoid of an inverse automaton is an inverse semigroup, whereas the transition monoid of a group automaton is a group.

The sets of states of an automaton is finite, and for this reason Proposition 1 can be used to construct group automata from inverse automata. To understand how, we need the following definition. Let $\mathbf{A} = (S, \Sigma, s_0, \delta, F)$ be an inverse automaton where $\Sigma = X \cup X^{-1}$. A group automaton **B** is said to be a *prolongation* of **A** if $\mathbf{B} = (S, \Sigma, s_0, \gamma, F)$ where γ is an extension of δ. The following result is now immediate by Proposition 1.

Proposition 3 *Every inverse automaton* **A** *can be prolonged to a group automaton* **B** *such that* $L(\mathbf{A}) \subseteq L(\mathbf{B})$. ■

We can now turn to the connection between combinatorial group theory and inverse semigroup theory. In what follows, the set X will always be finite and non-empty.

The free group $FG(X)$ has two basic properties. First of all, it is *residually finite*; this means that for each $w \in FG(X)$ there is a homomorphism θ to a finite group H such that $\theta(w) \neq 1$. Secondly, every subgroup of finite index contains a normal subgroup of finite index. An automata-theoretic proof of the second result may be found in [127]; we prove the first result below using inverse automata.

Theorem 4 *A finitely generated free group is residually finite.*

Proof Let $w \in FG(X)$ be a non-empty reduced string where $w = w_1 \ldots w_n$ and $w_i \in X \cup X^{-1}$. Construct an inverse automaton $\mathbf{LA}(w)$ as follows: there are $n+1$ states $s_0, \ldots, s_n$ and an edge labelled w_i from s_{i-1} to s_i for $i = 0, \ldots, n$. The initial state is s_0 and the final state is s_n. The string w is recognised by $\mathbf{LA}(w)$. Prolong this automaton to a group automaton **B** by

Proposition 3. There is therefore a homomorphism from $FG(X)$ to the transition group H of $\mathbf{B}$. Under this homomorphism the string w is mapped to a non-identity element of H because w is recognised by $\mathbf{B}$ and the initial state of $\mathbf{B}$ is not a terminal state. ■

The automaton $\mathbf{LA}(w)$ constructed in the proof of Theorem 4 is called the *linear inverse automaton* of w.

In 1950, Marshall Hall showed how to define a topology on any finitely generated free group [114]. Before we define it, we recall some definitions from topology.

A topological space is said to be *discrete* if every subset is open. If τ_1 and τ_2 are topologies on a set X then τ_1 is said to be *coarser* than τ_2 if $\tau_1 \subseteq \tau_2$. A *basis* for a topological space (X, τ) is a set of open sets β such that every element of τ is a union of elements of β.

Hall defined the *profinite topology* on a finitely generated free group to be the coarsest topology which makes every homomorphism onto a finite discrete group continuous. Using the properties of subgroups of finite index, it is easy to show that this topology can be described as the topology having as a basis all cosets of subgroups of finite index. Observe that if H is a subgroup of finite index, then it is not only open but also closed since the complement of H is the union of all open sets of the form Hg where $g \notin H$.

Although the definition of the topology is straightforward, deciding which sets are open (or closed) turns out to involve deep mathematical ideas. The first result in this area was obtained by Hall himself (Theorem 5 below) who proved that every finitely generated subgroup of a free group is closed in the profinite topology [113], [114]. Subsequently, Stallings reformulated Hall's work in terms of graph theory [396]. Meakin and Margolis [218] showed that Stallings' approach could be interpreted in terms of inverse automata. We now prove Hall's result using inverse automata.

Theorem 5 *Let H be a finitely generated subgroup of a finitely generated free group $FG(X)$. Then for every $w \in FG(X)$ such that $w \notin H$ there exists a subgroup of finite index K such that $H \subseteq K$ and $w \notin K$. In particular, H is closed in the profinite topology.*

Proof Let H be a finitely generated subgroup of the free group $FG(X)$ having reduced generators $Y = \{g_1, \ldots, g_n\}$. The first step in the proof is to construct an inverse automaton which recognises a reduced string precisely when it belongs to H.

We begin by constructing an automaton $\mathbf{FA}(Y)$ from Y called a *flower automaton*. This automaton has a unique initial and terminal state s_0 and one *petal* for each $g_i \in Y$ consisting of transitions which spell out g_i in a loop from s_0 to s_0. This automaton need not be inverse, but there is a simple procedure for converting it into one that is. Successively identify edges with the same label which either start or finish at the same state until no further identifications are possible. This process, called *folding*, produces an inverse automaton $\mathbf{A}(H)$ which can be shown to be independent of the foldings carried out and of the choice of reduced generators of H. The automaton $\mathbf{A}(H)$ has the desired property that a reduced string is recognised if, and only if, it belongs to H. We can use this automaton to prove Hall's theorem.

Let w be a reduced string which does not belong to H, so that w is not recognised by $\mathbf{A}(H)$. Then either w does not label a loop in $\mathbf{A}(H)$ based at s_0, or w does not label any path in $\mathbf{A}(H)$. In the first case, let $\mathbf{A}' = \mathbf{A}(H)$; in the second case, let $\mathbf{A}'$ be the automaton obtained from $\mathbf{A}(H)$ by glueing on a path reading w from the initial state adding new states where necessary. Then in either case $\mathbf{A}'$ is an inverse automaton which does not recognise w but which contains a sequence of edges labelled by the letters of w.

Let $\mathbf{B}$ be a group automaton which prolongs $\mathbf{A}'$, and let H_w be the language recognised by $\mathbf{B}$. Clearly, $H \subseteq H_w$ and $w \notin H_w$ by the construction of $\mathbf{A}'$. The language H_w is a subgroup of $FG(X)$, which has finite index in $FG(X)$ because H_w contains the kernel of the induced homomorphism from $FG(X)$ to the transition group of $\mathbf{B}$. Hence $K = H_w$.

It follows that H is the intersection of all subgroups of the form H_w where $w \in FG(X) \setminus H$. We observed earlier that subgroups of finite index are closed (as well as open). Thus H is closed in the profinite topology. ■

It is evident that Theorem 4 is a special case of Theorem 5. The next result generalises Theorem 5. We state it without proof, and then discuss its significance.

Theorem 6 *Let $H_1, \ldots, H_n$ be finitely generated subgroups of a finitely generated free group $FG(X)$. If $h \notin H_1 \ldots H_n$ then there are subgroups $K_1, \ldots, K_n$ each of which has finite index such that $H_i \subseteq K_i$ for each i and $h \notin K_1 \ldots K_n$. In particular, the subset $H_1 \ldots H_n$ is closed in the profinite topology.* ■

Theorem 6 was conjectured by Pin and Reutenauer in 1991 [322], who proved that it implied the type II conjecture. A direct proof of Theorem 6 was first provided by Ribes and Zalesskiĭ in 1993 [355] using profinite groups acting on graphs. Recently, an alternative proof has been found by Herwig and Lascar from a model-theoretic point of view [130]. Their approach fits in well with the main thesis of this section, so we briefly discuss their results.

A *cycle-free n-partitioned graph* is a graph satisfying the following conditions: the vertices may be partitioned into n subsets $U_1, \ldots, U_n$ in such a way that there is an edge joining the vertices u and v only when either $u \in U_i$ and $v \in U_{i+1}$ for $i = 1, \ldots, n-1$ or $u \in U_n$ and $v \in U_1$, together with the condition that there is no sequence of vertices $v_1, \ldots, v_n$ such that there are edges joining v_1 and v_2, v_2 and v_3, ..., v_{n-1} and v_n, and v_n and v_1. Herwig and Lascar prove the following result.

Theorem 7 *Let $\mathcal{G}$ be a finite cycle-free n-partitioned graph and let P be a set of partial automorphisms of $\mathcal{G}$. If $\mathcal{G}$ can be embedded in a cycle-free n-partitioned graph $\mathcal{G}'$ in such a way that every element of P can be extended to an automorphism of $\mathcal{G}'$, then $\mathcal{G}$ can be embedded in a finite cycle-free n-partitioned graph having the same property.* ■

It is clear that Proposition 1(1) is a trivial consequence of the above theorem. And just as Proposition 1(1), mediated by Proposition 3, is enough to establish Theorem 5, so too Theorem 7 can be used to establish Theorem 6; in fact, Herwig and Lascar show that Theorem 6 and Theorem 7 are equivalent.

Theorem 6 is not the most general result known about the profinite topology on free groups. To formulate this, we need some further definitions. Let $FG(X)$ be the free group on the finite set X. If H is a subgroup of $FG(X)$ then we write $x \equiv y \bmod H$ to mean that $xH = yH$. Let n be a positive integer, and let Y be a finite set (of unknowns). A *left-system*, which we denote by (E), is a finite set of equations of the form $x \equiv_i yg$ or $x \equiv_i g$, where $1 \leq i \leq n$, $x, y \in Y$ and $g \in FG(X)$. Let $\mathcal{H} = (H_1, \ldots, H_n)$ be a sequence of subgroups of $FG(X)$. A *solution of* (E) *modulo* $\mathcal{H}$ is a family $\{g_y \colon y \in Y\}$ of elements of $FG(X)$ such that for each equation $x \equiv_i yg$ in (E) the congruence $g_x \equiv g_y g$ holds mod H_i, and for each equation $x \equiv_i g$ in (E) the congruence $g_x \equiv g$ holds mod H_i.

Herwig and Lascar prove the following result.

Theorem 8 *Let n be a positive integer, and let (E) be a left-system in a finitely generated free group. Let $\mathcal{H} = (H_1, \ldots, H_n)$ be a sequence of finitely generated subgroups. If (E) does not have a solution modulo $\mathcal{H}$ then there is a sequence $\mathcal{K} = (K_1, \ldots, K_n)$ of subgroups of finite index such that $H_i \subseteq K_i$ and (E) has no solution modulo $\mathcal{K}$.* ■

Clearly, Theorem 8 generalises Theorem 6. Herwig and Lascar also prove a generalisation of Theorem 7 concerning the extension properties of partial automorphisms of a class of relational structures; this, in turn, is equivalent to Theorem 8.

At the beginning of this section we mentioned Rhodes' type II conjecture. This was proved by Chris Ash [11] as a corollary of a much more general result. Recently, Almeida and Delgado [5] have proved that Ash's main theorem is equivalent to Theorem 8.

Concluding remarks

We argued in Chapter 1 that partial symmetries were intrinsically important because they captured an aspect of our intuitions about symmetry which global symmetries did not. The examples above add another dimension to this. They suggest that even when our primary interest may be focussed on symmetries, we may still need to study them by means of partial symmetries. One reason for this can be simply stated: partial information about a symmetry can be represented by a partial symmetry. Thus the first step in constructing a symmetry may well be to construct a partial symmetry and then to extend it. A final example makes the point: in the construction of the Cayley graph of a group presentation using the Todd–Coxeter algorithm [74], a sequence of unambiguous partial Cayley graphs is constructed each of which defines an inverse semigroup approximation to the group specified by the presentation [219].

In the next section, we show how the Witt-property and the HNN-property lead to definitions in the theory of inverse semigroups.

2.2 E-unitary covers

An inverse monoid is said to be *factorisable* if every element lies beneath an element of the group of units with respect to the natural partial order. Thus structures satisfying the Witt-property are precisely those in which the semigroup of partial automorphisms is factorisable.

Proposition 1 *Let S be an inverse monoid.*

(1) *Let G be a subgroup of the group of units of S and put $T = [G]$. Then T is a factorisable inverse monoid with G as its group of units.*

(2) *Let T be an inverse subsemigroup of S such that for each $t \in T$ there exists an element $g \in U(S)$ such that $t \leq g$. Then T can be embedded in a factorisable inverse monoid.*

Proof (1) Firstly, T is closed under multiplication, because if $s \leq g$ and $t \leq h$ where $g, h \in G$ then $st \leq gh \in G$. From $s \leq g$ we have $s^{-1} \leq g^{-1} \in G$, and so T is closed under inversion. Thus T is an inverse semigroup. It is now immediate that T is factorisable with group of units G.
(2) Put $S' = [U(S)]$. By (1) above, this is a factorisable inverse monoid. Clearly, $T \subseteq S'$. ■

The proof of the following result is immediate from the proof of Proposition 2.1.1 and Proposition 1(2).

Proposition 2 *Let X be a set.*

(1) *$I(X)$ is factorisable if, and only if, X is finite.*

(2) *If X is infinite then $I(X)$ can be embedded in a factorisable inverse monoid.* ■

If we combine the above result with the Wagner–Preston representation theorem we obtain the following.

Theorem 3 *Every (finite) inverse semigroup can be embedded in a (finite) factorisable inverse monoid.* ■

The notion which is dual to that of an embedding is that of a cover (Section 1.4). We shall show how embeddings into factorisable inverse semigroups give rise to coverings belonging to another important class of semigroups. We need some definitions first.

Let θ: $S \to T$ be a homomorphism between inverse semigroups. By Proposition 1.4.21, θ induces a homomorphism $(\theta \mid E(S))$ from $E(S)$ to $E(T)$. If this induced homomorphism is also injective, then θ is said to be *idempotent-separating*.

An inverse semigroup is *E-unitary* if, whenever e is an idempotent and $e \leq s$, then s is an idempotent.

The result below is known as the McAlister covering theorem.

Theorem 4 *For every (finite) inverse semigroup S there is a (finite) E-unitary inverse semigroup P and a surjective, idempotent-separating homomorphism θ: $P \to S$.*

Proof Every (finite) inverse semigroup can be embedded in a (finite) factorisable inverse monoid by Theorem 3. Let ι: $S \to F$ be an embedding of the inverse semigroup S in the factorisable inverse monoid F, where F is finite if S is finite. Put

$$P = \{(s, g) \in S \times U(F) \colon \iota(s) \leq g\}.$$

We show first that P is an inverse subsemigroup of the direct product inverse semigroup $S \times U(F)$. Let $(s, g), (t, h) \in P$. Then $\iota(s) \leq g$ and $\iota(t) \leq h$. Thus $\iota(st) \leq gh$, so that $(st, gh) \in P$. Hence P is closed under multiplication. Let $(s, g) \in P$. Then $\iota(s) \leq g$, so that $\iota(s^{-1}) \leq g^{-1}$. Hence $(s^{-1}, g^{-1}) \in P$, and so P is closed under inverses. It follows that P is an inverse semigroup. To locate the idempotents of P, suppose that $(s, g)^2 = (s, g)$. Then $s^2 = s$ and $g = 1$. Thus $(s, g) = (e, 1)$ where e is an idempotent of S. Conversely, let e be any idempotent of S. Then $\iota(e) \leq 1$, the identity of F. Thus $(e, 1) \in P$ and $(e, 1)$ is an idempotent. Hence $E(P) = E(S) \times \{1\}$. To prove that P is E-unitary, let $(e, 1) \leq (s, g)$. Then $(e, 1) = (e, 1)(s, g)$. Thus $e = es$ and $1 = 1g$. It follows that $g = 1$. But by assumption $\iota(s) \leq g$ and so s is an idempotent. Hence (s, g) is an idempotent.

Define $\theta: P \to S$ by $\theta(s, g) = s$. Then θ is surjective, for if $s \in S$ then $\iota(s) \leq g$ for some $g \in U(F)$, since F is factorisable. Thus $(s, g) \in P$ and $\theta(s, g) = s$. That θ is a homomorphism is immediate. Finally, θ is idempotent-separating, for suppose $\theta(e, 1) = \theta(f, 1)$. Then $e = f$ and so $(e, 1) = (f, 1)$. ■

We say that P is an *E-unitary cover* of S. We shall develop the theory of E-unitary covers further in Section 2.4, before we can do that, however, we need to understand something about congruences on inverse semigroups.

2.3 Congruences

Congruences are the semigroup-theoretic way of handling homomorphisms. In the case of groups, congruences may be replaced by normal subgroups; something similar can be carried out for inverse semigroups as we show in Chapter 5. However, much of our work will use congruences themselves, and in this section we develop what we need.

Definition and properties

Let $\theta: S \to T$ be a homomorphism of semigroups. The *kernel* of θ is the relation $\ker\theta$ defined on S by:

$$\ker\theta = \{(a, b) \in S \times S: \theta(a) = \theta(b)\}.$$

It is straightforward to check that $\ker\theta$ is an equivalence relation on S which possesses the following additional property:

$$(a, b), (c, d) \in \ker\theta \Rightarrow (ac, bd) \in \ker\theta.$$

More generally, a *congruence* on a semigroup S is an equivalence relation ρ on S such that $(a, b), (c, d) \in \rho$ implies $(ac, bd) \in \rho$. It is useful to have one-sided versions of this definition as well: an equivalence relation is said to be a *left congruence* if $(a, b) \in \rho$ implies $(ca, cb) \in \rho$ for any $c \in S$. *Right congruences* are defined dually. It is left as an exercise to show that an equivalence relation is a congruence precisely when it is a left and right congruence. A congruence is said to be *proper* if $(a, b) \in \rho$ for some distinct elements $a, b \in S$.

We conclude that every semigroup homomorphism determines a congruence on its domain. The reverse is also true. Let ρ be an arbitrary congruence on the semigroup S. We denote the set of ρ-equivalence classes (or *congruence classes*) by S/ρ. A binary operation may be defined on the set S/ρ by mapping $(\rho(a), \rho(b))$ to $\rho(ab)$. This is well-defined because ρ is a congruence. It is usual to denote the binary operation by $\rho(a)\rho(b)$, but it is important to note that here we do not mean the product of subsets of a semigroup; indeed, as subsets we have that $\rho(a)\rho(b) \subseteq \rho(ab)$. Those congruences for which the binary operation on S/ρ coincides with the product of subsets are called *perfect*. For example, all congruences on groups are perfect. The set S/ρ is a semigroup with respect to this binary operation, called the *quotient* of S by ρ. The function which assigns to each element of S its congruence class in S/ρ is a homomorphism called the *associated natural homomorphism*, often denoted $\rho^\natural$. The following theorem is the first isomorphism theorem for semigroup homomorphisms.

Theorem 1 *Let* θ: $S \to T$ *be a homomorphism of semigroups and put* $\rho = \ker\theta$.

(1) ρ *is a congruence on* S, *and* $\rho^\natural$: $S \to S/\rho$ *is a surjective homomorphism.*

(2) *There is a unique injective homomorphism* ϕ: $S/\rho \to T$ *such that* $\phi\rho^\natural = \theta$. *If* θ *is surjective then* ϕ *is an isomorphism.*

Proof We have already dealt with the proof of (1).
(2) Define ϕ: $S/\rho \to T$ by $\phi(\rho(s)) = \theta(s)$. It is straightforward to check that ϕ is a well-defined injective homomorphism, and that it is the unique homomorphism such that $\phi\rho^\natural = \theta$. It is immediate that if θ is surjective then ϕ is surjective and so an isomorphism. ∎

There is a generalisation of the above result which is often useful.

Theorem 2 *Let* θ: $S \to T$ *be a homomorphism and let* ϕ: $S \to U$ *be a surjective homomorphism such that* $\ker\phi \subseteq \ker\theta$. *Then there is a unique homomorphism* ψ: $U \to T$ *such that* $\theta = \psi\phi$. *If* θ *is surjective then* ψ *is surjective.*

Proof Define ψ: $U \to T$ by $\psi(u) = \theta(s)$ where $\phi(s) = u$. It is easy to check that ψ is a well-defined homomorphism and the unique homomorphism such that $\theta = \psi\phi$. It is clear that if θ is surjective then ψ is surjective. ∎

The set of all congruences on a semigroup forms a lattice: the intersection of any non-empty set of congruences is again a congruence and the join of a non-empty set of congruences is the intersection of all congruences which contain their union. We may therefore refer to the *congruence lattice* of a semigroup.

If ρ is any relation on a semigroup S, then the intersection of all congruences containing ρ is a congruence, denoted $\rho^\sharp$, called the *congruence generated by* ρ. A more explicit description of $\rho^\sharp$ can be obtained in the following way. Denote by ρ^{-1} the relation defined by

$$(x, y) \in \rho^{-1} \Leftrightarrow (y, x) \in \rho.$$

We call ρ^{-1} the *converse* of ρ. Let $a, b \in S$. We say that *a is connected to b by an elementary ρ-transition* if there exists a pair of elements $c, d \in S$ and elements $x, y \in S^1$ such that $a = xcy$, $b = xdy$ and $(c, d) \in \rho \cup \rho^{-1}$. We write $a \to b$ to mean that there is an elementary ρ-transition connecting a to b. The proof of the following is straightforward.

Proposition 3 *Let ρ be a relation on the semigroup S. Then $(a, b) \in \rho^\sharp$ if, and only if, either $a = b$ or for some $n \in \mathbb{N}$ there is a sequence*

$$a = a_1 \to a_2 \to \ldots \to a_n = b$$

of elementary ρ-transitions connecting a to b. ■

Congruences on inverse semigroups have some additional properties (see also Proposition 4.3.2).

Proposition 4 *Let ρ be a congruence on an inverse semigroup S.*

(1) *If $(s, t) \in \rho$ then*

$$(s^{-1}, t^{-1}) \in \rho,\ (s^{-1}s, t^{-1}t) \in \rho \text{ and } (ss^{-1}, tt^{-1}) \in \rho.$$

(2) *If $(s, e) \in \rho$, where e is an idempotent, then*

$$(s, s^{-1}) \in \rho,\ (s, s^{-1}s) \in \rho \text{ and } (s, ss^{-1}) \in \rho.$$

Proof (1) Let $(s, t) \in \rho$. Then $\rho(s) = \rho(t)$ in the inverse semigroup S/ρ. By Proposition 1.4.21(1), $\rho(s^{-1}) = \rho(s)^{-1}$ and $\rho(t^{-1}) = \rho(t)^{-1}$. Thus $\rho(s^{-1}) = \rho(t^{-1})$, that is $(s^{-1}, t^{-1}) \in \rho$. The remaining proofs are straightforward. ■

Rees quotients

Although ideals are useful in semigroup theory, it is important to realise that the connection between ideals and congruences is a good deal more tenuous for semigroups than it is for rings. If ρ is a congruence on a semigroup S with zero, then the set $I = \rho(0)$ is an ideal of S; however, examples show that the congruence is not determined by this ideal.

Nevertheless, ideals can be used to construct rather simple congruences on semigroups. Let I be an ideal in the semigroup S. Define a relation ρ_I on S by:

$$(s, t) \in \rho_I \Leftrightarrow \text{either } s, t \in I \text{ or } s = t.$$

Then ρ_I is a congruence. The quotient semigroup S/ρ_I is isomorphic to the set $S \setminus I \cup \{0\}$ (we may assume that $0 \notin S \setminus I$) equipped with the following product: if $s, t \in S \setminus I$ then their product is st if $st \in S \setminus I$, all other products are defined to be 0. Such quotients are called *Rees quotients*. If I is an ideal in a semigroup S then we say that S is an *ideal extension of I by S/I*.

Syntactic congruences

Let S be a monoid and $L \subseteq S$. Define a relation ρ_L on S by:

$$(s,t) \in \rho_L \Leftrightarrow (\forall a, b \in S)(asb \in L \Leftrightarrow atb \in L).$$

It is left as an exercise to check that ρ_L is a congruence on the monoid S; it is called the *syntactic congruence* of L. This congruence occurs most frequently in connection with subsets of free monoids, or languages as we have defined them. To every language L over the alphabet Σ we may construct its syntactic congruence ρ_L on Σ^*. The monoid Σ^*/ρ_L is called the *syntactic monoid* of the language L.

Semigroup presentations

One of the ways in which semigroups can be defined is by means of presentations.

Let S be a semigroup. Regarding $X = S$ as a set, we can form the free semigroup $X^\dagger$. Define $\theta: X^\dagger \to S$ by $\theta(s_1, \ldots, s_n) = s_1 \ldots s_n$, where the right-hand product is calculated in S. Then θ is a surjective homomorphism and so S is isomorphic to the semigroup $X^\dagger/\ker\theta$. Consequently, every semigroup is a quotient of a free semigroup.

More generally, let X be a set and let ρ be a relation on $X^\dagger$. Then the semigroup $X^\dagger/\rho^\sharp$ is denoted by $\langle X \colon \rho\rangle$. A semigroup isomorphic to such a semigroup is said to be *generated by X subject to the relations ρ*. A *semigroup presentation* for S is a semigroup of the form $\langle X \colon \rho\rangle$ isomorphic to S. A semigroup is said to be *finitely generated* if it is isomorphic to a quotient of some free semigroup $X^\dagger$ for X finite, and it is said to be *finitely presented* if it is isomorphic to a semigroup of the form $\langle X \colon \rho\rangle$ where both X and ρ are finite. The quotient semigroup $X^\dagger/\rho^\sharp$ is usually denoted by $\langle x_i, i \in I \colon u_j = v_j, j \in K\rangle$, where $X = \{x_i \colon i \in I\}$ and $\rho = \{(u_j, v_j) \colon j \in K\}$.

It can be very difficult to determine the structure of a semigroup S described by means of a presentation $\langle X \colon \rho\rangle$. In particular, different strings in $X^\dagger$ may belong to the same congruence class modulo $\rho^\sharp$ and so determine the same element in S. A pair of strings in $X^\dagger$ are said to be *equivalent* if they are $\rho^\sharp$-related; if the strings u and v are equivalent we shall often write $u \equiv v$, when the relation $\rho^\sharp$ is clear, whereas if u and v are the same string we shall write $u = v$. The problem of finding an algorithm to decide when two strings are equivalent is called the *word problem*; it may not be solvable.

We now describe a result which is often helpful in solving word problems. Let S be a semigroup generated by $\{s_i \colon i \in I\}$, and let X be the set $\{x_i \colon i \in I\}$. Then S is a homomorphic image of $X^\dagger$ under the homomorphism θ which maps x_i to s_i. Let $u, v \in X^\dagger$. If $\theta(u) = \theta(v)$ in S then we say that S *satisfies the relation $u = v$ via θ*.

Proposition 5 *Let S be a semigroup generated by $\{s_i \colon i \in I\}$, and let X be the set $\{x_i \colon i \in I\}$. Let $\theta \colon X^\dagger \to S$ be the homomorphism which maps x_i to s_i. Let ρ be a relation on $X^\dagger$ such that $\theta(u) = \theta(v)$ for every $(u, v) \in \rho$. Then S is a homomorphic image of $\langle X \colon \rho \rangle$.*

Proof By assumption, $\rho \subseteq \ker \theta$, so that $\rho^\sharp \subseteq \ker \theta$. By Theorem 2, there is a surjective homomorphism from $X^\dagger / \rho^\sharp$ to S. ∎

Let $S = \langle X \colon \rho \rangle$ and put $\tau = \rho^\sharp$. A *transversal*, T, for τ is a subset of S which contains exactly one element from each τ-class. Let $\kappa \colon X^\dagger \to X^\dagger$ be the function which maps each element $u \in X^\dagger$ to the element $T \cap \tau(u)$. We may define a binary operation $\cdot$ on the set T by $u \cdot v = \kappa(uv)$. Then S is isomorphic to T with this product. If we can find such a transversal and an algorithm to compute τ then we can solve the word problem. In this case, the elements of the transversal are called *normal forms*.

Everything we have said in this section can be applied to other classes of semigroups. The category of monoids consists of monoids and monoid homomorphisms. Free monoids are the semigroups of the form X^*, and lead to *monoid presentations*. If we adjoin zeros to the free semigroup and the free monoid, then we obtain free semigroups with zero and free monoids with zero respectively, where homomorphisms are also required to map zeros to zeros. We obtain corresponding notions of presentation in each case.

2.4 The minimum group congruence

In this section, we return to the theory of E-unitary semigroups and E-unitary covers. The key idea is to relate inverse semigroups to groups by means of a special congruence.

Definition and characterisation

The relation σ is defined on the inverse semigroup S by

$$s \,\sigma\, t \Leftrightarrow \exists u \leq s, t$$

for all $s, t \in S$.

Theorem 1 *Let S be an inverse semigroup.*

(1) *σ is the smallest congruence on S containing the compatibility relation.*

(2) *S/σ is a group.*

(3) *If ρ is any congruence on S such that S/ρ is a group then $\sigma \subseteq \rho$.*

Proof (1) We begin by showing that σ is an equivalence relation. Reflexivity and symmetry are immediate. To prove transitivity, let $(a,b),(b,c) \in \sigma$. Then there exist elements $u, v \in S$ such that $u \leq a, b$ and $v \leq b, c$. Thus $u, v \leq b$. The set $[b]$ is a compatible subset by Lemma 1.4.14, and so $u \wedge v$ exists by Lemma 1.4.11. But $u \wedge v \leq a, c$ and so $(a,c) \in \sigma$. The fact that σ is a congruence follows from the fact that the natural partial order is compatible with the multiplication. The inclusion of the compatibility relation in σ follows from Lemma 1.4.11.

Let ρ be any congruence containing $\sim$, and let $(a,b) \in \sigma$. Then $z \leq a, b$ for some z. Thus $z \sim a$ and $z \sim b$. By assumption $(z,a),(z,b) \in \rho$. But ρ is an equivalence and so $(a,b) \in \rho$. Thus $\sigma \subseteq \rho$.
(2) Clearly, all idempotents are contained in a single σ-class (possibly with non-idempotent elements). Consequently, S/σ is an inverse semigroup with a single idempotent. Thus S/σ is a group by Proposition 1.4.4.
(3) Let ρ be any congruence such that S/ρ is a group. Let $(a,b) \in \sigma$. Then $z \leq a, b$ for some z. Hence $\rho(z) \leq \rho(a), \rho(b)$. But S/ρ is a group and so its natural partial order is equality. Hence $\rho(a) = \rho(b)$. ■

The congruence σ is called the *minimum group congruence.* The properties of this congruence lead naturally to the following result on the category of inverse semigroups.

Theorem 2 *The category of groups is a reflective subcategory of the category of inverse semigroups.*

Proof Let S be an inverse semigroup and $\sigma^\natural\colon S \to S/\sigma$ the natural homomorphism. Let $\theta\colon S \to G$ be a homomorphism to a group G. Then $\ker\theta$ is a group congruence on S and so $\sigma \subseteq \ker\theta$ by Theorem 1. Thus by Theorem 2.3.2, there is a unique homomorphism θ^* from S/σ to G making the diagram commute. ■

It follows by standard category theory (Chapter IV, Section 3 of [210]) that there is a functor from the category of inverse semigroups to the category of groups which takes each inverse semigroup S to S/σ. If $\theta\colon S \to T$ is a homomorphism of inverse semigroups then the function $\psi\colon S/\sigma \to T/\sigma$ defined by $\psi(\sigma(s)) = \sigma(\theta(s))$ is the corresponding group homomorphism (this can be checked directly).

Groups and groupoids of germs

The minimum group congruence has an important application in pseudogroup theory. Let Γ be the inverse monoid of all partial homeomorphisms between the open subsets of the topological space X. Let x be an arbitrary element of X. Consider the subset S of Γ consisting of all those elements of Γ which fix x. It is straightforward to check that S is an inverse submonoid of Γ. In topology, one is often interested in the local behaviour about the point x, which means behaviour on an open set containing x. For this reason, two elements of S which agree when restricted to some open set containing x should be regarded as equal. However, this relation is none other than the minimum group congruence on S. The corresponding congruence classes are called *germs* and the group S/σ is called the *group of germs* at x.

The disadvantage of this construction is that we have to choose one specific point. We can get around this difficulty in the following way. Define the set of *pointed partial homeomorphisms* of Γ to be the set

$$\Gamma' = \{(\theta, x) \in \Gamma \times X \colon x \in \operatorname{dom}\theta\}.$$

Anticipating definitions to be found in Section 3.1, this set may be endowed with the structure of a category by defining $\mathbf{d}(\theta, x) = (1, x)$ and $\mathbf{r}(\theta, x) = (1, \theta(x))$, where 1 is the identity of Γ, and defining the partial product of (θ, x) and (ψ, y) to be defined and equal to $(\theta\psi, y)$ only when $\mathbf{d}(\theta, x) = \mathbf{r}(\psi, y)$. Observe that the endomorphism monoid at $(1, x)$ is none other than the inverse monoid S we defined above. The category Γ' is an *inverse category* in the sense that for every morphism x there is a unique morphism x' such that $x = xx'x$ and $x' = x'xx'$. In this case, $(\theta, x)' = (\theta^{-1}, \theta(x))$. It turns out that much of the theory we have developed for inverse semigroups can be extended to inverse categories. In particular, we can define a congruence σ (in the sense of category theory [210]) on any inverse category C, such that C/σ is a groupoid and such that if ρ is any congruence on C for which C/ρ is a groupoid then $\sigma \subseteq \rho$. In the case of Γ', the groupoid Γ'/σ is called the *groupoid of germs* of the pseudogroup Γ. Additionally, Γ'/σ may be equipped with a topology in such a way that it becomes a topological groupoid. Such groupoids are important in the theory of local structures described in Section 1.2. They are discussed in more detail in Section 9.2.

E-unitary semigroups

E-unitary semigroups were introduced in Section 2.2 in connection with factorisable embeddings. We now provide some alternative characterisations of this important class of semigroups. First, we account for the terminology E-unitary.

A subset A of an inverse semigroup S is said to be *left* (resp. *right*) *unitary* if $a \in A$, $s \in S$ and $as \in A$ (resp. $sa \in A$) imply $s \in A$. A subset is said to be *unitary* if it is both left and right unitary.

Proposition 3 *Let S be an inverse semigroup. Then the following are equivalent:*

(1) *$E(S)$ is left unitary.*

(2) *$E(S)$ is right unitary.*

(3) *If e is an idempotent and $e \leq s$ then s is an idempotent.*

Proof (1) $\Rightarrow$ (2). Suppose that $E(S)$ is left unitary and $se \in E(S)$ where $e \in E(S)$. Then $se = fs$ for some idempotent f by Lemma 1.4.2. Thus s is an idempotent, since S is left unitary. Hence S is right unitary.
(2) $\Rightarrow$ (3). Suppose that $E(S)$ is right unitary and $e \leq s$ for some idempotent e. Then $e = se$ and so s is an idempotent.
(3) $\Rightarrow$ (1). Let $es = f$ be an idempotent. Then $f \leq s$ and so s is an idempotent. ■

E-unitary inverse semigroups first came to light as a consequence of the following result.

Theorem 4 *Let S be an inverse semigroup. Then the compatibility relation is transitive if, and only if, S is E-unitary.*

Proof Suppose that $\sim$ is transitive. Let $e \leq s$, where e is an idempotent. Then se^{-1} is an idempotent because $e = se = se^{-1}$, and $s^{-1}e$ is an idempotent because $s^{-1}e \leq s^{-1}s$. Thus $s \sim e$. Clearly $e \sim s^{-1}s$, and so, by our assumption that the compatibility relation is transitive, we have that $s \sim s^{-1}s$. But $s(s^{-1}s)^{-1} = s$, so that s is an idempotent.

Conversely, suppose that S is E-unitary and that $s \sim t$ and $t \sim u$. Clearly $(s^{-1}t)(t^{-1}u)$ is an idempotent and

$$(s^{-1}t)(t^{-1}u) = s^{-1}(tt^{-1})u \leq s^{-1}u.$$

But S is E-unitary and so $s^{-1}u$ is an idempotent. Similarly, su^{-1} is an idempotent. Hence $s \sim u$. ■

A congruence ρ on an inverse semigroup S is said to be *idempotent pure* if $a \in S$ and $e \in E(S)$ and $(a, e) \in \rho$ then a is an idempotent.

Proposition 5 *Let S be an inverse semigroup. Then a congruence ρ is idempotent pure if, and only if, $\rho \subseteq \sim$.*

Proof Let ρ be idempotent pure and let $(a, b) \in \rho$. Then $(ab^{-1}, bb^{-1}) \in \rho$. But ρ is idempotent pure and bb^{-1} is an idempotent. Thus ab^{-1} is an idempotent. Similarly, $a^{-1}b$ is an idempotent. Thus $a \sim b$.

Conversely, let ρ be a congruence contained in the compatibility relation. Let $(a, e) \in \rho$, where e is an idempotent. Then $(aa^{-1}, a) \in \rho$ by Proposition 2.3.4 and so $a \sim aa^{-1}$. But $a(a^{-1}a)^{-1} = a$, and so a is an idempotent. Hence ρ is idempotent pure. ■

Idempotent pure congruences, the minimum group congruence and E-unitary semigroups are all linked by the following result.

Theorem 6 *Let S be an inverse semigroup. Then the following conditions are equivalent:*

(1) *S is E-unitary.*

(2) $\sim \; = \sigma$.

(3) *σ is idempotent pure.*

(4) *$\sigma(e) = E(S)$ for any idempotent e.*

Proof (1) $\Rightarrow$ (2). By Theorem 1, the compatibility relation is contained in σ. Let $(a, b) \in \sigma$. Then $z \leq a, b$ for some z. It follows that $z^{-1}z \leq a^{-1}b$ and $zz^{-1} \leq ab^{-1}$. But S is E-unitary and so $a^{-1}b$ and ab^{-1} are both idempotents. Hence $a \sim b$.
(2) $\Rightarrow$ (3). By Proposition 5, a congruence is idempotent pure precisely when it is contained in the compatibility relation.
(3) $\Rightarrow$ (4). This is immediate from the definition of an idempotent pure congruence.
(4) $\Rightarrow$ (1) Suppose that $e \leq a$ where e is an idempotent. Then $(e, a) \in \sigma$. But by (4), the element a is an idempotent. ■

E-unitary covers over a group

The notion of an E-unitary cover may be refined with the help of the minimum group congruence. To do this, we first refine our notion of an embedding of an inverse semigroup in a factorisable inverse monoid. Suppose $\iota \colon S \rightarrow F$ is such an embedding. Because F is factorisable, each element $\iota(s)$ lies beneath some element of $U(F)$. However, it is quite possible for there to exist elements of $U(F)$ which do not lie above any element in the image of ι. We single out the case where this does not occur. An embedding $\iota \colon S \rightarrow F$ of an inverse semigroup S in a factorisable inverse monoid F is said to be *strict* if for each $g \in U(F)$ there exists $s \in S$ such that $\iota(s) \leq g$.

If an embedding $\iota \colon S \rightarrow F$ is not strict, then it can easily be converted to one that is: let

$$H = \{g \in U(F) \colon \iota(s) \leq g \text{ for some } s \in S\}.$$

Then H is a subgroup of $U(F)$. Put $F' = [H]$, a factorisable inverse submonoid of F with group of units H, by Proposition 2.2.1. Then the embedding $\iota: S \to F'$ is strict.

Theorem 7 *Let S be an inverse semigroup, and let $\iota: S \to F$ be a strict factorisable embedding. Let P be the E-unitary cover of S constructed from the embedding by the method of Theorem 2.2.4. Then P/σ is isomorphic to $U(F)$.*

Proof By the construction in Theorem 2.2.4

$$P = \{(s, g) \in S \times U(F) \colon \iota(s) \leq g\}.$$

We first prove that

$$(s, g) \, \sigma \, (t, h) \Leftrightarrow g = h.$$

Let $(s, g) \, \sigma \, (t, h)$. Then there is an element $(a, k) \in P$ such that $(a, k) \leq (s, g), (t, h)$. From the form taken by the order on P, this implies that $k = g = h$. In particular, $g = h$.

Conversely, consider the elements $(s, g), (t, g) \in P$. By definition, $\iota(s) \leq g$ and $\iota(t) \leq g$. Thus by Lemma 1.4.14, the elements $\iota(s)$ and $\iota(t)$ are compatible. In particular, $\iota(s^{-1}t)$ and $\iota(st^{-1})$ are idempotents. But ι is an embedding and so $s^{-1}t$ and st^{-1} are idempotents. Hence $s \sim t$. Thus there exists an element $a \in S$ such that $a \leq s, t$. Clearly, $(a, g) \in P$ and $(a, g) \leq (s, g), (t, g)$. Hence $(s, g) \, \sigma \, (t, g)$.

Define ϕ: $P/\sigma \to U(F)$ by $\phi(\sigma(s, g)) = g$, which is well-defined by the above result. This function is clearly a homomorphism; it is surjective because $\iota: S \to F$ is strict; and it is injective because if $\phi(\sigma(s, g)) = \phi(\sigma(t, h))$ then $g = h$ and so $\sigma(s, g) = \sigma(t, h)$. ■

We likewise refine our definition of an E-unitary cover. Let P be an E-unitary semigroup and suppose that there is a surjective, idempotent-separating homomorphism θ: $P \to S$. Then we say that P is an *E-unitary cover of the semigroup S over the group P/σ.*

Thus every strict factorisable embedding $\iota: S \to F$ gives rise to an E-unitary cover of S over a group isomorphic to $U(F)$. In Chapter 8, we prove that every E-unitary cover of S over a group G gives rise to a strict factorisable embedding of S into a factorisable inverse monoid with group of units G. Consequently, the theory of E-unitary covers of S is equivalent to the theory of factorisable embeddings, and both theories are intimately connected with the problem of extending partial bijections to bijections.

The Ore embedding theorem

We now describe a simple application of extending partial bijections to bijections.

A monoid S is said to be *left* (resp. *right*) *cancellative* if $ax = ay$ implies $x = y$ (resp. $xa = ya$ implies $x = y$). A monoid which is both left and right cancellative is said to be *cancellative.* Every submonoid of a group is cancellative, but the converse is not true: there are cancellative monoids which cannot be embedded in groups [48]. The Ore embedding theorem provides sufficient conditions for a cancellative monoid to be embedded in a group. We shall prove this result by means of inverse semigroups using an approach due to Rees [335].

Let S be a cancellative monoid. For each element $a \in S$, define the function $\lambda_a: S \to S$ by $\lambda_a(s) = as$. Since S is cancellative, these functions are injective. Thus the function $\lambda: S \to I(S)$ given by $\lambda(a) = \lambda_a$ is well-defined. It is straightforward to check that λ is an injective homomorphism. Thus S is isomorphic to the submonoid $S' = \{\lambda_a: a \in S\}$ of $I(S)$. Let $\Sigma(S)$ be the inverse submonoid of $I(S)$ generated by S'. Clearly, S is embedded in $\Sigma(S)$. The inverse semigroup $\Sigma(S)$ will be the centre of our attentions. Using the terminology introduced in Section 1.5, every element of S' is a right S-isomorphism. Thus $\Sigma(S)$ is an inverse subsemigroup of $\hat{S}$, the inverse semigroup of all right S-isomorphisms. It is important to observe that, in general, $\Sigma(S)$ will contain the zero of $I(S)$.

A monoid S is said to be *left reversible* if $aS \cap bS \neq \emptyset$ for all $a, b \in S$.

Lemma 8 *If S is a left reversible monoid then the product of any two non-empty elements of $\hat{S}$ is non-empty. In particular, if S is also cancellative then $\Sigma(S)$ does not contain the empty function.*

Proof Let $\alpha, \beta \in \hat{S}$ be non-empty elements. Then

$$\mathrm{dom}(\alpha\beta) = \beta^{-1}(\mathrm{dom}\,\alpha \cap \mathrm{im}\,\beta).$$

By assumption, $\mathrm{dom}\,\alpha$ and $\mathrm{im}\,\beta$ are non-empty right ideals of S. Let $a \in \mathrm{dom}\,\alpha$ and $b \in \mathrm{im}\,\beta$. By assumption $aS \cap bS \neq \emptyset$, and so there exist elements $x, y \in S$ such that $ax = by$. Clearly $ax = by \in \mathrm{dom}\,\alpha \cap \mathrm{im}\,\beta$. Thus $\mathrm{dom}\,\alpha \cap \mathrm{im}\,\beta \neq \emptyset$, and so $\mathrm{dom}(\alpha\beta) \neq \emptyset$. Hence $\alpha\beta$ is a non-empty element.

The proof of the final assertion follows from the fact that S', the set of generators of $\Sigma(S)$, does not contain the empty function. ∎

The inverse semigroup $\Sigma(S)$ has a further property.

Proposition 9 *Let S be a left reversible cancellative monoid. Then $\Sigma(S)$ is E-unitary.*

Proof The idempotents of $\hat{S}$ are the identities on the right ideals of S. Let $1_I \subseteq \alpha$ where $1_I, \alpha \in \Sigma$ and 1_I is an idempotent. By Lemma 8, I is a non-empty right ideal of S. Let $x \in \operatorname{dom}\alpha$ and $a \in I$. By left reversibility, $xaS \cap aS$ is non-empty, so that there exist elements y and z such that $ay = xaz$. But $\operatorname{dom}\alpha$ is a right ideal of S so that $xaz \in \operatorname{dom}\alpha$, and I is a right ideal of S, so that $ay \in I$. In particular, $\alpha(xaz) = \alpha(x)az$, and $\alpha(ay) = 1_I(ay) = ay$ since, by assumption, α agrees with 1_I on I. However $\alpha(xaz) = \alpha(ay)$, so that $\alpha(x)az = ay = xaz$. Hence $\alpha(x) = x$ by cancellativity. Thus α is an idempotent. Consequently, $\Sigma(S)$ is E-unitary. ■

We may now prove the Ore embedding theorem.

Theorem 10 *Let S be a left reversible cancellative monoid. Then S is embedded in the group $\Sigma(S)/\sigma$.*

Proof Form the group $\Sigma(S)/\sigma$ and let θ: $S \to \Sigma(S)/\sigma$ be the function $\theta(a) = \sigma(\lambda_a)$. This is a homomorphism because the function $a \mapsto \lambda_a$ is a homomorphism. We show that θ is injective. Suppose that $\theta(a) = \theta(b)$. Then $\sigma(\lambda_a) = \sigma(\lambda_b)$. Thus there is an element α of $\Sigma(S)$ such that $\alpha \subseteq \lambda_a, \lambda_b$. Let α: $I \to J$ where I and J are non-empty right ideals of S. Then λ_a and λ_b agree on I. Let $x \in I$. Then $ax = bx$ and so $a = b$, as required. ■

2.5 Notes on Chapter 2

Section 2.1

The exact nature of the relationship between model theory and partial bijections is as follows. Two structures $\mathcal{A}$ and $\mathcal{B}$ are *elementarily equivalent* if the set of first order sentences true of $\mathcal{A}$ is the same as the set of first order sentences true of $\mathcal{B}$. On the other hand, the structures $\mathcal{A}$ and $\mathcal{B}$ are said to be *finitely isomorphic* if there is a sequence (I_n) such that the following three conditions hold:

(FI1) Each I_n is a non-empty set of partial isomorphisms from $\mathcal{A}$ to $\mathcal{B}$.

(FI2) For each $\alpha \in I_{n+1}$ and $a \in \mathcal{A}$ there exists $\beta \in I_n$ such that $\alpha \subseteq \beta$ and $a \in \operatorname{dom}\beta$.

(FI3) For each $\alpha \in I_{n+1}$ and $b \in \mathcal{B}$ there exists $\beta \in I_n$ such that $\alpha \subseteq \beta$ and $b \in \operatorname{im}\beta$.

Fraïssé proved that $\mathcal{A}$ and $\mathcal{B}$ are elementarily equivalent if, and only if, they are finitely isomorphic (see [63] for a proof). This result concerns partial isomorphisms rather than partial automorphisms and so concerns the inverse category of partial isomorphisms between structures.

The relationship between extending partial automorphisms and Galois theory is taken up by [110].

In writing the section on combinatorial group theory and inverse semigroups, I relied on the survey article by Meakin [250], the long report by Margolis, Meakin and Sapir [219], and the paper by Margolis and Meakin [218]; the latter paper provides the essential link between the classical theory of finitely generated subgroups of free groups, due to Hall and Stallings, and inverse semigroup theory. In particular, they show that immersions of graphs are classified by inverse semigroups in much the same way that coverings of graphs are classified by groups; see Phillip Higgins' book [133] for an exposition of the theory of coverings. Further applications of inverse semigroup theory to the theory of finitely generated subgroups of free groups may be found in [31].

In finitely generated free groups, the relationship between subgroups of finite index and the finitely generated subgroups can be explained in terms of automata theory: the subgroups of finite index are precisely the *recognisable* subgroups, whereas the finitely generated subgroups are the *rational* subgroups. A good discussion of the relationship between combinatorial group theory and automata theory may be found in [127].

The type II conjecture was first stated in [165], but it arose out of much earlier work on the complexity of finite semigroups deriving from automata theory. I shall not state the conjecture formally here, save to say that it concerns finite monoids and their relationship with finite groups; I refer the reader to [125] for a full discussion. Instead, I would like to outline the background to those applications which are most closely connected with inverse semigroup theory. For standard terms from automata theory which are not defined consult Eilenberg [73] and Pin [315].

Given the importance of inverse semigroups, it is perhaps surprising that the first investigation into languages whose syntactic monoids are inverse was not begun until 1974, when Keenan and Lallement [167] investigated those inverse monoids which arise as the syntactic monoids of finite prefix code (see also [177] and [307]). But it was Reutenauer's 1979 paper [351] which was to prove particularly influential in understanding the ramifications of the type II conjecture. This paper defined a topology on the free monoid induced by Hall's topology on the corresponding free group. Injective automata were introduced in this paper as an important tool in understanding this topology. Indeed, one of Reutenauer's main results was that a rational submonoid of a free monoid is closed precisely when its minimal automaton is injective.

The transition monoid of an injective automaton has commuting idempotents since it is a subsemigroup of the symmetric inverse monoid on the set of states of the automaton. Finite semigroups with commuting idempotents and the closed rational subsets of free monoids became the twin concerns of much subsequent work culminating in Ash's two ground-breaking papers [10] and [11]. I shall deal first with finite semigroups with commuting idempotents before turning to closed rational subsets.

To understand the significance of finite semigroups with commuting idempotents, I shall need to introduce some ideas from finite semigroup theory and formal languages. A *variety* of finite semigroups is a class of semigroups closed under subsemigroups, homomorphic images and finite direct products. Observe that this is different from the usual notion of a variety in universal algebra since all structures are finite and only finite direct products are allowed. To each variety of finite semigroups $\mathbf{V}$ we can construct a *variety* $\Sigma^*\mathcal{V}$ of regular languages over a fixed alphabet Σ consisting of all regular Σ-languages whose syntactic semigroups belong to $\mathbf{V}$. The book by Pin [315] is a good source for these ideas. The finite inverse semigroups do not form a variety because subsemigroups of a finite inverse semigroup need not be inverse, but they generate a variety **Inv**: the smallest variety containing all the finite inverse semigroups. T. E. Hall [117] proved that the variety of languages $\Sigma^*\mathcal{I}$ corresponding to **Inv** is described by its finite biprefix codes. Margolis and Pin [225] proved that $\Sigma^*\mathcal{I}$ is the Boolean algebra generated by all languages of the form L or KaL where K and L are group languages and $a \in \Sigma$. But the most important result concerning **Inv** was first formulated by Margolis. It is not hard to show that every semigroup in **Inv** has commuting idempotents. Margolis conjectured, on the basis of the type II conjecture, that the converse also held. The relationship between the type II conjecture and this special problem is discussed in [222] and [225]. Margolis' conjecture was proved by Ash [10]: thus every finite semigroup with commuting idempotents is a homomorphic image of a subsemigroup of a finite inverse semigroup. The proof makes essential use of injective automata; it is also noteworthy that Ramsey's theorem is used to obtain a handle on the non-regular elements of the semigroup. See [315] for information on applications of Ramsey's theorem to semigroup theory.

The closed rational subsets now enter the frame in the following way. Using Ash's application of Ramsey's theorem, Pin [316] proved that a rational language L is accepted by an injective automaton if, and only if, L is closed in the Reutenauer topology and the idempotents in its syntactic monoid commute. Pin also showed that for rational languages whose syntactic monoids had commuting idempotents, it was possible to decide algebraically whether the language was closed or not. This led to Pin's topological conjecture which provided a precise criterion for deciding whether an arbitrary rational language was closed in terms of its syntactic monoid [317], [319]. The equivalence of this conjecture with the type II conjecture was established in [317], [319] in one direction, and in the other direction in [226].

Theorem 6 was originally conjectured in [322] where it was shown to imply the topological conjecture. In [125], it is proved that the topological conjecture implies Theorem 6. However, this proof is now known to be incorrect; it depends on Theorem 4.11 of [319], parts (a) and (c) of which were shown to be erroneous by Mario Branco. I am grateful to Jean–Eric Pin for information on this point. Ash proved the type II conjecture in [11].

The work of Herwig and Lascar [130] has added a new dimension to the Hall topology by demonstrating its importance in model theory. It is remarkable that a conjecture which originally concerned finite semigroups has so many important consequences.

Another point of contact between inverse semigroups and automata theory centres on the dot-depth hierarchy of language theory. This concerns the classification of a star-free language in terms of the smallest number of applications of concatenation (denoted by a dot) needed to describe it. Although this hierarchy concerns arbitrary finite aperiodic semigroups, inverse semigroups in the class have played a prominent rôle, because the strictness of the hierarchy was established using languages whose syntactic monoids were inverse [36]. Some recent work on inverse monoids of dot-depth two may be found in [52], [400], [427], [428], [429].

The bicyclic monoid of Section 3.4 and the polycyclic monoids of Section 9.3 are all syntactic monoids. They arise as the syntactic monoids of 'correct bracketing languages'.

Inverse automata were first discussed by Stephen [397] where they are important in combinatorial inverse semigroup theory. Injective automata are also termed *reversible automata* [316]. The thermodynamics of reversible Turing machines is discussed in [76].

Two-way automata are used to study rational subsets of free inverse semigroups in [304].

The relationship between groups and inverse semigroups from an automata-theoretic perspective is described in [32].

Finally, we return to semigroups with commuting idempotents. Ash's theorem tells us that in the finite case such semigroups are homomorphic images of subsemigroups of inverse semigroups. The characterisation of subsemigroups of arbitrary inverse semigroups was accomplished in the early 60s. A good account of this work (with many references) may be found in [388].

Section 2.2

Factorisable inverse monoids were first studied by Joubert [162] and later, independently, by Chen and Hsieh [41].

The existence of E-unitary covers for inverse semigroups was first explicitly proved by McAlister [231], [232]. However, the fact that every inverse semigroup can be embedded in a factorisable inverse monoid and the construction of Theorem 4 were known to Joubert [162]. Thus Joubert had all the components of the covering theorem except the notion of E-unitary semigroups [184]. Joubert used his results to study topological foliations.

In [252], the variety of inverse semigroups which possess E-unitary covers over abelian groups is shown to have undecidable word problem.

The class of E-unitary semigroups, originally called proper inverse semigroups, was introduced by Saîto [363] in 1965. They were defined as those inverse semigroups in which the idempotent σ-class contained only idempotents. Substantial work on the theory of E-unitary semigroups was carried out by McAlister in [231], [232], and [237]. See also O'Carroll [290], [293].

McAlister's covering theorem has been the inspiration for many generalisations to broader classes of semigroups, the most substantial to date being [84].

Section 2.3

The material on homomorphisms and congruences is classical. See Howie [146], [148] for background references. The term perfect is due to Wagner [422].

Lattice-theoretic properties of the congruence lattices of inverse semigroups have been the subject of a number of papers. We simply refer the reader to [155] and [14], [15].

The structure of the lattices of all inverse subsemigroups and all full inverse subsemigroups of an inverse semigroup is also interesting. A useful guide to this work is [158].

Section 2.4

The minimum group congruence was first used by Rees [335] but first explicitly defined by Wagner [420] and also characterised by Munn [261].

Inverse categories were introduced by Kastl [166], and their theory developed by Grandis [104]. Grandis applies them to the theory of local structures (Section 1.2) in [105]; this work is related to, but different from, Ehresmann's approach via pseudogroups. It is based on ideas derived from Lawvere's influential paper [195]. Lawvere's paper [196] provides a very succinct categorical description of categories of local structures.

The relationship between pseudogroups and topological groupoids is discussed by Reinhart [349] and Jekel [152]. Early papers on this connection are Gray [106] and Uesugi [416], [417].

The theory of embedding cancellative monoids in groups is discussed in greater generality in [48].

Chapter 3

The natural partial order

We have characterised groups as being precisely those inverse semigroups in which the natural partial order is equality (Proposition 1.4.10). This implies that the natural partial order is an important ingredient in the structure of an inverse semigroup. In this, and the next, chapter we examine the interaction of the algebraic and order-theoretic aspects of inverse semigroups.

In this chapter, we concentrate on the more elementary features of this interaction. In Section 3.1, we show that the multiplication of an inverse semigroup is determined by the natural partial order together with a groupoid constructed by restricting the usual multiplication. This is the first step in the ordered groupoid approach to inverse semigroups which was touched on in Section 1.1, and which is developed completely in Chapter 4.

In Section 3.2, we show how the groupoid underlying an inverse semigroup gives rise to a number of equivalence relations on the semigroup; these are none other than Green's relations, $\mathcal{H}$, $\mathcal{L}$, $\mathcal{R}$, $\mathcal{D}$ and $\mathcal{J}$, from general semigroup theory. The relationship between the $\mathcal{D}$ and $\mathcal{J}$ relations is particularly interesting from the order-theoretic point of view. These investigations lead naturally to some important examples of inverse semigroups discussed in Sections 3.3 and 3.4.

In Section 3.3, every primitive inverse semigroup is shown to be nothing other than a groupoid with a zero adjoined. This result demonstrates the nature of the relationship between groupoid theory and inverse semigroup theory. Primitive inverse semigroups are also characterised as those inverse semigroups in which the natural partial order is equality when restricted to the non-zero elements. This result emphasises the extent to which the natural partial order determines the algebraic structure of an inverse semigroup.

In Section 3.4, we introduce the bicyclic monoid. This is an E-unitary inverse semigroup at the other order-theoretic extreme to primitive inverse semigroups in that its idempotents form a chain dually isomorphic to the natural numbers. A complete characterisation of the full inverse subsemigroups of the bicyclic monoid is carried out in Section 5.4, and this forms the basis for the structure theory of inverse ω-semigroups.

3.1 The associated groupoid

In this section, we take the first steps in the groupoid approach to inverse semigroups. The theory is developed in full generality in the next chapter.

The arrow diagram of an inverse semigroup

We begin by recalling the concept of a directed graph or digraph. A *digraph* is a pair of sets $\mathcal{G} = (V, E)$, called respectively *vertices* and *edges* (or, more descriptively, *arrows*) together with two functions $\partial_0, \partial_1 \colon E \to V$; if e is an edge then $\partial_0(e)$ is called the *source vertex* and $\partial_1(e)$ is called the *target vertex* of e. Sometimes α is used instead of ∂_0, and ω is used instead of ∂_1.

We can construct a digraph from every inverse semigroup using the case of symmetric inverse monoids as motivation. Each element f of the symmetric inverse monoid $I(X)$ is a partial bijection with the partial identity on its domain being $f^{-1}f$ and the partial identity on its image being ff^{-1} (Proposition 1.1.1). Thus f can be regarded as an arrow from the idempotent $f^{-1}f$ to the idempotent ff^{-1}. This idea can be generalised to arbitrary inverse semigroups as follows. With every inverse semigroup S, we associate a digraph called the *arrow diagram* of S, whose vertices are labelled by the idempotents of S and whose edges (or arrows) are labelled by arbitrary elements of S; the element s labels an arrow which begins at the idempotent $s^{-1}s$ and ends at the idempotent ss^{-1}. Thus the maps $\mathbf{d}, \mathbf{r} \colon S \to E(S)$, given by $\mathbf{d}(s) = s^{-1}s$ and $\mathbf{r}(s) = ss^{-1}$, describe how the edges are attached to the vertices. The arrow representing s^{-1} is simply the opposite of the one representing s since $\mathbf{d}(s^{-1}) = \mathbf{r}(s)$ and $\mathbf{r}(s^{-1}) = \mathbf{d}(s)$.

In Section 1.1, the restricted product, $f \cdot g$, of two partial bijections f and g was introduced. This definition can also be generalised to arbitrary inverse semigroups. Let S be an inverse semigroup, and let $s, t \in S$ be any two elements of S. Then the *restricted product* $s \cdot t$ exists only when $s^{-1}s = tt^{-1}$, in which case it is equal to st. Thus $s \cdot t$ exists precisely when $\mathbf{d}(s) = \mathbf{r}(t)$, and so the restricted product of two elements is defined only when the corresponding arrows match head-to-tail.

Lemma 1 *Let S be an inverse semigroup. If $s \cdot t$ exists then $\mathbf{d}(s \cdot t) = \mathbf{d}(t)$ and $\mathbf{r}(s \cdot t) = \mathbf{r}(s)$ for any $s, t \in S$.*

Proof By definition $\mathbf{d}(s \cdot t) = t^{-1}s^{-1}st$, and $t^{-1}s^{-1}st = t^{-1}t$ since $s^{-1}s = tt^{-1}$. The proof of the other case is similar. ■

The above result tells us to how to represent the restricted product in arrow terms.

The following result is the abstract version of Proposition 1.1.5; it shows how the usual product can be reconstructed from the restricted product and the natural partial order.

Theorem 2 *Let S be an inverse semigroup.*

(1) *Let $s \in S$ and e an idempotent such that $e \leq s^{-1}s$. Then $a = se$ is the unique element in S such that $a \leq s$ and $a^{-1}a = e$.*

(2) *Let $s \in S$ and e an idempotent such that $e \leq ss^{-1}$. Then $a = es$ is the unique element in S such that $a \leq s$ and $aa^{-1} = e$.*

(3) *Let $s, t \in S$. Then $st = s' \cdot t'$ where $s' = se$, $t' = et$ and $e = s^{-1}stt^{-1}$.*

Proof (1) From the definition of the natural partial order we have that $a \leq s$. Also, $a^{-1}a = (se)^{-1}se = es^{-1}se = e$. Now let $b \leq s$ be such that $b^{-1}b = e$. Then $b = sb^{-1}b$, so that $b = se = a$.
(2) Similar to (1).
(3) Put $s' = se$ and $t' = et$ where $e = s^{-1}stt^{-1}$. Then $s' \leq s$ and $t' \leq t$. It is easy to check that $\mathbf{d}(s') = e$ and $\mathbf{r}(t') = e$. Thus $s' \cdot t'$ exists. But $s' \cdot t' = set = st$. ■

It is worth noting that Theorem 2(3) can be generalised by induction to any finite product of elements.

Proposition 3 *Let $s_i \in S$ where $i = 1, \ldots, n$, and put $s = s_1 \ldots s_n$. Then there exist elements $t_1, \ldots t_n$ such that $t_i \leq s_i$, $s = t_1 \cdot \ldots \cdot t_n$ and $t_i \cdot t_{i+1}$ is a restricted product for $i = 1, \ldots, n-1$.* ■

The natural partial order can also be represented on the arrow diagram of S; if $s \leq t$ then we draw an unlabelled arrow from s to t coloured differently from the arrows representing elements. Using Theorem 2, calculations in an inverse semigroup can now be visualised using the appropriate portions of its arrow diagram; something we would recommend the reader do in following the more complex calculations in this book.

The associated groupoid

The restricted product is important because it endows every inverse semigroup with the structure of a groupoid; the arrow representation of an inverse semigroup is nothing other than a reflection of this underlying groupoid structure. To prove this, we recall some definitions from category theory.

Categories are usually regarded as categories of structures with morphisms. But they can also be regarded as algebraic structures no different from groups, rings and fields except that the binary operation is only partially defined. We now define categories from this purely algebraic point of view.

Let C be a set equipped with a partial binary operation which we shall denote by $\cdot$ or by concatenation. If $x, y \in C$ and the product $x \cdot y$ is defined we write $\exists x \cdot y$. An element $e \in C$ is called an *identity* if $\exists e \cdot x$ implies $e \cdot x = x$ and $\exists x \cdot e$ implies $x \cdot e = x$. The set of identities of C is denoted C_o; the subscript 'o' stands for 'object'. The pair $(C, \cdot)$ is said to be a *category* if the following axioms hold:

(C1) $x \cdot (y \cdot z)$ exists if, and only if, $(x \cdot y) \cdot z$ exists, in which case they are equal.

(C2) $x \cdot (y \cdot z)$ exists if, and only if, $x \cdot y$ and $y \cdot z$ exist.

(C3) For each $x \in C$ there exist identities e and f such that $\exists x \cdot e$ and $\exists f \cdot x$.

From axiom (C3), it follows that the identities e and f are uniquely determined by x. We write $e = \mathbf{d}(x)$ and $f = \mathbf{r}(x)$, where $\mathbf{d}(x)$ is the *domain* identity and $\mathbf{r}(x)$ is the *range* identity. Observe that $\exists x \cdot y$ if, and only if, $\mathbf{d}(x) = \mathbf{r}(y)$.

The elements of a category are called *morphisms* or *homomorphisms*. If C is a category and e and f identities in C then we put

$$\hom(e, f) = \{x \in C \colon \mathbf{d}(x) = e \text{ and } \mathbf{r}(x) = f\},$$

the set of *homomorphisms from e to f*. Subsets of C of the form $\hom(e, f)$ are called *hom-sets*. We also put $\operatorname{end}(e) = \hom(e, e)$, the *endomorphism monoid at e*.

A category C is said to be a *groupoid* if for each $x \in C$ there is an element x^{-1} such that $x^{-1}x = \mathbf{d}(x)$ and $xx^{-1} = \mathbf{r}(x)$. It is easy to check that x^{-1} is unique with these properties.

Two elements x and y of a groupoid are said to be *connected* if there is a sequence of composable elements starting at $\mathbf{d}(x)$ and ending at $\mathbf{d}(y)$. This is an equivalence relation whose equivalence classes are called the *connected components* of the groupoid. A groupoid with one connected component is said to be *connected*.

Let C and D be categories. Then a function $F\colon C \to D$ is a *functor* if

$$F(\mathbf{d}(x)) = \mathbf{d}(F(x)) \text{ and } F(\mathbf{r}(x)) = \mathbf{r}(x)$$

for all $x \in C$, and $F(x \cdot y) = F(x) \cdot F(y)$ for all $x, y \in C$ for which $\exists x \cdot y$.

There is some important terminology concerning functors which we introduce now, although we do not need it until Chapter 4 and Chapter 8. Let e be an identity in the category C. Then the set of all elements x of C such that $\mathbf{d}(x) = e$ is called the *star of C at e*. Every functor $F\colon C \to D$ induces a function F_e from the star of C at e to the star of D at $F(e)$. If all the functions F_e are injective, then F is said to be *star injective*. Likewise, if all the functions F_e are surjective, then F is said to be *star surjective*. A functor which is both star injective and star surjective is said to be *star bijective* or a *covering functor*.

Proposition 4 *Every inverse semigroup S is a groupoid with respect to its restricted product.*

Proof We begin by showing that all idempotents of S are identities of $(S, \cdot)$. Let $e \in S$ be an idempotent and suppose that $e \cdot x$ is defined. Then $e = xx^{-1}$ and $e \cdot x = ex$. But $ex = (xx^{-1})x = x$. Similarly, if $x \cdot e$ is defined then it is equal to x. We now check that the axioms (C1), (C2) and (C3) hold.

Axiom (C1) holds: suppose that $x \cdot (y \cdot z)$ is defined. Then

$$x^{-1}x = (y \cdot z)(y \cdot z)^{-1} \text{ and } y^{-1}y = zz^{-1}.$$

But

$$(y \cdot z)(y \cdot z)^{-1} = yzz^{-1}y^{-1} = yy^{-1}.$$

Hence $x^{-1}x = yy^{-1}$, and so $x \cdot y$ is defined. Also $(xy)^{-1}(xy) = y^{-1}y = zz^{-1}$. Thus $(x \cdot y) \cdot z$ is defined. It is clear that $x \cdot (y \cdot z)$ is equal to $(x \cdot y) \cdot z$. A similar argument shows that if $(x \cdot y) \cdot z$ exists then $x \cdot (y \cdot z)$ exists and they are equal.

Axiom (C2) holds: suppose that $x \cdot y$ and $y \cdot z$ are defined. We show that $x \cdot (y \cdot z)$ is defined. We have that $x^{-1}x = yy^{-1}$ and $y^{-1}y = zz^{-1}$. Now

$$(yz)(yz)^{-1} = y(zz^{-1})y^{-1} = y(y^{-1}y)y^{-1} = yy^{-1} = x^{-1}x.$$

Thus $x \cdot (y \cdot z)$ is defined. The proof of the converse is straightforward.

Axiom (C3) holds: for each element x we have that $x \cdot (x^{-1}x)$ is defined, and we have seen that idempotents of S are identities. Thus we put $\mathbf{d}(x) = x^{-1}x$. Similarly, we put $xx^{-1} = \mathbf{r}(x)$. It is now clear that $(S, \cdot)$ is a category. The fact that it is a groupoid is immediate. ∎

We call $(S, \cdot)$ the *associated groupoid* of S.

Prehomomorphisms

The importance of the restricted product and the natural partial order to the structure of an inverse semigroup leads to a generalisation of semigroup homomorphisms. A function $\theta\colon S \to T$ between inverse semigroups is said to be a *prehomomorphism* if $\theta(st) \leq \theta(s)\theta(t)$ for all $s, t \in S$. It is called a *dual prehomomorphism* if $\theta(s)\theta(t) \leq \theta(st)$. We shall be solely interested in prehomomorphisms, but consult the notes at the end of this chapter for some remarks on dual prehomomorphisms.

Theorem 5 *Let $\theta\colon S \to T$ be a function between inverse semigroups. Then θ is a prehomomorphism if, and only if, it preserves the restricted product and the natural partial order; it is a homomorphism if, and only if, it is a prehomomorphism which satisfies $\theta(ef) = \theta(e)\theta(f)$ for all idempotents $e, f \in S$. In addition, the composition of prehomomorphisms is a prehomomorphism.*

Proof Let $\theta\colon S \to T$ be a prehomomorphism. We first prove that $\theta(s^{-1}) = \theta(s)^{-1}$ for each $s \in S$. By definition

$$\theta(s) = \theta(s(s^{-1}s)) \leq \theta(s)\theta(s^{-1}s) \leq \theta(s)\theta(s^{-1})\theta(s).$$

Similarly,

$$\theta(s^{-1}) \leq \theta(s^{-1})\theta(s)\theta(s^{-1}).$$

Put $a = \theta(s)$ and $b = \theta(s^{-1})$. Then $a \leq aba$ and $b \leq bab$. Now $ab \leq abab$, so that

$$ab = (ab)^2(ab)^{-1}(ab) = (ab)^2.$$

Similarly, $ba = (ba)^2$. Thus $a(ba) \leq a \leq aba$, and so $a = aba$. Similarly, $b = bab$. Hence $\theta(s^{-1}) = \theta(s)^{-1}$.

Next we show that if e is an idempotent then $\theta(e)$ is an idempotent. Let e be an idempotent. Then $\theta(e) = \theta(e^{-1}) = \theta(e)^{-1}$ by the result above. Thus

$$\theta(e) = \theta(ee) \leq \theta(e)\theta(e) = \theta(e)\theta(e^{-1}e) \leq \theta(e)\theta(e)^{-1}\theta(e) = \theta(e),$$

and so $\theta(e) = \theta(e)\theta(e)$.

We can now prove that θ is order-preserving. Let $s \leq t$. Then $s = te$ for some idempotent e. Thus

$$\theta(s) = \theta(te) \leq \theta(t)\theta(e) \leq \theta(t)$$

since $\theta(e)$ is an idempotent.

A key ingredient in proving that θ preserves restricted product is the following:

$$\theta(ss^{-1}) = \theta(s)\theta(s)^{-1} \text{ and } \theta(s^{-1}s) = \theta(s)^{-1}\theta(s)$$

for every $s \in S$. We show that $\theta(ss^{-1}) = \theta(s)\theta(s)^{-1}$; the proof of the other case is similar. Clearly, $\theta(ss^{-1}) \leq \theta(s)\theta(s)^{-1}$, so that

$$\theta(ss^{-1})\theta(s) \leq \theta(s)\theta(s)^{-1}\theta(s) = \theta(s).$$

But then

$$\theta(s) = \theta((ss^{-1})s) \leq \theta(ss^{-1})\theta(s) \leq \theta(s).$$

Thus $\theta(s) = \theta(ss^{-1})\theta(s)$, and so

$$\theta(s)\theta(s)^{-1} = \theta(ss^{-1})(\theta(s)\theta(s)^{-1}).$$

Hence $\theta(s)\theta(s)^{-1} \leq \theta(ss^{-1})$. But $\theta(ss^{-1}) \leq \theta(s)\theta(s)^{-1}$. It follows that $\theta(s)\theta(s)^{-1} = \theta(ss^{-1})$.

We can now prove that θ preserves restricted products. Suppose that $s \cdot t$ is defined. Then by the result above so too is $\theta(s) \cdot \theta(t)$. It remains to show that $\theta(s \cdot t) = \theta(s) \cdot \theta(t)$. Clearly $\theta(s \cdot t) \leq \theta(s) \cdot \theta(t)$. Now

$$\theta(s \cdot t)^{-1}\theta(s \cdot t) = \theta((s \cdot t)^{-1}(s \cdot t)) = \theta(t^{-1}s^{-1}st) = \theta(t^{-1}t),$$

and

$$[\theta(s) \cdot \theta(t)]^{-1}[\theta(s) \cdot \theta(t)] = \theta(t)^{-1}\theta(t) = \theta(t^{-1}t).$$

Hence $\theta(s \cdot t) = \theta(s) \cdot \theta(t)$ as required.

To prove the converse, suppose that θ preserves the restricted product and the natural partial order. We show that it is a prehomomorphism. Let st be a full product in S. Then $st = (se) \cdot (et)$ where $e = s^{-1}stt^{-1}$ by Theorem 2(3). Thus, by assumption, $\theta(st) = \theta(se) \cdot \theta(et)$. But $se \leq s$ and $et \leq t$ so that $\theta(se) \leq \theta(s)$ and $\theta(et) \leq \theta(t)$. Hence $\theta(st) \leq \theta(s)\theta(t)$ as required. This completes the proof that a function is a prehomomorphism if, and only if, it is order-preserving and preserves the restricted product.

We now prove that if $\theta\colon S \to T$ is a prehomomorphism satisfying $\theta(ef) = \theta(e)\theta(f)$ for all idempotents $e, f \in S$, then θ is a homomorphism. Let st be a full product in S. Then $st = (se) \cdot (et)$ where $e = s^{-1}stt^{-1}$. Thus $\theta(st) = \theta(se) \cdot \theta(et)$. We show that $\theta(se) = \theta(s)\theta(e)$. Clearly, $\theta(se) \leq \theta(s)\theta(e)$. Now

$$\theta(se)^{-1}\theta(se) = \theta((se)^{-1}(se)) = \theta(es^{-1}se) = \theta(e),$$

and

$$[\theta(s)\theta(e)]^{-1}\theta(s)\theta(e) = \theta(e)^{-1}\theta(s)^{-1}\theta(s)\theta(e) = \theta(e)^{-1}\theta(s^{-1}s)\theta(e) = \theta(e)$$

since $\theta(e) = \theta(s^{-1}s)\theta(tt^{-1})$ by assumption. Thus $\theta(se) = \theta(s)\theta(e)$. Similarly, $\theta(et) = \theta(e)\theta(t)$. It now follows that

$$\theta(st) = \theta(s)\theta(e)\theta(e)\theta(t) = \theta(s)\theta(s^{-1}s)\theta(tt^{-1})\theta(t) = \theta(s)\theta(t).$$

It is straightforward to check that the composition of prehomomorphisms is a prehomomorphism. ■

We can easily construct examples of prehomomorphisms which are not homomorphisms. Let L and M be meet semilattices and let $\theta: L \to M$ be an order-preserving function. Let $e, f \in L$. Then $e \wedge f \leq e, f$ and so $\theta(e \wedge f) \leq \theta(e), \theta(f)$ since θ is order-preserving. Thus

$$\theta(e \wedge f) \leq \theta(e) \wedge \theta(f)$$

since M is a meet semilattice. It follows that θ is a prehomomorphism from the inverse semigroup $(L, \wedge)$ to the inverse semigroup $(M, \wedge)$, but not in general a homomorphism.

Theorem 5 implies that every prehomomorphism between inverse semigroups induces a functor between their associated groupoids.

3.2 Green's relations

The arrow representation of the elements of an inverse semigroup leads naturally to a number of equivalence relations. These relations, called *Green's relations*, can be defined on any semigroup, but assume particularly simple forms on inverse semigroups. The relations we shall introduce are denoted by $\mathcal{L}$, $\mathcal{R}$, $\mathcal{H}$, $\mathcal{D}$ and, finally, $\mathcal{J}$. Tradition dictates that if $\mathcal{K}$ is one of Green's relations then the $\mathcal{K}$-equivalence class containing the element s is denoted by K_s. If $\mathcal{K}$ is one of Green's relations on a semigroup S, then we sometimes write $\mathcal{K}(S)$ rather than just $\mathcal{K}$ if we need to avoid ambiguity.

The relations $\mathcal{L}$, $\mathcal{R}$ and $\mathcal{H}$

Define relations $\mathcal{L}$ and $\mathcal{R}$ on the inverse semigroup S by

$$(s,t) \in \mathcal{L} \Leftrightarrow s^{-1}s = t^{-1}t \text{ and } (s,t) \in \mathcal{R} \Leftrightarrow ss^{-1} = tt^{-1}.$$

Both $\mathcal{L}$ and $\mathcal{R}$ are equivalence relations. It is easy to check that $\mathcal{L}$ is a right congruence and $\mathcal{R}$ is a left congruence.

There are natural interpretations of these relations in terms of arrow diagrams. Two elements are $\mathcal{L}$-related if they begin at the same vertex, and are $\mathcal{R}$-related if they end at the same vertex. In terms of the underlying groupoid structure, two elements are $\mathcal{L}$-related if they have the same domain and $\mathcal{R}$-related if they have the same range.

Put $\mathcal{H} = \mathcal{L} \cap \mathcal{R}$. This is again an equivalence relation. In arrow terms, elements which are $\mathcal{H}$-related begin and end at the same vertex; in groupoid terms the elements belong to the same hom-set. An inverse semigroup is said to be *combinatorial* if $\mathcal{H}$ is the equality relation; thus any two vertices are joined by *at most one* arrow.

The $\mathcal{H}$-relation is closely associated with the presence of *subgroups* in a semigroup, by which we mean a subsemigroup which is also a group. For every idempotent e in S, the subset eSe is a subsemigroup with identity e, called a *local submonoid.*

Proposition 1 *Let S be an inverse semigroup and e an idempotent in S.*

(1) *eSe is an inverse monoid with identity e.*

(2) *$H_e = U(eSe)$. In particular, H_e is a group with identity e.*

(3) *Every subgroup of S is contained in an $\mathcal{H}$-class.*

Proof (1) We have already noted that eSe is a monoid with identity e. Let $s \in eSe$. Then $es = s = se$. Thus $ss^{-1}, s^{-1}s \leq e$, so that $s^{-1} \in eSe$.
(2) Let $a \in H_e$. Then $a^{-1}a = e = aa^{-1}$. Thus $ea = a = ae$ and $a^{-1}e = a^{-1} = ea^{-1}$, so that $a, a^{-1} \in eSe$. But e is the identity of eSe. Thus $a \in U(eSe)$. The converse is clear.
(3) Let G be a subgroup with identity e. Then for each $a \in G$ there exists $b \in G$ such that $ab = e = ba$. Hence $a \,\mathcal{H}\, e$ and so $G \subseteq H_e$. ■

The characterisation of Green's relations given below in terms of principal ideals provides a link with the usual way of defining Green's relations in arbitrary semigroups.

Proposition 2 *Let S be an inverse semigroup and let $a, b \in S$.*

(1) *$a \,\mathcal{L}\, b$ if, and only if, $Sa = Sb$.*

(2) *$a \,\mathcal{R}\, b$ if, and only if, $aS = bS$.*

Proof We prove (1), the proof of (2) is similar. Suppose that $a^{-1}a = b^{-1}b$. Then $a = (ab^{-1})b$ and $b = (ba^{-1})a$. Thus $Sa = Sb$. Conversely, suppose that $Sa = Sb$. Then $a = xb$ and $b = ya$ for some $x, y \in S$. Now $a^{-1}a = (xb)^{-1}xb = b^{-1}x^{-1}xb \leq b^{-1}b$. Similarly $b^{-1}b \leq a^{-1}a$. Hence $a^{-1}a = b^{-1}b$. ■

The relationship between the natural partial order and the three Green's relations we have so far introduced is quite straightforward.

Proposition 3 *Let S be an inverse semigroup and let $s, t \in S$.*

(1) *$s \,\mathcal{L}\, t$ and $s \leq t$ implies $s = t$.*

(2) $s\,\mathcal{R}\,t$ *and* $s \leq t$ *implies* $s = t$.

(3) $s\,\mathcal{H}\,t$ *and* $s \leq t$ *implies* $s = t$.

Proof We prove (1), the proof of (2) is similar, and (3) follows from (1) and (2). By assumption, $s^{-1}s = t^{-1}t$ and $s = ts^{-1}s$. Thus $s = ts^{-1}s = tt^{-1}t = t$. ■

The $\mathcal{D}$-relation

Every arrow diagram of an inverse semigroup partitions the elements of S into larger classes: the connected components of the underlying groupoid. We show how this partition can be described in inverse semigroup-theoretic terms.

The relation $\mathcal{D}$ is defined to be the join of $\mathcal{L}$ and $\mathcal{R}$, that is, the smallest equivalence relation which contains both $\mathcal{L}$ and $\mathcal{R}$. Thus $\mathcal{D}$ is equal to the intersection of all equivalence relations containing the union of $\mathcal{L}$ and $\mathcal{R}$. This relation has a much more succinct description. First recall that if ρ and τ are relations on a set X then the relation $\rho \circ \tau$ is given by

$$\rho \circ \tau = \{(x, y) \in X \times X \colon \exists z \in X \text{ such that } (x, z) \in \rho \text{ and } (z, y) \in \tau\}.$$

With respect to this operation the set of all binary relations on X is a monoid which is partially ordered with respect to subset inclusion. If ρ is an equivalence relation on X then $\rho \circ \rho = \rho$ by reflexivity and transitivity.

Proposition 4 *Let S be an inverse semigroup. Then*

$$\mathcal{D} = \mathcal{L} \circ \mathcal{R} = \mathcal{R} \circ \mathcal{L}.$$

Proof We prove first that $\mathcal{L} \circ \mathcal{R} \subseteq \mathcal{R} \circ \mathcal{L}$. The reverse inclusion is proved similarly. Let $(a, b) \in \mathcal{L} \circ \mathcal{R}$. By definition there exists $c \in S$ such that

$$a^{-1}a = c^{-1}c \text{ and } cc^{-1} = bb^{-1}.$$

Put $d = ac^{-1}b$. Then

$$dd^{-1} = ac^{-1}b(ac^{-1}b)^{-1} = aa^{-1}$$

and so $(a, d) \in \mathcal{R}$. Also

$$d^{-1}d = (ac^{-1}b)^{-1}ac^{-1}b = b^{-1}b$$

and so $(d, b) \in \mathcal{L}$. Thus $\mathcal{L} \circ \mathcal{R} \subseteq \mathcal{R} \circ \mathcal{L}$.

Put $\mathcal{K} = \mathcal{L} \circ \mathcal{R}$. Clearly, $\mathcal{L}, \mathcal{R} \subseteq \mathcal{K}$. Using what we proved above, it is easy to check that $\mathcal{K}$ is an equivalence relation. Thus $\mathcal{D} \subseteq \mathcal{K}$. Now let ρ be any equivalence relation containing both $\mathcal{L}$ and $\mathcal{R}$. Then $\mathcal{L} \subseteq \rho$ and $\mathcal{R} \subseteq \rho$, so that $\mathcal{K} \subseteq \rho \circ \rho = \rho$. Thus $\mathcal{K} = \mathcal{D}$. ■

The next result shows, amongst other things, how to relate the $\mathcal{D}$-relation to the arrow diagram of the semigroup.

Proposition 5 *Let S be an inverse semigroup.*

(1) *Let $e, f \in E(S)$. Then $e \mathcal{D} f$ if, and only if, there is an element $a \in S$ such that $a^{-1}a = f$ and $aa^{-1} = e$.*

(2) *Let $s, t \in S$. Then $(s,t) \in \mathcal{D}$ if, and only if, there exist elements $a, b \in S$ such that*

$$\mathbf{d}(a) = \mathbf{d}(t), \quad \mathbf{r}(a) = \mathbf{d}(s), \quad \mathbf{d}(b) = \mathbf{r}(t), \quad \mathbf{r}(b) = \mathbf{r}(s)$$

and $t = b^{-1} \cdot s \cdot a$.

(3) *If $s = a_1 \cdot \ldots \cdot a_n$, then $(s, a_i) \in \mathcal{D}$ for each $i = 1, \ldots, n$.*

Proof (1) Suppose that $e \mathcal{D} f$. Then $e \mathcal{R} a \mathcal{L} f$ for some a by Proposition 4. But then $e = aa^{-1}$ and $f = a^{-1}a$. Conversely, if $a^{-1}a = f$ and $aa^{-1} = e$ then $f \mathcal{L} a$ and $a \mathcal{R} e$. Hence $e \mathcal{D} f$ by Proposition 4.
(2) Let $(s,t) \in \mathcal{D}$. Then $(s^{-1}s, t^{-1}t) \in \mathcal{D}$. By (1), there exists $a \in S$ such that $aa^{-1} = s^{-1}s$ and $a^{-1}a = t^{-1}t$. Put $b = sat^{-1}$. It is now routine to check that a and b have the required properties. The converse is clear.
(3) The result is true for $n = 2$, because if $a \cdot b$ is any restricted product then by definition $\mathbf{d}(a) = \mathbf{r}(b)$. Thus $a \mathcal{L} a^{-1}a = bb^{-1} \mathcal{R} b$. Hence $a \mathcal{D} b$ by Proposition 4. But $\mathbf{r}(a \cdot b) = \mathbf{r}(a)$ by Lemma 3.1.1. Thus $(a \cdot b, a) \in \mathcal{D}$ and $(a \cdot b, b) \in \mathcal{D}$. The result for arbitrary n now follows because $(a_i, a_{i+1}) \in \mathcal{D}$ for $i = 1, \ldots, n-1$, and $(s, a_1) \in \mathcal{D}$. ■

The result of (2) above suggests that elements *within* a single $\mathcal{D}$-class are 'isomorphic' to each other, although it is important to emphasise that when viewed within the whole inverse semigroup they may behave quite differently.

An inverse semigroup having a single $\mathcal{D}$-class is said to be *bisimple*; this is equivalent to the underlying groupoid being connected.

Let S be an inverse semigroup with zero. If $a \mathcal{L} 0$ then $a^{-1}a = 0$ and so $a = a(a^{-1}a) = a0 = 0$. Similarly, if $a \mathcal{R} 0$ then $a = 0$. We conclude that the zero forms a $\mathcal{D}$-class on its own. For inverse semigroups with zero, there is a modification of the definition of bisimplicity which is more appropriate. An inverse semigroup with zero is said to be *0-bisimple* if it contains exactly two $\mathcal{D}$-classes. The need to treat inverse semigroups and inverse semigroups with zero slightly differently in this way is a feature of the subject.

An important class of 0-bisimple inverse semigroups can be constructed as follows. Let I be a non-empty set with cardinality n. Put $B_n = (I \times I) \cup \{0\}$ and define a product on $I \times I \cup \{0\}$ by

$$(i,j)(k,l) = \begin{cases} (i,l) & \text{if } j = k \\ 0 & \text{otherwise.} \end{cases}$$

Then B_n is a combinatorial, 0-bisimple inverse semigroup in which the non-zero idempotents are the elements of the form (i,i), and in which the inverse of (i,j) is (j,i). Our notation B_n is justified, because the algebraic structure of the semigroup is entirely determined by the cardinality of the set I.

The inverse semigroups B_n arise naturally in ring theory. Let R be a ring with identity. Then for any natural number n, consider the set of all $n \times n$-matrices with at most one non-zero entry which must be 1. Under matrix multiplication these matrices form a semigroup isomorphic to B_n. We can think of these matrices as forming a representation of B_n in the matrix ring $M_n(R)$. More generally, any injective homomorphism of B_n into the multiplicative monoid of a ring R which maps the zero of B_n to the zero of the ring is said to be a *strong representation* of B_n if the sum of the images of the idempotents of B_n is 1. The image of B_n is called a *set of matrix units* in R. It can be proved that a ring R is isomorphic to the full ring of $n \times n$-matrices over some ring S if, and only if, R admits a strong embedding of B_n (Proposition 3.18 of [75]).

The semigroup B_2 has an unexpected application in deciding the equality of left and right compatibility relations. First we need a preliminary result.

Proposition 6 *The following are equivalent:*

(1) *The left and right compatibility relations are equal.*

(2) *For all $s,t \in S$, we have that $st \in E(S)$ if, and only if, $ts \in E(S)$.*

Proof (1) $\Rightarrow$ (2). Suppose $st \in E(S)$. Then $s(t^{-1})^{-1} \in E(S)$. Thus $s \sim_l t^{-1}$, and so by assumption, $s \sim_r t^{-1}$. Thus $s^{-1}t^{-1} \in E(S)$. Hence $ts \in E(S)$. The converse is proved similarly.
(2) $\Rightarrow$ (1). Suppose $s \sim_l t$. Then $st^{-1} \in E(S)$. Thus by assumption $t^{-1}s \in E(S)$. Hence $s^{-1}t \in E(S)$ and so $s \sim_r t$. The converse is proved similarly. ■

An inverse semigroup S is said to be *E-reflexive* if for all $s,t \in S$

$$st \in E(S) \Leftrightarrow ts \in E(S).$$

Thus the E-reflexive semigroups are precisely those in which the left and right compatibility relations are equal.

An example of an inverse semigroup which is not E-reflexive is the inverse semigroup B_2, with elements

$$e = (1,1), \quad s = (1,2), \quad s^{-1} = (2,1), \quad f = (2,2)$$

together with a zero 0. Clearly, $E(S) = \{0, e, f\}$ and $ss^{-1} = e$ and $s^{-1}s = f$. The semigroup B_2 is not E-reflexive because $fs = 0$ is an idempotent, but $sf = s$ is not.

The general situation is described by the following result.

Theorem 7 *Let S be an inverse semigroup. Then the following are equivalent:*

(1) *There exists an element $s \in S$ such that $s^2 < s$.*

(2) *S contains an isomorphic copy of B_2.*

(3) *S is not E-reflexive.*

Proof (1) $\Rightarrow$ (2). Let s be an element such that $s^2 < s$. Put

$$T = \{s, s^{-1}, e = ss^{-1}, f = s^{-1}s, 0 = s^2\}.$$

We show that T is a subsemigroup of S isomorphic to B_2. From $s^2 \leq s$, we obtain $s^2 = s^2 s^{-2} s$ and so $s^3 = s^2 s^{-2} s^2$. Thus $s^3 = s^2$. It follows that $s^4 = s^3 = s^2$. Thus s^2 is an idempotent and so $s^{-3} = s^2 = s^{-2}$. It is straightforward to show that all elements of T are distinct. The Cayley table of T can now easily be constructed and yields a semigroup isomorphic to B_2.
(2) $\Rightarrow$ (3). We have already observed that B_2 is not E-reflexive.
(3) $\Rightarrow$ (1). Let $s, t \in S$ be such that st is an idempotent but $a = ts$ is not. Observe that

$$a^3 = tststs = t(stst)s = t(st)^2 s = t(st)s = a^2.$$

We show that $a^3 = a^2$ together with $a \neq a^2$ implies that $a^2 < a$. From $a^3 = a^2$ we obtain

$$a^3 a^{-1} = a^2 a^{-1} \leq a.$$

Thus $a^3 a^{-2} a \leq a$. But $a^3 = a^2$ implies $a^4 = a^2$ and so a^2 is an idempotent. Thus $a^{-2} = a^2$, which implies that $a^6 \leq a$. However, $a^6 = a^2$, and so $a^2 < a$. ■

The $\mathcal{J}$-relation

The remaining Green's relation, $\mathcal{J}$, is defined in terms of principal ideals. Let S be an inverse semigroup. Then

$$(s, t) \in \mathcal{J} \Leftrightarrow SsS = StS.$$

By Proposition 5(2), $\mathcal{D} \subseteq \mathcal{J}$. The meaning of the $\mathcal{J}$-relation for inverse semigroups is clarified by the following result.

Proposition 8 *Let S be an inverse semigroup.*

(1) *$a \in SbS$ if, and only if, there exists $u \in S$ such that $a \mathcal{D} u \leq b$.*

(2) *$a \mathcal{J} b$ if, and only if, $a \mathcal{D} b' \leq b$ and $b \mathcal{D} a' \leq a$ for some $a', b' \in S$.*

Proof (1) Let $a \in SbS$. Then $a = xby$ for some $x, y \in S$. By Proposition 3.1.3, there exist elements x', y' and b' such that $a = x' \cdot b' \cdot y'$ is a restricted product where $x' \leq x$, $b' \leq b$ and $a' \leq a$. Hence $a \mathcal{D} b'$ by Proposition 5(3) which, together with $b' \leq b$, gives $a \mathcal{D} b' \leq b$.

Conversely, suppose that $a \mathcal{D} b' \leq b$. From $a \mathcal{D} b'$ we have that $a \mathcal{J} b'$, and from $b' \leq b$ we have that $Sb'S \subseteq SbS$. Thus $a \in SbS$.
(2) Immediate from (1). ■

As an example, we determine all of Green's relations on the symmetric inverse monoid. By Proposition 1.1.1(1), two partial bijections are $\mathcal{L}$-related when they have the same domain, and $\mathcal{R}$-related when they have the same image. If two partial bijections are $\mathcal{D}$-related then by Proposition 5(2) their images (or, equally well, their domains) will be in bijective correspondence and so have the same cardinalities; conversely, if two partial functions have images with the same cardinalities then they must be $\mathcal{D}$-related, again by Proposition 5(2). Thus two partial bijections are $\mathcal{D}$-related precisely when their images (or domains) have the same cardinality. It remains to calculate the $\mathcal{J}$-relation; as we now show, it turns out that in this case $\mathcal{J} = \mathcal{D}$. Let $f, g \in I(X)$ such that $f \mathcal{J} g$. Then by Proposition 8(2), there exist $h, k \in I(X)$ such that $f \mathcal{D} h \leq g$ and $g \mathcal{D} k \leq f$. It follows that the partial bijections f and g have domains with the same cardinality and so are $\mathcal{D}$-related.

The absence of ideals leads to notions of simplicity. A semigroup S is said to be *simple* if S is the only non-empty ideal. An inverse semigroup with zero is said to be 0-*simple* if it contains at least one non-zero element and the only ideals are $\{0\}$ and S. Observe that (0-)bisimple inverse semigroups are always (0-)simple.

Proposition 9 *Let S be an inverse semigroup.*

(1) *S is simple if, and only if, the only $\mathcal{J}$-class is S.*

(2) *If S has a zero then it is 0-simple if, and only if, $S \neq \{0\}$ and the only $\mathcal{J}$-classes are $\{0\}$ and $S \setminus \{0\}$.*

Proof We prove (1); the proof of (2) is similar. Let S be simple and let $s, t \in S$. Both SsS and StS are ideals of S and so must be equal. Thus $(s, t) \in \mathcal{J}$. Conversely, suppose that the only $\mathcal{J}$-class is S. Let I be any non-empty ideal of S with $s \in I$ and let $t \in S$. By assumption, $(s, t) \in \mathcal{J}$. Thus $t = asb$ for some $a, b \in S$. But then $t \in I$. Hence $I = S$. ■

Using Proposition 8, we can obtain alternative characterisations of simple and 0-simple inverse semigroups.

Proposition 10 *Let S be an inverse semigroup (with zero).*

(1) *S is simple (resp. 0-simple) if, and only if, for any two (non-zero) elements s and t in S there exists an element s' such that $s \mathcal{D} s' \leq t$.*

(2) *S is simple (resp. 0-simple) if, and only if, for any two (non-zero) idempotents e and f in S there exists an idempotent i such that $e \mathcal{D} i \leq f$.*

Proof (1) We prove the simple case; the 0-simple case is similar. By Proposition 9(1), an inverse semigroup is simple if it consists of one $\mathcal{J}$-class. Thus any two elements of S are $\mathcal{J}$-related. The result is now immediate by Proposition 8.
(2) We prove the simple case; the 0-simple case is similar. Suppose the condition on the idempotents holds. Let $s, t \in S$. Then $e = ss^{-1}$ and $f = tt^{-1}$ are idempotents and so, by assumption, there is an idempotent i such that $e \mathcal{D} i \leq f$. Put $u = it$. Then $u \leq t$, and $uu^{-1} = it(it)^{-1} = itt^{-1} = if = i$. Thus $s \mathcal{D} u \leq t$. The proof of the converse is straightforward. ■

As an example of a 0-simple inverse semigroup, consider the inverse semigroup $\Gamma(\mathbb{R}^n)$ of all partial homeomorphisms between the open subsets of $\mathbb{R}^n$ [18]. Every non-empty open subset U of $\mathbb{R}^n$ contains an open ball B. It is well-known that every open ball in $\mathbb{R}^n$ is homeomorphic to $\mathbb{R}^n$. Thus $1 \mathcal{D} 1_B \leq 1_U$, where 1 is the identity function on $\mathbb{R}^n$. On the other hand, $1_U \leq 1$. Thus every non-zero idempotent in $\Gamma(\mathbb{R}^n)$ is $\mathcal{J}$-related to the identity, and so $\Gamma(\mathbb{R}^n)$ is 0-simple by Proposition 10(2). The inverse semigroup $\Gamma(\mathbb{R}^n)$ is not 0-bisimple since it is easy to find examples of non-empty open subsets of $\mathbb{R}^n$ which are not homeomorphic.

A semigroup is said to be *congruence-free* if the only congruences are equality and the universal congruence. Clearly all congruence-free inverse semigroups are simple and all congruence-free inverse semigroups with zero (which contain non-zero elements) are 0-simple. An important family of congruence-free inverse semigroups, the polycyclic monoids, is discussed in Section 9.3.

The relationships between Green's relations on a subsemigroup and in the original semigroup are described below. The proofs are all straightforward.

Proposition 11 *Let S be an inverse subsemigroup of the inverse semigroup T.*

(1) $\mathcal{L}(T) \cap (S \times S) = \mathcal{L}(S)$.

(2) $\mathcal{R}(T) \cap (S \times S) = \mathcal{R}(S)$.

(3) $\mathcal{H}(T) \cap (S \times S) = \mathcal{H}(S)$.

(4) $\mathcal{D}(S) \subseteq \mathcal{D}(T) \cap (S \times S)$.

(5) $\mathcal{J}(S) \subseteq \mathcal{J}(T) \cap (S \times S)$. ■

Idempotent-separating homomorphisms

A congruence is said to be *idempotent-separating* if its associated natural homomorphism is idempotent-separating. The following result is the analogue of Proposition 2.4.5 which characterises idempotent pure congruences.

Proposition 12 *Let S be an inverse semigroup. Then a congruence ρ is idempotent-separating if, and only if, $\rho \subseteq \mathcal{H}$.*

Proof Let ρ be idempotent-separating and let $(a,b) \in \rho$. Then

$$(a^{-1}a, b^{-1}b), (aa^{-1}, bb^{-1}) \in \rho$$

by Proposition 2.3.4. But ρ is idempotent-separating and so $a^{-1}a = b^{-1}b$ and $aa^{-1} = bb^{-1}$. Hence $(a,b) \in \mathcal{H}$. Conversely, let $\rho \subseteq \mathcal{H}$, and suppose that $(e,f) \in \rho$ where e and f are idempotents. Then $(e,f) \in \mathcal{H}$. But an $\mathcal{H}$-class containing an idempotent is a group by Proposition 3.2.1(2), and so $e = f$. ■

Homomorphisms between groups are always idempotent-separating. In Chapter 5, we show that there are further grounds for regarding idempotent-separating homomorphisms as the most natural generalisations of group homomorphisms.

If $\theta\colon S \to T$ is a homomorphism between inverse semigroups then all of Green's relations are *preserved* by θ in the sense that if $(a,b) \in \mathcal{K}(S)$ then $(\theta(a), \theta(b)) \in \mathcal{K}(T)$. For surjective, idempotent-separating homomorphisms the converse is also true.

Proposition 13 *Let $\theta\colon S \to T$ be a surjective, idempotent-separating homomorphism.*

(1) *If $\mathcal{K}$ is any one of Green's relations then*

$$(s,t) \in \mathcal{K}(S) \Leftrightarrow (\theta(s), \theta(t)) \in \mathcal{K}(T).$$

(2) *S is (0-)bisimple if, and only if, T is (0-)bisimple.*

(3) *S is (0-)simple if, and only if, T is (0-)simple.*

Proof (1) We have already observed that $(s,t) \in \mathcal{K}(S)$ implies $(\theta(s), \theta(t)) \in \mathcal{K}(T)$. We prove the converses. We begin with $\mathcal{K} = \mathcal{L}$. Let $(\theta(a), \theta(b)) \in \mathcal{L}$. Then $\theta(a)^{-1}\theta(a) = \theta(b)^{-1}\theta(b)$ in T. Thus $\theta(a^{-1}a) = \theta(b^{-1}b)$. But θ is idempotent-separating and so $a^{-1}a = b^{-1}b$. Thus $(a,b) \in \mathcal{L}$. The proof for $\mathcal{R}$ is, of course, similar and the result for $\mathcal{H}$ then follows. The result for $\mathcal{D}$ follows from the fact that $\mathcal{D} = \mathcal{L} \circ \mathcal{R}$. This only leaves the case where $\mathcal{K} = \mathcal{J}$.

Let $(\theta(a), \theta(b)) \in \mathcal{J}$ in T. Then

$$\theta(a) = \theta(x)\theta(b)\theta(y) \text{ and } \theta(b) = \theta(u)\theta(a)\theta(v)$$

for some $x, y, u, v \in S$, since θ is surjective. But θ is idempotent-separating, and so $\ker\theta$ is contained in $\mathcal{H}$ by Proposition 12. Thus $(a, xby) \in \mathcal{H}$ and $(b, uav) \in \mathcal{H}$. But $\mathcal{H} \subseteq \mathcal{J}$ and so $(a, b) \in \mathcal{J}$.

The proofs of (2) and (3) now follow from (1) above. ■

Idempotent pure prehomomorphisms

Prehomomorphisms preserve the $\mathcal{L}$ and $\mathcal{R}$ relations by the proof of Theorem 3.1.5. Let θ: $S \to T$ be a prehomomorphism between inverse semigroups. Then for each idempotent e in S, θ induces a function by restriction from L_e to $L_{\theta(e)}$. We say that θ is *$\mathcal{L}$-injective* (resp. *$\mathcal{L}$-surjective*) if all of these induced functions are injective (resp. surjective). If θ is both $\mathcal{L}$-injective and $\mathcal{L}$-surjective then it is said to be *$\mathcal{L}$-bijective*.

Notions of *$\mathcal{R}$-injective*, *$\mathcal{R}$-surjective* and *$\mathcal{R}$-bijective* are defined similarly.

Proposition 14 *Let θ: $S \to T$ be a prehomomorphism between inverse semigroups. Then the following are equivalent:*

(1) *θ is $\mathcal{L}$-injective.*

(2) *θ is $\mathcal{R}$-injective.*

(3) *If $\theta(s)$ is an idempotent then s is an idempotent.*

Proof (1) $\Rightarrow$ (2). Let $(s, t) \in \mathcal{R}$ and $\theta(s) = \theta(t)$. Then $(s^{-1}, t^{-1}) \in \mathcal{L}$ and $\theta(s^{-1}) = \theta(t^{-1})$ by Theorem 3.1.5. Thus by (1), $s^{-1} = t^{-1}$, and so $s = t$.
(2) $\Rightarrow$ (3). Let $\theta(s)$ be an idempotent. Then $\theta(s) = \theta(s^{-1})$ and so $\theta(s) = \theta(s)^{-1}$ by the proof of Theorem 3.1.5. Now

$$\theta(ss^{-1}) = \theta(s)\theta(s^{-1}) = \theta(s)\theta(s)^{-1} = \theta(s)^2 = \theta(s),$$

by Theorem 3.1.5. But $ss^{-1} \,\mathcal{R}\, s$, and so, by assumption, $ss^{-1} = s$.
(3) $\Rightarrow$ (1). Let $(s, t) \in \mathcal{L}$ and $\theta(s) = \theta(t)$. Then $\theta(s)\theta(s^{-1}) = \theta(t)\theta(s^{-1})$. Thus $\theta(ts^{-1})$ is an idempotent and so, by assumption, ts^{-1} is an idempotent. Now $ts^{-1}ts^{-1} = ts^{-1}$. Thus $ts^{-1}ts^{-1}s = ts^{-1}s$. But $(s, t) \in \mathcal{L}$ and so $s^{-1}s = t^{-1}t$. Hence $ts^{-1}t = t$. By symmetry, $st^{-1}s = s$, and so $s^{-1}ts^{-1} = s^{-1}$. Thus $s^{-1} = t^{-1}$, which gives $s = t$. ■

$\mathcal{L}$-injective prehomomorphisms are also said to be *idempotent pure*. Intuitively, they are something like immersions in topology in that they are 'locally injective functions'.

The $\mathcal{D}$ and $\mathcal{J}$ relations and the natural partial order

In Proposition 3, we saw that the natural partial order interacts in a straightforward way with Green's relations $\mathcal{L}$, $\mathcal{R}$ and $\mathcal{H}$. The relationship between the natural partial order and the $\mathcal{D}$-relation is more interesting. The key to understanding this relationship is the following result.

Proposition 15 *Let S be an inverse semigroup. If $x \leq y \,\mathcal{D}\, v$ then there exists $u \in S$ such that $u \leq v$ and $u \,\mathcal{D}\, x$. If $x \neq y$ then $u \neq v$.*

Proof From $(y, v) \in \mathcal{D}$ it follows that $(yy^{-1}, vv^{-1}) \in \mathcal{D}$. Thus by Proposition 5(1), there is an element $a \in S$ such that $yy^{-1} = aa^{-1}$ and $vv^{-1} = a^{-1}a$. Put $b = xx^{-1}a$ and $u = b^{-1}bv$. Clearly, $b \leq a$ and $u \leq v$. We shall show that $x \,\mathcal{D}\, u$. Firstly,

$$bb^{-1} = xx^{-1}a(xx^{-1}a)^{-1} = xx^{-1}aa^{-1}$$

but $aa^{-1} = yy^{-1}$ and $x \leq y$ so that $bb^{-1} = xx^{-1}$. Secondly,

$$uu^{-1} = b^{-1}bv(b^{-1}bv)^{-1} = b^{-1}bvv^{-1}$$

but $vv^{-1} = a^{-1}a$ and $b \leq a$ so that $b^{-1}b = uu^{-1}$. Thus $x \,\mathcal{D}\, u$ by Proposition 5(1).

Now suppose that $u = v$. We showed above that $b^{-1}b = uu^{-1}$ and $a^{-1}a = vv^{-1}$, so that $b^{-1}b = a^{-1}a$. But $b \leq a$ so that $a = b$ by Proposition 3. Now $aa^{-1} = yy^{-1}$ and $bb^{-1} = xx^{-1}$. Thus $yy^{-1} = xx^{-1}$. But $x \leq y$ so that $x = y$ by Proposition 3. Thus $x \neq y$ implies $u \neq v$. ■

An inverse semigroup is said to be *completely semisimple* when the natural partial order is equality when restricted to any $\mathcal{D}$-class. As a corollary of the above result we have the following important theorem.

Theorem 16 *Let S be an inverse semigroup. If D is a $\mathcal{D}$-class of S, then the restriction of the natural partial order to D is either equality or for every $b \in D$ there exists $a \in D$ such that $a < b$. In particular, the inverse semigroup S is either completely semisimple or there exists a pair of distinct, $\mathcal{D}$-related elements $a, b \in S$ such that $a < b$.* ■

Theorem 16 is important in understanding the relationship between the $\mathcal{D}$- and $\mathcal{J}$-relations.

Proposition 17 *Let S be an inverse semigroup and let $y \in S$. If the $\mathcal{D}$-class D_y has a minimal element then $D_y = J_y$.*

Proof Suppose that D_y contains a minimal element. Clearly, $D_y \subseteq J_y$ so, to prove the reverse inclusion, let $x \in J_y$. By Proposition 8(2), there exists an element y' such that $x \mathcal{D} y' \leq y$. Since $x \mathcal{D} y'$ it is also true that $y' \in J_y$, so that by Proposition 8(2) again there exists an element y'' such that $y \mathcal{D} y'' \leq y'$. Hence $y'' \leq y' \leq y$ and $y'' \mathcal{D} y$. Since D_y has a minimal element, every element of D_y is minimal by Theorem 16. Consequently $y'' = y' = y$. It therefore follows that $x \in D_y$. Hence $J_y \subseteq D_y$. Consequently $D_y = J_y$. ∎

We can now easily obtain necessary and sufficient conditions for the $\mathcal{D}$- and $\mathcal{J}$-relations to coincide.

Theorem 18 *Let S be an inverse semigroup. Then $\mathcal{J} = \mathcal{D}$ if, and only if, from $x \mathcal{D} y$ and $x \leq u \leq y$ we may deduce that $x \mathcal{D} u$.*

Proof Suppose that $\mathcal{J} = \mathcal{D}$. Let $x \mathcal{D} y$ and $x \leq u \leq y$. Then $y \mathcal{D} x \leq u$, so that $SyS \subseteq SuS$ by Proposition 8. But $u \leq y$ implies that $SuS \subseteq SyS$. Thus $u \mathcal{J} y$. By assumption $u \mathcal{D} y$, and so $x \mathcal{D} u$.

Conversely, suppose that $(x, y) \in \mathcal{J}$. Then by Proposition 8, there exist elements x' and y' such that $y \mathcal{D} x' \leq x$ and $x \mathcal{D} y' \leq y$. We thus have $y' \leq y \mathcal{D} x'$. By Proposition 15, there exists an element x'' such that $x'' \leq x'$ and $x'' \mathcal{D} y'$. Thus $x'' \mathcal{D} x$ and $x'' \leq x' \leq x$, so that $x'' \mathcal{D} x'$ by assumption. Hence $x \mathcal{D} x'$ and so $x \mathcal{D} y$ giving $\mathcal{J} = \mathcal{D}$. ∎

A special case of the above result is worth noting.

Corollary 19 *Let S be an inverse semigroup. If $x \mathcal{D} y$ and $x \leq y$ implies $x = y$, then $\mathcal{J} = \mathcal{D}$.* ∎

3.3 Primitive inverse semigroups

This section concerns the relationship between groupoids and inverse semigroups. In Proposition 1.4.4, we proved that an inverse semigroup in which the natural partial order is equality is just a group. In this section, we prove an analogous result for inverse semigroups with zero. This is another example of the need to treat inverse semigroups with zero differently from plain inverse semigroups; we saw it in the distinction between bisimple and 0-bisimple semigroups and between simple and 0-simple semigroups.

Let S be an inverse semigroup with zero. Then the zero is the smallest element in the semigroup with respect to the natural partial order. Thus if we required the order to be equality then the semigroup would consist of just a zero. Instead, we shall consider the case where the natural partial order is equality when restricted to the non-zero elements.

A non-zero idempotent in a semigroup with zero is said to be *primitive* if it is minimal relative to the natural partial order on the set of non-zero idempotents. An inverse semigroup S with zero is said to be *primitive* if every non-zero idempotent is primitive. In the case of an inverse semigroup without zero, a primitive idempotent will just be a minimal idempotent, and a primitive inverse semigroup will be an inverse semigroup in which every idempotent is minimal.

Proposition 1 *Let S be an inverse semigroup with zero. Then the natural partial order is equality when restricted to $S\backslash\{0\}$ if, and only if, S is primitive.*

Proof If the natural partial order is equality when restricted to $S\backslash\{0\}$ then it is immediate that every non-zero idempotent is primitive. To prove the converse, suppose that every idempotent is minimal in $E(S)\backslash\{0\}$. Let $s \leq t$. Then $s^{-1}s \leq t^{-1}t$. By assumption, either $s^{-1}s = 0$ or $s^{-1}s = t^{-1}t$. In the former case, we have that $s = ss^{-1}s = 0$, and in the latter case we have that $s = ts^{-1}s = tt^{-1}t = t$. ∎

The existence of just one primitive idempotent in a simple or 0-simple semigroup has important implications for the structure of that semigroup.

Proposition 2 *Let S be an inverse semigroup. If S is 0-simple, then S has a primitive idempotent if, and only if, S is a primitive, 0-bisimple, inverse semigroup. If S is simple, then S has a primitive idempotent if, and only if, S is a group.*

Proof Let S be a 0-simple inverse semigroup with a primitive idempotent e. Any element beneath an idempotent is also an idempotent, and so e is a minimal element in its $\mathcal{D}$-class. Thus the natural partial order is equality on D_e by Theorem 3.2.16. Consequently, $D_e = J_e$ by Proposition 3.2.17. By assumption, S is 0-simple and so $S\backslash\{0\} = J_e$ by Proposition 3.2.9. Thus S is 0-bisimple and the natural partial order is equality when restricted to $S\backslash\{0\}$. The converse is immediate.

Now suppose that S is an inverse semigroup with a primitive idempotent. A similar argument to the above shows that the natural partial order on S is equality. Thus S is a group by Proposition 1.4.4. ∎

A semigroup with zero is said to be *completely 0-simple* if it is 0-simple and has a primitive idempotent. Inverse, completely 0-simple semigroups are called *Brandt semigroups*. Thus Proposition 2 says, in particular, that Brandt semigroups are 0-bisimple.

There is an easy way of constructing examples of primitive inverse semigroups.

Proposition 3 *Let G be a groupoid. Let $0 \notin G$ and put $G^0 = G \cup \{0\}$. Define a binary operation on G^0 as follows: if $x, y \in G$ and $\exists x \cdot y$ in the groupoid G*

then $xy = x \cdot y$; *all other products in* G^0 *are* 0. *With this operation* G^0 *is a primitive inverse semigroup.*

Proof Let $x, y, z \in G^0$. It is easy to check that $x(yz) = 0$ precisely when $(xy)z = 0$. Thus G^0 is a semigroup since the category product is associative. The idempotents of G^0 are the identities of G together with 0. If $e, f \in E(G^0)$ then

$$ef \neq 0 \Leftrightarrow e = f \neq 0.$$

Thus the idempotents form a commutative subsemigroup. Clearly for each $x \in G$ we have that $x = xx^{-1}x$ and $x^{-1} = x^{-1}xx^{-1}$ where x^{-1} is the inverse of x in the groupoid G. Also, $0 = 000$. Thus G^0 is an inverse semigroup. Let e be a non-zero idempotent of G^0. Let $s \leq e$ for some $s \in G^0$. Then $s = ss^{-1}e$. If $s \neq 0$, then $ss^{-1} = e$, so that $s = e$. Thus G^0 is a primitive inverse semigroup. ■

The semigroup G^0 is said to arise from the groupoid G by *adjoining a zero.*

Theorem 4 *Let* S *be an inverse semigroup with zero. Then* S *is primitive if, and only if, it is isomorphic to a groupoid with zero adjoined.*

Proof We proved in Proposition 3 that every groupoid with a zero adjoined is a primitive inverse semigroup. To prove the converse, let S be a primitive inverse semigroup with zero. By Proposition 3.1.4, $(S, \cdot)$ is a groupoid. In this groupoid, the zero of S is a connected component on its own, since $a \mathcal{D} 0$ implies $a = 0$. Thus $G = S \setminus \{0\}$ is a well-defined groupoid with respect to the restricted product. We shall prove that the semigroup S is isomorphic to the semigroup obtained by adjoining a zero to G.

Let $\iota\colon S \to G^0$ be the obvious bijection. We show that ι is a semigroup homomorphism. Consider the case where x and y are non-zero elements of S. Then $xy = (xe) \cdot (ey)$ where $e = x^{-1}xyy^{-1}$ by Theorem 3.1.2. By assumption, S is primitive. Thus $e = x^{-1}x = yy^{-1}$ or $e = 0$. In the former case, $xy = x \cdot y$, a restricted product. But then $\iota(xy) = \iota(x) \cdot \iota(y)$. In the latter case, the product of x and y in S is 0, and the restricted product of x and y does not exist. Thus the product of x and y in G^0 is 0, and so $\iota(xy) = 0 = \iota(x)\iota(y)$. If either of x or y is zero then it is immediate that ι preserves the product. ■

We can analyse the above result further. First we need a definition. Let $\{S_i\colon i \in I\}$ be a family of disjoint semigroups with zero, where the zero of S_i is denoted by 0_i. Let S be the union of the sets $S_i \setminus \{0_i\}$ together with a new symbol 0. Define a product on S as follows. If $x, y \in S \setminus \{0\}$ then $xy = 0$ unless there is an $i \in I$ such that $x, y \in S_i$ and the product xy in S_i is non-zero, in which case the product of x and y is defined to be xy. The element 0 is defined to act as a zero in all other cases. It is easy to show that S is a semigroup; S is said to be a *0-direct union* of the S_i. It is clear that each of the semigroups S_i is embedded in S.

Every groupoid is a disjoint union of its connected components. In Section 3.2, we saw that the $\mathcal{D}$-relation in an inverse semigroup corresponds to the equivalence relation determined by the connected components of the associated groupoid. From these observations and Proposition 2 and Theorem 4, the proof of the following is now immediate.

Theorem 5 *Brandt semigroups are precisely the connected groupoids with a zero adjoined, and every primitive inverse semigroup with zero is a 0-direct union of Brandt semigroups.* ■

The terminology Brandt semigroup arises because connected groupoids used to be known as Brandt groupoids. Readers should also be aware that groupoids are sometimes termed Croisot groupoids in older papers.

To determine the structure of Brandt semigroups in more detail, we look first at a simple way of constructing connected groupoids.

Proposition 6 *Let G be a group and I a non-empty set. Define a partial product on $I \times G \times I$ by $(i,g,j)(j,h,k) = (i,gh,k)$ and undefined in all other cases. Then $I \times G \times I$ is a connected groupoid, and every connected groupoid is isomorphic to one constructed in this way.*

Proof The first part of the proposition is straightfoward.

Let H be a connected groupoid. Let e be any identity of H, and put $G = \hom(e,e)$ and $I = G_o$. We shall show that H is isomorphic to the groupoid $I \times G \times I$. Since H is connected, for each $f \in H_0$ we may choose an element $x_f \in H$ such that

$$\mathbf{d}(x_f) = e \text{ and } \mathbf{r}(x_f) = f.$$

Define a function $\theta : I \times G \times I \to H$ by $\theta(i,g,j) = x_i g x_j^{-1}$. This is well-defined, since

$$\mathbf{d}(x_i) = e,\ \mathbf{r}(g) = e,\ \mathbf{d}(g) = e, \text{ and } \mathbf{r}(x_j^{-1}) = \mathbf{d}(x_j) = e.$$

We show that θ is an isomorphism of groupoids. To begin with θ is a functor. The identities of $I \times G \times I$ are the elements of the form $(i,1,i)$ where i is an identity of G. Clearly $\theta(i,1,i)$ is an identity of H. Let

$$(i,g,j), (j,h,l) \in I \times G \times I$$

so that $(i,g,j)(j,h,l)$ is defined. Then $\theta(i,g,j) = x_i g x_j^{-1}$ and $\theta(j,h,l) = x_j h x_l^{-1}$. A simple calculation shows that $\theta(i,g,j)\theta(j,h,l) = \theta(i,gh,l)$. Thus

θ is a functor. We next show that θ is injective. Suppose that $\theta(i,g,j) = \theta(k,h,l)$. Then $x_i g x_j^{-1} = x_k h x_l^{-1}$. Now $\mathbf{r}(x_i g x_j^{-1}) = \mathbf{r}(x_i) = i$. Similarly $\mathbf{r}(x_k h x_l^{-1}) = \mathbf{r}(x_k) = k$. Thus $i = k$, and so $x_i = x_k$ from which we obtain $gx_j^{-1} = hx_l^{-1}$. Also $\mathbf{d}(gx_j^{-1}) = \mathbf{r}(x_j) = j$ and $\mathbf{d}(hx_l^{-1}) = \mathbf{r}(x_l) = l$. Thus $j = l$ and so $x_j = x_l$. It follows that $g = h$ and so $(i,g,j) = (k,h,l)$. Finally we show that θ is surjective. Let $x \in H$, where $\mathbf{d}(x) = j$ and $\mathbf{r}(x) = i$. Then $x_i^{-1} x x_j \in G$. Put $g = x_i^{-1} x x_j$, so that $x = x_i g x_j^{-1}$. Then $\theta(i,g,j) = x$. ■

Let G be a group and I a non-empty set. Let $B(G,I)$ be the groupoid $I \times G \times I$ with a zero adjoined. If G and H are isomorphic groups and I and J have the same cardinality then $B(G,I)$ is isomorphic to $B(H,J)$. When I has cardinality n we write $B(G,n)$ instead of $B(G,I)$. When the group is trivial we write $B_n = B(\{1\},n)$; these are the semigroups introduced in Section 3.2.

Combining Theorem 5 and Proposition 6, we obtain the following.

Theorem 7 *Every Brandt semigroup is isomorphic to a semigroup of the form $B(G,I)$ for some group G and for some non-empty set I.* ■

3.4 The bicyclic monoid

We proved in Theorem 3.2.16 that an inverse semigroup is either completely semisimple or there exists a pair of distinct, comparable, $\mathcal{D}$-related elements. We now examine the implications for the structure of an inverse semigroup if the latter situation occurs.

Definition and basic properties

Let S be an inverse semigroup possessing a $\mathcal{D}$-class D on which the natural partial order is not trivial. Then for every idempotent $e \in D$ there exists an idempotent $f \in D$ such that $f < e$ by Theorem 3.2.16. However, since $e \mathcal{D} f$, there exists an element $a \in D$ such that $a^{-1}a = e$ and $aa^{-1} = f$ by Proposition 3.2.5. We focus our attention on the inverse subsemigroup of S generated by a. Observe that this is a monoid with identity e, because $ea = a = ae$ and $ea^{-1} = a^{-1} = a^{-1}e$. We show below that this inverse monoid is determined uniquely up to isomorphism by the fact that it is generated as a monoid by a and a^{-1} and the fact that $a^{-1}a$ is the identity.

To obtain a concrete representation of this monoid, suppose that S were a symmetric inverse monoid. Then e would be a partial identity defined on a set isomorphic to a proper subset of itself, namely the domain of f. Such sets are said to be *Dedekind infinite*. Thus the simplest example of idempotents e and f satisfying the above conditions arises in the symmetric inverse monoid $I(\mathbb{N})$; the set $\mathbb{N}$ is isomorphic to the proper subset $\mathbb{N} \setminus \{0\}$ by means of the successor function α: $\mathbb{N} \to \mathbb{N}$ defined by $\alpha(n) = n + 1$. Since $\alpha \in I(\mathbb{N})$, it generates an inverse submonoid of $I(\mathbb{N})$ which we denote by B. Observe that $\alpha^{-1}\alpha = 1$, the identity on $\mathbb{N}$.

The above observations motivate the following definition. The *bicyclic monoid* is the monoid given by the (monoid) presentation

$$P_1 = \langle p, q \colon pq = 1 \rangle.$$

Define θ: $P_1 \to B$ by $\theta(p) = \alpha^{-1}$ and $\theta(q) = \alpha$. By the monoid version of Proposition 2.3.5, B is a homomorphic image of P_1. We shall show that they are in fact isomorphic.

In what follows we shall use the convention introduced in Section 2.3: if u and v are strings then '$u = v$' means the strings are identically equal, whereas '$u \equiv v$' means that the strings are equivalent in the presentation.

Theorem 1 *The bicyclic monoid is isomorphic to the inverse monoid generated by the successor function on the natural numbers.*

Proof It is immediate from $pq = 1$, that every element of P_1 is equivalent to a string of the form $q^m p^n$ for some $m, n \in \mathbb{N}$. Now suppose that $q^m p^n \equiv q^r p^s$. Then $\theta(q^m p^n) = \theta(q^r p^s)$, so that $\alpha^m \alpha^{-n} = \alpha^r \alpha^{-s}$. But $(\alpha^m \alpha^{-n})(n) = m$ and so $(\alpha^r \alpha^{-s})(n) = m$. However,

$$(\alpha^r \alpha^{-s})(n) = (n - s) + r$$

and $0 \leq n - s$. Thus $m - n = r - s$. Similarly, $(\alpha^r \alpha^{-s})(s) = r$ and so $(\alpha^m \alpha^{-n})(s)$ is defined. In particular, $0 \leq s - n$. Thus $n = s$ and so $m = r$, which gives $q^m p^n = q^r p^s$. Consequently, the elements $q^m p^n$ form a transversal for the congruence classes of the presentation and so form a set of normal forms. Thus the homomorphism from P_1 to B is an injection, and consequently the semigroups P_1 and B are isomorphic. ■

The transversal obtained above can be used to obtain a more convenient representation for the bicyclic monoid. The operation '$\dot{-}$', sometimes called *monus*, is defined on the set of natural numbers as follows:

$$a \dot{-} b = \begin{cases} a - b & \text{if } a \geq b \\ 0 & \text{otherwise.} \end{cases}$$

On the set $\mathbb{N} \times \mathbb{N}$ define a binary operation by

$$(m, n)(r, s) = (m + (r \dot{-} n), s + (n \dot{-} r)).$$

Proposition 2 *The bicyclic monoid is isomorphic to the set $\mathbb{N} \times \mathbb{N}$ equipped with the above binary operation.*

Proof Consider the product $(q^m p^n)(q^r p^s)$. If $n = r$ then this product is $q^m p^s$; if $n > r$ then it is $q^m p^{(n-r)+s}$; and if $n < r$ then it is $q^{(m+(r-n))}p^s$. Define the function θ: $P_1 \to \mathbb{N} \times \mathbb{N}$ which maps $q^m p^n$ to (m, n). This is a bijection, and it is now easy to check that it is a homomorphism of the respective binary operations. ■

From now on, we shall usually assume that P_1 is given in terms of ordered pairs of natural numbers.

The main inverse semigroup-theoretic properties of the bicyclic monoid are as follows.

Theorem 3 *The bicyclic monoid is a combinatorial, bisimple, E-unitary inverse monoid.*

Proof It is straighforward to check that the idempotents are the elements of the form (m, m). If (m, m) and (n, n) are idempotents then a direct computation shows that

$$(m, m)(n, n) = (\max(m, n), \max(m, n)) = (n, n)(m, m).$$

Thus the idempotents form a commutative subsemigroup. To show that the bicyclic monoid is inverse, observe that $(m, n)(n, s) = (m, s)$, so that $(m, n) = (m, n)(n, m)(m, n)$ for every element (m, n). Thus the bicyclic monoid is regular with commuting idempotents and so inverse. The inverse of (m, n) is (n, m). Observe that

$$(m, n)^{-1}(m, n) = (n, n) \text{ and } (m, n)(m, n)^{-1} = (m, m).$$

Thus the bicylic monoid is combinatorial.

If (m, m) and (n, n) are any two idempotents then the element (m, n) satisfies $(m, m)\,\mathcal{R}\,(m, n)$ and $(m, n)\,\mathcal{L}\,(n, n)$. Thus $(m, m)\,\mathcal{D}\,(n, n)$, and so the bicyclic monoid is bisimple.

We now characterise the natural partial order. Suppose that $(m, n) \leq (p, q)$. Then $(m, n) = (p, q)(n, n)$. By definition

$$(p, q)(n, n) = (p + (n \dot{-} q), n + (q \dot{-} n)).$$

Thus

$$m = p + (n \dot{-} q) \text{ and } n = n + (q \dot{-} n).$$

But $n = n + (q \dot{-} n)$ implies that $0 = q \dot{-} n$ so that $q \leq n$. Put $a = n - q$. Then $m = a + p$ and $n = a + q$. Conversely, suppose that $m = a + p$ and $n = a + q$ for some natural number a. Then

$$(p, q)(n, n) = (p + (n \dot{-} q), n + (q \dot{-} n)) = (m, n).$$

We have therefore proved that the natural partial order is given by:

$$(m,n) \leq (p,q) \Leftrightarrow m = a+p \text{ and } n = a+q$$

for some natural number a.

It is now immediate that any element above an idempotent is also an idempotent. Thus the bicyclic monoid is E-unitary. ■

By Theorem 3, the natural partial order on the semilattice of idempotents is given by $(m,m) \leq (n,n)$ precisely when $m \geq n$, which is just the dual of the usual ordering on the natural numbers. We denote by $(\omega, \leq)$ the poset consisting of the natural numbers under the dual of the usual partial order. Clearly, the semilattice of idempotents of the bicyclic monoid is isomorphic to this semilattice. More generally, an *inverse ω-semigroup* is any inverse semigroup whose semilattice of idempotents is isomorphic to the poset $(\omega, \leq)$.

The following lemma will be used below; it is also of independent interest because it shows how conjugation of idempotents can be used to recapture the addition and monus operations on the natural numbers. The proofs are by direct computation.

Lemma 4 *Let* $(a,0), (m,m) \in P_1$.

(1) $(a,0)(m,m)(a,0)^{-1} = (a+m, a+m)$.

(2) $(a,0)^{-1}(m,m)(a,0) = (m \dot{-} a, m \dot{-} a)$. ■

All homomorphic images of the bicyclic monoid can be described. The key to the solution of this problem turns out to be the minimum group congruence.

Theorem 5 *Any proper congruence on the bicyclic monoid is a group congruence. In particular,* P_1/σ *is the group of integers.*

Proof Let ρ be a proper congruence on P_1. We first show that $(0,0)\,\rho\,(1,1)$. Since ρ is proper, there are at least two distinct elements $(m,n),(p,q) \in P_1$ such that $(m,n)\,\rho\,(p,q)$. By taking inverses of elements if necessary, we can assume that $m \neq p$. Without loss of generality, assume that $m < p$. By Proposition 2.3.4, we have that $(m,m)\,\rho\,(p,p)$. By Lemma 4,

$$(0,m)(m,m)(m,0) = (0,0)$$

and

$$(0,m)(p,p)(m,0) = (p-m, p-m).$$

Thus $(0,0)\,\rho\,(p-m,p-m)$. Put $t = p-m$ and let $0 \leq k \leq t$. Then $(k,k) = (0,0)(k,k)$, since $(0,0)$ is the identity, and $(t,t) = (t,t)(k,k)$ from the proof of Theorem 3. Thus

$$(k,k) = (0,0)(k,k)\,\rho\,(t,t)(k,k) = (t,t).$$

It follows that $(k,k)\,\rho\,(t,t)$ for every $0 \leq k \leq t$. Since $t \geq 1$, we have that $(0,0)\,\rho\,(t,t)$ and $(1,1)\,\rho\,(t,t)$. Hence $(0,0)\,\rho\,(1,1)$.

We can now show that all idempotents are identified by ρ. For any $k \geq 0$, we have that $(k+1,k+1) = (k,0)(1,1)(0,k)$ and $(k,k) = (k,0)(0,0)(0,k)$ by Lemma 4. Thus

$$(k+1,k+1) = (k,0)(1,1)(0,k)\,\rho\,(k,0)(0,0)(0,k) = (k,k).$$

Hence all idempotents are ρ-related to $(0,0)$. Consequently, every proper congruence is a group congruence.

We now characterise the minimum group congruence; we prove that

$$(m,n)\,\sigma\,(p,q) \Leftrightarrow m-n = p-q.$$

Suppose that $(m,n)\,\sigma\,(p,q)$. Then there exists (e,f) such that $(e,f) \leq (m,n)$ and $(e,f) \leq (p,q)$. Thus there exist natural numbers a and b such that

$$e = m+a,\ f = n+a,\ e = p+b,\ \text{and}\ f = q+b$$

from the proof of Theorem 3. Thus $m-n = e-f = p-q$.

Conversely, suppose that $m-n = p-q$. Put $e = \max\{m,p\}$ and $f = \max\{n,q\}$. We shall prove that $(e,f) \leq (m,n),(p,q)$. Let $a = e-m$. It is zero if $p \leq m$ and $p-m$ otherwise. Now consider $f-n$. It is zero if $q \leq n$ and $q-n$ otherwise. But $p-m = q-n$. Thus in all cases, $a = e-m = f-n$. Hence $e = m+a$ and $f = n+a$, so that $(e,f) \leq (m,n)$. We may similarly show that $(e,f) \leq (p,q)$.

Define $\theta\colon P_1 \to \mathbb{Z}$ by $\theta(\sigma(m,n)) = m-n$. By the above result, this is a well-defined bijection. It is straightforward to check that it is a homomorphism. ∎

The bicyclic monoid as a subsemigroup

We use the above result to obtain a *semigroup* presentation of the bicyclic monoid.

Theorem 6 *The semigroup given by the semigroup presentation*

$$S = \langle a,b\colon aba = a, bab = b, a = a^2b \text{ and } b = ab^2\rangle$$

is a monoid with identity ab, which is isomorphic to the bicyclic monoid.

Proof Put $e = ab$. Then $ea = aba = a$ and $ae = a^2b = a$. Similarly, $eb = ab^2 = b$ and $be = bab = b$. Every element of S is a product of a's and b's. Thus e is an identity (and therefore the identity) for S. It follows that S is a homomorphic image of $\langle p,q\colon pq = 1\rangle$ under the homomorphism mapping

p to a and q to b. Consider the congruence induced on $\langle p, q \colon pq = 1 \rangle$ by this homomorphism. If it were proper, then by Theorem 5, S would be a group. However, we may define a surjective homomorphism from the free semigroup on $\{a, b\}$ to P_1 which maps a to p and b to q. Furthermore, the images under this homomorphism of each of the relations in the presentation of S also hold in P_1. Thus P_1 is a homomorphic image of S by Proposition 2.3.5. Thus if S were a group, then so too would P_1. Hence the congruence cannot be proper and so the semigroups are isomorphic. ■

We now investigate the occurrence of the bicyclic semigroup as a subsemigroup.

Proposition 7 *Let S be an inverse semigroup. If a $\mathcal{J}$-class J contains elements x and y such that $x < y$, then J contains a copy of the bicyclic monoid.*

Proof From $x < y$ we have that $x^{-1}x \leq y^{-1}y$. Observe that if $x^{-1}x = y^{-1}y$ then we would have $x = y$. Thus, in fact, $x^{-1}x < y^{-1}y$. Clearly, $x^{-1}x, y^{-1}y \in J$. So, without loss of generality, we may assume that x and y are idempotents e and f such that $e < f$ and $e, f \in J$. Since $e \mathcal{J} f$ there exist elements $u, v \in S$ such that $f = uev$. Put $a = fue$ and $b = evf$. Then

$$aba = (fue)(evf)(fue) = f(uev)fue = fue = a$$

and

$$bab = (evf)(fue)(evf) = evf(uev)f = evf = b.$$

Thus a and b are mutually inverse, and so ab and ba are idempotents. Now $ab = (fue)(evf) = fuevf = f$ and $ba = (evf)(fue) = evfue \leq e < f$. Thus $a^2b = af = aef = ae = a$ and $ab^2 = fb = b$. Thus a and b generate a subsemigroup T of S which is a homomorphic image of the bicyclic monoid by Theorem 6. However, $ab \neq ba$ and so this subsemigroup is in fact isomorphic to the bicyclic monoid. By Theorem 3, the bicyclic monoid is bisimple, and so T is a subsemigroup of J. ■

The above result enables us to refine the statement of Proposition 3.3.2.

Theorem 8 *Let S be a 0-simple inverse semigroup. Then either S is a Brandt semigroup or for every (resp. for some) non-zero idempotent e, there exists a bicyclic subsemigroup of S with identity e.*

Proof Let S be a 0-simple inverse semigroup which is not completely 0-simple. Then it contains an idempotent f which is not primitive. (In fact, by Proposition 3.3.2 every non-zero idempotent is non-primitive.) Thus there exists a non-zero idempotent e such that $e < f$. Since S is 0-simple, $(e, f) \in \mathcal{J}$ by

Proposition 3.2.9(2). The result now follows by Proposition 7. ■

More generally, we have the following dichotomy in the class of inverse semigroups; the proof follows from Theorem 3.2.16 and Proposition 7.

Theorem 9 *An inverse semigroup is either completely semisimple or contains a copy of the bicyclic monoid (but not both).* ■

Applications of the bicyclic monoid

One of the earliest applications of the bicyclic monoid was in ring theory. Jacobson [151] investigated those rings R containing elements $p, q \in R$ such that $pq = 1$ and $qp \neq 1$; thus in our terms, he was considering an embedding of the bicyclic monoid in the multiplicative monoid of the ring. The idempotents of the bicyclic monoid are of the form $q^i p^i$ where $i \in \mathbb{N}$. Define

$$e_{i,j} = q^{i-1}p^{j-1} - q^i p^j$$

for all non-zero natural numbers i and j. It can be shown that none of the $e_{i,j}$ is zero and that the set of all $e_{i,j}$ together with the zero forms a subsemigroup of R isomorphic to $B_{\aleph_0}$. This has important consequences for the structure of the ring R. In particular, neither the ascending nor descending chain conditions hold for principal right ideals generated by idempotents.

Another important application of the bicyclic monoid is in formal language theory. Let $\Sigma = \{(,)\}$, and let L be the language over Σ consisting of all strings which are correct bracketings. Then the syntactic monoid of L is the bicyclic monoid [176].

A final application comes from λ-calculus. Hofmann and Mislove [141], [142] show that any homomorphic image of the bicyclic semigroup in a compact topological semigroup is a group; this is intuitively plausible, because compactness is a finiteness condition and the only finite images of the bicyclic monoid are groups. They then use their result to investigate the existence of certain models of the λ-calculus.

3.5 Notes on Chapter 3

Section 3.1

The arrow representation of inverse semigroups is essentially the same as the inductive groupoid approach to inverse semigroups discussed in Chapter 4.

The origin of the work on the associated groupoid and the natural partial order lies in the papers of Ehresmann [66], Schein [373] and Nambooripad [280]. Most semigroup theorists only became aware of this work in 1979 with the translation [381] of Schein's paper [373] which had originally been published in 1965.

Prehomomorphisms were introduced by McAlister and Reilly [246], though they called them $\vee$-prehomomorphisms and included as part of the definition the property that $\theta(x^{-1}) = \theta(x)^{-1}$. This property was shown to be redundant by McAlister [240]. What we have called dual prehomomorphisms were originally termed $\wedge$-prehomomorphisms by McAlister and Reilly, and are what Petrich [312] means by what he calls a prehomomorphism. The theory of prehomomorphisms was further developed by McAlister [233], [240].

Section 3.2

Green's relations are a classical tool of semigroup theory. For more information about them and further references consult [148].

Combinatorial inverse semigroups are an important class of inverse semigroups. Ash [8], [9] and Meakin [249] proved that any downward directed partially ordered set arises as the partially ordered set of $\mathcal{J}$-classes of some combinatorial, completely semisimple inverse semigroup. A description of all combinatorial inverse semigroups in terms of category-theoretic constructions was given by Rajan [334]. Families of combinatorial inverse semigroups are studied in the deep papers of Kadourek [163], [164].

(0-)bisimple semigroups and monoids also form an important class of inverse semigroups. Bisimple inverse monoids were introduced in Clifford's 1953 paper [47]. Clifford may have been motivated by his own work on lattice-ordered groups. Bisimple inverse semigroups were described by Reilly [339], and 0-bisimple inverse semigroups by McAlister [230] in terms of generalised RP-systems. This classical work is generalised to arbitrary inverse semigroups in Chapter 10. Other descriptions of these semigroups were obtained by Munn [267] and Reilly and Clifford [345]. The structure of an important class of bisimple inverse monoids, the ω-semigroups, is determined in Chapter 5.

Although bisimple inverse semigroups have only one $\mathcal{D}$-class they are not otherwise elementary in structure: Reilly proved [337] that every inverse semigroup could be embedded in a bisimple inverse monoid.

Congruence-free inverse semigroups were studied by Munn [271], [272]. Section IV.3 of Petrich [312] discusses necessary and sufficient conditions for congruence-freeness.

O'Carroll [294] defined E-reflexive semigroups. Petrich [312] uses the term E-reflexive in a stronger sense. He also states of E-reflexive semigroups that 'this class seems to play no appreciable role' (page 164). But Theorem 7, due to McAlister [237], is sufficiently interesting on its own to warrant the inclusion of E-reflexive semigroups. See also the notes to Section 1.4.

Idempotent-separating congruences were introduced by Preston [328]. In

[261] Munn showed that they were precisely the congruences contained in the $\mathcal{H}$-relation. Idempotent pure congruences were introduced by Green [107].

The material in this section concerning the relationship between $\mathcal{D}$ and $\mathcal{J}$ is adapted from Nambooripad's paper [281].

Section 3.3

Brandt groupoids, now termed connected groupoids, go back to the work of Brandt [34]. Clifford [46] showed that Brandt groupoids with a zero adjoined were semigroups isomorphic to $B(G, I)$. Munn [257] made the connection between Brandt semigroups and completely 0-simple semigroups.

Section 3.4

Jacobson's paper [151] is the earliest application I could find of the bicyclic monoid, but Karl Hofmann has suggested (private communication) that it would have been effectively known in functional analysis before this date because of its connection with the shift mapping. The bicyclic monoid appears to have been first explicitly defined within semigroup theory by Ljapin [205].

Chapter 4

Ordered groupoids

The results of this chapter will only be used in Chapter 8.

In Section 3.1, we showed that a groupoid could be constructed from an inverse semigroup, and that this groupoid, together with the natural partial order, could be used to reconstruct the original inverse semigroup. The aim of this chapter is to place these results in their correct conceptual framework. This is done by axiomatising the properties of groupoids equipped with partial orders arising in this way from inverse semigroups. The resulting ordered groupoids are said to be inductive, and the main result of this chapter (Theorem 4.1.8) shows that the category of inverse semigroups is isomorphic to the category of inductive groupoids. Such groupoids were the basis of Ehresmann's work on pseudogroups (Section 1.1).

It is tempting to think that, because we have established an isomorphism between inverse semigroups and inductive groupoids, we may therefore dispense with inductive groupoids. However, this misses an important point: the category of inductive groupoids sits inside the larger category of ordered groupoids. It is possible that constructions concerning inverse semigroups may best be carried out by translating them to the category of inductive groupoids, making use of the extra space afforded by the category of ordered groupoids and then interpreting the results in terms of inverse semigroups. Just such a procedure is realised in Chapter 8, where we use ordered groupoids to study E-unitary semigroups, E-unitary covers, and idempotent pure congruences.

The idea of using categories (and, more generally, semigroupoids) to help us study semigroups is now an established feature of the subject. In addition to Ehresmann's approach via ordered groupoids, the work of Tilson [412] has been particularly influential. The general idea of using categories as algebraic devices for studying other mathematical domains is argued in [195].

4.1 Inductive groupoids

In this section, we obtain an abstract characterisation of inverse semigroups equipped with their restricted product and natural partial orders.

From inverse semigroups to inductive groupoids

Let $(G, \cdot)$ be a groupoid, and let $\leq$ be a partial order defined on G. Then $(G, \cdot, \leq)$ is an *ordered groupoid* if the following axioms hold:

(OG1) $x \leq y$ implies $x^{-1} \leq y^{-1}$ for all $x, y \in G$.

(OG2) For all $x, y, u, v \in G$, if $x \leq y, u \leq v, \exists xu$ and $\exists yv$ then $xu \leq yv$.

(OG3) Let $x \in G$ and let e be an identity such that $e \leq \mathbf{d}(x)$. Then there exists a unique element $(x \mid e)$, called the *restriction of x to e*, such that $(x \mid e) \leq x$ and $\mathbf{d}(x \mid e) = e$.

(OG3)* Let $x \in G$ and let e be an identity such that $e \leq \mathbf{r}(x)$. Then there exists a unique element $(e \mid x)$, called the *corestriction of x to e*, such that $(e \mid x) \leq x$ and $\mathbf{r}(e \mid x) = e$.

An ordered groupoid is said to be *inductive* if the partially ordered set of identities forms a meet-semilattice. The justification for the definition of inductive groupoid is provided by the following proposition.

Proposition 1 *Let S be an inverse semigroup. Then $(S, \cdot, \leq)$ is an inductive groupoid.*

Proof By Proposition 3.1.4, $(S, \cdot)$ is a groupoid. Axioms (OG1) and (OG2) hold by Lemma 1.4.6 and Proposition 1.4.7, and axioms (OG3) and (OG3)* hold by Theorem 3.1.2. ■

The inductive groupoid associated with S is denoted by $\mathbf{G}(S)$.

A functor between two ordered groupoids is said to be *ordered* if it is order-preserving. An ordered functor between two inductive groupoids is said to be *inductive* if it preserves the meet operation on the set of identities. An *isomorphism* of ordered groupoids is a bijective ordered functor whose inverse is an ordered functor. Ordered functors can also be star injective, star surjective and star bijective (or covering functors).

Proposition 2 *Let $\theta: G \to H$ be an ordered functor between ordered groupoids.*

(1) *If $(x \mid e)$ is defined in G then $(\theta(x) \mid \theta(e))$ is defined in H and $\theta(x \mid e) = (\theta(x) \mid \theta(e))$.*

(2) *If* $(e \,|\, x)$ *is defined in* G *then* $(\theta(e) \,|\, \theta(x))$ *is defined in* H *and* $\theta(e \,|\, x) = (\theta(e) \,|\, \theta(x))$.

Proof We shall prove (1); the proof of (2) is similar. By definition $(x \,|\, e) \leq x$ and so $\theta(x \,|\, e) \leq \theta(x)$ since θ is an ordered functor. But

$$\mathbf{d}(\theta(x \,|\, e)) = \theta(\mathbf{d}(x \,|\, e)) = \theta(e)$$

since θ is a functor. But by axiom (OG3), $(\theta(x) \,|\, \theta(e))$ is the unique element less than $\theta(x)$ and with domain $\theta(e)$. Thus $\theta(x \,|\, e) = (\theta(x) \,|\, \theta(e))$. ■

We now establish some of the basic properties of ordered groupoids.

Proposition 3 *Let* $(G, \cdot, \leq)$ *be an ordered groupoid.*

(1) *If* $x \leq y$ *then* $\mathbf{d}(x) \leq \mathbf{d}(y)$ *and* $\mathbf{r}(x) \leq \mathbf{r}(y)$.

(2) *The order* $\leq$ *restricted to* hom-sets *is trivial.*

(3) *If* $\exists xy$ *and* e *is an identity such that* $e \leq \mathbf{d}(xy)$ *then*

$$(xy \,|\, e) = (x \,|\, \mathbf{r}(y \,|\, e))(y \,|\, e).$$

(4) *If* $\exists xy$ *and* e *is an identity such that* $e \leq \mathbf{r}(xy)$ *then*

$$(e \,|\, xy) = (e \,|\, x)(\mathbf{d}(e \,|\, x) \,|\, y).$$

(5) *If* $z \leq xy$ *then there exist elements* x' *and* y' *such that* $\exists x'y'$, $x' \leq x$, $y' \leq y$ *and* $z = x'y'$.

(6) *Axiom* (OG3)* *is a consequence of the axioms* (OG1) *and* (OG3).

(7) *The set of identities* G_o *is an order ideal of* G.

(8) *If* $f \leq e \leq \mathbf{d}(x)$ *then* $(x \,|\, f) \leq (x \,|\, e) \leq x$.

(9) *If* $f \leq e \leq \mathbf{r}(x)$ *then* $(f \,|\, x) \leq (e \,|\, x) \leq x$.

(10) *Let* $x, y, e, f \in G$ *such that*

$$x \leq y,\ f \leq e,\ f \leq \mathbf{d}(x) \text{ and } e \leq \mathbf{d}(y).$$

Then $(x \,|\, f) \leq (y \,|\, e)$.

Proof (1) This is immediate from axioms (OG1) and (OG2).
(2) Suppose that

$$\mathbf{d}(x) = \mathbf{d}(y),\ \mathbf{r}(x) = \mathbf{r}(y) \text{ and } x \leq y.$$

In particular, $\mathbf{d}(x) \leq \mathbf{d}(y)$ and so by axiom (OG3) there is a unique element $(y \,|\, \mathbf{d}(x))$ such that

$$(y \,|\, \mathbf{d}(x)) \leq y \text{ and } \mathbf{d}(y \,|\, \mathbf{d}(x)) = \mathbf{d}(x).$$

But the element x also has the property that $x \leq y$ and $\mathbf{d}(x) = \mathbf{d}(x)$. Thus by uniqueness $(y \,|\, \mathbf{d}(x)) = x$. However, $(y \,|\, \mathbf{d}(y)) = x$ since $\mathbf{d}(x) = \mathbf{d}(y)$. But $(y \,|\, \mathbf{d}(y)) = y$. Hence $x = y$.
(3) Since $e \leq \mathbf{d}(xy) = \mathbf{d}(y)$ the restriction $(y \,|\, e)$ is defined. Since $(y \,|\, e) \leq y$ we have that $\mathbf{r}(y \,|\, e) \leq \mathbf{r}(y) = \mathbf{d}(x)$. Thus $(x \,|\, \mathbf{r}(y \,|\, e))$ exists and the product $(x \,|\, \mathbf{r}(y \,|\, e))(y \,|\, e)$ exists. Clearly, $(x \,|\, \mathbf{r}(y \,|\, e))(y \,|\, e) \leq xy$. But

$$\mathbf{d}((x \,|\, \mathbf{r}(y \,|\, e))(y \,|\, e)) = e,$$

and so $(x \,|\, \mathbf{r}(y \,|\, e))(y \,|\, e) = (xy \,|\, e)$.
(4) Similar to the proof of (3).
(5) Let $z \leq xy$. Then $\mathbf{d}(z) \leq \mathbf{d}(xy)$ by (1). Thus $(xy \,|\, \mathbf{d}(z))$ exists. Now $\mathbf{d}(xy \,|\, \mathbf{d}(z)) = \mathbf{d}(z)$ and $(xy \,|\, \mathbf{d}(z)) \leq xy$, so that $z = (xy \,|\, \mathbf{d}(z))$. By (3),

$$(xy \,|\, \mathbf{d}(z)) = (x \,|\, \mathbf{r}(y \,|\, \mathbf{d}(z)))(y \,|\, \mathbf{d}(z)).$$

Put $x' = (x \,|\, \mathbf{r}(y \,|\, \mathbf{d}(z)))$ and $y' = (y \,|\, \mathbf{d}(z))$, and we have the result.
(6) Suppose that the axioms (OG1) and (OG3) hold. We show that axiom (OG3)* holds. Let $e \leq \mathbf{r}(x)$. Then $e \leq \mathbf{d}(x^{-1})$, so that $(x^{-1} \,|\, e)$ exists by axiom (OG3). Define $(e \,|\, x) = (x^{-1} \,|\, e)^{-1}$. Then $(e \,|\, x) \leq x$ by axiom (OG1), and $\mathbf{r}(e \,|\, x) = \mathbf{d}(x^{-1} \,|\, e) = e$. Now for uniqueness. Suppose that $y \leq x$ and $\mathbf{r}(y) = e$. Then $y^{-1} \leq x^{-1}$ by axiom (OG1) and $\mathbf{d}(y^{-1}) = e$. Thus $y^{-1} = (x^{-1} \,|\, e)$ by (OG3), and so $y = (e \,|\, x)$ by (OG1).
(7) Let $x \leq e$ where e is an identity. Then $\mathbf{d}(x) \leq e$. But $x, \mathbf{d}(x) \leq e$ and $\mathbf{d}(x) = \mathbf{d}(\mathbf{d}(x))$. Thus $x = \mathbf{d}(x)$.
(8) Let $f \leq e \leq \mathbf{d}(x)$. Both $(x \,|\, e)$ and $(x \,|\, f)$ exist. Now $f \leq \mathbf{d}(x \,|\, e)$ and so the element $((x \,|\, e) \,|\, f)$ exists. But $\mathbf{d}(x \,|\, f) = f$ and $(x \,|\, f) \leq x$. Thus $((x \,|\, e) \,|\, f) = (x \,|\, f)$, and so $(x \,|\, f) \leq (x \,|\, e)$.
(9) Similar to (8).
(10) By (8) we have that $(y \,|\, f) \leq (y \,|\, e)$. However, $(x \,|\, f), (y \,|\, f) \leq y$ and $\mathbf{d}(x \,|\, f) = \mathbf{d}(y \,|\, f)$. Thus by axiom (OG3), we have that $(x \,|\, f) = (y \,|\, f)$. Hence $(x \,|\, f) \leq (y \,|\, e)$. ■

The terminology 'ordered groupoid' is slightly misleading since, by result (2) above, ordered groupoids with one identity are just groups rather than ordered groups.

The term 'inductive groupoid' was used by Ehresmann to refer to a much more restricted class of ordered groupoids than we have defined, but the terminology is well-established.

Let G be an ordered groupoid and H a subset of G. Then we say that H is an *ordered subgroupoid* if it is a subgroupoid of G and an ordered groupoid with respect to the induced order. This is equivalent to the condition that H be a subgroupoid of G and that if $x \in H$ and $e \in H_o$ and $e \leq \mathbf{d}(x)$ then $(x \mid e) \in H$.

Let θ: $G \to K$ be an injective ordered functor. The image of θ is a subgroupoid of K, because if $\theta(x)\theta(y)$ is defined in K then $\theta(\mathbf{d}(x)) = \theta(\mathbf{r}(y))$ and so xy is defined in G; this gives $\theta(x)\theta(y) = \theta(xy)$. However, the image of θ need not be an ordered subgroupoid of K. A stronger notion than an injective ordered functor is what we term an *ordered embedding*; this is an ordered functor θ: $G \to K$ such that for all $g, h \in G$

$$g \leq h \Leftrightarrow \theta(g) \leq \theta(h).$$

The image of θ is an ordered subgroupoid of K which is isomorphic to G.

In verifying that a structure is an ordered groupoid, it is sometimes more convenient to use the following characterisation. We shall need the following two axioms. Let G be a groupoid and $\leq$ a partial order defined on G. The axioms (OI) and (OG4) are defined as follows:

(OI) G_o is an order ideal of G.

(OG4) For all $x \in G$ and $e \in G_o$, if $e \leq \mathbf{d}(x)$ then there exists $y \in G$ such that $y \leq x$ and $\mathbf{d}(y) = e$.

Proposition 4 *Let $(G, \cdot)$ be a groupoid and $\leq$ a partial order defined on G. Then $(G, \cdot, \leq)$ is an ordered groupoid if, and only if, the axioms* (OG1), (OG2), (OI) *and* (OG4) *hold.*

Proof If G is an ordered groupoid then axioms (OG1) and (OG2) hold by definition, axiom (OI) holds by Proposition 3(7) and axiom (OG3) implies that axiom (OG4) holds.

To prove the converse, it is enough to show that axiom (OG3) holds because axiom (OG3)* follows from the other axioms for an ordered groupoid by Proposition 3(6). Let $u, v \leq x$ be such that $\mathbf{d}(u) = \mathbf{d}(v) = e$. We shall show that $u = v$ which, together with axiom (OG4), will imply that axiom (OG3) holds. Clearly $\mathbf{d}(u) = \mathbf{r}(v^{-1}) = e$. Thus uv^{-1} is defined. By axiom (OG1), we have that $v^{-1} \leq x^{-1}$. Thus by axiom (OG2), we have that $uv^{-1} \leq xx^{-1}$.

Now xx^{-1} is an identity and so by axiom (OI), the element uv^{-1} is an identity. Thus $u = v$, as required. ■

From inductive groupoids to inverse semigroups

We have shown how to construct an inductive groupoid from an inverse semigroup, it remains to show how to construct an inverse semigroup from an inductive groupoid.

Let G be an ordered groupoid and let $x, y \in G$ be such that $e = \mathbf{d}(x) \wedge \mathbf{r}(y)$ exists. Put

$$x \otimes y = (x \mid e)(e \mid y),$$

and call $x \otimes y$ the *pseudoproduct* of x and y. It is immediate from the definition that the pseudoproduct is everywhere defined in an inductive groupoid. The next result provides a neat, order-theoretic way of viewing the pseudoproduct.

Lemma 5 *Let G be an ordered groupoid. For each pair $x, y \in G$ put*

$$< x, y > = \{(x', y') \in G \times G \colon \mathbf{d}(x') = \mathbf{r}(y') \text{ and } x' \leq x \text{ and } y' \leq y\},$$

regarded as a subset of the ordered set $G \times G$. Then $x \otimes y$ exists if, and only if, there is a maximum element (x', y') of $< x, y >$. In which case, $x \otimes y = x'y'$.

Proof Suppose that $x \otimes y$ exists. Then $e = \mathbf{d}(x) \wedge \mathbf{r}(y)$ exists, and

$$((x \mid e), (e \mid y)) \in < x, y > .$$

Let $(u, v) \in < x, y >$. Then

$$u \leq x,\ v \leq y \text{ and } \mathbf{d}(u) = \mathbf{r}(v) = f,$$

say. Thus by axioms (OG3) and (OG3)* we have that $u = (x \mid f)$ and $v = (f \mid y)$. By Proposition 3(1), $\mathbf{d}(u) \leq \mathbf{d}(x)$ and $\mathbf{r}(v) \leq \mathbf{r}(y)$. Thus $f \leq \mathbf{d}(x), \mathbf{r}(y)$, and so, by assumption, $f \leq e$. By Proposition 3(8),(9), it follows that

$$u = (x \mid f) \leq (x \mid e) \text{ and } v = (f \mid y) \leq (e \mid x).$$

Thus $((x \mid e), (e \mid y))$ is the maximum element of $< x, y >$.

Conversely, suppose that the maximum element of $< x, y >$ exists and equals (x', y'). Put $e = \mathbf{d}(x') = \mathbf{r}(y')$. Clearly, $e \leq \mathbf{d}(x), \mathbf{r}(y)$. Now let f be any identity such that $f \leq \mathbf{d}(x), \mathbf{r}(y)$. Then

$$(x \mid f) \leq x,\ (f \mid y) \leq y \text{ and } \mathbf{d}(x \mid f) = f = \mathbf{r}(f \mid y).$$

Thus $((x \mid f), (f \mid y)) \in < x, y >$ and so $((x \mid f), (f \mid y)) \leq (x', y')$. Hence $f \leq e$, which implies that $e = \mathbf{d}(x) \wedge \mathbf{r}(y)$. Thus $x \otimes y$ exists. The proof of the last assertion is now immediate. ■

It will be an immediate consequence of the following result that the pseudoproduct on an inductive groupoid is associative.

Lemma 6 *Let G be an ordered groupoid. Then for all $x, y, z \in G$ if $x \otimes (y \otimes z)$ and $(x \otimes y) \otimes z$ both exist then they are equal.*

Proof Let $(x \otimes y) \otimes z = az'$ where (a, z') is the maximum element of $< x \otimes y, z >$. Let $x \otimes y = x'y'$ where (x', y') is the maximum element of $< x, y >$. Then we have that

$$a \leq x \otimes y,\ z' \leq z,\ x' \leq x \text{ and } y' \leq y.$$

By Proposition 3(5), $a \leq x'y'$ implies that there are elements $x'' \leq x'$ and $y'' \leq y'$ such that $a = x''y''$. Thus

$$(x \otimes y) \otimes z = (x''y'')z' = x''(y''z').$$

Now, $y'' \leq y' \leq y$ and $z' \leq z$, so that $(y'', z') \in\ < y, z >$. Thus $y''z' \leq y \otimes z$. Similarly, $(x'', y''z') \in\ < x, y \otimes z >$ and so $x''(y''z') \leq x \otimes (y \otimes z)$. Hence $(x \otimes y) \otimes z \leq x \otimes (y \otimes z)$. The reverse inequality follows by symmetry. ■

If $(G, \cdot, \leq)$ is an inductive groupoid, then the inverse semigroup $(G, \otimes)$ will be denoted by $\mathbf{S}(G)$. We can now show how to construct an inverse semigroup from an inductive groupoid.

Proposition 7 *Let $(G, \cdot, \leq)$ be an inductive groupoid.*

(1) *$(G, \otimes)$ is an inverse semigroup.*

(2) $\mathbf{G}(\mathbf{S}(G, \cdot, \leq)) = (G, \cdot, \leq)$.

(3) *For any inverse semigroup S we have that $\mathbf{S}(\mathbf{G}(S)) = S$.*

Proof (1) By Lemma 6, $(G, \otimes)$ is a semigroup. If $x, y \in G$ and $\exists x \cdot y$ in the groupoid G then $x \cdot y = x \otimes y$. But for each element $x \in G$ we have that $x = x \cdot x^{-1} \cdot x$ and $x^{-1} = x^{-1} \cdot x \cdot x^{-1}$. Thus $(G, \otimes)$ is a regular semigroup. It is easy to check that the idempotents of $(G, \otimes)$ are precisely the identities of $(G, \cdot)$. Let e and f be two idempotents of $(G, \otimes)$. Then

$$e \otimes f = (e \,|\, e \wedge f)(e \wedge f \,|\, f) = e \wedge f = (f \,|\, e \wedge f)(e \wedge f \,|\, e) = f \otimes e$$

so that the idempotents commute. It follows that $(G, \otimes)$ is an inverse semigroup.

(2) We show first that the natural partial order on $(G, \otimes)$ is just $\leq$. Suppose

that $x = e \otimes y$ in $(G, \otimes)$ for some idempotent e. Then $x = (e \wedge \mathbf{r}(y) \,|\, y)$ and so $x \leq y$ in $(G, \cdot, \leq)$. Conversely, suppose that $x \leq y$ in $(G, \cdot, \leq)$. Then $x = (\mathbf{r}(x) \,|\, y)$. But $(\mathbf{r}(x) \,|\, y) = \mathbf{r}(x) \otimes y$. Thus $x \leq y$ in $(G, \otimes)$. Now we turn to the restricted product. The restricted product of x and y is defined in $(G, \otimes)$ precisely when $x^{-1} \otimes x = y \otimes y^{-1}$. But from the properties of the pseudoproduct, we have that $x^{-1} \otimes x = x^{-1} \cdot x$ and $y \otimes y^{-1} = y \cdot y^{-1}$ in $(G, \cdot)$. Thus the restricted product of x and y exists in $(G, \otimes)$ precisely when the product $x \cdot y$ exists in $(G, \cdot)$. Thus $\mathbf{G}(\mathbf{S}(G, \cdot, \leq)) = (G, \cdot, \leq)$.
(3) The pseudoproduct in $\mathbf{G}(S)$ is given by

$$s \otimes t = (s \,|\, e) \cdot (e \,|\, t),$$

where $e = \mathbf{d}(s) \wedge \mathbf{r}(t)$ and the product on the right is the restricted product in S. But $(s \,|\, e) = se$ and $(e \,|\, t) = et$ and $e = s^{-1}stt^{-1}$ by Theorem 3.1.2. Thus $s \otimes t = st$. Hence $\mathbf{S}(\mathbf{G}(S)) = S$. ∎

The Ehresmann–Schein–Nambooripad theorem

Our main theorem is the following; we have called it the Ehresmann–Schein–Nambooripad theorem to reflect the diverse origins of its various components.

Theorem 8 *The category of inverse semigroups and prehomomorphisms is isomorphic to the category of inductive groupoids and ordered functors; and the category of inverse semigroups and homomorphisms is isomorphic to the category of inductive groupoids and inductive functors.*

Proof Define a function $\mathbf{G}$ from the category of inverse semigroups and prehomomorphisms to the category of inductive groupoids and ordered functors as follows: for each inverse semigroup S we define $\mathbf{G}(S) = (S, \cdot, \leq)$, an inductive groupoid by Proposition 1, and if $\theta \colon S \to T$ is a prehomomorphism then $\mathbf{G}(\theta) \colon \mathbf{G}(S) \to \mathbf{G}(T)$ is defined to be the same function on the underlying sets; this is an ordered functor by Theorem 3.1.5. It is easy to check that $\mathbf{G}$ defines a functor.

Define a function $\mathbf{S}$ from the category of inductive groupoids and ordered functors to the category of inverse semigroups and prehomomorphisms as follows: for each inductive groupoid G we define $\mathbf{S}(G) = (G, \otimes)$, an inverse semigroup by Proposition 7, and if $\phi \colon G \to H$ is an ordered functor between inductive groupoids then $\mathbf{S}(\phi) \colon \mathbf{S}(G) \to \mathbf{S}(H)$ is defined to be the same function on the underlying sets; this is a prehomomorphism by Theorem 3.1.5 since it preserves the restricted product of $\mathbf{S}(G)$ and the natural partial order. It is easy to check that $\mathbf{S}$ is a functor.

By Proposition 7, we have that $\mathbf{G}(\mathbf{S}(G, \cdot, \leq)) = (G, \cdot, \leq)$ and $\mathbf{S}(\mathbf{G}(S)) = S$. It is now immediate that the category of inverse semigroups and prehomomorphisms is isomorphic to the category of inductive groupoids and ordered functors. By Theorem 3.1.5 this isomorphism restricts to an isomorphism between the category of inverse semigroups and semigroup homomorphisms and the category of inductive groupoids and inductive functors. ∎

We have already met a special case of the above theorem in Proposition 1.4.9. Let S be a commutative, idempotent semigroup; in other words, S is an inverse semigroup in which every element is an idempotent. The associated groupoid consists entirely of identities. The order on these identities is just the natural partial order of S. Thus the inductive groupoid associated with S is a meet semilattice.

There is a final observation to be made, which is of crucial importance in Chapter 8. Star injective ordered functors between inductive groupoids correspond to idempotent pure prehomomorphisms between the respective inverse semigroups.

A generalisation of the Wagner–Preston theorem

There is a generalisation of the Wagner–Preston representation theorem which holds for ordered groupoids. First, we need to refine the notion of an ordered functor.

Let $\theta\colon G \to H$ be an ordered functor between ordered groupoids. We say that θ is *pseudoproduct preserving* if $x \otimes y$ exists in G then $\theta(x) \otimes \theta(y)$ exists in H and $\theta(x \otimes y) = \theta(x) \otimes \theta(y)$. It is straightforward to check (using Proposition 2) that an ordered functor $\theta\colon G \to H$ is pseudoproduct preserving precisely when for all identities $e, f \in G_o$ if $e \wedge f$ exists in G then $\theta(e) \wedge \theta(f)$ exists in H and $\theta(e \wedge f) = \theta(e) \wedge \theta(f)$. Thus θ preserves meets of pairs of identities whenever they exist.

The composition of pseudoproduct preserving ordered functors is again a pseudoproduct preserving ordered functor. It is clear that inductive functors between inductive groupoids are pseudoproduct preserving.

In the result which follows, $I(X)$ is regarded as an inductive groupoid.

Theorem 9 *Let $(G, \cdot, \leq)$ be an ordered groupoid. Then there is an injective pseudoproduct preserving ordered embedding $\theta\colon G \to I(X)$ for some set X.*

Proof The proof is essentially a translation of the proof of Theorem 1.5.1 into the language of ordered groupoids. Let $X = G$. For each $a \in G$ define a partial function θ_a of $I(X)$ as follows:

$$\operatorname{dom}\theta_a = \{x \in G\colon \mathbf{r}(x) \leq \mathbf{d}(a)\} \text{ and } \operatorname{im}\theta_a = \{y \in G\colon \mathbf{r}(y) \leq \mathbf{r}(a)\}$$

and $\theta_a(x) = a \otimes x = (a \,|\, \mathbf{r}(x)) \cdot x$. It is straightforward to check that θ_a is a well-defined function. To show that θ_a is injective, let $x, x' \in \operatorname{dom}\theta_a$ such that

$\theta_a(x) = \theta_a(x')$. Then $(a \,|\, \mathbf{r}(x))\cdot x = (a \,|\, \mathbf{r}(x'))\cdot x'$. But this implies $\mathbf{r}(a \,|\, \mathbf{r}(x)) = \mathbf{r}(a \,|\, \mathbf{r}(x'))$, and $(a \,|\, \mathbf{r}(x')), (a \,|\, \mathbf{r}(x')) \leq a$. Hence $(a \,|\, \mathbf{r}(x)) = (a \,|\, \mathbf{r}(x'))$, which gives $x = x'$. To show that θ_a is surjective let $y \in \operatorname{im}\theta_a$. Put $x = a^{-1} \otimes y$, which is well-defined. Clearly, $x \in \operatorname{dom}\theta_a$ so that $\theta_a(x) = a \otimes (a^{-1} \otimes y)$ is defined. However, it is clear that $(a \otimes a^{-1}) \otimes y$ is also defined. Thus by Lemma 6, we have that

$$\theta_a(x) = a \otimes (a^{-1} \otimes y) = (a \otimes a^{-1}) \otimes y = y.$$

Define $\theta\colon G \to I(X)$ by $\theta(a) = \theta_a$. This is well-defined by the above arguments. Clearly, θ maps identities to identities. Suppose that $a\cdot b$ is defined in G. Then $\operatorname{dom}\theta_a = \operatorname{im}\theta_b$, and it is easy to check that $\theta_a\theta_b = \theta_{ab}$ again by Lemma 6. Thus θ is a functor.

We now show that θ is an ordered embedding. Suppose that $a \leq b$. Then $a = b \otimes \mathbf{d}(a)$. Clearly, $\operatorname{dom}\theta_a \subseteq \operatorname{dom}\theta_b$. Let $x \in \operatorname{dom}\theta_a$. Then $\mathbf{r}(x) \leq \mathbf{d}(a)$, and so $x = \mathbf{d}(a) \otimes x$. Both $\theta_a(x)$ and $\theta_b(x)$ are defined. But

$$\theta_b(x) = b \otimes x = b \otimes (\mathbf{d}(a) \otimes x)$$

and

$$\theta_a(x) = a \otimes x = (b \otimes \mathbf{d}(a)) \otimes x.$$

By Lemma 6, we have that $\theta_b(x) = \theta_a(x)$. Hence $\theta_a \subseteq \theta_b$.

Now suppose that $\theta_a \subseteq \theta_b$. Then $\mathbf{d}(a) \leq \mathbf{d}(b)$ and $\theta_a(\mathbf{d}(a)) = \theta_b(\mathbf{d}(a))$. Thus $a = a \otimes \mathbf{d}(a) = b \otimes \mathbf{d}(a)$. Hence $a \leq b$.

Finally, we prove that θ is pseudoproduct preserving. Let $e, f \in G_o$ such that $e \wedge f$ exists. From the definition, θ_e is the identity function on the set $\operatorname{dom}\theta_e$, θ_f is the identity function on the set $\operatorname{dom}\theta_f$, and $\theta_{e\wedge f}$ is the identity function on the set $\operatorname{dom}\theta_{e\wedge f}$. However, it is easy to check that $\operatorname{dom}\theta_{e\wedge f} = \operatorname{dom}\theta_e \cap \operatorname{dom}\theta_f$. Thus $\theta(e \wedge f) = \theta(e) \wedge \theta(f)$. ■

4.2 Ordered groupoids from $*$-semigroups

Our main applications of ordered groupoids will be in Chapter 8, but here we present a class of examples of ordered groupoids which arise in any semigroup with involution.

Partial isometries in $*$-semigroups

In Section 1.4, we defined semigroups with involution. Here we shall denote the involution by $s \mapsto s^*$. In this context, semigroups with involution are also termed *$*$-semigroups*.

A *partial isometry* in a $*$-semigroup is an element s such that $s = ss^*s$. We denote the set of partial isometries of a $*$-semigroup S by $PI(S)$. A *projection* in a $*$-semigroup is an element e such that $e^2 = e$ and $e^* = e$. Every projection is a partial isometry, since $ee^*e = eee = e^3 = e$.

Lemma 1 *Let S be a $*$-semigroup.*

(1) *Let s be a partial isometry in S. Then s^*s and ss^* are projections.*

(2) *Let e and f be projections. Then $e = ef$ if, and only if, $e = fe$.*

Proof (1) We prove that s^*s is a projection; the proof that ss^* is a projection is similar. First s^*s is an idempotent since $(s^*s)^2 = s^*(ss^*s) = s^*s$, using $ss^*s = s$. Next $(s^*s)^* = s^*s^{**} = s^*s$, using axioms (IS1) and (IS2).
(2) Suppose that $e = ef$. Then $e^* = (ef)^*$. But $e^* = e$ and $(ef)^* = f^*e^* = fe$, using the fact that e and f are projections and axiom (IS2). Thus $e = fe$. The converse is similar. ■

Define a relation $\leq$ on $PI(S)$ by:

$$s \leq t \Leftrightarrow s = ss^*t \text{ and } ss^* = ss^*tt^*.$$

Lemma 2 *Let S be a $*$-semigroup, and let $s, t \in S$ be partial isometries.*

(1) *$s \leq t$ if, and only if, $s = et$ and $e = ett^*$ for some projection e.*

(2) *$s \leq t$ if, and only if, $s = tf$ and $f = t^*tf$ for some projection f. In particular, $s \leq t$ if, and only if, $s = ts^*s$ and $s^*s = s^*st^*t$.*

(3) *The relation $\leq$ is a partial order on $PI(S)$.*

(4) *If e and f are projections then $e \leq f$ if, and only if, $e = ef$.*

Proof (1) Suppose $s = et$ and $e = ett^*$ where e is a projection. Now $es = s$ so that $ess^* = ss^*$. Thus $ss^* = ss^*e$ by Lemma 1(2). Hence $s = ss^*s = ss^*(et) = ss^*t$. Also $ss^* = ss^*e = ss^*ett^* = ss^*tt^*$. The converse is clear.
(2) Suppose $s \leq t$. Then $s = et$ and $e = ett^*$ for some projection e by (1). By Lemma 1(2), we have that $e = tt^*e$. Thus $s = et = t(t^*et)$. It is easy to check that t^*et is a projection. Put $f = t^*et$. Clearly, $f = t^*tf$. Thus $s = tf$ and $f = t^*tf$ as required. The converse is proved similarly.

To prove the last assertion, suppose that $s = tf$ and $f = t^*tf$ for some projection f. Then $sf = s$ and so $s^*sf = s^*s$. Thus $s^*s = fs^*s$ by Lemma 1(2). It follows that $s = ts^*s$. Also $s^*s = t^*ts^*s$ and so $s^*s = s^*st^*t$ by Lemma 1(2).
(3) Reflexivity: we have that $s = (ss^*)s$ and $(ss^*)(ss^*) = ss^*$ by Lemma 1.

Anti-symmetry: let $s \leq t$ and $t \leq s$. Then $s = ss^*t$, $ss^* = ss^*tt^*$ and $t = tt^*s$, $tt^* = tt^*ss^*$. By Lemma 1, $ss^* = ss^*tt^* = tt^*ss^* = tt^*$. Thus $s = ss^*t = tt^*t = t$.

Transitivity: let $s \leq t$ and $t \leq u$. Then $s = ss^*t$ and $ss^* = ss^*tt^*$, and $t = tt^*u$ and $tt^* = tt^*uu^*$. But $s = ss^*t = ss^*tt^*u = ss^*u$ and $ss^*uu^* = ss^*tt^*uu^* = ss^*tt^* = ss^*$. Thus $s \leq u$.
(4) Suppose that $e \leq f$. Then $e = ee^*f$ and $ee^* = ee^*ff^*$. But e is a projection and so $ee^* = ee = e$. Thus $e = ef$. Conversely, suppose that $e = ef$. Then $e = ee^*$ and $f = ff^*$ since e and f are projections. Thus $e = ee^*f$ and $ee^* = ee^*ff^*$. Hence $e \leq f$. ■

Define a partial binary operation on $PI(S)$ as follows: $s \cdot t$ is defined only when $s^*s = tt^*$ in which case we put $s \cdot t = st$. We write $\exists s \cdot t$ if $s \cdot t$ is defined.

Theorem 3 *Let S be a $*$-semigroup. Then the triple $(PI(S), \cdot, \leq)$ is an ordered groupoid.*

Proof We begin by showing that the set of partial isometries is closed under the partial binary operation. Let $s, t \in PI(S)$ be such that $\exists s \cdot t$. Then

$$st(st)^*st = st(t^*s^*)st = s(tt^*)(s^*s)t.$$

But by assumption, $s^*s = tt^*$, and so $st(st)^*st = st$. Thus $s \cdot t \in PI(S)$. We now show that $(PI(S), \cdot)$ is a groupoid by verifying that the axioms in Section 3.1 hold.

Axiom (C1) holds: suppose that $(s \cdot t) \cdot u$ is defined. We shall prove that $s \cdot (t \cdot u)$ is defined and that they are equal. By assumption, we have that $s^*s = tt^*$ and $(st)^*st = uu^*$. From the latter equality, we have that $t^*s^*st = uu^*$. But $s^*s = tt^*$ and so $t^*t = uu^*$. Thus $\exists t \cdot u$. Now $tu(tu)^* = tuu^*t^* = tt^* = s^*s$. Thus $\exists s \cdot (t \cdot u)$. Clearly $(s \cdot t) \cdot u = s \cdot (t \cdot u)$. The converse can be proved similarly.

Axiom (C2) holds: suppose that $\exists s \cdot t$ and $\exists t \cdot u$. Then $s^*s = tt^*$ and $t^*t = uu^*$, so that

$$tu(tu)^* = t(uu^*)t^* = t(t^*t)t^* = tt^* = s^*s.$$

Thus $s \cdot (t \cdot u)$ is defined. The proof of the converse is straightforward.

Axiom (C3) holds: we first show that all projections are identities. Let e be a projection and suppose that $\exists s \cdot e$. Then $s^*s = ee^*$. But $ee^* = ee = e$ so that $s^*s = e$. Thus $s \cdot e = se = ss^*s = s$. Similarly, if $\exists e \cdot s$ then $e \cdot s = s$. Now we show that all identities are projections. Let e be an identity. The product $e \cdot (e^*e)$ is defined but, by assumption, e is an identity and so $e \cdot (e^*e) = e^*e$. But $e \cdot (e^*e) = ee^*e = e$. Thus $e = e^*e$. But by Lemma 1, e^*e is a projection and so e is a projection.

Observe that $s = (ss^*)s = s(s^*s)$. It follows that $PI(S)$ is a groupoid if we define $s^{-1} = s^*$.

We now show that the partial isometries form an ordered groupoid by showing that axioms (OG1), (OG2) and (OG3) hold.

Axiom (OG1) holds: let $s \leq t$. Then $s = ss^*t$ and $ss^* = ss^*tt^*$. Thus $s^* = t^*ss^*$ by axiom (IS2), and $(t^*)^*t^*ss^* = tt^*ss^* = ss^*$ by Lemma 1(2). Hence $s^* \leq t^*$ by Lemma 2(2).

Axiom (OG2) holds: let

$$s \leq t,\ u \leq v,\ \exists s \cdot u \text{ and } \exists t \cdot v.$$

Then

$$s = ss^*t,\ ss^* = ss^*tt^* \text{ and } u = uu^*v,\ uu^* = uu^*vv^*,$$

and

$$s^*s = uu^* \text{ and } t^*t = vv^*.$$

Firstly,

$$(su)(su)^*tv = s(uu^*)s^*tv = (ss^*)(ss^*)tv = ss^*tv = sv,$$

but

$$sv = ss^*sv = suu^*v = su.$$

Thus $su = (su)(su)^*tv$. Also

$$(su)(su)^*(tv)(tv)^* = suu^*s^*tvv^*t^* = s(s^*s)s^*t(t^*t)t^* = ss^*tt^* = ss^*,$$

but

$$ss^* = ss^*ss^* = suu^*s^* = su(su)^*.$$

Thus $su(su)^* = (su)(su)^*(tv)(tv)^*$. Hence $su \leq tv$.

Axiom (OG3) holds: let $e \leq s^*s$ where e is a projection. Define $(s \mid e) = se$. First we show that $(s \mid e) \leq s$. We have that

$$(se)(se)^*ss^* = ses^*ss^* = ses^* = (se)es^* = (se)(se)^*,$$

and

$$(se)(se)^*s = sees^*s = ses^*s = se.$$

Hence $(s \mid e) \leq s$. Now suppose that $t \leq s$ and $t^*t = e$. Then $t = st^*t$ and $t^*t = t^*ts^*s$. But then $t = se = (s \mid e)$. ∎

Application

We apply Theorem 3 to C^*-algebras. Underlying every C^*-algebra is a $*$-semigroup, so we may define partial isometries in C^*-algebras; these are precisely the partial isometries in the usual sense [426]. Thus the partial isometries in an arbitrary C^*-algebra form an ordered groupoid.

In certain cases, we can say more. An important class of C^*-algebras are those of the form $M = M_n(\mathbb{C})$, algebras of $n \times n$-matrices over the complex numbers with the star operation being the conjugate transpose of the matrix. The projections in M are closely related to the geometric properties of the vector space $\mathbb{C}^n$: if E and F are projections then it can be shown, using standard linear algebra [118], that

$$E \leq F \Leftrightarrow \operatorname{im} E \subseteq \operatorname{im} F.$$

Now let E and F be any two projections. Put $V = \operatorname{im} E \cap \operatorname{im} F$. Then $\mathbb{C}^n = V \oplus V^\perp$. Let J be the unique projection with image V and kernel $V^\perp$. Then $J = E \wedge F$ in the partially ordered set of projections, and so the projections form a meet semilattice with respect to the order. We have thus proved the following result.

Theorem 4 *Let $M_n(\mathbb{C})$ be the C^*-algebra of $n \times n$ complex matrices. Then the set of all partial isometries forms an inverse semigroup when the product of two partial isometries A and B is defined by:*

$$A \otimes B = A(A^*A \wedge BB^*)B,$$

where $A^*A \wedge BB^*$ is the projection defined above. ■

Special partial isometries

The definition of the partial order on the set of partial isometries of a $*$-semigroup is somewhat complex. However there is at least one situation where it can be simplified. A $*$-semigroup S is said to be *ordered* if there is a partial order $\subseteq$ defined on S such that:

(OIS1) If $a \subseteq b$ then $a^* \subseteq b^*$.

(OIS2) If $a \subseteq b$ and $c \subseteq d$ then $ac \subseteq bd$.

In what follows, we assume that S is a monoid with identity 1. In ordered $*$-monoids, we shall only be interested in a subset of partial isometries: a *special* partial isometry is a partial isometry $a \in S$ such that

$$1 \subseteq a^*a \text{ and } 1 \subseteq aa^*.$$

We shall denote the set of special partial isometries in S by $SPI(S)$.

Theorem 5 *Let S be an ordered $*$-monoid. Then the set of all special partial isometries is an ordered subgroupoid of the ordered groupoid of partial isometries. The groupoid order on $SPI(S)$ is the reverse of the semigroup order.*

Proof We begin by showing that $SPI(S)$ is closed under the partial binary operation. Let $a, b \in SPI(S)$ such that $\exists a \cdot b$. Then

$$1 \subseteq a^*a, b^*b, \text{ and } 1 \subseteq aa^*, bb^*, \text{ and } a^*a = bb^*.$$

From $1 \subseteq a^*a$, we obtain $b^*1b \subseteq b^*(a^*a)b$ by axiom (OIS2). But $1 \subseteq b^*b$. Hence $1 \subseteq b^*(a^*a)b = (ab)^*ab$. We may similarly show that $1 \subseteq ab(ab)^*$. Thus $SPI(S)$ is closed under the partial product. Clearly, if $a \in SPI(S)$ then $a^* \in SPI(S)$. Thus $SPI(S)$ is closed under the involution.

To show that $SPI(S)$ is an ordered subgroupoid of $PI(S)$, let $a \in SPI(S)$ and e a projection in $SPI(S)$ such that $e \leq a^*a$. From the proof of Theorem 3, $(a \,|\, e) = ae$. Thus we need to prove that $ae \in SPI(S)$. We know that $(ae)^*ae = e$ and so $1 \subseteq (ae)^*ae$ since $1 \subseteq e$. But then $1 \subseteq e$ implies $a1a^* \subseteq aea^*$ and so $1 \subseteq aea^* = (ae)(ae)^*$.

It remains to prove the assertion concerning the order. Suppose $a \leq b$ in the groupoid order. Then $a = aa^*b$ and $aa^* = aa^*bb^*$. From $1 \subseteq aa^*$ we obtain $b = 1b \subseteq aa^*b = a$, so that $b \subseteq a$.

Conversely, suppose that $b \subseteq a$. Then $aa^*b \subseteq aa^*a = a$. By (OIS1), $b^* \subseteq a^*$. Thus $1 \subseteq b^*b \subseteq a^*b$ and so $a \subseteq aa^*b$. Hence $a = aa^*b$. Also, $aa^*bb^* \subseteq aa^*aa^* = aa^*$, and $aa^* = aa^*1 \subseteq aa^*bb^*$, which gives $aa^* = aa^*bb^*$. Hence $a \leq b$. ■

We now consider two applications of Theorem 5.

Group coset semigroups

Let G be a group and let M be the monoid of all *non-empty* subsets of G with the binary operation being the product of subsets; the identity is $\{1\}$. Then M is an ordered $*$-semigroup when we take the star to be the inverse of a subset and the order to be subset inclusion. A partial isometry in M is any non-empty subset A of G such that $A = AA^{-1}A$. By Proposition 1.4.26, such subsets are precisely the cosets of subgroups of G. In particular, for each partial isometry A, the sets $A^{-1}A$ and AA^{-1} are subgroups of G and so for every partial isometry $\{1\} \subseteq A^{-1}A, AA^{-1}$. Thus every partial isometry is a special partial isometry. The set of all partial isometries, denoted by $K(G)$, is an ordered groupoid with the groupoid ordering being the reverse of set inclusion by Theorem 5. The projections are clearly just the subgroups of G; it follows that the projections form a semilattice under the groupoid ordering because given any two subgroups A and B of G the subgroup they generate is the meet of A and B in the ordered groupoid $K(G)$. Thus $K(G)$ is an inductive groupoid. We have therefore proved the following theorem; it is also a special case of Theorem 1.4.28 (which can be proved using similar ideas).

Theorem 6 *Let G be a group. Then the set of all cosets of subgroups of G, $K(G)$, forms an inverse semigroup when we define the product by*

$$A \otimes B = A(A^{-1}A \vee BB^{-1})B$$

where $A^{-1}A \vee BB^{-1}$ is the subgroup generated by $A^{-1}A$ and BB^{-1}. ■

The semigroup $K(G)$, which we termed a group coset semigroup in Section 1.4, is a factorisable inverse monoid with group of units isomorphic to G. These semigroups are, in some sense, prototypes of all factorisable inverse monoids, as we now show.

Theorem 7 *Let S be a factorisable inverse monoid with group of units G. Define $\phi: S \to K(G)$ by $\phi(s) = \{g \in G: s \leq g\}$. Then ϕ is an $\mathcal{L}$-bijective prehomomorphism.*

Proof The function ϕ is well-defined, because if $g, h, k \in \phi(s)$, then $s = ss^{-1}s \leq gh^{-1}k$. Thus $gh^{-1}k \in \phi(s)$. It follows that $\phi(s) \in K(G)$. To prove that ϕ is a prehomomorphism, we have to show that $\phi(st) \leq \phi(s)\phi(t)$. But the natural partial order in $K(G)$ is just the reverse of set-inclusion, so we have to prove that $\phi(s)\phi(t) \subseteq \phi(st)$, which is straightforward.

It remains to prove that ϕ is an $\mathcal{L}$-bijection. Suppose first that $\phi(a) = \phi(b)$ and $a^{-1}a = b^{-1}b$. Then there exists $g \in G$ such that $a, b \leq g$. Thus $a = ga^{-1}a = gb^{-1}b = b$. Hence ϕ is $\mathcal{L}$-injective. To prove that ϕ is $\mathcal{L}$-surjective, let $e \in E(S)$ and let $A \in K(G)$ such that $\phi(e) = A^{-1} \otimes A = A^{-1}A$. Let $g \in A$ and put $s = ge$. Observe that $s^{-1}s = (ge)^{-1}ge = e$. Because ϕ is a prehomomorphism, we have that $\phi(s)^{-1}\phi(s) = \phi(s^{-1}s) = \phi(e) = A^{-1}A$. Thus $\phi(s) = gA^{-1}A = A$ by the dual of Proposition 1.4.26. ■

The dual symmetric inverse monoid

Let X be a set and let $B(X)$ be the monoid of all binary relations under composition of relations. Then $B(X)$ is a $*$-semigroup when we take the star operation to be the converse operation. The partial isometries in this $*$-semigroup are those relations ρ on X such that $\rho = \rho\rho^{-1}\rho$. Such relations are called *di-functional.*

Lemma 8 *The projections in $B(X)$ are precisely the partial equivalence relations.*

Proof From the definition every projection in $B(X)$ is symmetric and transitive and so is a partial equivalence relation.

Suppose now that ρ is symmetric and transitive. Then $\rho^{-1} = \rho$ and $\rho\rho \subseteq \rho$. To prove that the reverse inclusion holds let $(y, x) \in \rho$. Then $(x, y) \in \rho^{-1} - \rho$, and so $(x, x) \in \rho$ by transitivity. It follows that $(y, x), (x, x) \in \rho$, which implies $(y, x) \in \rho\rho$. Hence $\rho \subseteq \rho\rho$, and so $\rho = \rho\rho$. ■

The identity of $B(X)$ is the binary relation $\Delta_X = X \times X$. The set $Be(X)$ of special partial isometries is given by

$$Be(X) = \{\rho \in B(X)\colon \Delta_X \subseteq \rho^{-1}\rho,\ \rho\rho^{-1} \text{ and } \rho = \rho\rho^{-1}\rho\}.$$

Elements of $Be(X)$ are called *bi-equivalences.* By Lemma 8, the projections in $Be(X)$ are precisely the equivalence relations on X.

Proposition 9 *There is a bijective correspondence between bi-equivalences in $B(X)$ and bijections between quotients of X.*

Proof Let ρ be a bi-equivalence in $B(X)$. The relations $\rho^{-1}\rho$ and $\rho\rho^{-1}$ are equivalence relations on X: they are clearly both symmetric, reflexivity follows from the fact that both contain Δ_X, and transitivity is a consequence of di-functionality. Put $\sigma = \rho^{-1}\rho$ and $\tau = \rho\rho^{-1}$. Define a function

$$f\colon X/\sigma \to X/\tau,$$

by $f(\sigma(x)) = \tau(y)$, where $(y, x) \in \rho$. This is well-defined, for suppose that $\sigma(x) = \sigma(x')$ and $(y', x') \in \rho$. We show that $\tau(y) = \tau(y')$. There exists $z \in X$ such that

$$(z, x) \in \rho \text{ and } (z, x') \in \rho.$$

But then

$$(y, x) \in \rho,\ (x, z) \in \rho^{-1},\ (z, x') \in \rho \text{ and } (x', y') \in \rho^{-1},$$

and so $(y, y') \in \rho\rho^{-1} = \tau$. Thus $\tau(y) = \tau(y')$. It remains to show that f is a bijection. Suppose that $f(\sigma(u)) = f(\sigma(v))$. Then $\tau(w) = \tau(z)$ where $(w, u) \in \rho$ and $(z, v) \in \rho$. It is easy to show that $(u, v) \in \rho^{-1}\rho = \sigma$, and so $\sigma(u) = \sigma(v)$. Thus f is an injection. Finally, f is a surjection. To see this, let $\tau(y) \in X/\tau$. Then $(y, y) \in \rho\rho^{-1}$, and so $(y, x) \in \rho$ for some $x \in X$. But then $f(\sigma(x)) = \tau(y)$.

Conversely, let σ and τ be two equivalences on X, and let $f\colon X/\sigma \to X/\tau$ be a bijection. Define a relation ρ on X by

$$(y, x) \in \rho \Leftrightarrow \tau(y) = f(\sigma(x)).$$

We show first that $\rho = \rho\rho^{-1}\rho$. Let $(u, v) \in \rho\rho^{-1}\rho$. Then there exist $x, y \in X$ such that $(u, x) \in \rho$, $(x, y) \in \rho^{-1}$ and $(y, v) \in \rho$. Thus $(u, x), (y, x), (y, v) \in \rho$.

By definition, $\tau(u) = f(\sigma(x))$, $\tau(y) = f(\sigma(x))$ and $\tau(y) = f(\sigma(v))$. But f is a bijective function and so $\tau(u) = \tau(y)$ and $\sigma(x) = \sigma(v)$. Thus $\tau(u) = f(\sigma(v))$. Hence $(u, v) \in \rho$. It follows that $\rho\rho^{-1}\rho \subseteq \rho$. The reverse inclusion is immediate. Next we show that $\rho^{-1}\rho = \sigma$. Let $(u, v) \in \rho^{-1}\rho$. Then $(u, x) \in \rho^{-1}$ and $(x, v) \in \rho$ for some $x \in X$. Thus $(x, u), (x, v) \in \rho$. By definition, $\tau(x) = f(\sigma(u))$ and $\tau(x) = f(\sigma(v))$. But f is bijective and so $\sigma(u) = \sigma(v)$, so that $(u, v) \in \sigma$. Thus $\rho^{-1}\rho \subseteq \sigma$. To prove the reverse inclusion let $(u, v) \in \sigma$. Then $\sigma(u) = \sigma(v)$. Let $\tau(y) = f(\sigma(u)) = f(\sigma(v))$. Then $(y, u), (y, v) \in \rho$, so that $(u, v) \in \rho^{-1}\rho$. Hence $\rho^{-1}\rho = \sigma$. The proof that $\rho\rho^{-1} = \tau$ is similar. ∎

The projections of $Be(X)$ are the equivalence relations on X, and these form a meet semilattice under the reverse of subset inclusion: the meet of two equivalences being the intersection of all equivalence relations which contain their union. Thus $Be(X)$ is an inductive groupoid. We have proved the following theorem.

Theorem 10 *The set of all bi-equivalences on X, $Be(X)$, forms an inverse semigroup when we define*

$$\rho \otimes \tau = \rho(\rho^{-1}\rho \vee \tau\tau^{-1})\tau,$$

where $\rho^{-1}\rho \vee \tau\tau^{-1}$ is the smallest equivalence relation containing $\rho^{-1}\rho$ and $\tau\tau^{-1}$. ∎

The inverse semigroup $Be(X)$ is called the *dual symmetric inverse monoid*, because its elements are bijections between quotients rather than bijections between subobjects.

The construction of $Be(X)$ can be generalised to algebras. Let A be an algebra and let ρ be a subalgebra of $A \times A$ which is also a bi-equivalence. Then ρ is called a *bi-congruence* on A. The set of all bi-congruences on A is denoted by $Bc(A)$. As before, $Bc(A)$ becomes an inverse semigroup. In certain cases, the form of the product can be simplified. An algebra A is said to be *congruence permutable* if $\rho\tau = \tau\rho$ for any congruences ρ and τ on A. For example, any group is congruence permutable, because congruences on groups are determined by normal subgroups and normal subgroups commute under the usual product of subsets of a group. In such algebras, the join of two congruences is just their product. It follows that if A is congruence permutable then the product in $Bc(A)$ reduces to the usual product of binary relations.

4.3 Ordered congruences

We have seen that inverse semigroups can be regarded as special kinds of ordered groupoids: namely the inductive ordered groupoids. It transpires, as we demonstrate in Chapter 8, that more general kinds of ordered groupoids are important in inverse semigroup theory. Intuitively, the category of ordered groupoids provides us with more room for carrying out calculations, even when we are primarily interested in inverse semigroups. In this section, we shall prove a first isomorphism theorem (Theorem 6) for a class of ordered functors and a corresponding class of congruences. We begin with a result which motivates what follows.

Proposition 1 *Let S and T be inverse semigroups and θ: $S \to T$ a homomorphism.*

(1) *θ preserves and reflects the natural partial orders.*

(2) *If $\theta(s) \cdot \theta(t)$ is a restricted product in T, then there exist elements $s', t' \in S$ such that:*

$$\exists s' \cdot t', \quad \theta(s') = \theta(s), \quad \theta(t') = \theta(t).$$

Proof (1) This was proved in Proposition 1.4.21.
(2) By assumption, $\theta(s^{-1}s) = \theta(tt^{-1})$. By Theorem 3.1.2, $st = s' \cdot t'$ where $s' = se$ and $t' = et$ and $e = s^{-1}stt^{-1}$. Thus

$$\theta(s') = \theta(se) = \theta(stt^{-1}) = \theta(ss^{-1}s) = \theta(s).$$

Similarly, $\theta(t') = \theta(t)$. ■

With the above result as a model we make the following definition. An ordered functor θ: $G \to H$ between ordered groupoids will be called a *special functor* if the following two axioms hold:

(SF1) θ reflects partial orders.

(SF2) If $\theta(x)\theta(y)$ is defined in H, then there exist elements x' and $y' \in G$ such that $x'y'$ is defined, $\theta(x') = \theta(x)$ and $\theta(y') = \theta(y)$.

If θ: $G \to H$ is an ordered functor, define the *kernel of* θ, denoted $\ker\theta$, by

$$\ker\theta = \{(x,y) \in G \times G\colon \theta(x) = \theta(y)\}.$$

The proof of the following result is immediate from the definition of a special functor and the properties of ordered groupoids.

Proposition 2 *Let θ: $G \to H$ be a special functor. Then $\rho = \ker\theta$ is an equivalence relation on G satisfying the following axioms:*

(OCON1) *If $\exists xy$ and $\exists x'y'$ and $(x,x'), (y,y') \in \rho$, then $(xy, x'y') \in \rho$.*

(OCON2) *If* $(x, y) \in \rho$, *then* $(\mathbf{d}(x), \mathbf{d}(y)), (\mathbf{r}(x), \mathbf{r}(y)) \in \rho$.

(OCON3) *If* $(\mathbf{d}(x), \mathbf{r}(y)) \in \rho$, *then there exist elements* $x', y' \in G$ *such that* $(x', x), (y', y) \in \rho$ *and* $\mathbf{d}(x') = \mathbf{r}(y')$.

(OCON4) *If* $x, y, y' \in G, x \leq y$ *and* $(y, y') \in \rho$, *then there exists an element* $x' \in G$ *such that* $(x, x') \in \rho$ *and* $x' \leq y'$.

(OCON5) *If* $x, y \in G, x \leq y, e \leq \mathbf{d}(y)$ *and* $(e, \mathbf{d}(x)) \in \rho$, *then there exists an element* $x' \in G$ *such that* $x' \leq y$, $\mathbf{d}(x') = e$ *and* $(x, x') \in \rho$.

(OCON6) *For all identities*, $e, f, f' \in G$, *if* $f \leq e \leq f'$ *and* $(f, f') \in \rho$, *then* $(e, f) \in \rho$. ∎

We shall use the properties listed in Proposition 2 to define what we mean by an ordered congruence, but before we can do this we need to define one further property.

Let ρ be an equivalence relation on a groupoid G which satisfies axiom (OCON1). On the set G/ρ of equivalence classes define a partial product as follows:

$$\rho(x)\rho(y) = \begin{cases} \rho(x'y') & \text{if } x' \,\rho\, x,\ y' \,\rho\, y \text{ and } \exists x'y' \\ \text{undefined} & \text{otherwise.} \end{cases}$$

Axiom (OCON1) ensures that is a well-defined partial binary operation. An equivalence ρ on a groupoid G satisfying axiom (OCON1) is said to satisfy the *associativity condition* (AC) if the product $\rho(x)(\rho(y)\rho(z))$ is defined if, and only if, the product $(\rho(x)\rho(y))\rho(z)$ is defined and they are equal.

We can now state the key definition. An *ordered congruence* on an ordered groupoid G is an equivalence ρ on G satisfying the axioms (OCON1) to (OCON6) of Proposition 2 together with (AC).

Before showing that quotients of ordered groupoids by ordered congruences are again ordered groupoids, we deal with the unordered case.

Lemma 3 *Let* ρ *be an equivalence relation on a groupoid* G *satisfying axioms* (OCON1), (OCON2), (OCON3) *and* (AC). *Then* G/ρ *is a groupoid and the map* $\rho^\natural\colon G \to G/\rho$ *is a functor satisfying axiom* (SF2).

Proof We have already defined a partial product on G/ρ which is well-defined by axiom (OCON1). We begin by proving that the identities of G/ρ are the elements of the form $\rho(e)$, where e is an identity of G. First of all, let e be an identity of G and suppose that the product $\rho(x)\rho(e)$ is defined. By definition, there exist elements x' and a of G such that

$$x' \,\rho\, x,\ a \,\rho\, e,\ \exists x'a \text{ and } \rho(x)\rho(e) = \rho(x'a).$$

By axiom (OCON2), $a \,\rho\, e$ implies that $\mathbf{r}(a) \,\rho\, e$. But $x'a$ defined implies that $x'\mathbf{r}(a)$ is defined, so that

$$\rho(x)\rho(e) = \rho(x'\mathbf{r}(a)) = \rho(x') = \rho(x).$$

We may similarly show that whenever $\rho(e)\rho(x)$ is defined it equals $\rho(x)$. Thus $\rho(e)$ is an identity of G/ρ. Conversely, let $\rho(a)$ be an identity of G/ρ. Then $\rho(a)\rho(\mathbf{d}(a))$ is defined and $\rho(a)\rho(\mathbf{d}(a)) = \rho(\mathbf{d}(a))$, since $\rho(a)$ is an identity. But $\rho(\mathbf{d}(a))$ is an identity, so that $\rho(a)\rho(\mathbf{d}(a)) = \rho(a)$. Thus $\rho(a) = \rho(\mathbf{d}(a))$.

Now define the functions $\mathbf{d}$ and $\mathbf{r}$ on G/ρ by

$$\mathbf{d}(\rho(a)) = \rho(\mathbf{d}(a)) \text{ and } \mathbf{r}(\rho(a)) = \rho(\mathbf{r}(a)),$$

which are well-defined by axiom (OCON2). We shall now show that the product $\rho(x)\rho(y)$ is defined if, and only if, $\mathbf{d}(\rho(x)) = \mathbf{r}(\rho(y))$. Suppose that $\rho(x)\rho(y)$ is defined. By the definition of the product there exist elements x' and y' such that $x' \,\rho\, x, y' \,\rho\, y$ and $x'y'$ is defined. Thus $\rho(x)\rho(y) = \rho(x'y')$. We have that

$$\mathbf{d}(\rho(x)) = \mathbf{d}(\rho(x')) = \rho(\mathbf{d}(x')) \text{ and } \mathbf{r}(\rho(y)) = \mathbf{r}(\rho(y')) = \rho(\mathbf{r}(y')).$$

Thus $\mathbf{d}(\rho(x)) = \mathbf{r}(\rho(y))$. The converse holds by axiom (OCON3).

It is now straightforward to show that

$$\mathbf{d}(\rho(x)\rho(y)) = \mathbf{d}(\rho(y)) \text{ and } \mathbf{r}(\rho(x)\rho(y)) = \mathbf{r}(\rho(x)).$$

We may now prove that G/ρ is a groupoid.

Axiom (C1) holds: by (AC) the product $\rho(x)(\rho(y)\rho(z))$ is defined if, and only if, the product $(\rho(x)\rho(y))\rho(z)$ is defined and they are equal.

Axiom (C2) holds: suppose that the products $\rho(x)\rho(y)$ and $\rho(y)\rho(z)$ are defined. Then $\mathbf{d}(\rho(x)) = \mathbf{r}(\rho(y))$ and $\mathbf{d}(\rho(y)) = \mathbf{r}(\rho(z))$. Thus

$$\mathbf{r}(\rho(y)\rho(z)) = \mathbf{r}(\rho(y)) = \mathbf{d}(\rho(x)),$$

so that the product $\rho(x)(\rho(y)\rho(z))$ is defined. The proof of the converse is similar.

Axiom (C3) holds: it is clear that $\mathbf{d}(\rho(x))$ is the right identity of $\rho(x)$, and that $\mathbf{r}(\rho(x))$ is the left identity.

Observe that the products $\rho(x)\rho(x^{-1})$ and $\rho(x^{-1})\rho(x)$ are defined and are identities. Thus $\rho(x)$ is invertible with inverse $\rho(x^{-1})$. It follows that G/ρ is a groupoid.

The verification that $\rho^{\natural}$ is a functor is now straightforward, and axiom (SF2) holds from the definition of the partial binary operation. ■

It is a consequence of the above result that if $x \,\rho\, y$ then $x^{-1} \,\rho\, y^{-1}$; a fact that we shall use on a number of occasions below.

Lemma 4 *Let ρ be an ordered congruence on an ordered groupoid G. Then the following conditions hold:*

(1) *Let $x, y \in G, x \leq y$ and $\mathbf{d}(x) \,\rho\, \mathbf{d}(y)$. Then $x \,\rho\, y$.*

(2) *Let $x, y, y' \in G$ such that $y \leq x \leq y'$ and $y \,\rho\, y'$. Then $x \,\rho\, y$.*

(3) *Let $x, y \in G$, $x \leq y$, $e \leq \mathbf{r}(y)$ and $e \,\rho\, \mathbf{r}(x)$. Then there exists an element x' such that $x' \leq y$, $\mathbf{r}(x') = e$ and $x' \,\rho\, x$.*

Proof (1) Let $x, y \in G$, $x \leq y$ and $\mathbf{d}(x) \,\rho\, \mathbf{d}(y)$. Then in particular $\mathbf{d}(y) \leq \mathbf{d}(y)$ and $\mathbf{d}(y) \,\rho\, \mathbf{d}(x)$. By axiom (OCON5), there exists an element $x' \in G$ such that $x' \leq y$, $\mathbf{d}(x') = \mathbf{d}(y)$ and $x \,\rho\, x'$. But these conditions imply that $x' = y$ and so $x \,\rho\, y$, as required.
(2) Let $x, y, y' \in G$ such that $y \leq x \leq y'$ and $y \,\rho\, y'$. By axiom (OCON2), we have that $\mathbf{d}(y) \,\rho\, \mathbf{d}(y')$. We also have that $\mathbf{d}(y) \leq \mathbf{d}(x) \leq \mathbf{d}(y')$. By axiom (OCON6), we obtain $\mathbf{d}(y) \,\rho\, \mathbf{d}(x)$. But then (1) gives $x \,\rho\, y$.
(3) Let $x \leq y$, $e \leq \mathbf{r}(y)$ and $e \,\rho\, \mathbf{r}(x)$. Then $x^{-1} \leq y^{-1}$, $e \leq \mathbf{d}(y^{-1})$ and $e \,\rho\, \mathbf{d}(x^{-1})$. By axiom (OCON5), there exists an element $z \in G$ such that $z \leq y^{-1}$, $\mathbf{d}(z) = e$ and $x^{-1} \,\rho\, z$. Thus $z^{-1} \leq y$, $\mathbf{r}(z^{-1}) = e$ and $x \,\rho\, z^{-1}$, so that we may put $x' = z^{-1}$. ■

Let ρ be an ordered congruence on the ordered groupoid G. Define a relation $\leq$ on G/ρ by $\rho(x) \leq \rho(y)$ if, and only if, for each $y' \in \rho(y)$ there exists $x' \in \rho(x)$ such that $x' \leq y'$.

Lemma 5 *The relation $\leq$ defined above is a partial order.*

Proof Reflexivity of $\leq$ is immediate. Transitivity is obtained directly from the definition. It remains to show that $\leq$ is antisymmetric. Let $\rho(x) \leq \rho(y)$ and $\rho(y) \leq \rho(x)$. Then there exists $x' \in \rho(x)$ such that $x' \leq y$ and there exists $y' \in \rho(y)$ such that $y' \leq x'$. Thus $y' \leq x' \leq y$ and $y' \,\rho\, y$. By Lemma 4(2), we have that $y' \,\rho\, x'$ and so $\rho(y) = \rho(x)$. ■

We now come to the main theorem of this section. It amounts to a first isomorphism theorem for special functors and ordered congruences.

Theorem 6 *Let G be an ordered groupoid.*

(1) *Let ρ be an ordered congruence on G. With respect to the product and order defined above, G/ρ is an ordered groupoid and $\rho^{\natural}\colon G \to G/\rho$ is a special ordered functor.*

(2) *Let $\theta\colon G \to H$ be a special ordered functor between ordered groupoids. Then $\rho = \ker\theta$ is an ordered congruence, and there is a unique, injective, special ordered functor $\psi\colon G/\rho \to H$ such that $\theta = \psi\rho^{\natural}$.*

Proof (1) By Lemma 3, $\rho^\natural: G \to G/\rho$ is a functor between groupoids satisfying axiom (SF2). That $\rho^\natural$ preserves the order is proved by axiom (OCON4). The fact that $\rho^\natural$ reflects partial orders is immediate from the definition of the order. Thus axiom (SF1) holds. It remains to show that G/ρ is an ordered groupoid. By Proposition 4.1.4, it is enough to check that the axioms (OG1), (OG2), (OG4) and (OI) hold.

Axiom (OG1) holds: this follows from the fact that $x \,\rho\, y$ implies $x^{-1} \,\rho\, y^{-1}$.

Axiom (OG2) holds: suppose that the products $\rho(x)\rho(y)$ and $\rho(u)\rho(v)$ exist and that $\rho(u) \leq \rho(x)$ and $\rho(v) \leq \rho(y)$. By definition there exist elements x' and $y' \in G$ such that the product $x'y'$ exists, $\rho(x') = \rho(x)$ and $\rho(y') = \rho(y)$. Thus $\rho(x)\rho(y) = \rho(x'y')$. There also exist elements $u' \leq x'$ and $v' \leq y'$ such that $u' \,\rho\, u$ and $v' \,\rho\, v$. Since the product $\rho(u)\rho(v)$ exists, we have that $\mathbf{d}(u') \,\rho\, \mathbf{r}(v')$. In particular, we have that

$$u' \leq x', \ \mathbf{r}(v') \leq \mathbf{d}(x') \ \text{ and } \mathbf{r}(v') \,\rho\, \mathbf{d}(u').$$

Thus by axiom (OCON5) there exists an element u'' such that

$$u'' \leq x', \ \mathbf{d}(u'') = \mathbf{r}(v') \text{ and } u'' \,\rho\, u'.$$

Thus $u''v' \leq x'y'$, and so $\rho(u)\rho(v) \leq \rho(x)\rho(y)$.

Axiom (OG4) holds: let $\rho(e) \leq \mathbf{d}(\rho(x))$, where $\rho(e)$ is an identity. Then there exists an identity e' such that $\rho(e') = \rho(e)$ and $e' \leq \mathbf{d}(x)$. But then $\rho(x \mid e') \leq \rho(x)$ and $\mathbf{d}(\rho(x \mid e')) = \rho(e)$.

Axiom (OI) holds: let $\rho(x)$ be an identity in G/ρ. Then $\rho(x) = \rho(e)$ for some identity e of G by Lemma 3. If $\rho(y) \leq \rho(x)$, then there exists an identity $f \in \rho(y)$ such that $f \leq e$. Thus $\rho(y)$ is an identity.

(2) Define $\psi: G/\rho \to H$ by $\psi(\rho(x)) = \theta(x)$. This is a well-defined function since $\rho = \ker\theta$. To show that ψ is a functor observe first that if $\rho(a)$ is an identity in G/ρ then $\rho(a) = \rho(e)$ for some identity e by Lemma 3. Thus $\psi(\rho(a)) = \theta(e)$, an identity. Next, suppose that the product $\rho(x)\rho(y)$ is defined in G/ρ. Then $\rho(x)\rho(y) = \rho(x'y')$ where $x' \in \rho(x)$, $y' \in \rho(y)$ and $\exists x'y'$. Thus $\psi(\rho(x)\rho(y)) = \theta(x'y')$. On the other hand $\psi(\rho(x)) = \theta(x')$ and $\psi(\rho(y)) = \theta(y')$ and $\theta(x')$ and $\theta(y')$ are both defined and $\theta(x')\theta(y') = \theta(x'y')$.

To show that ψ is an ordered functor, let $\rho(x) \leq \rho(y)$. Then there exists $x' \in \rho(x)$ such that $x' \leq y$. But then $\psi(\rho(x)) = \theta(x')$, $\psi(\rho(y)) = \theta(y)$ and $\theta(x') \leq \theta(y)$.

We now show that ψ is a special ordered functor. To show that axiom (SF1) holds, suppose that $\psi(\rho(x)) \leq \psi(\rho(y))$. Then $\theta(x) \leq \theta(y)$. But, by assumption, θ is a special ordered functor and so there exists an element $x' \in G$ such that $x' \leq y$ and $\theta(x') = \theta(x)$. Thus $\rho(x) \leq \rho(y)$.

To show that axiom (SF2) holds suppose that $\psi(\rho(x))\psi(\rho(y))$ is defined. Then $\theta(x)\theta(y)$ is defined and so, by assumption, there exist elements x' and y'

such that $\exists x'y'$ and $\theta(x') = \theta(x)$ and $\theta(y') = \theta(y)$. Thus $\rho(x)\rho(y)$ is defined. It is clear that ψ is injective, and by definition we have that $\theta = \psi\rho^\natural$. It is also straightforward to see that if $\psi'\colon G/\rho \to H$ is any special ordered functor such that $\theta = \psi'\rho^\natural$, then $\psi' = \psi$. ∎

One special class of ordered congruences is important in Chapter 8.

Lemma 7 *Let ρ be an equivalence on a groupoid G which satisfies axioms* (OCON1), (OCON2) *and the following:*

(OCON3)* *If $\mathbf{d}(x)\,\rho\,\mathbf{r}(y)$, then there exists an element x' such that $x'\,\rho\,x$ and $\mathbf{d}(x') = \mathbf{r}(y)$.*

Then ρ satisfies axiom (OCON3) *and* (AC).

Proof It is clear that axiom (OCON3)* implies that axiom (OCON3) holds. To see that (AC) holds, suppose that $\rho(x)(\rho(y)\rho(x))$ is defined. By definition of the product, this implies that $\mathbf{d}(y)\,\rho\,\mathbf{r}(z)$. Thus there exists an element $y'\,\rho\,y$ such that $y'z$ exists. Hence $\rho(y)\rho(z) = \rho(y'z)$. Again $\mathbf{d}(x)\,\rho\,\mathbf{r}(y')$ so that there exists an element $x' \in \rho(x)$ such that $x'(y'z)$ exists. Thus

$$\rho(x)(\rho(y)\rho(z)) = \rho(x'(y'z)) = \rho((x'y')z) = (\rho(x)\rho(y))\rho(z).$$

It is now clear that (AC) holds. ∎

4.4 Notes on Chapter 4

Section 4.1

Ordered groupoids were introduced by Charles Ehresmann to provide an algebraic model of pseudogroups of transformations. The theory of ordered groupoids and pseudogroups was initiated in [66] and subsequently developed in a number of papers notably [67], [68], and [69]. The *Oeuvres complètes* [71] contains not only reprints of all Ehresmann's papers but also detailed annotations.

The pseudoproduct on classes of ordered groupoids was defined in [66] and [67], where Ehresmann states that it is associative but does not provide a proof. A proof is given in some unpublished notes from a course delivered in 1961 at the Université de Montréal and later in the 1964 paper [69], although by this time in the context of ordered categories. We follow Ehresmann's proof of the associativity of the pseudoproduct in Lemmas 5 and 6.

As I mentioned above, Ehresmann was also interested in ordered categories not just ordered groupoids. Basing itself on Ehresmann's work, my paper [181] attempts to answer the following question: just what order structure does one need on a category in order to define an associative pseudoproduct.

Ehresmann's work is particularly interesting because it is based on the intuitions he gained whilst working in differential geometry. It is true that his papers can be quite difficult to follow on occasion, but the annotations provided by Prof. Andrée Ehresmann in the *Oeuvres complètes* clear up many points.

In Chapter 8, I show how a theory that Ehresmann developed as part of the categorical foundations of differential geometry (see Section 1.2) provides exactly what we need to understand idempotent pure extensions of inverse semigroups; as a corollary, we obtain the structure theory of E-unitary inverse semigroups. This result alone shows that Ehresmann's work must be regarded as belonging to inverse semigroup theory.

The equivalence of inverse semigroups and inductive groupoids was known to Ehresmann, but Boris Schein [373] deserves the credit for making explicit the connection between Wagner's school of inverse semigroup theory and Ehresmann's. Ehresmann's papers [66], [67] are the basis for Schein's paper published in Russian in 1965 [373] and translated with notes as [381]. Schein recognised the connection between Ehresmann's work and Wagner's and, by dropping the completion properties needed by Ehresmann, was able to show that inverse semigroups could be realised as inductive groupoids (in Schein's sense, the one adopted here).

The idea of relating a class of semigroups to a class of ordered groupoids was taken up by Nambooripad in 1979 [280]. Using ordered groupoids as defined here, he extended Schein's result to arbitrary regular semigroups. Since Nambooripad also considered homomorphisms, the inverse semigroup-theoretic version of Nambooripad's theorem yields the isomorphism between the category of inverse semigroups and semigroup homomorphisms and the category of inductive groupoids and inductive functors. The connection between prehomomorphisms and ordered functors is essentially to be found in [283].

The inductive groupoid approach to inverse semigroups was further pursued in Germany, perhaps because Germany was the home of groupoid theory (see [89]). The first paper of the German school of ordered groupoid theory was [357]. The book [120] contains a chapter on ordered groupoids and categories and one on Ehresmann's work on species of structures. Recent papers in this School are [139], [140].

The idea of applying groupoid theory (as developed in [133]) to inverse semigroups via ordered groupoids has been developed in [180], [182], [185] and [189], and [121], [282], and [301].

Meakin's structure mapping approach to inverse semigroups [248] is a variant of the inductive groupoid approach. He applies this technique to the problem of constructing $\mathcal{H}$-coextensions of inverse semigroups (an inverse semigroup T is an $\mathcal{H}$-*coextension* of an inverse semigroup S if there is a surjective idempotent-separating homomorphism from T to S [109]).

Section 4.2

The important rôle of partial isometries in the theory of C^*-algebras is discussed in [426]. Leech [198] discusses partial isometries in matrix C^*-algebras.

The dual symmetric inverse monoid and the inverse semigroup of bi-congruences was introduced by FitzGerald [81], [82] and anticipated by Leech [197]. Bi-equivalances and bi-congruences had been considered earlier by Schein [385], but without the pseudoproduct multiplication.

My presentation of this section was influenced by Hoehnke's paper [140].

Section 4.3

The results of this section are based on work to be found in Ehresmann's book [70], Joubert's long paper [162] and my paper [183].

Chapter 5

Extensions of inverse semigroups

In this chapter, we depart from our main interest in partial symmetries to take a closer look at the way inverse semigroups can be constructed from simpler building blocks. Our starting point is the fact that congruences on groups have a much simpler and more familiar description. If ρ is the congruence in question, then $\rho(1)$, the congruence-class containing the identity, is a normal subgroup. On the other hand, if N is a normal subgroup, then a congruence ρ can be defined by

$$(a, b) \in \rho \Leftrightarrow ab^{-1} \in N.$$

Something similar can be carried out for inverse semigroups; the essential difference is that the presence of more than one idempotent means that two pieces of information are required: the union of the congruence classes containing the idempotents, called the Kernel, and the restriction of the congruence to the semilattice of idempotents, called the trace.

When a congruence is idempotent-separating, it may be entirely described in terms of its Kernel which is, in a precise sense, a normal subsemigroup. This is a situation which directly generalises the group case, and suggests the problem of constructing the inverse semigroup from the Kernel of the congruence and the quotient semigroup. Even in the group case, this is a difficult problem. So, rather than trying to obtain an exact solution, we look for something looser. This leads us to what are termed λ-semidirect products. With the help of these we may embed the original semigroup in one constructed from the normal subsemigroup and the quotient.

A particularly pleasant situation in group theory is where an extension splits. This may also be generalised to inverse semigroups, and leads to a structure theory for the class of inverse ω-semigroups.

5.1 The Kernel–trace description of a congruence

In this section, we show that every congruence on an inverse semigroup is determined by a pair consisting of a normal subsemigroup and a normal congruence on the semilattice of idempotents.

The Kernel and the trace

Let ρ be a congruence on an inverse semigroup S. The *Kernel* of ρ, denoted $Ker\,\rho$, is the union of the ρ-classes containing idempotents. The *trace* of ρ, denoted $\mathrm{tr}\,\rho$, is the restriction of ρ to $E(S)$. We follow Howie and make a typographical distinction between the 'kernel', which is a congruence, and the 'Kernel', which is an inverse subsemigroup. To characterise abstractly pairs of the form $(\mathrm{Ker}\,\rho, \mathrm{tr}\,\rho)$ we shall need the following definitions.

An inverse subsemigroup N of an inverse semigroup S is said to be *self-conjugate* if $s^{-1}Ns \subseteq N$ for all $s \in S$. A full, self-conjugate inverse subsemigroup of S is called a *normal subsemigroup* of S. A congruence ρ on the semilattice $E(S)$ is said to be *normal* if $(e, f) \in \rho$ implies $(a^{-1}ea, a^{-1}fe) \in \rho$ for all $a \in S$.

Essentially, every homomorphism from an inverse semigroup determines and is determined by a normal subsemigroup *and* a normal congruence on the semilattice of idempotents. The remainder of this section is devoted to proving this assertion.

Proposition 1 *Let ρ be a congruence on an inverse semigroup S. Put $N =$ $\mathrm{Ker}\,\rho$ and $\nu = \mathrm{tr}\,\rho$.*

(1) *N is a normal subsemigroup of S.*

(2) *ν is a normal congruence on $E(S)$.*

(3) *For any $a \in S$ and $e \in E(S)$, if $ae \in N$ and $(e, a^{-1}a) \in \nu$ then $a \in N$.*

(4) *$a \in N$ implies $(aa^{-1}, a^{-1}a) \in \nu$.*

Proof (1) Clearly N is a full, inverse subsemigroup of S. It is self-conjugate because the conjugate of an idempotent is also an idempotent.
(2) Clearly ν is a congruence on $E(S)$. The fact that ν is normal follows from the fact that the conjugate of an idempotent is an idempotent.
(3) From $(e, a^{-1}a) \in \nu$ we obtain $(ae, a) \in \rho$. Thus $\rho(ae) = \rho(a)$. But

$ae \in N$ means $\rho(ae)$ contains an idempotent and so $\rho(a)$ contains an idempotent. Hence $a \in N$.
(4) Let $a \in N$. Then $(a, e) \in \rho$ for some idempotent e. Thus $(a^{-1}a, aa^{-1}) \in \rho$ by Proposition 2.3.4, and so $(a^{-1}a, aa^{-1}) \in \nu$. ■

On the basis of the above results, we now make the following definition. A *congruence pair* (N, ν) on an inverse semigroup S consists of a normal subsemigroup N and a normal congruence ν satisfying the following two axioms for any $a \in S$ and $e \in E(S)$:

(CP1) $ae \in N$ and $(e, a^{-1}a) \in \nu$ imply $a \in N$.

(CP2) $a \in N$ implies $(aa^{-1}, a^{-1}a) \in \nu$.

Proposition 1 shows that $(\operatorname{Ker}\rho, \operatorname{tr}\rho)$ is a congruence pair for every congruence ρ.

Let (N, ν) be an arbitrary congruence pair on a semigroup S. Define a relation $\rho_{(N,\nu)}$ by

$$(a, b) \in \rho_{(N,\nu)} \Leftrightarrow ab^{-1} \in N \text{ and } (a^{-1}a, b^{-1}b) \in \nu.$$

We can now prove that there is a bijection between the set of congruences on S and the set of congruence pairs on S.

Theorem 2 *Let S be an inverse semigroup.*

(1) *Let ρ be a congruence on S. Then $(\operatorname{Ker}\rho, \operatorname{tr}\rho)$ is a congruence pair and the congruence associated with $(\operatorname{Ker}\rho, \operatorname{tr}\rho)$ is ρ.*

(2) *Let (N, ν) be a congruence pair. Then $\rho_{(N,\nu)}$ is a congruence on S such that $\operatorname{Ker}\rho_{(N,\nu)} = N$ and $\operatorname{tr}\rho_{(N,\nu)} = \nu$.*

Proof (1) Put $N = \operatorname{Ker}\rho$ and $\nu = \operatorname{tr}\rho$. By Proposition 1, (N, ν) is a congruence pair. Thus it only remains to show that

$$(a, b) \in \rho \Leftrightarrow ab^{-1} \in N \text{ and } (a^{-1}a, b^{-1}b) \in \nu.$$

Suppose that $ab^{-1} \in N$ and $(a^{-1}a, b^{-1}b) \in \nu$. Then $(a^{-1}a, b^{-1}b) \in \rho$ and $(ab^{-1}, e) \in \rho$ for some idempotent e. By Proposition 2.3.4,

$$(ab^{-1}, ba^{-1}) \in \rho \text{ and } (ab^{-1}, ab^{-1}ba^{-1}) \in \rho.$$

From $(ab^{-1}, ba^{-1}) \in \rho$ we obtain $(ab^{-1}a, ba^{-1}a) \in \rho$. But $(a^{-1}a, b^{-1}b) \in \rho$ and so $(ab^{-1}a, b) \in \rho$. Similarly, from $(ab^{-1}, ab^{-1}ba^{-1}) \in \rho$ we obtain $(ab^{-1}a, ab^{-1}b) \in \rho$. But $(a^{-1}a, b^{-1}b) \in \rho$ and so

$$(ab^{-1}a, a) \in \rho.$$

Thus $(a,b) \in \rho$. The converse is straightforward.
(2) We show first that $\rho_{(N,\nu)}$ is an equivalence relation. Reflexivity and symmetry hold because N is a full inverse subsemigroup and ν is an equivalence. To prove transitivity suppose that $(a,b),(b,c) \in \rho_{(N,\nu)}$. Then

$$ab^{-1}, bc^{-1} \in N \text{ and } (a^{-1}a, b^{-1}b), (b^{-1}b, c^{-1}c) \in \nu.$$

Thus $a(b^{-1}b)c^{-1} \in N$ and $(a^{-1}a, c^{-1}c) \in \nu$. By the proof of Lemma 1.4.2,

$$a(b^{-1}b)c^{-1} = ac^{-1}(cb^{-1}bc^{-1}).$$

From $(a^{-1}a, b^{-1}b) \in \nu$ we obtain $(ca^{-1}ac^{-1}, cb^{-1}bc^{-1}) \in \nu$ since ν is normal. Hence

$$ac^{-1}(cb^{-1}bc^{-1}) \in N \text{ and } (cb^{-1}bc^{-1}, (ac^{-1})^{-1}ac^{-1}) \in \nu.$$

Thus $ac^{-1} \in N$ by axiom (CP1). Hence $(a,c) \in \rho_{(N,\nu)}$.

Next we show that $\rho_{(N,\nu)}$ is a congruence. Let $(a,b) \in \rho_{(N,\nu)}$ and $c \in S$. By assumption, $ab^{-1} \in N$ and $(a^{-1}a, b^{-1}b) \in \nu$. We prove first that $\rho_{(N,\nu)}$ is a right congruence by showing that $(ac, bc) \in \rho_{(N,\nu)}$. Observe that

$$(ac)^{-1}ac = c^{-1}a^{-1}ac \text{ and } (bc)^{-1}bc = c^{-1}b^{-1}bc.$$

Thus $((ac)^{-1}ac, (bc)^{-1}bc) \in \nu$ since ν is normal. Also

$$ac(bc)^{-1} = a(cc^{-1})b^{-1} = ab^{-1}(bcc^{-1}b^{-1})$$

by the proof of Lemma 1.4.2. But N is a full inverse subsemigroup so that $bcc^{-1}b^{-1} \in N$. Thus $ac(bc)^{-1} = ab^{-1}(bcc^{-1}b^{-1}) \in N$. Hence $(ac, bc) \in \rho_{(N,\nu)}$.

We prove now that $\rho_{(N,\nu)}$ is a left congruence by showing that $(ca, cb) \in \rho_{(N,\nu)}$. Firstly, $ca(cb)^{-1} = c(ab^{-1})c^{-1}$, but $ab^{-1} \in N$ and N is self-conjugate so that $ca(cb)^{-1} \in N$. Our main problem therefore is to show that the elements

$$(ca)^{-1}ca = a^{-1}c^{-1}ca \text{ and } (cb)^{-1}cb = b^{-1}c^{-1}cb$$

are ν-related. To make the presentation of this part of the proof easier to read, we shall write $x \equiv y$ instead of $(x,y) \in \nu$ throughout. Put $e = c^{-1}c$. We shall show that $a^{-1}ea \equiv b^{-1}eb$, which will conclude the proof that $\rho_{(N,\nu)}$ is a congruence. Write

$$a^{-1}ea = (a^{-1}ea)(a^{-1}a)(a^{-1}ea).$$

But $a^{-1}a \equiv b^{-1}b$ and so

$$a^{-1}ea \equiv (a^{-1}ea)(b^{-1}b)(a^{-1}ea).$$

Now

$$(a^{-1}ea)(b^{-1}b)(a^{-1}ea) = (a^{-1}e)(ab^{-1})(ab^{-1})^{-1}(ea).$$

By axiom (CP2), $ab^{-1} \in N$ implies that

$$ab^{-1}(ab^{-1})^{-1} \equiv (ab^{-1})^{-1}ab^{-1}.$$

Thus

$$(a^{-1}e)(ab^{-1})(ab^{-1})^{-1}(ea) \equiv (a^{-1}e)(ab^{-1})^{-1}(ab^{-1})(ea),$$

and so

$$a^{-1}ea \equiv (a^{-1}e)(ba^{-1}ab^{-1})(ea).$$

Now

$$(a^{-1}e)(ba^{-1}ab^{-1})(ea) = a^{-1}(ab^{-1}e)^{-1}(ab^{-1}e)a.$$

But $ab^{-1} \in N$, and N is a full subsemigroup, so that $ab^{-1}e \in N$. From axiom (CP2) and the fact that ν is normal we obtain

$$a^{-1}(ab^{-1}e)^{-1}(ab^{-1}e)a \equiv a^{-1}(ab^{-1}e)(ab^{-1}e)^{-1}a.$$

Thus

$$a^{-1}ea \equiv a^{-1}(ab^{-1}e)(ab^{-1}e)^{-1}a.$$

But $a^{-1}(ab^{-1}e)(ab^{-1}e)^{-1}a = a^{-1}ab^{-1}eb$, so that we in fact have

$$a^{-1}ea \equiv a^{-1}ab^{-1}eb.$$

But then from $a^{-1}a \equiv b^{-1}b$ we obtain $a^{-1}ea \equiv b^{-1}eb$ as required.

We now calculate $\operatorname{Ker}\rho_{(N,\nu)}$. Let $a \in \operatorname{Ker}\rho_{(N,\nu)}$. Then there is an idempotent $e \in S$ such that $(a,e) \in \rho_{(N,\nu)}$. But then $ae \in N$ and $(a^{-1}a, e) \in \nu$. Thus $a \in N$ by axiom (CP1). Hence $\operatorname{Ker}\rho_{(N,\nu)} \subseteq N$. To prove the reverse inclusion, suppose that $a \in N$. Then $a(a^{-1}a) \in N$ and $(a^{-1}a, a^{-1}a) \in \nu$. Thus $(a, a^{-1}a) \in \rho_{(N,\nu)}$. Hence $a \in \operatorname{Ker}\rho_{(N,\nu)}$, and so $N \subseteq \operatorname{Ker}\rho_{(N,\nu)}$.

The fact that $\operatorname{tr}\rho_{(N,\nu)} = \nu$ is immediate. ∎

The preceding theorem demonstrates that congruence pairs are the correct analogues of normal subgroups.

The idempotent-separating and idempotent pure congruences have natural interpretations in terms of the Kernel-trace description. A congruence is idempotent-separating just when its trace is the equality congruence, whereas a congruence is idempotent pure just when its Kernel is the semilattice of idempotents.

Normal extensions

We now return to the group case for further motivation. Let K and F be groups. An *extension of K by F* is a group G together with an embedding

ι: $K \to G$ and a surjection ω: $G \to F$ such that $\iota(K) = \operatorname{Ker}\omega$. The analogous definition for inverse semigroups runs as follows.

Let K and F be arbitrary inverse semigroups, and let π: $K \to E(F)$ be a surjective homomorphism. The triple (K, π, F) is called a *normal extension triple*. An inverse semigroup S is said to be a *normal extension of* K *by* F *along* π if there is an embedding ι: $K \to S$ and a surjection ω: $S \to F$ such that the following two axioms hold:

(NE1) $\iota(K) = \operatorname{Ker}\omega$.

(NE2) $\omega\iota = \pi$.

Axiom (NE1) says that K is the Kernel of the surjection ω, whereas axiom (NE2) says that the trace of ω is the trace of π. The triple (ι, S, ω) is called a *solution of the normal extension problem for the triple* (K, π, F).

Two solutions (ι, S, ω) and (ι', S', ω') are said to be *equivalent* if there is an isomorphism η: $S \to S'$ such that $\omega'\eta = \omega$ and $\eta\iota = \iota'$. The *extension problem* consists in classifying all normal extensions of the triple (K, π, F) up to equivalence. Classically, two cases of the extension problem have received considerable attention: if π is an idempotent-separating homomorphism then normal extensions of (K, ι, F) are called *idempotent-separating extensions*; if K is a semilattice then the corresponding extensions are called *normal extensions of semilattices* or *idempotent pure extensions*.

5.2 Idempotent-separating extensions

The structure of normal extensions of semilattices is described in Chapter 8. In this section, we concentrate on some of the basic results which emerge from the theory of idempotent-separating extensions.

Idempotent-separating congruences

Of all the types of congruences on inverse semigroups, it is the idempotent-separating congruences which behave most like group congruences, in that they are entirely determined by their Kernels. This is because their traces are just the equality relation on the semilattice of idempotents.

Lemma 1 *Let N be a normal subsemigroup of an inverse semigroup S. Then N is the Kernel of an idempotent-separating congruence on S if, and only if, $a \in N$ implies $a^{-1}a = aa^{-1}$.*

Proof Let N be the Kernel of an idempotent-separating congruence on S. Then since the trace is the equality relation, $a \in N$ implies $a^{-1}a = aa^{-1}$, by axiom (CP2). Conversely, let N be a normal subsemigroup of S such that $a \in N$ implies $a^{-1}a = aa^{-1}$. Then N together with the equality relation on the semilattice of idempotents is a congruence pair. Thus N is the Kernel of an idempotent-separating congruence. ■

For every inverse semigroup S, we define the set $Z(E(S))$, the *centraliser of the idempotents*, to consist of all elements of S which commute with every idempotent. It is easy to check that $Z(E(S))$ is a full inverse subsemigroup of S. If $Z(E(S)) = S$ the semigroup is said to be *Clifford*. It is easy to see that $Z(E(S))$ is always a Clifford semigroup, possibly consisting only of idempotents.

Lemma 2 *Let N be a normal subsemigroup of an inverse semigroup S. Then N is the Kernel of an idempotent-separating homomorphism if, and only if, N is contained in $Z(E(S))$.*

Proof Let N be the Kernel of the idempotent-separating congruence ρ. Let $a \in N$, and let e be any idempotent in S. Since N is a normal subsemigroup, we have that $ae \in N$. Thus $(ae)^{-1}ae = ae(ae)^{-1}$ by Lemma 1, and so $ea^{-1}a = aea^{-1}$. But $a^{-1}a = aa^{-1}$ by Lemma 1, so that $eaa^{-1} = aea^{-1}$. Thus $ea = ae$. Consequently, $a \in Z(E(S))$.

Conversely, let N be a normal subsemigroup contained in $Z(E(S))$, and let $a \in N$. Then $a = a(a^{-1}a) = (a^{-1}a)a$ and so $aa^{-1} = (a^{-1}a)aa^{-1}$. Thus $aa^{-1} \leq a^{-1}a$. We may similarly show that $a^{-1}a \leq aa^{-1}$. Hence $a^{-1}a = aa^{-1}$. Thus N is the Kernel of an idempotent-separating congruence by Lemma 1. ■

The above result tells us two important things: firstly, that Kernels of idempotent-separating congruences are always Clifford semigroups, and secondly, that the centraliser of the idempotents of an inverse semigroup is the Kernel of an idempotent-separating congruence. We shall now describe that congruence.

Define the relation μ on the inverse semigroup S by:

$$(a, b) \in \mu \Leftrightarrow (\forall e \in E(S))\, aea^{-1} = beb^{-1}.$$

Proposition 3 *Let S be an inverse semigroup. Then μ is the unique idempotent-separating congruence such that* $\operatorname{Ker} \mu = Z(E(S))$.

Proof It is clear that μ is an equivalence relation. To show that μ is a congruence let $(a, b), (c, d) \in \mu$ and let $e \in E(S)$. Then

$$(ac)e(ac)^{-1} = a(cec^{-1})a^{-1} \text{ and } (bd)e(bd)^{-1} = b(ded^{-1})b^{-1}.$$

But from $(c,d) \in \mu$, we obtain $cec^{-1} = ded^{-1}$, an idempotent, and from $(a,b) \in \mu$ we obtain $a(cec^{-1})a^{-1} = b(ded^{-1})b^{-1}$. Thus $(ac,bd) \in \mu$. It is immediate that μ is idempotent-separating.

We now show that $\operatorname{Ker}\mu = Z(E(S))$. Let $a \in \operatorname{Ker}\mu$. Then $(a,f) \in \mu$ for some idempotent f. For any idempotent e we have that $(aea^{-1}, ef) \in \mu$ so that $aea^{-1} = ef$ since μ is idempotent-separating. But by Proposition 3.2.12, $\mu \subseteq \mathcal{H}$ and so $fa = a$. Thus $ae = efa = ea$, and so $a \in Z(E(S))$.

Conversely, let $a \in Z(E(S))$. Then $a = a(a^{-1}a) = (a^{-1}a)a$ and so $aa^{-1} = (a^{-1}a)(aa^{-1})$. Thus $aa^{-1} \leq a^{-1}a$. Similarly, $a^{-1}a \leq aa^{-1}$. Thus $a^{-1}a = aa^{-1}$. But then for any idempotent e we have that

$$aea^{-1} = eaa^{-1} = ea^{-1}a = a^{-1}ae(a^{-1}a)^{-1}.$$

Thus $(a, a^{-1}a) \in \mu$, and so $a \in \operatorname{Ker}\mu$. Uniqueness follows from Theorem 5.1.2. ■

In Proposition 3.2.12, we proved that a congruence is idempotent-separating precisely when it is contained in the $\mathcal{H}$-relation. This result can be refined as follows.

Proposition 4 *Let S be an inverse semigroup. Then μ is the maximum idempotent-separating congruence on S.*

Proof Let ρ be any idempotent-separating congruence on S. Let $(s,t) \in \rho$ and $e \in E(S)$. Then $(s^{-1}, t^{-1}) \in \rho$ by Proposition 2.3.4, and $(e,e) \in \rho$, so that $(ses^{-1}, tet^{-1}) \in \rho$. But ρ is idempotent-separating and so $ses^{-1} = tet^{-1}$. Hence $(s,t) \in \mu$. Thus $\rho \subseteq \mu$. ■

An inverse semigroup is said to be *fundamental* if μ is the equality relation.

Proposition 5 *Let S be an inverse semigroup.*

(1) *S is fundamental if, and only if, $Z(E(S)) = E(S)$.*

(2) *S/μ is fundamental.*

Proof (1) Suppose that S is fundamental. Let $a \in Z(E(S))$. By Proposition 3, $\operatorname{Ker}\mu = Z(E(S))$. Thus $(a,e) \in \mu$ for some $e \in E(S)$. But then $a = e$, since μ is equality, and so a is an idempotent. Thus $Z(E(S)) = E(S)$.

Conversely, suppose that $Z(E(S)) = E(S)$. Let $(a,b) \in \mu$. Then

$$(ab^{-1}, bb^{-1}) \in \mu,$$

and so $ab^{-1} \in \operatorname{Ker}\mu$. But $\operatorname{Ker}\mu = Z(E(S))$ by Proposition 3, and so $ab^{-1} \in Z(E(S))$. Thus ab^{-1} is an idempotent, by assumption. But then $ab^{-1} = bb^{-1}$

since μ is idempotent-separating, which gives $ab^{-1}b = b$. But $(a, b) \in \mathcal{H}$ by Proposition 3.2.12, and so $a = b$.
(2) Let $\mu(s)$ and $\mu(t)$ be μ-related in S/μ. By Lallement's lemma, every idempotent in S/μ is of the form $\mu(e)$ where $e \in E(S)$. Thus

$$\mu(s)\mu(e)\mu(s)^{-1} = \mu(t)\mu(e)\mu(t)^{-1}$$

so that $\mu(ses^{-1}) = \mu(tet^{-1})$. But both ses^{-1} and tet^{-1} are idempotents, so that $ses^{-1} = tet^{-1}$ for every $e \in E(S)$. Thus $(s, t) \in \mu$. ■

Combining Lemma 2, Proposition 3 and Proposition 5 we obtain the following decomposition of an arbitrary inverse semigroup.

Theorem 6 *Every inverse semigroup is an idempotent-separating extension of a Clifford semigroup by a fundamental inverse semigroup.* ■

Fundamental inverse semigroups

Theorem 6 throws the spotlight on the classes of fundamental and Clifford semigroups. We shall describe these semigroups in more detail beginning with fundamental semigroups.

Let $(E, \leq)$ be a meet semilattice, and let T_E be the set of all order isomorphisms between principal order ideals of E. Clearly, T_E is a subset of $I(E)$. The proof of the following is straightforward.

Theorem 7 *The set T_E is an inverse subsemigroup of $I(E)$ whose semilattice of idempotents is isomorphic to E.* ■

T_E is called the *Munn semigroup* of the semilattice E. This inverse semigroup is second only to the symmetric inverse monoid in its importance in inverse semigroup theory. It is particularly useful in constructing examples of inverse semigroups with a given semilattice of idempotents.

As we show next, every inverse semigroup can be represented in the Munn semigroup of its semilattice of idempotents. Unlike the Wagner–Preston representation, however, the Munn representation, as it is termed, is not injective. At first sight, this might appear to be a disadvantage but, as we show in Section 5.3, it leads to a natural way of decomposing any inverse semigroup into simpler pieces. The following is the Munn representation theorem.

Theorem 8 *Let S be an inverse semigroup. Then there is an idempotent-separating homomorphism $\delta\colon S \to T_{E(S)}$ such that* $\ker \delta = \mu$, *and whose image is a full inverse subsemigroup of $T_{E(S)}$.*

Proof For each $s \in S$ define the function

$$\delta_s\colon [s^{-1}s] \to [ss^{-1}]$$

by $\delta_s(e) = ses^{-1}$. We first show that δ_s is well-defined. Let $e \leq s^{-1}s$. Then $ss^{-1}\delta_s(e) = \delta_s(e)$, and so $\delta_s(e) \leq ss^{-1}$. To show that δ_s is order-preserving, let $e \leq f \in [s^{-1}s]$. Then

$$\delta_s(e)\delta_s(f) = ses^{-1}sfs^{-1} = sefs^{-1} = \delta_s(e).$$

Thus $\delta_s(e) \leq \delta_s(f)$.

Consider now the function $\delta_{s^{-1}}\colon [ss^{-1}] \to [s^{-1}s]$. This is order-preserving by the argument above. For each $e \in [s^{-1}s]$, we have that

$$\delta_{s^{-1}}(\delta_s(e)) = \delta_{s^{-1}}(ses^{-1}) = s^{-1}ses^{-1}s = e.$$

Similarly, $\delta_s(\delta_{s^{-1}}(f)) = f$ for each $f \in [ss^{-1}]$. Thus δ_s and $\delta_{s^{-1}}$ are mutually inverse, and so δ_s is an order isomorphism.

Define $\delta\colon S \to T_{E(S)}$ by $\delta(s) = \delta_s$. To show that δ is a homomorphism, we begin by calculating $\mathrm{dom}(\delta_s\delta_t)$ for any $s, t \in S$. We have that

$$\mathrm{dom}(\delta_s\delta_t) = \delta_t^{-1}([s^{-1}s] \cap [tt^{-1}]) = \delta_t^{-1}([s^{-1}stt^{-1}]).$$

But $\delta_t^{-1} = \delta_{t^{-1}}$ and so

$$\mathrm{dom}(\delta_s\delta_t) = [(st)^{-1}st] = \mathrm{dom}(\delta_{st}).$$

If $e \in \mathrm{dom}\,\delta_{st}$ then

$$\delta_{st}(e) = (st)e(st)^{-1} = s(tet^{-1})s^{-1} = \delta_s(\delta_t(e)).$$

Hence $\delta_s\delta_t = \delta_{st}$.

To show that δ is idempotent-separating, suppose that $\delta(e) = \delta(f)$ where e and f are idempotents of S. Then $\mathrm{dom}\,\delta(e) = \mathrm{dom}\,\delta(f)$. Thus $e = f$.

Because δ is idempotent-separating its kernel is contained in the maximum idempotent-separating congruence μ. Let $(s, t) \in \mu$. The elements s and t are $\mathcal{H}$-related, and so the domains of δ_s and δ_t are the same. Let $e \in \mathrm{dom}\,\delta_s$. Then

$$\delta_s(e) = ses^{-1} = tet^{-1} = \delta_t(e).$$

Thus $\delta_s = \delta_t$, and so $(s, t) \in \ker\delta$. Hence $\ker\delta = \mu$.

The image of δ is a full inverse subsemigroup of $T_{E(S)}$ because every idempotent in $T_{E(S)}$ is of the form $1_{[e]}$ for some $e \in E(S)$, and $\delta_e = 1_{[e]}$. ∎

The homomorphism $\delta\colon S \to T_E$ is called the *Munn representation* of S.

Just as arbitrary inverse semigroups are the inverse subsemigroups of symmetric inverse semigroups, so the following theorem tells us that fundamental inverse semigroups are the *full* inverse subsemigroups of Munn semigroups.

Theorem 9 *Let S be an inverse semigroup. Then S is fundamental if, and only if, S is isomorphic to a full inverse subsemigroup of the Munn semigroup $T_{E(S)}$.*

Proof Let S be a fundamental inverse semigroup. By Theorem 8, there is a homomorphism $\delta: S \to T_{E(S)}$ such that $\ker\delta = \mu$. By assumption, μ is the equality congruence, and so δ is an injective homomorphism. Thus S is isomorphic to its image in $T_{E(S)}$, which is a full inverse subsemigroup.

Conversely, let S be a full inverse subsemigroup of a Munn semigroup T_E. Clearly, we can assume that $E = E(S)$. We calculate the maximum idempotent-separating congruence of S. Let $\alpha, \beta \in S$ and suppose that $(\alpha, \beta) \in \mu$ in S. Then $(\alpha, \beta) \in \mathcal{H}(S)$ and so $\operatorname{dom}\alpha = \operatorname{dom}\beta$. Let $e \in \operatorname{dom}\alpha$. Then $1_{[e]} \in S$, since S is a full inverse subsemigroup of $T_{E(S)}$. By assumption $\alpha 1_{[e]}\alpha^{-1} = \beta 1_{[e]}\beta^{-1}$. It is easy to check that $1_{[\alpha(e)]} = \alpha 1_{[e]}\alpha^{-1}$ and $1_{[\beta(e)]} = \beta 1_{[e]}\beta^{-1}$. Thus $\alpha(e) = \beta(e)$. Hence $\alpha = \beta$, and so S is fundamental. ■

There is another way of representing fundamental inverse semigroups which is of interest in view of our discussion of pseudogroups in Chapter 1. We need a few definitions from topology.

A topological space X is said to be T_0 if for each pair of elements $x, y \in X$ there exists an open set which contains one but not both of x and y. A *basis* for a topological space is a set of open sets β such that every open set of the topology is a union of elements of β. Let X be an arbitrary set and β a set of subsets of X whose union is X and with the property that the intersection of any two elements of β is a union of elements of β. Then a topology can be defined on X by defining the open sets to be the unions of elements of β.

The inverse semigroup of all homeomorphisms between open subsets of (X, τ) will be denoted by $\Gamma(X, \tau)$. An inverse subsemigroup S of $\Gamma(X, \tau)$ is said to be *topologically complete* if the set-theoretic domains of the elements of S form a basis for τ.

Theorem 10 *An inverse semigroup is fundamental if, and only if, it is isomorphic to a topologically complete inverse semigroup on a T_0-space.*

Proof Let S be a fundamental inverse semigroup. We can assume by Theorem 9, that S is a full inverse subsemigroup of a Munn semigroup T_E. Put $\beta = \{[e]: e \in E\}$. Clearly, E is the union of the elements of β, and β is closed under finite intersections. Thus β is the basis of a topology on the set E. It is easy to check that each element of S is a homeomorphism between open subsets of E. It remains to show that this topology is T_0. Let $e, f \in E$ be distinct idempotents. If $f \leq e$ then $[f]$ is an open set containing f but not e.

If $f \not\leq e$ then $[e]$ is an open set containing e but not f. Thus the topology is T_0.

Conversely, let S be a topologically complete inverse subsemigroup of the inverse semigroup $\Gamma(X, \tau)$ where τ is T_0 and $\beta = \{\mathrm{dom}\,\alpha\colon \alpha \in S\}$ is a basis for τ. We shall prove that S is fundamental by showing that the centraliser of the idempotents of S contains only idempotents (Proposition 5). Let $\phi \in S \setminus E(S)$. Then there exists $x \in \mathrm{dom}\,\phi$ such that $\phi(x) \neq x$, because ϕ is not an idempotent. Since τ is T_0, there exists an open set U such that

$$\text{either } (\phi(x) \in U \text{ and } x \notin U) \text{ or } (\phi(x) \notin U \text{ and } x \in U).$$

Since β is a basis for the topology, $U = \bigcup B_i$ for some $B_i \in \beta$. It follows that there is a $B = B_i \in \beta$ such that

$$\text{either } (\phi(x) \in B \text{ and } x \notin B) \text{ or } (\phi(x) \notin B \text{ and } x \in B).$$

Observe that $1_B \in S$ since $B = \mathrm{dom}\,\alpha$ for some $\alpha \in S$. Thus the elements $\phi 1_B$ and $1_B\phi$ belong to S. In the first case, $\phi(x) \in B$ and $x \notin B$, so that whereas $(\phi 1_B)(x)$ is not defined, $(1_B\phi)(x)$ is defined. Thus $\phi \notin Z(E(S))$. In the second case, $(\phi 1_B)(x)$ is defined and $(1_B\phi)(x)$ is not defined. Thus once again $\phi \notin Z(E(S))$. Hence in either case $\phi \notin Z(E(S))$. ■

Clifford semigroups

We now obtain some information about the structure of Clifford semigroups. The basic construction we need is the following.

Let $(E, \leq)$ be a meet semilattice, and let $\{G_e\colon e \in E\}$ be a family of disjoint groups indexed by the elements of E, the identity of G_e being denoted by 1_e. For each pair e, f of elements of E where $e \geq f$ let $\phi_{e,f}\colon G_e \to G_f$ be a group homomorphism, such that the following two axioms hold:

(PG1) $\phi_{e,e}$ is the identity homomorphism on G_e.

(PG2) If $e \geq f \geq g$ then $\phi_{f,g}\phi_{e,f} = \phi_{e,g}$.

We call such a family

$$(G_e, \phi_{e,f}) = (\{G_e\colon e \in E\}, \{\phi_{e,f}\colon e, f \in E, f \leq e\})$$

a *presheaf of groups (over the semilattice E).*

Proposition 11 *Let $(G_e, \phi_{e,f})$ be a presheaf of groups. Let $S = S(G_e, \phi_{e,f})$ be the union of the G_e equipped with the product $\otimes$ on S defined by:*

$$x \otimes y = \phi_{e,e\wedge f}(x)\phi_{f,e\wedge f}(y),$$

where $x \in G_e$ and $y \in G_f$. Then $(S, \otimes)$ is a Clifford semigroup.

Proof Clearly $\otimes$ is well-defined. To prove associativity, let $x \in G_e$, $y \in G_f$ and $z \in G_g$ and put $i = e \wedge f \wedge g$. By definition

$$(x \otimes y) \otimes z = \phi_{e\wedge f,i}(\phi_{e,e\wedge f}(x)\phi_{f,e\wedge f}(y))\phi_{g,i}(z).$$

But

$$\phi_{e\wedge f,i}(\phi_{e,e\wedge f}(x)\phi_{f,e\wedge f}(y)) = \phi_{e\wedge f,i}(\phi_{e,e\wedge f}(x))\phi_{e\wedge f,i}(\phi_{f,e\wedge f}(y)).$$

By axiom (PG2) this simplifies to $\phi_{e,i}(x)\phi_{f,i}(y)$. Thus

$$(x \otimes y) \otimes z = \phi_{e,i}(x)\phi_{f,i}(y)\phi_{g,i}(z).$$

A similar argument shows that $x \otimes (y \otimes z)$ likewise reduces to the right-hand side of the above equation. Thus $(S, \otimes)$ is a semigroup.

Observe that if $x, y \in G_e$ then $x \otimes y = xy$. Thus if $x \in G_e$ and x^{-1} is the inverse of x in the group G_e then

$$x = x \otimes x^{-1} \otimes x \text{ and } x^{-1} = x^{-1} \otimes x \otimes x^{-1}$$

by axiom (PG1). Thus $(S, \otimes)$ is a regular semigroup.

The idempotents of S are just the identities of the groups G_e, again by axiom (PG1). It is easy to check that $1_e \otimes 1_f = 1_{e\wedge f}$. Thus the idempotents commute. We have thus shown that $(S, \otimes)$ is an inverse semigroup.

To finish off, let $x \in G_f$. Then

$$1_e \otimes x = \varphi_{e,e\wedge f}(1_e)\varphi_{f,e\wedge f}(x) = 1_{e\wedge f}\varphi_{f,e\wedge f}(x) = \varphi_{f,e\wedge f}(x),$$

and similarly, $x \otimes 1_e = \varphi_{f,e\wedge f}(x)$. Consequently, the idempotents of S are central. ∎

The product $\otimes$ would usually be written as concatenation.

The semigroup S constructed in Proposition 11 is called a *strong semilattice of groups*. We now prove, amongst other things, that every Clifford semigroup is isomorphic to a strong semilattice of groups.

Theorem 12 *Let S be an inverse semigroup. Then the following are equivalent.*

(1) *S is a Clifford semigroup.*

(2) *For every $s \in S$ we have that $s^{-1}s = ss^{-1}$.*

(3) *Every $\mathcal{H}$-class is a group.*

(4) *Every μ-class is a group.*

(5) *S is isomorphic to a strong semilattice of groups.*

Proof (1) $\Rightarrow$ (2). Let S be a Clifford semigroup and let $s \in S$. Since the idempotents are central $s = s(s^{-1}s) = (s^{-1}s)s$. Thus $ss^{-1} \leq s^{-1}s$. We may similarly show that $s^{-1}s \leq ss^{-1}$, from which we obtain $s^{-1}s = ss^{-1}$.
(2) $\Rightarrow$ (3). Let $s \in S$. By assumption $s^{-1}s = ss^{-1} = e$, say. Hence $s\,\mathcal{H}\,e$. Thus each $\mathcal{H}$-class contains an idempotent, and so each $\mathcal{H}$-class is a group by Proposition 3.2.1.
(3) $\Rightarrow$ (2). Let $s \in S$. By assumption, H_s is a group and so contains an idempotent e. But then $e = s^{-1}s = ss^{-1}$.
(3) $\Rightarrow$ (4). We shall show that $\mu = \mathcal{H}$ using the fact that (2) is equivalent to (3). Let $(a,b) \in \mathcal{H}$. Then for every idempotent e we have that

$$aea^{-1} = (ae)(ae)^{-1} = (ae)^{-1}(ae) = ea^{-1}a$$

and

$$beb^{-1} = (be)(be)^{-1} = (be)^{-1}(be) = eb^{-1}b.$$

By assumption, $a^{-1}a = b^{-1}b$, so that $(a,b) \in \mu$. Hence $\mu = \mathcal{H}$.
(4) $\Rightarrow$ (5). Let S be an inverse semigroup in which each μ-class is a group. Put $G_e = \mu(e)$ for each idempotent $e \in E(S)$. Then $\{G_e : e \in E(S)\}$ is a family of disjoint groups. If $e \geq f$ define $\phi_{e,f}\colon G_e \to G_f$ by $\phi_{e,f}(a) = af$. This is a well-defined function, because if $a\,\mu\,e$ then $af\,\mu\,f$. We now show that $(G_e, \phi_{e,f})$ is a presheaf of groups over the semilattice $E(S)$.

Axiom (PG1) holds: let $e \in E(S)$ and $a \in G_e$. Then $\phi_{e,e}(a) = ae$. But $a \in G_e = \mu(e)$, so that $ae = a$.

Axiom (PG2) holds: let $e \geq f \geq g$ and $a \in G_e$. Then

$$(\phi_{f,g}\phi_{e,f})(a) = \phi_{f,g}(\phi_{e,f}(a)) = afg = ag = \phi_{e,g}(a).$$

Let T be the strong semilattice of groups constructed from this presheaf of groups as in Proposition 11. Define a function $\theta\colon S \to T$ by $\theta(a) = a$, where $\theta(a) \in G_e$ if $a \in \mu(e)$. This function is a bijection, so it only remains to show that it is a homomorphism. We calculate $a \otimes b$ in T, where $a \in G_e$ and $b \in G_f$. By definition

$$a \otimes b = \phi_{e,ef}(a)\phi_{f,ef}(b) = aefbef.$$

However $a \in \mu(e)$ and $b \in \mu(f)$ imply that $ae = a$, $bf = b$ and $ab\,\mu\,ef$. Hence $aefbef = ab$, and so $a \otimes b = ab$. Thus θ is a homomorphism.
(5) $\Rightarrow$ (1). This was proved in Proposition 11. ■

There is a natural class of inverse semigroups which are automatically Clifford.

Proposition 13 *Let S be an inverse semigroup whose idempotents form a finite chain. Then S is a Clifford semigroup.*

Proof Let $s \in S$. The function δ_s: $[s^{-1}s] \to [ss^{-1}]$ is defined as in the proof of Theorem 8; it is an order isomorphism. Thus the sets $[s^{-1}s]$ and $[ss^{-1}]$ have the same cardinality. But the idempotents form a finite chain, and so neither $s^{-1}s < ss^{-1}$ nor $ss^{-1} < s^{-1}s$ is a possibility. Thus $s^{-1}s = ss^{-1}$. Consequently, S is a Clifford semigroup by Theorem 12. ∎

The semigroups with finite chains of idempotents are often called *finite chains of groups*.

Concluding remarks

Theorem 6, combined with Theorems 9 and 12, provides an overview of the class of inverse semigroups. To a certain extent, it describes inverse semigroups in terms of groups and semilattices. It is possible to make the result more precise by developing the theory of idempotent-separating extensions along the lines of group extensions; references to this work are discussed in the notes at the end of this chapter. However, this leads to a theory which is complex even when the inverse semigroups are groups. For this reason, we prefer to aim for something less ambitious but sufficient for our purposes. Consequently, we now turn to the theory of semidirect products appropriate for inverse semigroups.

5.3 λ-semidirect products

Extension theory is concerned with the problem of constructing inverse semigroups from simpler pieces. Even in the group case, this is difficult. For this reason, we shall not attempt to describe the general situation here, instead we shall concentrate on some special cases based on semidirect products. Before we can do this, however, we have to deal with a fundamental stumbling-block: the classical semidirect product of two inverse semigroups need not be inverse. Bernd Billhardt showed how to get around this difficulty by modifying the definition of semidirect products in the inverse case to obtain what he termed λ-semidirect products. The λ-semidirect product of inverse semigroups is again inverse and so the path is clear to develop a general theory.

Throughout this section, it will ease notation if we recall the definitions made in Chapter 1: $\mathbf{r}(s) = ss^{-1}$ and $\mathbf{d}(s) = s^{-1}s$ for each s in the inverse semigroup S. In particular, the idempotent $\mathbf{r}(s)$ was denoted $\lambda(s)$ by Petrich [312], which accounts for Billhardt's terminology.

One of our main results will be a proof that every inverse semigroup can be embedded in a λ-wreath product of a Clifford semigroup by a fundamental inverse semigroup.

Definition and basic properties

Let K and T be inverse semigroups. The semigroup T is said to *act on K by endomorphisms* if for every $t \in T$ there is a map $a \mapsto t \cdot a$ from K to itself satisfying the following two axioms for all $t, u \in T$ and for all $a, b \in K$:

(AE1) $t \cdot (ab) = (t \cdot a)(t \cdot b)$.

(AE2) $(tu) \cdot a = t \cdot (u \cdot a)$.

If T is a monoid with identity 1 then we may also require the additional condition:

(AE3) $1 \cdot a = a$ for all $a \in K$.

The axioms (AE1) and (AE2) (resp. (AE1), (AE2) and (AE3)) are equivalent to the existence of a homomorphism (resp. monoid homomorphism) from T to the monoid of endomorphisms of K.

There are a couple of useful deductions to be made from these axioms. Firstly,

$$t \cdot a^{-1} = (t \cdot a)^{-1}$$

for all $t \in T$ and $a \in K$, because inverses are mapped to inverses by endomorphisms; secondly, if e is an idempotent in K then $t \cdot e$ is an idempotent because idempotents are mapped to idempotents by endomorphisms.

Let T act on K by endomorphisms. The *(classical) semidirect product*, $K * T$, of K by T is the set $K \times T$ equipped with the multiplication

$$(a, t)(b, u) = (a(t \cdot b), tu).$$

As we noted above, the semidirect product of inverse semigroups need not be inverse. For this reason, we modify the definition in the following way.

Let T and K be inverse semigroups and let T act on K by endomorphisms. Put

$$K *^{\lambda} T = \{(a, t) \in K \times T \colon a = \mathbf{r}(t) \cdot a\}$$

and define a binary operation on $K * T$ by

$$(a, t)(b, u) = ((\mathbf{r}(tu) \cdot a)(t \cdot b), tu).$$

The result which justifies this definition is the following.

Proposition 1 *Let K and T be inverse semigroups such that T acts on K by endomorphisms. Then $K *^{\lambda} T$ is an inverse semigroup. If, in addition, K and T are both monoids and axiom* (AE3) *holds then $K *^{\lambda} T$ is an inverse monoid.*

Proof First we have to show that the multiplication is well-defined. By definition

$$(a,t)(b,u) = ((\mathbf{r}(tu)\cdot a)(t\cdot b), tu).$$

To show that this ordered pair actually belongs to the set $K *^\lambda T$ we have to compute

$$\mathbf{r}(tu)\cdot((\mathbf{r}(tu)\cdot a)(t\cdot b)).$$

By axiom (AE1) this is just

$$(\mathbf{r}(tu)\cdot(\mathbf{r}(tu)\cdot a))(\mathbf{r}(tu)\cdot(t\cdot b)).$$

By axiom (AE2) the first term in this product is $(\mathbf{r}(tu)\cdot a)$ and the second term is $(\mathbf{r}(tu)t)\cdot b$. Now $\mathbf{r}(tu)t = t\mathbf{r}(u)$. Thus by axiom (AE2), the second term is $t\cdot(\mathbf{r}(u)\cdot b)$. But $(b,u)\in K * T$ and so $\mathbf{r}(u)\cdot b = b$. Thus

$$\mathbf{r}(tu)\cdot((\mathbf{r}(tu)\cdot a)(t\cdot b)) = ((\mathbf{r}(tu)\cdot a)(t\cdot b)$$

as required.

Next we show that the binary operation is associative. Let

$$(a,t),(b,u),(c,v)\in K *^\lambda T.$$

First we calculate $[(a,t)(b,u)](c,v)$; only its first component need be considered. This is just

$$[\mathbf{r}(tuv)\cdot((\mathbf{r}(tu)\cdot a)(t\cdot b))][(tu)\cdot c].$$

The first term of this product quickly simplifies to

$$(\mathbf{r}(tuv)\cdot a)((t\mathbf{r}(uv))\cdot b)$$

by axioms (AE1) and (AE2) and using the fact that

$$\mathbf{r}(tuv)\mathbf{r}(tu) = \mathbf{r}(tuv) \text{ and } \mathbf{r}(tuv)t = t\mathbf{r}(uv).$$

Thus the first component is just

$$(\mathbf{r}(tuv)\cdot a)((t\mathbf{r}(uv))\cdot b)((tu)\cdot c).$$

It is easy to check that the first component of $(a,t)[(b,u)(c,v)]$ is the same.

To show that the semigroup is regular, define

$$(a,t)^{-1} = (t^{-1}\cdot a^{-1}, t^{-1}).$$

This is a well-defined element of $K *^\lambda T$ because

$$\mathbf{r}(t^{-1})\cdot(t^{-1}\cdot a^{-1}) = (\mathbf{r}(t^{-1})t^{-1})\cdot a^{-1} = t^{-1}\cdot a^{-1}.$$

Observe that

$$(a,t)(a,t)^{-1} = (\mathbf{r}(a),\mathbf{r}(t))$$

using the fact that $t^{-1} \cdot (a^{-1}) = (t^{-1} \cdot a)^{-1}$. The following two equations are now easy to check:

$$(a,t)(a,t)^{-1}(a,t) = (\mathbf{r}(a),\mathbf{r}(t))(a,t) = (a,t)$$

and

$$(a,t)^{-1}(a,t)(a,t)^{-1} = (a,t)^{-1}(\mathbf{r}(a),\mathbf{r}(t)) = (a,t)^{-1}.$$

Next we locate the idempotents. Let $(a,t)^2 = (a,t)$. Then $t^2 = t$ and $(\mathbf{r}(t^2)\cdot a)(t\cdot a) = a$. Thus t is an idempotent, so that $a = t\cdot a$, which implies that a is an idempotent. It is now easy to check that $(a,t) \in K*^{\lambda}T$ is an idempotent precisely when a and t are idempotents. The fact that idempotents commute follows from the fact that if a is an idempotent in K and i an idempotent in T then $i \cdot a$ is an idempotent. We have therefore proved that $K *^{\lambda} T$ is an inverse semigroup.

Now suppose that T and K are monoids and axiom (AE3) holds. Denote the identities of both T and K by 1. Then $(1,1)$ is a well-defined idempotent of $K *^{\lambda} T$ such that

$$(1,1)(a,t) = ((\mathbf{r}(t)\cdot 1)(1\cdot a),t) = ((\mathbf{r}(t)\cdot 1)(\mathbf{r}(t)\cdot a),t) = (\mathbf{r}(t)\cdot(1a),t) = (a,t)$$

and

$$(a,t)(1,1) = ((\mathbf{r}(t)\cdot a)(t\cdot 1),t) = ((t\cdot(t^{-1}\cdot a))(t\cdot 1),t) = (t\cdot((t^{-1}\cdot a)1),t) = (a,t).$$

Thus $(1,1)$ is the identity. ■

The semigroup $K *^{\lambda} T$ is called a *λ-semidirect product of K by T*. There are two special cases of this construction which are worth making explicit.

Proposition 2 *Let K and T be inverse semigroups and let T act on K by endomorphisms.*

(1) *If $K = E$ is a semilattice and T is a group and the action also satisfies axiom* (AE3), *then $E*^{\lambda}T$ is the classical semidirect product of a semilattice by a group.*

(2) *If K and T are both groups and the action also satsifies axiom* (AE3), *then $K *^{\lambda} T$ is a group and is the classical semidirect product of the group K by the group T.*

Proof (1) Because T is a group and axiom (AE3) holds, the underlying set of $E*^{\lambda}T$ is just $E \times T$. The product of (e,g) and (f,h) simplifies to $(e \wedge g\cdot f, gh)$, which is just the usual muliplication in the classical semidirect product.
(2) The only idempotent in $K *^{\lambda} T$ is $(1,1)$. Thus it is a group. The claim is now clear. ■

λ-wreath products

We now describe a way of constructing examples of λ-semidirect products. Let S and T be arbitrary inverse semigroups. Let $K = S^T$ be the set of all functions from T to S. Define a binary operation on K by $(f \oplus g)(a) = f(a)g(a)$. With respect to this operation K is an inverse semigroup: f^{-1} is defined by $f^{-1}(a) = f(a)^{-1}$; of course, here f^{-1} does not mean the inverse of the function f in the usual sense.

The proofs of the following are straightforward.

Proposition 3 *Let S and T be inverse semigroups.*

(1) *If S is a semilattice then S^T is a semilattice.*

(2) *If S is a group then S^T is a group.*

(3) *If S is a Clifford semigroup then S^T is a Clifford semigroup.*

■

Define an action of T on $K = S^T$ as follows: if $f \in K$ and $t \in T$ then $t \cdot f$ is defined by $(t \cdot f)(a) = f(at)$ for all $a \in T$.

Proposition 4 *With the notation above T acts on K by endomorphisms. If T is a monoid then axiom* (AE3) *holds.*

Proof Axiom (AE1) holds: $(t \cdot (f \oplus g))(a) = (f \oplus g)(at) = f(at)g(at) = (t \cdot f)(a)(t \cdot g)(a) = ((t \cdot f) \oplus (t \cdot g))(a)$.
Axiom (AE2) holds: $((tu) \cdot f)(a) = f(atu) = f((at)u) = (u \cdot f)(at) = (t \cdot (u \cdot f))(a)$.

If T is a monoid then it is clear that axiom (AE3) holds. ■

The semigroup $S^T *^{\lambda} T$, constructed from the above action, is denoted by $S \operatorname{Wr}^{\lambda} T$ and called the (*standard*) *λ-wreath product of S by T*.

Billhardt congruences

It is now time to connect congruences with λ-semidirect products. A congruence ρ on an inverse semigroup S is called a *Billhardt congruence* if for each $s \in S$ the set $\{t^{-1}t\colon t \in \rho(s)\}$ contains a maximum element with respect to the natural partial order. A *Billhardt transversal* for a Billhardt congruence ρ is a transversal $\{s_r\}$ of the ρ-equivalence classes such that $s_r^{-1}s_r$ is the largest element of $\{t^{-1}t\colon t \in \rho(s_r)\}$. For congruences of this type, the Kaloujnine–Krasner theorem [402] for groups may be generalised.

Theorem 5 *Let ρ be a Billhardt congruence on an inverse semigroup S. Then S can be embedded in* $\operatorname{Ker}\rho \operatorname{Wr}^{\lambda} S/\rho$.

Proof Choose a Billhardt transversal for ρ. For each $s \in S$, define a function $f_s\colon S/\rho \to \operatorname{Ker}\rho$ by

$$f_s(\rho(a)) = (ass^{-1})_r s(as)_r^{-1}.$$

This is well-defined because

$$((ass^{-1})_r, ass^{-1}) \in \rho \text{ and } ((as)_r^{-1}, (as)^{-1}) \in \rho$$

so that

$$(f_s(\rho(a)), as(as)^{-1}) \in \rho,$$

and $as(as)^{-1}$ is an idempotent. Thus $f_s(\rho(a)) \in \operatorname{Ker}\rho$.

Form the λ-wreath product $\operatorname{Ker}\rho \operatorname{Wr}^{\lambda} S/\rho$. Define a function

$$\phi\colon S \to \operatorname{Ker}\rho \operatorname{Wr}^{\lambda} S/\rho \text{ by } \phi(s) = (f_s, \rho(s)).$$

This is well-defined because $\mathbf{r}(\rho(s)) \cdot f_s = f_s$, which can be easily verified.

To show that ϕ is injective, suppose that $\phi(s) = \phi(t)$. Then $f_s = f_t$ and $\rho(s) = \rho(t)$. By definition

$$s^{-1}s \leq s_r^{-1}s_r \text{ and } ss^{-1} \leq (ss^{-1})_r^{-1}(ss^{-1})_r$$

so that

$$s = (ss^{-1})_r^{-1}(ss^{-1})_r ss_r^{-1}s_r.$$

But $f_s(\rho(ss^{-1})) = (ss^{-1})_r ss_r^{-1}$, and so

$$s = (ss^{-1})_r^{-1} f_s(\rho(ss^{-1}))s_r.$$

We may similarly show that

$$t = (tt^{-1})_r^{-1} f_t(\rho(tt^{-1}))t_r.$$

But

$$(ss^{-1})_r^{-1} = (tt^{-1})_r^{-1},\ f_s(\rho(ss^{-1})) = f_t(\rho(tt^{-1})) \text{ and } s_r = t_r.$$

Hence $s = t$.

To show that ϕ is a homomorphism let $s, t \in S$. We need to show that $\phi(st) = \phi(s)\phi(t)$, but this boils down to showing that

$$f_{st} = (\mathbf{r}(\rho(st)) \cdot f_s) \oplus (\rho(s) \cdot f_t).$$

Let $\rho(a)$ be any element of S/ρ. Then

$$(\mathbf{r}(\rho(st)) \cdot f_s) \oplus (\rho(s) \cdot f_t)(\rho(a)) = f_s(\rho(a\mathbf{r}(st)))f_t(\rho(as)).$$

By definition

$$f_s(\rho(a\mathbf{r}(st)))f_t(\rho(as)) = [(a\mathbf{r}(st))_r s]\mathbf{d}((as\mathbf{r}(t))_r)t(ast)_r^{-1}.$$

But $((a\mathbf{r}(st))_r s, as\mathbf{r}(t)) \in \rho$ and so $\mathbf{d}(a\mathbf{r}(st))_r s) \leq \mathbf{d}((as\mathbf{r}(t))_r)$. Thus the product simplifies to $(a\mathbf{r}(st))_r st(ast)_r^{-1}$ which is just $f_{st}(\rho(a))$. ∎

The structure of arbitrary inverse semigroups

The following result provides us with a natural class of Billhardt congruences.

Proposition 6 *Every idempotent-separating congruence is a Billhardt congruence.*

Proof Let ρ be an idempotent-separating congruence on an inverse semigroup S. Let $s \in S$ and consider the set $\{t^{-1}t\colon t \in \rho(s)\}$. Let $a^{-1}a$ and $b^{-1}b$ be elements of this set where $a, b \in \rho(s)$. Then $a\,\rho\,b$, so that $a^{-1}a\,\rho\,b^{-1}b$, which gives $a^{-1}a = b^{-1}b$ since ρ is idempotent-separating. Thus ρ is a Billhardt congruence because the set $\{t^{-1}t\colon t \in \rho(s)\}$ contains exactly one element. ■

We now have the following; it tells us, in a general way, how to construct inverse semigroups from Clifford semigroups and fuundamental semigroups.

Theorem 7 *Every inverse semigroup can be embedded in a λ-semidirect product of a Clifford semigroup by a fundamental inverse semigroup.*

Proof Let S be an inverse semigroup. By Proposition 6, the maximum idempotent-separating congruence μ on S is a Billhardt congruence. Thus S can be embedded in $\operatorname{Ker}\mu \operatorname{Wr}^{\lambda} S/\mu$ by Theorem 5. By Proposition 5.2.5, the semigroup S/μ is fundamental. From the definition of the λ-wreath product, $\operatorname{Ker}\mu \operatorname{Wr}^{\lambda} S/\mu$ is the λ-semidirect product $(\operatorname{Ker}\mu)^{S/\mu} *^{\lambda} S/\mu$. By Proposition 5.2.3, $\operatorname{Ker}\mu$ is a Clifford semigroup, and so by Proposition 3, $(\operatorname{Ker}\mu)^{S/\mu}$ is a Clifford semigroup. Thus S can be embedded in a λ-semidirect product of a Clifford semigroup by a fundamental inverse semigroup. ■

Split Billhardt congruences

Theorem 5 enables us to obtain embedding theorems, but in order to obtain isomorphism theorems we need to restrict further the class of congruences considered. The motivation for this again comes from group theory. A group extension G of H by F is said to *split* if there is a subgroup K of G such that $HK = G$ and $H \cap K = \{1\}$. It can be shown that G is a semidirect product of H by K.

In this section, we shall define what it means for a Billhardt congruence on an inverse semigroup S to split, and show that S is then isomorphic to a special kind of semidirect product. We begin by introducing a corresponding special class of actions by endomorphisms.

Let T act on K by endomorphisms. Consider the following axioms:

(AE4) For each $a \in K$ there exists $\epsilon(a) \in E(T)$ such that

$$e \cdot a = a \Leftrightarrow \epsilon(a) \leq e$$

for all $e \in E(T)$.

(AE5) $\epsilon(ab) = \epsilon(a)\epsilon(b)$ for all $a, b \in K$.

(AE6) For each $e \in E(T)$ there exists $a \in K$ such that $e = \epsilon(a)$.

Axioms (AE5) and (AE6) simply mean that the function ϵ: $K \to E(T)$ is a surjective homomorphism.

Lemma 8 *Let T act on K by endomorphisms and satisfy in addition the axioms* (AE4), (AE5) *and* (AE6).

(1) $\mathbf{r}(t) \cdot a = a$ *implies* $\epsilon(t^{-1} \cdot a) = t^{-1}\epsilon(a)t$ *for all* $a \in K$ *and* $t \in T$.

(2) $\epsilon(a^{-1}) = \epsilon(a)^{-1} = \epsilon(a)$ *for all* $a \in K$.

(3) $[(\epsilon(b)s) \cdot a][\epsilon(s \cdot a) \cdot b] = (s \cdot a)b$ *for all* $a, b \in K$ *and* $t \in T$.

(4) *If* $\mathbf{r}(t) = \epsilon(a)$ *then*

$$t \cdot b = t \cdot [(t^{-1}\epsilon(a)t) \cdot b] = t \cdot (\epsilon(t^{-1} \cdot a) \cdot b)$$

for all $a, b \in K$ *and* $t \in T$.

Proof (1) Let $u \in T$ and suppose that

$$\mathbf{r}(u) \cdot (t^{-1} \cdot a) = t^{-1} \cdot a.$$

Multiplying the equation on the left by t and using $\mathbf{r}(t) \cdot a = a$ we obtain $\mathbf{r}(tu) \cdot a = a$. Thus $\epsilon(a) \leq \mathbf{r}(tu)$ by axiom (AE4). Multiplying on the left by t^{-1} and on the right by t, we obtain

$$t^{-1}\epsilon(a)t \leq \mathbf{r}(u)\mathbf{d}(t) \leq \mathbf{r}(u).$$

In particular, $\epsilon(t^{-1} \cdot a) \cdot (t^{-1} \cdot a) = t^{-1} \cdot a$ by axiom (AE4). Thus

$$t^{-1}\epsilon(a)t \leq \epsilon(t^{-1} \cdot a).$$

Clearly, $(t^{-1}\epsilon(a)t) \cdot (t^{-1} \cdot a) = t^{-1} \cdot a$, so that

$$\epsilon(t^{-1} \cdot a) \leq t^{-1}\epsilon(a)t.$$

Hence $\epsilon(t^{-1} \cdot a) = t^{-1}\epsilon(a)t$.
(2) Immediate since ϵ is a homomorphism of inverse semigroups by axiom

(AE5).
(3) Consider first the term $\epsilon(s \cdot a) \cdot b$. By axiom (AE4), we have that $\epsilon(b) \cdot b = b$. Thus

$$\epsilon(s \cdot a) \cdot b = \epsilon(s \cdot a) \cdot (\epsilon(b) \cdot b).$$

But $\epsilon(s \cdot a)$ and $\epsilon(b)$ are idempotents and so

$$\epsilon(s \cdot a) \cdot b = \epsilon(b) \cdot (\epsilon(s \cdot a) \cdot b).$$

For the other term we have that

$$(\epsilon(b)s) \cdot a = \epsilon(b) \cdot (s \cdot a).$$

Thus the product of the two terms is equal to

$$\epsilon(b) \cdot ((s \cdot a)(\epsilon(s \cdot a) \cdot b)).$$

Now

$$s \cdot a = \epsilon(s \cdot a) \cdot (s \cdot a).$$

Thus the product is equal to

$$\epsilon(b) \cdot (\epsilon(s \cdot a) \cdot ((s \cdot a)b)).$$

But

$$\epsilon(b)\epsilon(s \cdot a) = \epsilon(s \cdot a)\epsilon(b) = \epsilon((s \cdot a)b).$$

Thus the product is equal to $(s \cdot a)b$.
(4) Straightforward calculation using (1). ∎

Let T act on K by endomorphisms and satisfy in addition the axioms (AE4), (AE5) and (AE6). Put

$$K \bowtie T = \{(a,t) \in K \times T \colon \mathbf{r}(t) = \epsilon(a)\}.$$

Theorem 9 *$K \bowtie T$ is an inverse subsemigroup of $K *^{\lambda} T$ in which the product takes the form $(a,t)(b,u) = (a(t \cdot b), tu)$.*

Proof First observe that $K \bowtie T$ is a subset of $K *^{\lambda} T$, for if $(a,t) \in K \bowtie T$ then $\mathbf{r}(t) = \epsilon(a)$ and so $\mathbf{r}(t) \cdot a = a$. Thus $(a,t) \in K *^{\lambda} T$. Next we show that $K \bowtie T$ is closed under inverses. Let $(a,t) \in K \bowtie T$. Then $(t^{-1} \cdot a^{-1}, t^{-1}) \in K *^{\lambda} T$. Now $\mathbf{r}(t) \cdot a = a$ implies that $\mathbf{r}(t) \cdot (a^{-1}) = a^{-1}$. By Lemma 8(1), we have that

$$\epsilon(t^{-1} \cdot a^{-1}) = t^{-1}\epsilon(a^{-1})t = t^{-1}\epsilon(a)t.$$

But $t^{-1}\epsilon(a)t = t^{-1}\mathbf{r}(t)t = \mathbf{r}(t^{-1})$. Thus $\mathbf{r}(t^{-1}) = \epsilon(t^{-1} \cdot a^{-1})$.

Next we show that the form of the product of two elements from $K \bowtie T$ simplifies. Let $(a,t),(b,u) \in K \bowtie T$, so that $\mathbf{r}(t) = \epsilon(a)$ and $\mathbf{r}(u) = \epsilon(b)$. By definition

$$(a,t)(b,u) = ((\mathbf{r}(tu)\cdot a)(t\cdot b), tu).$$

Now $\mathbf{r}(tu) = t\mathbf{r}(u)t^{-1} = t\epsilon(b)t^{-1}$. Thus the product is equal to

$$(((t\epsilon(b)t^{-1})\cdot a)(t\cdot b), tu).$$

Considering only the first component,

$$((t\epsilon(b)t^{-1})\cdot a)(t\cdot b) = t\cdot[((\epsilon(b)t^{-1})\cdot a)(\epsilon(t^{-1}\cdot a)\cdot b)]$$

by Lemma 8(1). But this simplifies to $t\cdot[(t^{-1}\cdot a)b]$ by Lemma 8(3) and $t\cdot[(t^{-1}\cdot a)b] = a(t\cdot b)$.

It remains to show that $K \bowtie T$ is closed under multiplication. Let

$$(a,t),(b,u) \in K \bowtie T.$$

We shall show that

$$\epsilon(a(t\cdot b)) = \mathbf{r}(tu).$$

Observe that

$$a(t\cdot b) = ((t\epsilon(b)t^{-1})\cdot a)(t\cdot b)) = t\cdot \mathbf{a},$$

where

$$\mathbf{a} = [(\epsilon(b)t^{-1})\cdot a][(t^{-1}\epsilon(a)t)\cdot b],$$

using Lemma 8(4). But $\mathbf{r}(t^{-1})\cdot\mathbf{a} = \mathbf{a}$. Thus $\epsilon(t\cdot\mathbf{a}) = t\epsilon(\mathbf{a})t^{-1}$ by Lemma 8(1). But $\mathbf{a} = (t^{-1}\cdot a)b$ by Lemma 8(1) and Lemma 8(3), so that

$$\epsilon(t\cdot\mathbf{a}) = t\epsilon((t^{-1}\cdot a)b)t^{-1}.$$

Hence

$$\epsilon(a(t\cdot b)) = t\epsilon((t^{-1}\cdot a)b)t^{-1}.$$

But ϵ is a homomorphism and $\epsilon(t^{-1}\cdot a) = t^{-1}\epsilon(a)t$ by Lemma 8(1) so that

$$t\epsilon((t^{-1}\cdot a)b)t^{-1} = \epsilon(a)t\epsilon(b)t^{-1}.$$

This quickly reduces to $\mathbf{r}(tu)$ since $\epsilon(a) = \mathbf{r}(t)$ and $\epsilon(b) = \mathbf{r}(u)$. ■

$K \bowtie T$ is called a *full restricted semidirect product* of K by T.

Billhardt congruences are intimately connected with λ-semidirect products; we now isolate those Billhardt congruences which are likewise connected with full restricted semidirect products.

Let ρ be a Billhardt congruence on an inverse semigroup S. We say that ρ *splits* if there exists a Billhardt transversal $\{s_r\}$ such that $(st)_r = s_r t_r$. Observe that such a transversal forms an inverse subsemigroup of S. Conversely, if there exists an inverse subsemigroup T of S satisfying the following two conditions:

(1) each ρ-class contains exactly one element from T;

(2) for each $t \in T$, the element $t^{-1}t$ is the maximum element of

$$\{a^{-1}a\colon a \in \rho(t)\};$$

then T is a Billhardt transversal. In particular, a Billhardt transversal is isomorphic to S/ρ.

Lemma 10 *Let ρ be a split Billhardt congruence on an inverse semigroup S.*

(1) $s_r^{-1} = (s^{-1})_r$.

(2) *If $k \in \operatorname{Ker}\rho$ then k_r is an idempotent.*

(3) *If $k \in \operatorname{Ker}\rho$ then $(s_r k)^{-1} s_r k \leq s_r^{-1} s_r$ for any $s \in S$.*

Proof The proofs of (1) and (2) follow from the fact that the s_r form a semigroup isomorphic to S/ρ, which is inverse.
(3) Now $(s_r k)^{-1} s_r k = k^{-1} s_r^{-1} s_r k$. But $k^{-1} s_r^{-1} s_r k \,\rho\, s_r^{-1} s_r k^{-1} k$ since $\rho(k)$ is an idempotent. Thus

$$(s_r k)^{-1} s_r k \leq (s_r^{-1} s_r k^{-1} k)_r^{-1} (s_r^{-1} s_r k^{-1} k)_r.$$

But

$$(s_r^{-1} s_r k^{-1} k)_r^{-1} (s_r^{-1} s_r k^{-1} k)_r = (k^{-1} k s_r^{-1} s_r)_r (s_r^{-1} s_r k^{-1} k)_r = k_r^{-1} k_r s_r^{-1} s_r$$

and k_r is an idempotent by (2), so that

$$k_r^{-1} k_r s_r^{-1} s_r = k_r s_r^{-1} s_r \leq s_r^{-1} s_r.$$

■

Split Billhardt congruences give rise to actions by endomorphisms satisfying axioms (AE4), (AE5) and (AE6).

Lemma 11 *Let ρ be a split Billhardt congruence on an inverse semigroup S. Then S/ρ acts on* Ker ρ *by endomorphisms on the left and satisfies in addition axioms* (AE4), (AE5) *and* (AE6).

Proof Define an action

$$S/\rho \times \operatorname{Ker}\rho \to \operatorname{Ker}\rho \text{ by } \rho(s) \cdot k = s_r k s_r^{-1}.$$

This is independent of the choice of representative from $\rho(s)$ because if $\rho(s) = \rho(t)$ then $s_r = t_r$. Furthermore, $\rho(s) \cdot k \in \operatorname{Ker} \rho$ because

$$\rho(s_r k s_r^{-1}) = \rho(s_r)\rho(k)\rho(s_r)^{-1}$$

and $\rho(k)$ contains an idempotent by assumption. To show that S/ρ acts by endomorphisms on $\operatorname{Ker}\rho$, we check that axioms (AE1) and (AE2) hold.

Axiom (AE1) holds: by definition

$$(\rho(s) \cdot k)(\rho(s) \cdot l) = s_r k(s_r^{-1} s_r) l s_r^{-1}.$$

But by Lemma 10(3), $(s_r k)^{-1}(s_r k) \leq s_r^{-1} s_r$. Thus

$$(\rho(s) \cdot k)(\rho(s) \cdot l) = s_r k l s_r^{-1} = \rho(s) \cdot (kl).$$

It is straightforward to check that (AE2) holds.

Define ϵ: $\operatorname{Ker}\rho \to E(S/\rho)$ by $\epsilon(k) = \rho(k)$. Recall that for such a k the element k_r is an idempotent by Lemma 10(2).

To show that axiom (AE4) holds, recall that by Lallement's lemma the idempotents of S/ρ are the elements of the form $\rho(e)$ where e is an idempotent. Suppose that $\rho(e) \cdot k = k$. Then $e_r k e_r^{-1} = k$. Thus $\rho(k) = \rho(e_r)\rho(k)\rho(e_r)^{-1}$. But $\rho(e_r)$ and $\rho(k)$ are idempotents and so $\rho(k) = \rho(e_r)\rho(k)$. Hence $\epsilon(k) \leq \rho(e)$.

Conversely, suppose that $\epsilon(k) \leq \rho(e)$ where e is an idempotent. Then $\rho(k) \leq \rho(e)$. But $\rho(k) = \rho(ke)$ and so $k_r = (ke)_r = k_r e_r$. Thus $k_r \leq e_r$. By definition $\rho(e) \cdot k = e_r k e_r^{-1}$. By Lemma 10(2),

$$k^{-1}k \leq k_r = k_r^{-1} \geq kk^{-1}.$$

Thus $k^{-1}k, kk^{-1} \leq e_r$. It follows that $\rho(e) \cdot k = k$.

Axiom (AE5) holds: this is straightforward to check.

Axiom (AE6) holds: let $\rho(e) \in S/\rho$ be an idempotent, where e is an idempotent (using Lallement's lemma again). Then $e \in \operatorname{Ker}\rho$ and $\epsilon(e) = \rho(e)$. ■

The relationship between split Billhardt congruences and full restricted semidirect products can now be spelt out; it generalises the theory of split extensions of groups.

Theorem 12 *Let ρ be a split Billhardt congruence on an inverse semigroup S. Then S is isomorphic to* $\operatorname{Ker}\rho \bowtie S/\rho$.

Proof Using the definitions in the proof of Lemma 11, the semigroup S/ρ acts on $\operatorname{Ker}\rho$ by endomorphisms in such a way that axioms (AE4), (AE5) and

(AE6) hold. We may therefore construct the full restricted semidirect product $\operatorname{Ker}\rho \bowtie S/\rho$ by Theorem 9. Define a function

$$\psi\colon S \to \operatorname{Ker}\rho \bowtie S/\rho \text{ by } \psi(s) = (ss_r^{-1}, \rho(s)).$$

Observe that from $s\,\rho\,s_r$ we obtain $ss_r^{-1}\,\rho\,s_r s_r^{-1}$, so that $ss_r^{-1} \in \operatorname{Ker}\rho$; and

$$\epsilon(ss_r^{-1}) = \rho(ss_r^{-1}) = \mathbf{r}(\rho(s)).$$

Thus ψ is well-defined. We prove that ψ is an isomorphism.

To show that ψ is injective, suppose that $\psi(s) = \psi(t)$. Then $ss_r^{-1} = tt_r^{-1}$ and $\rho(s) = \rho(t)$. In particular, $s_r = t_r$. Now

$$s = ss^{-1}s = s(s^{-1}s)(s_r^{-1}s_r) = s(s_r^{-1}s_r)$$

since $s^{-1}s \leq s_r^{-1}s_r$. But

$$(ss_r^{-1})s_r = tt_r^{-1}t_r = t.$$

To show that ψ is surjective, let $(k, \rho(s)) \in \operatorname{Ker}\rho \bowtie S/\rho$. Then $\mathbf{r}(\rho(s)) = \epsilon(k)$. Thus $k_r = s_r s_r^{-1}$. Now

$$(ks_r)(ks_r)_r^{-1} = (ks_r)(s_r^{-1}k_r^{-1}) = kk_rk_r^{-1} = k$$

since $k^{-1}k \leq k_r$, and

$$\rho(ks_r) = \rho(k)\rho(s_r) = \rho(s).$$

Thus $\psi(ks_r) = (k, \rho(s))$.

To show that ψ is a homomorphism let $s, t \in S$. Then

$$\psi(st) = ((st)(st)_r^{-1}, \rho(st))$$

and

$$\psi(s) = (ss_r^{-1}, \rho(s)) \text{ and } \psi(t) = (tt_r^{-1}, \rho(t)).$$

Thus

$$\psi(s)\psi(t) = ((ss_r^{-1})(\rho(s)\cdot(tt_r^{-1})), \rho(st)),$$

but the first component reduces as follows:

$$(ss_r^{-1})(\rho(s)\cdot(tt_r^{-1})) = (ss_r^{-1})s_r(tt_r^{-1})s_r^{-1} = ss_r^{-1}s_rtt_r^{-1}s_r^{-1} = st(s_rt_r)^{-1}$$

since $s^{-1}s \leq s_r^{-1}s_r$. ∎

In the next section, we show how split Billhardt congruences arise naturally in the theory of inverse ω-semigroups.

5.4 Inverse ω-semigroups

Recall that an inverse ω-semigroup is an inverse semigroup whose semilattice of idempotents is isomorphic to the set of natural numbers with the reverse of their usual order; the bicyclic monoid is an ω-semigroup (Section 3.4). In this section, we show how the theory of split Billhardt congruences can be used to obtain a structure theory for ω-semigroups. The key to this theory is the bicyclic monoid.

Fundamental ω-semigroups

The bicyclic monoid is combinatorial and so fundamental. Thus by Theorem 5.2.9, it is isomorphic to a full inverse subsemigroup of the Munn semigroup on its semilattice of idempotents, which we shall denote by T_ω. In fact, a much stronger result is true.

Theorem 1 *The Munn semigroup T_ω is isomorphic to the bicyclic monoid.*

Proof We show that T_ω is isomorphic to the set $\mathbb{N} \times \mathbb{N}$ equipped with the binary operation defined in Proposition 3.4.2.

The principal order ideals of $(\omega, \leq)$ are just the sets of the form $[n]$ for each $n \in \omega$. Clearly, any two principal order ideals are isomorphic and, moreover, there is exactly one isomorphism between them. We now obtain an explicit description of these isomorphisms. Let the unique isomorphism from $[m]$ to $[n]$ be denoted by $\theta_{m,n}$. Then it is easy to check that

$$\theta_{m,n}(x) = x + (n - m).$$

We now compute the product $\theta_{m,n}\theta_{p,q}$. Put $t = \max\{m, q\}$. Then

$$\mathrm{dom}(\theta_{m,n}\theta_{p,q}) = \theta_{p,q}^{-1}([t]) = [t + p - q] = [p + (m \dot{-} q)]$$

and

$$\mathrm{im}(\theta_{m,n}\theta_{p,q}) = \theta_{m,n}([t]) = [t + n - m] = [n + (q \dot{-} m)].$$

If we define the function ι: $T_\omega \to P_1$ by $\iota(\theta_{m,n}) = (n, m)$, then it is easy to check that it is an isomorphism. ■

Because of Theorem 1, every fundamental ω-semigroup is isomorphic to a full inverse subsemigroup of the bicyclic monoid. We now obtain explicit descriptions of such inverse subsemigroups. To do this we need a number of definitions.

Let m be a natural number. The set E_m is defined to be the empty set when $m = 0$ and for $m \geq 1$ we define

$$E_m = \{(0, 0), \ldots, (m - 1, m - 1)\}.$$

Thus E_m consists of the top m idempotents of the bicyclic monoid. Let d be a non-zero natural number. Define the set

$$I_{(m,d)} = \{(a,b) \in P_1 \colon m \leq a, b \text{ and } a \equiv b \bmod d\}.$$

Finally, we put

$$B_{(m,d)} = E_m \cup I_{(m,d)}$$

and

$$B_d = B_{(0,d)} = \{(a,b) \in P_1 \colon a \equiv b \bmod d\}.$$

Theorem 2 *$B_{(m,d)}$ is a full inverse subsemigroup of the bicyclic monoid such that:*

(1) *B_1 is the bicyclic monoid and bisimple.*

(2) *For $d \geq 2$, B_d is a simple inverse monoid with d $\mathcal{D}$-classes.*

(3) *For $m \geq 1$, $B_{(m,d)}$ is non-simple having $I_{(m,d)}$ as a proper ideal. Furthermore, $I_{(m,d)}$ is isomorphic to B_d.*

Proof By definition $B_{(m,d)} = E_m \cup I_{(m,d)}$. We show first that $I_{(m,d)}$ is an inverse subsemigroup of P_1. Let $(a,b),(g,h) \in I_{(m,d)}$. By definition $m \leq a, b, g, h$ where $a \equiv b \bmod d$ and $g \equiv h \bmod d$. Also by definition

$$(a,b)(g,h) = (a + (g \dot{-} b), h + (b \dot{-} g)).$$

Since $m \leq a, h$ the entries of the product are both greater than or equal to m. The difference of the two entries is

$$a - h + ((g \dot{-} b) - (b \dot{-} g)) = (a - b) + (g - h)$$

which is divisible by d. Thus $I_{(m,d)}$ is closed under multiplication. It is immediate that $I_{(m,d)}$ is closed under inverses and so $I_{(m,d)}$ is an inverse subsemigroup. The largest idempotent in $I_{(m,d)}$ is (m,m); it is clear that all smaller idempotents of P_1 belong to $I_{(m,d)}$. The elements of E_m are all the idempotents of P_1 larger than any of the idempotents of $I_{(m,d)}$. Thus $B_{(m,d)}$ is a full inverse subsemigroup.
(1) Immediate from the definition and Theorem 3.4.3.
(2) First observe that $(a,b)\,\mathcal{D}\,(g,h)$ precisely when $b \equiv g \bmod d$. Thus the idempotents $(0,0),\ldots,(d-1,d-1)$ form an idempotent transversal of the $\mathcal{D}$-classes. Given a pair of idempotents (a,a) and (b,b) from this transversal we may easily find a natural number c such that $c \geq b$ and $c \equiv a \pmod d$. Consequently, $(a,a)\,\mathcal{D}\,(c,c) \leq (b,b)$ and the semigroup is simple by Proposition 3.2.10.

(3) The only elements of $B_{(m,d)}$ not contained in $I_{(m,d)}$ are the idempotents $(0,0), \ldots, (m-1, m-1)$, each of which is larger than any idempotent in $I_{(m,d)}$. It is now straightforward to check that $I_{(m,d)}$ is a (proper) ideal.

Define ϕ: $I_{(m,d)} \to B_d$ by $\phi(a,b) = (a-m, b-m)$. It is immediate that ϕ is a well-defined bijection, so it only remains to prove that it is a homomorphism. By definition

$$\phi((a,b)(g,h)) = (a - m + (g \dot{-} b), h - m + (b \dot{-} g))$$

and

$$\phi(a,b)\phi(g,h) = (a - m + ((g-m) \dot{-} (b-m)), h - m + ((b-m) \dot{-} (g-m)))$$

which are equal from the properties of the monus operation. Thus ϕ is a homomorphism, and so an isomorphism. ∎

We prove in Theorem 4, that every full inverse subsemigroup of the bicyclic monoid is of the form $B_{(m,d)}$ for some m and d. The following lemma will be useful.

Lemma 3 *In the bicyclic monoid the following results hold.*

(1) *Let $(a, a+d)$ be an element of the bicyclic monoid and let $(b,b) \leq (a,a)$. Then $(b,b)(a,a+d)^k = (b, b+kd)$ for all $k \geq 0$.*

(2) *The elements of $I_{(m,d)}$ are precisely the elements of the form*

$$(m, m+d)^{-k}(n,n)(m,m+d)^l$$

where $(n,n) \leq (m,m)$ and $0 \leq k, l$.

Proof (1) It can be proved by induction that $(a, a+d)^k = (a, a+kd)$ for all $k \geq 1$, and a direct calculation shows that $(b,b)(a,a+d)^k = (b, b+kd)$ for $k \geq 1$. It is immediate that the result holds when $k = 0$.
(2) We show first that every element of $I_{(m,d)}$ is of the form $(n+kd, n+ld)$ where $m \leq n$ and $0 \leq k, l$. Let $(a,b) \in I_{(m,d)}$. Without loss of generality we may assume that $b \leq a$. Then $a = b + qd$ for some q since d divides $a - b$. By assumption, $m \leq b$ so that $b - m = q'd + r$ for some q' and r. Thus

$$(a,b) = ((m+r) + (q'+q)d, (m+r) + q'd),$$

which is of the form $(n+kd, n+ld)$ where $m \leq n$ and $0 \leq k, l$. Conversely, it is clear that every element of this form belongs to $I_{(m,d)}$. But by (1) above, we have that

$$(n+kd, n+ld) = (m, m+d)^{-k}(n,n)(m,m+d)^l$$

as required. ∎

Let U be a full inverse subsemigroup of the bicyclic monoid which contains non-idempotent elements. Put

$$m_0 = \min\{m \in \mathbb{N}\colon (m,n) \in U \text{ for some } n \neq m\},$$

and

$$d_0 = \min\{d \in \mathbb{N} \setminus \{0\}\colon (m_0, m_0 + d) \in U\}.$$

Theorem 4 *Let U be a full inverse subsemigroup of the bicyclic monoid. Then U is either the semilattice of idempotents or $U = B_{(m_0,d_0)}$ for some m_0 and d_0.*

Proof Suppose that U is not the semilattice of idempotents. Then there exists $(m,n) \in U$ for some $m,n \in \mathbb{N}$ such that $m \neq n$. Thus m_0 and d_0 are defined and $(m_0, m_0 + d_0) \in U$. By assumption, U is a full inverse subsemigroup of P_1 so that in particular $(m_0, m_0) \in U$. If $(n,n) \leq (m_0, m_0)$, then

$$(n,n)(m_0, m_0 + d_0) = (n, n + kd_0) \in U$$

for all $k \geq 0$ by Lemma 3(1). Hence $I_{(m_0,d_0)} \subseteq U$ by Lemma 3(2), and so $B_{(m_0,d_0)} \subseteq U$ since $E_{m_0} \subseteq U$.

To prove the reverse inclusion, let $(m,n) \in U$ be a non-idempotent. By taking inverses if necessary, we can assume that $m \leq n$. By assumption, $m,n \geq m_0$. Thus we can write, $(m,n) = (m, m + kd_0 + r)$, where $0 \leq r < d_0$. We shall prove that $r = 0$. Now

$$(m, m + kd_0 + r)(m + kd_0, m) = (m, m + r) \in U.$$

If $m = m_0$ then $r = 0$ and $(m,n) \in B_{(m,d)}$, so we may suppose that $m > m_0$. Then $(m - 1 + d_0, m - 1) \in U$ and so

$$(m - 1 + d_0, m - 1)^{-1}(m, m + r)(m - 1 + d_0, m - 1) = (m - 1, m - 1 + r)$$

giving $(m-1, m-1+r) \in U$. If $m-1 = m_0$ then $r = 0$ and $(m,n) \in B_{(m,d)}$; otherwise, we repeat the above argument. Ultimately, we obtain $(m_0, m_0+r) \in U$. Thus $r = 0$, and so $(m,n) \in B_{(m_0,d_0)}$. ∎

Theorems 2 and 4 constitute a complete classification of fundamental inverse ω-semigroups. The following special cases are worth noting.

Theorem 5 *The only bisimple, fundamental ω-semigroup is the bicyclic monoid; and the only simple, fundamental ω-semigroup with d $\mathcal{D}$-classes, where $d \geq 2$, is B_d.*

Proof Let S be a bisimple, fundamental ω-semigroup. By Theorem 5.2.9, S is isomorphic to a full inverse subsemigroup of T_ω, and so by Theorem 1, it is isomorphic to a full inverse subsemigroup of the bicyclic monoid. But the only bisimple, full inverse subsemigroup of the bicyclic monoid is the bicyclic monoid itself by Theorem 2 and Theorem 4. The result on simple fundamental ω-semigroups is proved similarly. ∎

A bisimple inverse ω-semigroup is called a *Reilly semigroup*.

The structure of inverse ω-semigroups

The description of fundamental ω-semigroups obtained in Theorem 4 is the basis for a classification of arbitrary inverse ω-semigroups.

Theorem 6 *Let S be an inverse ω-semigroup.*

(1) *The semigroup S is a Clifford semigroup if, and only if, S/μ is a semilattice.*

(2) *The semigroup S is bisimple if, and only if, S/μ is isomorphic to the bicyclic monoid.*

(3) *The semigroup S is simple if, and only if, S/μ is isomorphic to some B_d, where $d \geq 2$.*

(4) *If the semigroup S is none of the above then it is an ideal extension of a simple ω-semigroup by a Clifford semigroup with a finite chain of idempotents with a zero adjoined.*

Proof (1) If S is a Clifford semigroup then S/μ is a semilattice by Theorem 5.2.12. Conversely, if S/ρ is a semilattice then each μ-class of S contains an idempotent by Lallement's lemma and so S is a Clifford semigroup by Theorem 5.2.12.
(2) If S is bisimple then its image under the Munn representation is isomorphic to a full bisimple inverse ω-semigroup by Theorem 5.2.9. But the only such inverse subsemigroup of the bicyclic monoid is the bicyclic monoid itself by Theorem 5. Conversely, if S/μ is isomorphic to the bicyclic monoid then it is bisimple. Thus S is bisimple by Proposition 3.2.13.
(3) If S is simple then its image under the Munn representation is isomorphic to a full simple inverse ω-semigroup by Theorem 5.2.9. But the only such inverse subsemigroups of the bicyclic monoid are those of the form B_d for some $d \geq 1$ by Theorem 5. Conversely, if S/μ is isomorphic to a semigroup of the form B_d where $d \geq 1$ then it is simple and non-bisimple. Thus S is simple and not bisimple by Proposition 3.2.13.

(4) By Theorem 4, S/μ is isomorphic to some $B_{(m,d)}$ where $m \geq 1$. By Theorem 2(3), this semigroup is of the form $E_m \cup I_{(m,d)}$ where $I_{(m,d)}$ is a simple ω-semigroup and ideal of $B_{(m,d)}$ and E_m consists of the top m idempotents of P_1. The inverse image of $I_{(m,d)}$ under the Munn representation is a simple inverse subsemigroup I of S by Proposition 3.2.13. Thus I is a simple ω-semigroup, and it is straightforward to check that I is an ideal of S. The complement $S \setminus I$ is mapped to the finite chain of idempotents E_m. Thus μ induces a surjective idempotent-separating homomorphism from the Rees quotient S/I to the Rees quotient $B_{(m,d)}/I_{(m,d)}$ with the property that the inverse image of the zero is just zero. But $B_{(m,d)}/I_{(m,d)}$ is isomorphic to the chain E_m with a zero adjoined. Thus the inverse semigroup S/I has a finite chain of idempotents and so it is a Clifford semigroup by Proposition 5.2.13; by construction, it is isomorphic to a finite chain of groups with a zero adjoined. ■

We have already noted that idempotent-separating congruences are automatically Billhardt congruences. We now determine precisely when idempotent-separating congruences split.

Proposition 7 *Let ρ be an idempotent-separating congruence on an inverse semigroup S. Then ρ is a split Billhardt congruence if, and only if, there is an inverse subsemigroup T of S which meets each ρ-class exactly once.*

Proof Let T be an inverse subsemigroup which meets each ρ-class exactly once. For each $s \in S$ let s_r be the unique element of $\rho(s) \cap T$. It is immediate that $(st)_r = s_r t_r$. By Proposition 5.3.6, the sets $\{t^{-1}t \colon t \in \rho(s)\}$ contain exactly one element. Consequently, T is automatically a Billhardt transversal for ρ. The converse is immediate. ■

The crucial result which will lead to a structure theorem for inverse ω-semigroups is the following.

Theorem 8 *Let S be an inverse ω-semigroup. Then $\mathcal{H} = \mu$ and μ is a split Billhardt congruence.*

Proof To show that $\mathcal{H} = \mu$ it is enough to to prove that $\mathcal{H} \subseteq \mu$. Let $a \,\mathcal{H}\, b$. With the notation of Theorem 5.2.8, the composite $\delta_b^{-1}\delta_a$ is an order automorphism of the ω-semilattice $[a^{-1}a] = [b^{-1}b]$. But this implies that $\delta_b^{-1}\delta_a$ is the identity on $[a^{-1}a] = [b^{-1}b]$. Similarly, $\delta_b\delta_a^{-1}$ is the identity on $[aa^{-1}] = [bb^{-1}]$. Thus $\delta_a^{-1} = \delta_b^{-1}$ and so $\delta_a = \delta_b$. Now let e be any idempotent. Put $f = a^{-1}ae = b^{-1}be$. Then

$$afa^{-1} = \delta_a(f) = \delta_b(f) = bfb^{-1}$$

and so $aea^{-1} = beb^{-1}$. Hence $a \,\mu\, b$.

We now show that μ is a split Billhardt congruence using Theorem 6. By Proposition 7, we have to show that there is an inverse subsemigroup T of S which meets each μ-class exactly once.

If S/μ is a semilattice then each μ-class contains an idempotent and, since μ is idempotent-separating, exactly one idempotent. Thus a suitable semigroup transversal in this case is the semilattice of idempotents.

Next observe that it is enough to prove the result for simple ω-semigroups; by Theorem 6(4), if S is not simple then it is an ideal extension of a simple ω-semigroup S' by a Clifford semigroup with a finite chain of idempotents with a zero adjoined. Let the maximum idempotent in S' be e. If T' is a suitable semigroup transversal for the μ-relation on S', then we can obtain a suitable semigroup transversal T for the μ-relation on S by defining

$$T = T' \cup \{f \in E(S) \colon e \leq f\}.$$

We may therefore suppose that S is a simple inverse ω-semigroup. Let 1 be the identity of S. Since S is not a Clifford semigroup, there exist elements a such that $aa^{-1} = 1$ and $a^{-1}a \neq 1$. Consequently, there is an element a such that $aa^{-1} = 1$ and $a^{-1}a$ is as large as possible subject to the condition that $a^{-1}a \neq 1$. By assumption, S/μ is isomorphic to a semigroup of the form B_d. Let $\theta \colon S/\mu \to B_d$ be the isomorphism. Then $\theta(\mu(a)) = (0, d)$. We show that the subset T of S defined by

$$T = \{a^{-k}ea^l \colon \text{ where } e \in E(S) \text{ and } 0 \leq k, l\}$$

is an inverse semigroup transversal of μ. By Lemma 3,

$$B_d = \{(0,d)^{-k}(n,n)(0,d)^l \colon 0 \leq k, l\}.$$

Thus every μ-class contains an element of T. Next, suppose that

$$(a^{-k}ea^l, a^{-u}fa^v) \in \mathcal{H}$$

where $e, f \in E(S)$. Then

$$a^{-k}ea^k = a^{-u}fa^u \text{ and } a^{-l}ea^l = a^{-v}fa^v.$$

Using the homomorphism from S onto B_d, we deduce that $k - l = u - v$. But then

$$a^{-k}ea^l = a^{-k}e(a^ka^{-k})a^l = a^{-u}fa^ua^{-k}a^l$$

and from $u - k = v - l$ we obtain

$$a^{-u}fa^ua^{-k}a^l = a^{-u}fa^{u-k}a^l = a^{-u}fa^{v-l}a^l = a^{-u}fa^v.$$

Hence $a^{-k}ea^l = a^{-u}fa^v$. Thus each μ-class contains exactly one element of T. It is straightforward to check that T forms an inverse subsemigroup of S. ■

An immediate consequence of Theorem 5.3.12 and Theorem 8 is the following isomorphism theorem for ω-semigroups.

Theorem 9 *Every inverse ω-semigroup is isomorphic to a full restricted semidirect product of a Clifford semigroup by a fundamental inverse semigroup.* ■

5.5 Notes on Chapter 5

Section 5.1

The first attempt to describe inverse semigroup congruences in terms inspired by group theory was Preston's kernel-normal systems [325]. The Kernel–trace description of congruences on inverse semigroups was introduced by Scheiblich [369], and developed by Green [108] and Petrich [308]. Chapter III of [312] is a thorough presentation of the Kernel–trace theory of congruences.

The right framework for studying inverse semigroup extensions and their relations with congruences is the theory of normal extensions initiated by Petrich [311] and described in Chapter V of [312]. The *normal extension problem* amounts to constructing all possible solutions up to equivalence of an extension triple. This was first solved by Allouch [4], and later by Billhardt [26] using λ-semidirect products. Historically, two special cases of the normal extension problem received particular attention: idempotent-separating extensions and idempotent pure extensions.

The theory of idempotent-separating extensions is based on the fact that idempotent-separating congruences may be described entirely in terms of their Kernels. The theory of such extensions is a direct generalisation of the Schreier extension theory of groups. The details were worked out by Coudron [50], and D'Alarcao [57].

The theory of idempotent pure extensions is developed in Chapter 8. It has been most successful in describing the structure of E-unitary inverse semigroups.

The theory of normal extensions is still in its early stages, and is nowhere near as developed as the theory of group extensions. The first steps in the cohomology of inverse semigroups were taken by Lausch [179], who improves on the Coudron–D'Alarcao theory of idempotent-separating congruences, and Loganathan [206], who derives O'Carroll's theory of idempotent pure extensions (see also Chapter 8).

The extension theory of inverse semigroups is further complicated by the existence of other kinds of kernels. One such is the derived category of a homomorphism introduced by Tilson [412]; an early version of Tilson's work was applied to inverse semigroups by Margolis and Pin [223]. This categorical approach to semigroup extensions is described in Margolis [214] and Tilson [413].

Section 5.2

Howie [145] proved that there was a maximum idempotent-separating congruence μ, and obtained an expression for it.

Fundamental inverse semigroups were studied first by Munn [266], and also by Wagner [423], [424], [425] who called them *antigroups*, a terminology adopted by Petrich [312]. The Munn semigroup and the Munn representation were introduced by Munn in [263] (Howie had already observed in 1964 that every element of an inverse semigroup induces an order isomorphism between two principal order ideals of its semilattice of idempotents [145]).

The result that fundamental inverse semigroups could be represented by means of pseudogroups is due to Wagner [424]. It is the starting point for later work of Zhitomirskiĭ [440], [441]. He also described the topologically complete inverse semigroups of partial homeomorphisms of T_1- and T_2-spaces. (Recall that a topological space is T_1 if for each pair of points x and y there is an open set which contains x but not y, and T_2, or *hausdorff*, if there are disjoint open sets containing x and y respectively). He also showed that every inverse semigroup was isomorphic to an inverse semigroup of all local automorphisms of a sheaf.

Clifford semigroups were introduced and their structure determined by Clifford [45]. There is a categorical way of viewing the building blocks of strong semilattices of groups. A meet semilattice $(E, \leq)$ can be regarded as a category in the following way: there is an arrow from e to f precisely when $e \leq f$. Consider now the dual category $(E, \leq)^{\mathrm{op}}$ in which the arrows are turned round. Let G: $(E, \leq)^{\mathrm{op}} \to$ **Groups** be any functor to the category of groups. For every $e \in E$, the image $G(e)$ is a group. If $f \leq e$ then there is a group homomorphism $\phi_{e,f}$: $G(e) \to G(f)$. The fact that G is a functor implies that $\phi_{e,e}$ is the identity on $G(e)$ and if $i \leq f \leq e$ then $\phi_{f,i}\phi_{e,f} = \phi_{e,i}$. Thus G is precisely what we have called a *presheaf of groups over the meet semilattice* $(E, \leq)$. Classically, such presheaves are usually considered over the lattice of open subsets of a topological space (see [409]).

An application of abelian Clifford semigroups may be found in [111] and [112]. Languages with Clifford semigroups as syntactic monoids are described in [177].

Section 5.3

λ-semidirect products and λ-wreath products were introduced by Billhardt in [25], [26], [27], and [28]. The work of this section is very largely based on portions of this work.

Mária Szendrei has pointed out [407] that earlier work of Houghton's [144] is a precursor of Billhardt's. Houghton also defined a wreath product $W(K,T)$ of inverse semigroups K and T. Szendrei shows that there are embeddings from $W(K,T)$ into some $K^T *^\lambda T$, and any $K *^\lambda T$ into $W(K,T)$. Houghton proved that if S is an inverse semigroup and ρ an idempotent-separating congruence on S, then S is embeddable in $W(\mathrm{Ker}\,\rho, S/\rho)$ By means of Szendrei's embedding we see that this implies Theorem 7.

Billhardt and Szittyai [29] have strengthened Theorem 7 and earlier work of Cowan [51] on varieties. They have proved that if ρ is an idempotent-separating congruence on an inverse semigroup S then S can be embedded in a λ-semidirect product of a *group* K by S/ρ where K belongs to the variety of groups generated by the idempotent classes of ρ.

Congruences on (inverse) semigroups having the property that each congruence class contains a maximum element were considered by O'Carroll [287], [290].

The work on split Billhardt congruences is based entirely on [26].

Inverse semigroups in which μ has a transversal which is an inverse subsemigroup were considered by Munn [267] and Meakin [248].

Section 5.4

The theory of ω-inverse semigroups played an important rôle in the early development of inverse semigroup theory. The starting point was Rees' paper [336]. On the face of it, this has nothing to do with inverse semigroups; it dealt with the structure of left cancellative monoids in which the principal right ideals form an ω-semilattice (in our terminology). Rees proves that the structure of such a semigroup is completely determined by its group of units and an endomorphism of that group.

The connection between certain left (or right) cancellative monoids and what we now know as bisimple inverse monoids was established by Clifford [47]. He showed that right cancellative monoids in which the intersection of two principal left ideals is again a principal left ideal could be used to construct bisimple inverse monoids, and that, conversely, every bisimple inverse monoid arises in this way. This work is discussed and generalised in Chapter 10.

Using the dual of Clifford's results, it is apparent that the left cancellative monoids which Rees considered can be used to construct a class of bisimple inverse monoids: in fact, the bisimple ω-semigroups. This raises the problem of how to modify Rees' structure theory to describe these semigroups. This was achieved in Reilly's 1966 paper on the structure of bisimple ω-semigroups [338]. It was the first successful description of a class of semigroups which was not primitive.

Reilly's work used the following construction. Let S be an inverse monoid and let θ: $S \to U(S)$ be an endomorphism. The set $BR(S,\theta) = \mathbb{N} \times S \times \mathbb{N}$ is endowed with the binary operation

$$(m,a,n)(p,b,q) = (m + (p \dot{-} n), \theta^{p \dot{-} n}(a)\theta^{n \dot{-} p}(b), q + (n \dot{-} p)).$$

Then $BR(S,\theta)$ is a simple inverse monoid called the *Bruck–Reilly extension of S determined by θ*. It can be shown that $BR(S,\theta)$ is bisimple if, and only if, S is bisimple. Semigroups of the form $BR(S,\theta)$ are called *Bruck–Reilly semigroups.* Reilly proved that an inverse semigroup is a Reilly semigroup if, and only if, it is isomorphic to a Bruck-Reilly semigroup $BR(G,\theta)$ where G is a group.

Reilly's work opened up the possibility of describing all ω-semigroups, something which was achieved independently by Kochin [171] for simple ω-semigroups, and Munn [265] for arbitrary ω-semigroups.

It was natural to try to generalise this success to broader classes of semigroups. The property of ω-semigroups that was singled out as significant was that of uniformity; a semilattice is *uniform* if any two principal ideals are isomorphic. It turns out that such semilattices are precisely the semilattices of idempotents of bisimple inverse semigroups [263], [267]. The theory of uniform semilattices was pursued by Hickey [131], [132].

Petrich [312] devotes a chapter to ω-semigroups, Howie [148] describes the structure theorems for bisimple and simple ω-semigroups, and Grillet [109] derives the structure theorem for bisimple inverse ω-semigroups. For this reason, I have preferred to show how the theory of ω-semigroups is the result of a more general idea: the fact that the maximum idempotent-separating congruence is a split Billhardt congruence. However, let me briefly indicate the connection between this approach and the classical approach in the case of Reilly semigroups. Let S be a Reilly semigroup. Then by Theorem 9, there is an isomorphism from S to a semigroup of the form $Z \bowtie P_1$ where Z is a Clifford semigroup which acts by endomorphisms on the bicyclic monoid P_1. One can define an isomorphism Φ: $Z \bowtie P_1 \to BR(G,\theta)$ for a suitable G and θ constructed from $Z \bowtie P_1$.

The E-unitary Reilly semigroups are characterised by the associated group endomorphism being injective. It is interesting to note that groups equipped with such injective endomorphisms are useful in the investigation of self-similar tilings [91].

The analysis we have carried out for Reilly semigroups can also be generalised to simple ω-semigroups; they are isomorphic to Bruck–Reilly semigroups over finite chains of groups. The details may be found in [26].

Chapter 6

Free inverse semigroups

We begin by showing that free inverse semigroups exist; this is relatively straightforward and poses few problems. The next step, however, is more complex: the solution of the word problem. There are two basic approaches to this. The first is in the spirit of group theory: certain strings of the free inverse semigroup are singled out as normal forms; in addition, there is an algorithm which transforms each string into a (unique) normal form. This solution of the word problem leads rapidly to a structural description of free inverse semigroups which enables us to prove that they are E-unitary, combinatorial and completely semisimple.

The second approach to the word problem is completely different and rather unexpected: from every string in the free inverse semigroup we can construct a tree with two distinguished vertices, called a Munn tree. Two strings represent the same element in the free inverse semigroup precisely when their associated Munn trees are equal. Munn trees are easy to construct, and all the usual features of an inverse semigroup, such as the multiplication, the formation of inverses, and the natural partial order, have natural, tree-theoretic interpretations.

Although both approaches are valid, Munn's seems to provide the correct perspective on the word problem, and it is the basis of all deeper work in combinatorial inverse semigroup theory.

6.1 The existence of free inverse semigroups

Let X be a non-empty set. An inverse semigroup $FIS(X)$, equipped with a function ι: $X \to FIS(X)$, is said to be a *free inverse semigroup on* X if for every inverse semigroup S and function κ: $X \to S$ there exists a unique homomorphism θ: $FIS(X) \to S$ such that $\theta\iota = \kappa$. We may likewise define 'free inverse monoids', 'free inverse semigroups with zero' and 'free inverse monoids with zero'. We shall concentrate on 'free inverse semigroups' because the other types of free inverse semigroup may be obtained by adjoining an identity or a zero as appropriate.

Free semigroups with involution

To show that free inverse semigroups exist, we must first construct free semigroups with involution. Let X be a non-empty set. A semigroup with involution $FSI(X)$, equipped with a function $\iota: X \to FIS(X)$, is said to be a *free semigroup with involution on* X if for every semigroup with involution S and function $\kappa: X \to S$ there exists a unique involution homomorphism $\theta: FIS(X) \to S$ such that $\theta\iota = \kappa$. The construction of this semigroup is straightforward, as we now show.

Denote by X^{-1} a set disjoint from and bijective with the set X; the element of X^{-1} in bijective correspondence with $x \in X$ is denoted by x^{-1}. Form the free semigroup $FSI(X) = (X \cup X^{-1})^{\dagger}$, and define a unary operation on this semigroup as follows: for each $y \in X \cup X^{-1}$ put $y^{-1} = x^{-1}$ if $y = x \in X$ and put $y^{-1} = x$ if $y = x^{-1} \in X^{-1}$. If $y_1 \dots y_n \in (X \cup X)^{\dagger}$ then define $(y_1 \dots y_n)^{-1} = y_n^{-1} \dots y_1^{-1}$. This unary operation turns $FSI(X)$ into a semigroup with involution. Let $\iota': X \to FSI(X)$ be the insertion of the elements of X into $FSI(X)$.

Theorem 1 *The semigroup $(FSI(X), \iota')$, with the unary operation defined above, is the free semigroup with involution on X.*

Proof Let $\kappa: X \to T$ be any function to a semigroup with involution T. Define $\theta: X \cup X^{-1} \to T$ by

$$\theta(x) = \kappa(x) \text{ and } \theta(x^{-1}) = \kappa(x)^{-1}.$$

Because $FSI(X)$ is a free semigroup on the set $X \cup X^{-1}$, the function θ can be extended to a homomorphism $\theta: FSI(X) \to T$. It is straightforward to check that θ preserves the involution. Clearly, θ is the unique involution homomorphism such that $\theta\iota' = \kappa$. ■

Free inverse semigroups

The free inverse semigroup on X is constructed from $FSI(X)$ by factoring out by means of an appropriate congruence. To see what this congruence must be, we need the following alternative characterisation of inverse semigroups. Inverse semigroups can be viewed as algebras with one binary operation, multiplication, and one unary operation, inversion, satisfying associativity, the axioms (IS1) and (IS2) from Section 1.4, and the following two axioms:

(IS3) $s = ss^{-1}s$.

(IS4) $ss^{-1}tt^{-1} = tt^{-1}ss^{-1}$.

It now follows from universal algebra that free inverse semigroups exist, but we show this explictly.

The binary relation

$$\rho = \{(uu^{-1}u, u)\colon u \in FSI(X)\} \cup \{(uu^{-1}vv^{-1}, vv^{-1}uu^{-1})\colon u, v \in FSI(X)\}$$

defined on $FSI(X)$ generates a congruence $\rho^\sharp$ on $FSI(X)$. Put

$$FIS(X) = FSI(X)/\rho^\sharp$$

and let $\iota\colon X \to FIS(X)$ be the insertion of the set X given by $\iota = \rho^\natural\iota'$. If u and v are the same string we shall write $u = v$; if u and v belong to the same $\rho^\sharp$-class then we shall write $u \equiv v$ and say that they are 'equivalent'.

Theorem 2 $(FIS(X), \iota)$, *as defined above, is the free inverse semigroup on* X.

Proof We show first that $FIS(X)$ is an inverse semigroup. From the definition,

$$u \equiv uu^{-1}u \text{ and } u^{-1} \equiv u^{-1}uu^{-1}$$

for every string $u \in FIS(X)$. Thus the semigroup is regular. To show it is inverse we prove that its idempotents commute. To do this we shall prove that every idempotent in $FIS(X)$ is equivalent to a product of elements of the form uu^{-1} where $u \in FSI(X)$; commutativity of the idempotents is then clear. Let z be an idempotent. Then $z \equiv z^2 = zz$. By assumption, $z \equiv zz^{-1}z$ so that

$$z \equiv (zz^{-1}z)(zz^{-1}z).$$

Rebracketing this product we obtain

$$z \equiv z(z^{-1}z)(zz^{-1})z.$$

By assumption,

$$(z^{-1}z)(zz^{-1}) \equiv (zz^{-1})(z^{-1}z).$$

Thus

$$z \equiv z(zz^{-1})(z^{-1}z)z.$$

But z is an idempotent and so $z \equiv (zz^{-1})(z^{-1}z)$.

We may now prove that $FIS(X)$ is the free inverse semigroup on the set X. Let $\kappa\colon X \to T$ be any function to an inverse semigroup T. Of course,

T is a semigroup with involution and so by Theorem 1 there is a unique involution homomorphism θ: $FSI(X) \to T$ such that $\theta\iota' = \kappa$. By definition, $(uu^{-1}u, u) \in \rho$ for every $u \in FSI(X)$. But $\theta(uu^{-1}u) = \theta(u)\theta(u)^{-1}\theta(u)$, since θ is an involution homomorphism, and $\theta(u)\theta(u)^{-1}\theta(u) = \theta(u)$ since T is inverse. Thus $(uu^{-1}u, u) \in \ker\theta$. Similarly, $(uu^{-1}vv^{-1}, vv^{-1}uu^{-1}) \in \ker\theta$ for all $u, v \in FSI(X)$. Hence $\rho^\sharp \subseteq \ker\theta$, and so by Theorem 2.3.2 there is a unique homomorphism ϕ: $FIS(X) \to T$ such that $\phi(\rho^\sharp)^\natural = \theta$. Thus $\phi\iota = \kappa$. Suppose now that ϕ': $FIS(X) \to T$ is a homomorphism such that $\phi'\iota = \kappa$. The set $\iota(X)$ generates $FIS(X)$ as an inverse semigroup and ϕ and ϕ' agree on all elements of the form $\iota(x)$ where $x \in X$. Thus $\phi = \phi'$. ■

The free inverse monoid $FIM(X)$ on X is just $FIS(X)^1$ and the free inverse semigroup with zero is $FIS(X)^0$.

Presentations of inverse semigroups

The existence of free inverse semigroups enables us to define the notion of an inverse semigroup presentation.

An *inverse semigroup presentation* is a pair (X, ρ) where ρ is a relation on $FIS(X)$. The inverse semigroup $FIS(X)/\rho^\sharp$ is said to be *presented* by the *generators* X and the *relations* ρ and is denoted by $S = \text{Inv}\langle X\colon \rho\rangle$. If both X and ρ are finite we say that S is *finitely presented.* An inverse semigroup S is said to be *monogenic* if it has an inverse semigroup presentation with one generator.

In a similar way, we can define 'inverse monoid presentations', 'presentations of inverse semigroups with zero', and 'presentations of inverse semigroups with both a zero and an identity'. There is no separate notation for such presentations; the presence of a zero or a one in the presentation signals that the appropriate type of inverse semigroup presentation is intended.

6.2 Solving the word problem

We have shown that free inverse semigroups exist, but the construction of Theorem 6.1.2 does not tell us anything about its properties. The difficulty is that quite distinct strings may represent the same element in the free inverse semigroup; we give an example.

Let $X = \{a, b\}$, and consider the two strings

$$u = bb^{-1}aa^{-1}baaa^{-1}a^{-1}bb^{-1}a \text{ and } v = aa^{-1}bbb^{-1}aaa^{-1}$$

in the free semigroup with involution on X. We claim that they both represent the same element of the free inverse semigroup $FIS(\{a, b\})$. To see this, we argue as follows. Firstly,

$$u = (bb^{-1}aa^{-1})baaa^{-1}a^{-1}bb^{-1}a \equiv (aa^{-1}bb^{-1})baaa^{-1}a^{-1}bb^{-1}a$$

since $bb^{-1}aa^{-1} \equiv aa^{-1}bb^{-1}$. But

$$u \equiv aa^{-1}(bb^{-1}b)aaa^{-1}a^{-1}bb^{-1}a \equiv aa^{-1}baaa^{-1}a^{-1}bb^{-1}a$$

since $b \equiv bb^{-1}b$. Now

$$u \equiv aa^{-1}b[((aa)(aa)^{-1})bb^{-1}]a$$

and $(aa)(aa)^{-1}bb^{-1} \equiv bb^{-1}(aa)(aa)^{-1}$. Thus

$$u \equiv aa^{-1}bbb^{-1}aaa^{-1}a^{-1}a.$$

Also $aa^{-1}a^{-1}a \equiv a^{-1}aaa^{-1}$ and so

$$u \equiv aa^{-1}bbb^{-1}aa^{-1}aaa^{-1}.$$

But $a \equiv aa^{-1}a$, so that

$$u \equiv aa^{-1}bbb^{-1}aaa^{-1} = v.$$

Recall that the free group $FG(X)$ on X can be regarded as the set of all reduced strings equipped with the product given by $x \cdot y = \mathrm{red}(xy)$ where $x, y \in FG(X)$ (Section 2.1). To solve the word problem for free inverse semigroups, we shall invoke the solution of the word problem for groups. The reason for this stems from the following result.

Proposition 1 *Let X be a non-empty set. Then $FIS(X)/\sigma$ is the free group on X, and*

$$(u, v) \in \sigma \Leftrightarrow \mathrm{red}(u) = \mathrm{red}(v)$$

for any $u, v \in FIS(X)$.

Proof Define θ: $FIS(X) \to FG(X)$ by $\theta(w) = \mathrm{red}(w)$. This is a well-defined function since for every ordered pair $(u_1, u_2) \in \rho$, the relation used in the construction of $FIS(X)$, it is trivial that $\theta(u_1) = \theta(u_2)$. The function θ is a homomorphism because from the solution of the word problem in free groups we know that $\mathrm{red}(\mathrm{red}(u)\mathrm{red}(v)) = \mathrm{red}(uv)$. It follows that $\sigma \subseteq \ker\theta$ by Theorem 2.4.1. Now observe that for any two strings $v_1 = pxx^{-1}q$ and $v_2 = pq$ in the free inverse semigroup, we have that $\sigma(v_1) = \sigma(v_2)$, because xx^{-1} is an idempotent. Furthermore, $\mathrm{red}(v_1) = \mathrm{red}(v_2)$. Thus if $u, v \in FIS(X)$ are such that $\mathrm{red}(u) = \mathrm{red}(v)$ then $\sigma(u) = \sigma(v)$. Hence $\sigma = \ker\theta$. ■

Scheiblich normal form

We describe an algorithm which will transform every string in $FIS(X)$ into an equivalent string having a special form, which we call Scheiblich normal form. We shall later prove that two strings in Scheiblich normal form are equivalent precisely when they are equal in a suitably well-defined sense. Scheiblich normal forms are the appropriate generalisation to free inverse semigroups of the reduced strings used to solve the word problem in free groups.

The algorithm centres on strings of the form zz^{-1}; such strings represent the identity element in free groups but represent idempotents in free inverse semigroups. Although such strings cannot be erased in free inverse semigroups, they can be moved around using Lemma 1.4.2. The basic idea behind the algorithm we shall describe is the following: given any string w_0 from $FIS(X)$ use Lemma 1.4.2 to move certain 'obvious' idempotent substrings of w_0 to the front. The resulting string will be equivalent to w_0, but will consist of a product of idempotents followed by a non-idempotent element. It will turn out that the idempotents and the non-idempotent element will be quite special.

We now describe the algorithm. Let w_0 be a string in $FIS(X)$. If w_0 is reduced, then $w_0 \equiv (w_0 w_0^{-1})w_0$ and the algorithm terminates. If w_0 is not reduced, then let u be the longest prefix of w_0 which is reduced as it stands (i.e. $\text{red}(u) = u$). Then $w_0 = uv$ where both u and v are non-empty strings. Let z be the longest suffix of u such that z^{-1} is a prefix of v; the string z is non-empty because of the way we chose u. Then $u = u_1 z$ and $v = z^{-1}v_1$ for some strings u_1 and v_1. Thus $w_0 = uv = u_1 zz^{-1}v_1$. But $u_1 zz^{-1}v_1 \equiv (u_1 zz^{-1}u_1^{-1})u_1 v_1$. Hence $w_0 \equiv (uu^{-1})u_1 v_1$. Put $w_1 = u_1 v_1$. Clearly, the length of w_1 is strictly less than the length of w_0. If w_1 is reduced, then $w_0 \equiv (uu^{-1})(w_1 w_1^{-1})w_1$ and the algorithm terminates. Otherwise we may repeat the above procedure with w_1 instead of w_0. Continuing in this way, we obtain the following result.

Proposition 2 *Let w_0 be a string in $FIS(X)$. Then*

$$w_0 \equiv (a_1 a_1^{-1}) \dots (a_n a_n^{-1})a,$$

where the a_i and a are reduced, and $a = a_i$ for some i. ∎

Our normal form will be constructed from the string of Proposition 2 by deleting certain redundant strings. To do this we shall use the following result.

Lemma 3 *Let a and b be reduced strings. If b is a prefix of a then $aa^{-1} \equiv aa^{-1}bb^{-1}$ in $FIS(X)$.*

Proof By assumption $a = bu$ for some string u. But then $aa^{-1}bb^{-1} = (bu)(bu)^{-1}bb^{-1} \equiv (bu)(bu)^{-1} = aa^{-1}$. ∎

We now apply Lemma 3 to the string $(a_1a_1^{-1})\dots(a_na_n^{-1})a$ obtained in Proposition 2. Omit from the set $\{a_1,\dots,a_n\}$ any strings which are proper prefixes of other strings in the set. In this way, we obtain a set $\{b_1,\dots,b_m\}$ where no string is a proper prefix of another. Then $w_0 \equiv (b_1b_1^{-1})\dots(b_mb_m^{-1})a$ by Lemma 3. By assumption, $a = a_i$ for some i and consequently a will be a prefix of one of the b_j's.

The properties of the resulting string are the basis of the following definition. A string $w = (a_1a_1^{-1})\dots(a_na_n^{-1})a$ in $FSI(X)$ is said to be in *Scheiblich normal form* if the following conditions hold:

(SNF1) The strings a_i and a are all reduced.

(SNF2) No string a_i in the set $\{a_1,\dots,a_n\}$ is a proper prefix of any other string in the set.

(SNF3) The string a is a prefix of one of the a_i.

The strings $\{a_1,\dots,a_n\}$ are called the *components* of w, and a is called the *root* of w. We have therefore proved our first, basic result.

Theorem 4 *Every element of $FIS(X)$ is equivalent to a string in Scheiblich normal form.* ■

The algorithm for computing a Scheiblich normal form is straightforward, as the following example shows. Let $w_0 = a^2a^{-3}abb^{-1}ab^{-1}bcaa^{-1}cc^{-1}$. Then applying the algorithm we obtain the following; in each case, a_i is the longest prefix of w_{i-1} which is reduced as it stands.

1. $a_1 = aa, \quad w_1 = a^{-1}abb^{-1}ab^{-1}bcaa^{-1}cc^{-1}$,
2. $a_2 = a^{-1}, \quad w_2 = bb^{-1}ab^{-1}bcaa^{-1}cc^{-1}$,
3. $a_3 = b, \quad w_3 = ab^{-1}bcaa^{-1}cc^{-1}$,
4. $a_4 = ab^{-1}, \quad w_4 = acaa^{-1}cc^{-1}$,
5. $a_5 = aca, \quad w_5 = accc^{-1}$,
6. $a_6 = acc, \quad w_6 = ac$.

Thus the components of w_0 are

$$\{aa, a^{-1}, b, ab^{-1}, aca, acc\}$$

and the root of w_0 is ac.

It now remains to decide when two strings in Scheiblich normal form are equivalent. The clue to solving this problem is provided by the following two lemmas.

Lemma 5 *Let* $u = (a_1a_1^{-1})\dots(a_ma_m^{-1})a$ *and* $v = (b_1b_1^{-1})\dots(b_nb_n^{-1})b$ *be two Scheiblich normal forms. Then* $u \equiv v$ *implies* $a = b$.

Proof Suppose that $u \equiv v$. Then $\sigma(u) = \sigma(v)$. Thus $\mathrm{red}(u) = \mathrm{red}(v)$ by Proposition 1. But $\mathrm{red}(u) = a$ and $\mathrm{red}(v) = b$, and so $a = b$. ■

Lemma 6 *The Scheiblich normal form* $w = (a_1a_1^{-1})\dots(a_na_n^{-1})a$ *represents an idempotent if, and only if,* a *is empty.*

Proof If a is empty then w is a product of idempotents, and so w is an idempotent. Conversely, suppose that w is an idempotent. Then $\sigma(w)$ is the identity of the group $FIS(X)/\sigma$. But $\sigma(ww^{-1})$ is also the identity of this group. Thus $\sigma(w) = \sigma(ww^{-1})$, and so by Proposition 1 we have that $\mathrm{red}(w) = 1$. It follows that a is empty. ■

The above two lemmas suggest that the solution of the word problem will follow from a description of the idempotents of the free inverse semigroup.

The prefix order on reduced strings

To obtain further insight into the idempotent structure of free inverse semigroups we introduce some ideas which are suggested by Lemma 3.

Define a partial ordering $\leq_p$ on $FG(X)$ by $b \leq_p a$ if b is a prefix of a; this means that $a = bu$ for some string u where bu is reduced as it stands (i.e. $\mathrm{red}(bu) = bu$). We call this the *prefix order* on reduced strings. If $b \leq_p a$ and $b \neq a$ then we say that b is a *proper prefix* of a. Lemma 3 states that

$$b \leq_p a \Rightarrow aa^{-1} \leq bb^{-1}.$$

A pair of elements $x, y \in FG(X)$ are said to be *incomparable* with respect to the prefix order if neither $x \leq_p y$ nor $y \leq_p x$. A nonempty subset A of $FG(X)$ is said to be *incomparable* if any two distinct elements of A are incomparable. If $A \subseteq FG(X)$ is a finite, non-empty subset, define $\max(A)$ to be the set of maximal elements of A with respect to the prefix order. The subsets A which satisfy $\max(A) = A$ are just the finite incomparable subsets of $FG(X)$. Denote by $\mathcal{I}$ the set of all non-empty finite incomparable subsets of $FG(X) \setminus \{1\}$.

If $A \subseteq FG(X)$ is any non-empty subset then we denote by $[A]^{\downarrow}$ the order ideal of $(FG(X), \leq_p)$ generated by A. Let $\mathcal{Y}$ be the set of all finite, non-empty order ideals of $FG(X)$ which contain at least one non-identity element.

Proposition 7 *There is a bijection from $\mathcal{Y}$ to $\mathcal{I}$.*

Proof If $A \in \mathcal{Y}$ then $\max(A)$ is an element of $\mathcal{I}$. Thus $A \mapsto \max(A)$ is a well-defined function from $\mathcal{Y}$ to $\mathcal{I}$. If $B \in \mathcal{I}$ then $[B]^{\downarrow}$ is a finite non-empty order ideal which contains at least one non-identity element. Thus $B \mapsto [B]^{\downarrow}$ is a well-defined function from $\mathcal{I}$ to $\mathcal{Y}$. We now show that these functions are mutually inverse.

Let $A \in \mathcal{Y}$. Since $\max(A) \subseteq A$, we have that $[\max(A)]^{\downarrow} \subseteq A$. To prove the reverse inclusion, let $a \in A$. Then $a \leq_p a'$ for some maximal element a' of A. Hence $A \subseteq [\max(A)]^{\downarrow}$, and so $A = [\max(A)]^{\downarrow}$.

Let $B \in \mathcal{I}$. Clearly $B \subseteq [B]^{\downarrow}$. We show first that every element of B is maximal in the set $[B]^{\downarrow}$. Suppose $b \in B$ is not maximal in $[B]^{\downarrow}$. Then there exists $b' \in [B]^{\downarrow}$ such that $b <_p b'$. By definition $b' \leq_p b''$ for some $b'' \in B$. But this implies $b, b'' \in B$ are distinct comparable elements of B, which is a contradiction. Thus $B \subseteq \max([B]^{\downarrow})$. Conversely, let $a \in \max([B]^{\downarrow})$. By definition $a \leq_p b$ for some $b \in B$. But a is maximal, and so $a = b \in B$. Thus $B = \max([B]^{\downarrow})$. ■

Let $u = (a_1 a_1^{-1}) \ldots (a_n a_n^{-1})$ be an idempotent in Scheiblich normal form. Then $\{a_1, \ldots, a_n\}$ is a set of incomparable elements of $FG(X) \setminus \{1\}$. This set generates an order ideal $[\{a_1, \ldots, a_n\}]^{\downarrow}$ which is an element of $\mathcal{Y}$. The set of components of u is the set $\max([\{a_1, \ldots, a_n\}]^{\downarrow})$ by Proposition 7. Conversely, if $A \in \mathcal{Y}$, then $\max(A)$ is the set of components of an idempotent in Scheiblich normal form.

Thus it appears that idempotents in Scheiblich normal form correspond to finite non-empty order ideals of $FG(X)$ which contain at least one non-identity element.

This correspondence also extends to products of idempotents. In what follows we shall write $[a_1, \ldots, a_n]^{\downarrow}$ instead of $[\{a_1, \ldots, a_n\}]^{\downarrow}$.

Proposition 8 *Let $u = (a_1 a_1^{-1}) \ldots (a_m a_m^{-1})$ and $v = (b_1 b_1^{-1}) \ldots (b_n b_n^{-1})$ be two idempotents in Scheiblich normal form. Then a set of components of uv is the set* $\max([a_1, \ldots, a_m]^{\downarrow} \cup [b_1, \ldots, b_n]^{\downarrow})$.

Proof The product uv is equivalent to the concatenation of the Scheiblich normal form u and the Scheiblich normal form v. This fails to be in Scheiblich normal form only if any of the a_i or b_j is a proper prefix of any of the other strings. Thus the set

$$\max([a_1, \ldots, a_m]^{\downarrow} \cup [b_1, \ldots, b_n]^{\downarrow})$$

is a set of components of uv. ■

We have to say *a* set of components because we have yet to prove that the components are unique.

Since the union of order ideals is an order ideal, $(\mathcal{Y}, \cup)$ is a semilattice. The order in $(\mathcal{Y}, \cup)$ is given by $A \leq B$ if, and only if, $B \subseteq A$.

The main conclusion of these deliberations is the following: Proposition 8 suggests that the semilattice of idempotents of the free inverse semigroup on X is isomorphic to $(\mathcal{Y}, \cup)$. We shall prove this, and much more, in what follows.

Equivalence of Scheiblich normal forms

We now proceed to determine when two Scheiblich normal forms are equivalent. First, a remark about notation. Two partially ordered sets will figure in what follows: the partially ordered set $(FG(X), \leq_p)$ and the partially ordered set $(\mathcal{Y}, \cup)$. If $A \subseteq FG(X)$ then the order ideal it generates is denoted $[A]^{\downarrow}$, whereas if $I \in \mathcal{Y}$ then $[I]$ is the principal order ideal which it generates.

Let $T_{\mathcal{Y}}$ be the Munn semigroup of $(\mathcal{Y}, \cup)$. For each $y \in X \cup X^{-1}$ the set $\{1, y\}$ is an order ideal of $(FG(X), \leq_p)$ and so $\{1, y\} \in \mathcal{Y}$. The principal order ideal $[\{1, y\}]$ in $(\mathcal{Y}, \cup)$ can also be described by

$$[\{1, y\}] = \{I \in \mathcal{Y}: y \in I\}.$$

We shall define a function from X to $T_{\mathcal{Y}}$. The following observation tells us how to do this. Let $x \in X$. Then the idempotent $x^{-1}x$ corresponds to the principal order ideal $[\{1, x^{-1}\}]$ in $\mathcal{Y}$ and the idempotent xx^{-1} corresponds to the principal order ideal $[\{1, x\}]$ in $\mathcal{Y}$. If $e \leq x^{-1}x$ in the free inverse semigroup then $xex^{-1} \leq xx^{-1}$. We now make the following definition. For each $x \in X$ define a function ϕ_x from $[\{1, x^{-1}\}]$ to $[\{1, x\}]$ by

$$\phi_x(I) = x \cdot I = \{\text{red}(xw): w \in I\}.$$

The following lemma shows, in particular, that this function is well-defined.

Lemma 9 *Let $I \in \mathcal{Y}$, and let $a \in FG(X)$. Then $a \cdot I \in \mathcal{Y}$ if, and only if, $a^{-1} \in I$.*

Proof Suppose first that $a \cdot I \in \mathcal{Y}$. Then, in particular, $1 \in a \cdot I$. Thus there exists $w \in I$ such that $1 = a \cdot w$. But w is a reduced string and so $w = a^{-1} \in I$.

To prove the converse, let $a^{-1} \in I$. We shall prove that $a \cdot I \in \mathcal{Y}$. Clearly $1 \in a \cdot I$, and $a \in a \cdot I$ since $1 \in I$. Thus it only remains to show that $a \cdot I$ is an order ideal of $FG(X)$. Let $b \in a \cdot I$, and let c be a prefix of b. We show that $c \in a \cdot I$. By definition $b = a \cdot u$ for some $u \in I$. Let v be the longest suffix of a such that v^{-1} is a prefix of u. Then $a = a'v$ and $u = v^{-1}u'$ for some strings a' and u'. Thus $a \cdot u = a'u'$. Clearly, c is a prefix of $a'u'$. There are two cases to consider. Firstly, suppose that c is a prefix of a'. Then $a' = cp$ for some string p and so $a = a'v = cpv$. Thus $v^{-1}p^{-1}$ is a prefix of a^{-1} and so is an element of I since $a^{-1} \in I$. But then $a \cdot (v^{-1}p^{-1}) = a'v(v^{-1}p^{-1}) = cpv(v^{-1}p^{-1}) = c$. Hence $c \in a \cdot I$. Secondly, suppose that $c = a'p$ where p is a prefix of u'. Then $v^{-1}p$ is a prefix of u and $u \in I$, so that $v^{-1}p \in I$. Thus $a \cdot (v^{-1}p) = (a'v) \cdot (v^{-1}p) = a'p = c$. Hence $c \in a \cdot I$. ∎

The following result will be used to determine when two Scheiblich normal forms are equivalent.

Proposition 10 *There is a homomorphism*

$$\phi \colon FIS(X) \to T_{\mathcal{Y}}$$

such that $\phi(x) = \phi_x$ *for each* $x \in X$. *Furthermore, for each idempotent* $u = (a_1 a_1^{-1}) \dots (a_n a_n^{-1})$ *in Scheiblich normal form, the domain of* $\phi(u)$ *is* $[[a_1, \dots, a_n]^{\downarrow}]$.

Proof We have to show first that ϕ_x is a well-defined isomorphism of semilattices. By Lemma 9, we know that ϕ_x is a well-defined function. The semilattice operation in $\mathcal{Y}$ is union, which is clearly preserved by ϕ_x. That ϕ_x is injective is straightforward from the definitions. To prove that it is surjective, let $J \in [\{1, x\}]$. Then $I = x^{-1} \cdot J \in [\{1, x^{-1}\}]$ by Lemma 9, and $\phi_x(I) = J$.

We thus have a function $\phi \colon X \to T_{\mathcal{Y}}$. From the definition of the free inverse semigroup $FIS(X)$ on the set X, we may extend this function to a homomorphism $\phi \colon FIS(X) \to T_{\mathcal{Y}}$, where $\phi(x) = \phi_x$ for $x \in X$.

To prove the final assertion, we prove first the following special case. Let a be a non-empty reduced string. We prove that the domain of $\phi(aa^{-1})$ is $[[a]^{\downarrow}]$. The result is clearly true when $a \in X \cup X^{-1}$. So assume the result is true for all non-empty reduced strings of length n or less. Let a be a reduced string of length $n+1$. Then $a = by$ where $y \in X \cup X^{-1}$ and b is a reduced string of length n. It is straightforward to check that $\operatorname{dom} \phi(aa^{-1}) = \phi(b)([\{1, y\}])$. We prove that $\phi(b)([\{1, y\}]) = [[a]^{\downarrow}]$. Let $J \in \phi(b)([\{1, y\}])$. Then $J = b \cdot I$ where $I \in [\{1, y\}]$. Thus $y \in I$, and so $a = by \in J$. Hence $J \in [[a]^{\downarrow}]$. On the other hand, if $J \in [[a]^{\downarrow}]$ then $a \in J$ and so $b \in J$ because $b \leq_p a$. Thus $I = b^{-1} \cdot J \in \mathcal{Y}$ by Lemma 9, and since $b = ay \in J$, we have that $y \in I$. It follows that $I \in [\{1, y\}]$. But then $\phi(b)(I) = b \cdot I = J$, and so $J \in \phi(b)([\{1, y\}])$ Hence for all non-empty reduced strings a we have shown that the domain of $\phi(aa^{-1})$ is $[[a]^{\downarrow}]$.

Now let $u = (a_1 a_1^{-1}) \dots (a_n a_n^{-1})$ be an idempotent in Scheiblich normal form. Then

$$\phi(u) = \phi(a_1 a_1^{-1}) \dots \phi(a_n a_n^{-1}).$$

We have just proved that the domain of $\phi(a_i a_i^{-1})$ is $[[a_i]^{\downarrow}]$. Thus the domain of $\phi(u)$ is the union of the domains of the $\phi(a_i a_i^{-1})$, since the product of elements in $\mathcal{Y}$ is their union. Hence the domain of $\phi(u)$ is $[[a_1, \ldots, a_n]^{\downarrow}]$. ■

The following result together with Theorem 4 solves the word problem for free inverse semigroups.

Theorem 11 *Let* $u = (a_1 a_1^{-1}) \ldots (a_m a_m^{-1})a$ *and* $v = (b_1 b_1^{-1}) \ldots (b_n b_n^{-1})b$ *be two Scheiblich normal forms in the free inverse semigroup* $FIS(X)$. *Then* $u \equiv v$ *if, and only if,* $\{a_1, \ldots, a_m\} = \{b_1, \ldots b_n\}$ *and* $a = b$.

Proof Let $u \equiv v$. Then $a = b$ by Lemma 5. Thus

$$(a_1 a_1^{-1}) \ldots (a_m a_m^{-1}) a a^{-1} \equiv (b_1 b_1^{-1}) \ldots (b_n b_n^{-1}) a a^{-1}.$$

By the definition of Scheiblich normal form, a is a prefix of one of the a_i and $b = a$ is a prefix of one of the b_j. Thus

$$u' = (a_1 a_1^{-1}) \ldots (a_m a_m^{-1}) \equiv (b_1 b_1^{-1}) \ldots (b_n b_n^{-1}) = v'$$

by Lemma 3. Now $\phi(u') = \phi(v')$. Thus by Proposition 10, we have that

$$\{a_1, \ldots, a_m\} = \{b_1, \ldots, b_n\}.$$

We have therefore shown that $u \equiv v$ implies that the components of u are the same as the components of v and the root of u is the same as the root of v. The converse is clear. ■

Our results on the possible semilattice structure of the free inverse semigroup can now be made precise.

Theorem 12 *The semilattice of idempotents of the free inverse semigroup* $FIS(X)$ *is isomorphic to* $(\mathcal{Y}, \cup)$.

Proof By Lemma 6 and Theorem 11, the idempotents of $FIS(X)$ are in bijective correspondence with the finite non-empty incomparable subsets of $FG(X) \setminus \{1\}$, and thus by Proposition 7 with the elements of $\mathcal{Y}$. By Proposition 8 this correspondence is an isomorphism of semilattices. ■

We illustrate Theorem 11 by returning to the example we discussed at the beginning of this section. Let $X = \{a, b\}$, and consider the two strings

$$u_0 = u = bb^{-1}aa^{-1}baaa^{-1}a^{-1}bb^{-1}a \text{ and } v_0 = v = aa^{-1}bbb^{-1}aaa^{-1}.$$

We first compute the components and root of u:

1. $a_1 = b, \quad u_1 = aa^{-1}baaa^{-1}a^{-1}bb^{-1}a,$
2. $a_2 = a, \quad u_2 = baaa^{-1}a^{-1}bb^{-1}a,$
3. $a_3 = baa, \quad u_3 = bbb^{-1}a,$
4. $a_4 = bb, \quad u_4 = ba.$

Thus the set of components of u is $\{a, bb, baa\}$ and the root is ba. Now we compute the components and root of v:

1. $b_1 = a, \quad v_1 = bbb^{-1}aaa^{-1},$
2. $b_2 = bb, \quad v_2 = baaa^{-1},$
3. $b_3 = baa, \quad v_3 = ba.$

Thus the set of components of v is $\{a, bb, baa\}$ and the root is ba. We therefore deduce that u and v are equivalent.

Multiplication of Scheiblich normal forms

In order to obtain structural information about free inverse semigroups, we need to obtain a usable description of the product of two strings in Scheiblich normal form. First we prove a simple lemma.

Lemma 13 *Let $a \leq_p x$ and $b \leq_p y$ in $(FG(X), \leq_p)$. Then either* $\text{red}(ab) \leq_p x$ *or* $\text{red}(ab) \leq_p \text{red}(ay)$.

Proof Let u be the longest suffix of a such that u^{-1} is a prefix of b. There are two cases two consider. Suppose first that $u = b^{-1}$. Then $a = a_0 b^{-1}$ for some string a_0. Thus

$$\text{red}(ab) = a_0 \leq_p a \leq_p x.$$

Now suppose that $u \neq b^{-1}$. Then $a = a_0 u$ and $b = u^{-1} b_0$ where b_0 is non-empty. Thus $\text{red}(ab) = a_0 b_0$. By assumption $b \leq_p y$ and so $y = bv$, which is reduced as it stands. We claim that $a_0 b_0 v$ is reduced; this is because $a_0 b_0$ is reduced, and $b_0 v$ is reduced because bv is reduced, and finally b_0 is non-empty. Now $ay = abv$ and $\text{red}(abv) = a_0 b_0 v$, so that $\text{red}(ab) \leq_p \text{red}(ay)$. ■

The desired characterisation of the Scheiblich normal form of a product now follows.

Theorem 14 *Let $u = (a_1 a_1^{-1}) \ldots (a_m a_m^{-1}) a$ and $v = (b_1 b_1^{-1}) \ldots (b_n b_n^{-1}) b$ be two Scheiblich normal forms. Then the root of uv is $a \cdot b$ and the set of components of uv is*

$$\max(\{a_1, \ldots, a_m, a \cdot b_1, \ldots, a \cdot b_n\}).$$

Proof It is easy to check that

$$uv \equiv (a_1 a_1^{-1}) \dots (a_m a_m^{-1})(ab_1)(ab_1)^{-1} \dots (ab_n)(ab_n)^{-1} ab.$$

Next we show for each $i = 1, \dots, n$ that

$$(ab_i)(ab_i)^{-1} \equiv \text{red}(ab_i)(\text{red}(ab_i))^{-1} aa^{-1}.$$

If ab_i is not a reduced string, then we can write $a = cu$ and $b_i = u^{-1}d$ for some strings c, u, d such that $\text{red}(ab_i) = cd$. Thus

$$ab_i = c(uu^{-1})d \equiv cd(d^{-1}uu^{-1}d) \equiv \text{red}(ab_i) b_i^{-1} b_i.$$

Hence

$$(ab_i)(ab_i)^{-1} \equiv \text{red}(ab_i)(a_i b_i)^{-1}.$$

But

$$(ab_i)^{-1} = b_i^{-1} a^{-1} = d^{-1}uu^{-1}c^{-1} \equiv d^{-1}c^{-1}(cuu^{-1}c^{-1}) \equiv (\text{red}(ab_i))^{-1} aa^{-1}.$$

Thus we have proved the result.

Next we show that

$$ab \equiv aa^{-1}\text{red}(ab).$$

If ab is reduced then the result is clear. Otherwise, $a = cw$ and $b = w^{-1}d$ for some strings c, w, d. Then a simple calculation shows that

$$ab \equiv cww^{-1}c^{-1}cd \equiv aa^{-1}\text{red}(ab).$$

Hence

$$uv \equiv (a_1 a_1^{-1}) \dots (a_m a_m^{-1})(a \cdot b_1)(a \cdot b_1)^{-1} \dots (a \cdot b_n)(a \cdot b_n)^{-1}(a \cdot b)$$

since a is a prefix of one of the a_i, and consequently, by Lemma 3, aa^{-1} can be omitted from the product.

By assumption, $a \leq_p a_i$ for some i and $b \leq_p b_j$ for some j. Thus by Lemma 13, $\text{red}(ab) \leq_p a_i$ or $\text{red}(ab) \leq_p \text{red}(ab_j)$. Finally, the set of components is

$$\max(\{a_1, \dots, a_m, a \cdot b_1, \dots, a \cdot b_n\}).$$

■

6.3 The structure of free inverse semigroups

Now that we have solved the word problem for free inverse semigroups, we can obtain information about their structure. To do this we shall describe a model of free inverse semigroups derived from Scheiblich normal form and Theorem 6.2.14. Put

$$\mathcal{X} = \{g \cdot A \colon A \in \mathcal{Y} \text{ and } g \in FG(X)\}$$

and

$$P(FG(X), \mathcal{X}, \mathcal{Y}) = \{(A, g) \in \mathcal{Y} \times FG(X) \colon g \in A\}.$$

Define a product on $P(FG(X), \mathcal{X}, \mathcal{Y})$ by

$$(A, g)(B, h) = (A \cup g \cdot B, g \cdot h).$$

We use Lemma 6.2.9 to show that this is well-defined; $g^{-1} \cdot A$ is an order ideal because $g \in A$. Thus $g^{-1} \cdot A \cup B$ is an order ideal. Similarly, $g \cdot (g^{-1} \cdot A \cup B)$ is an order ideal because $g^{-1} \in g^{-1} \cdot A \cup B$. But $A \cup g \cdot B = g \cdot (g^{-1} \cdot A \cup B)$.

We now have the following representation of the free inverse semigroup $FIS(X)$. At the same time we prove that $P(FG(X), \mathcal{X}, \mathcal{Y})$ is inverse.

Theorem 1 *Let X be a non-empty set. Then the free inverse semigroup $FIS(X)$ is isomorphic to $P(FG(X), \mathcal{X}, \mathcal{Y})$.*

Proof Let $u \in FIS(S)$. By Theorems 6.2.4 and 6.2.11, we may assume that $u = (a_1 a_1^{-1}) \ldots (a_n a_n^{-1})a$ is in Scheiblich normal form. Define $\theta \colon FIS(X) \to P$ by $\theta(u) = ([a_1, \ldots, a_n]^{\downarrow}, a)$. It is clear that θ is well-defined. It is a homomorphism by Theorem 6.2.14. Injectivity of θ is an immediate consequence of Proposition 6.2.7 and Theorem 6.2.11, and surjectivity follows from Proposition 6.2.7. ■

The algebraic properties of the free inverse semigroup can now be very easily determined using the inverse semigroup $P(FG(X), \mathcal{X}, \mathcal{Y})$. The following results are straightfoward to prove.

Proposition 2 *In the inverse semigroup $P = P(FG(X), \mathcal{X}, \mathcal{Y})$ the following properties hold:*

(1) *The idempotents of P are the elements of the form $(A, 1)$.*

(2) $(A, g)^{-1} = (g^{-1} \cdot A, g^{-1})$.

(3) $(A, g) \leq (B, h) \Leftrightarrow B \subseteq A$ *and* $g = h$.

(4) $(A, g)\, \mathcal{L}\, (B, h) \Leftrightarrow g^{-1} \cdot A = h^{-1} \cdot B$.

(5) $(A, g) \mathcal{R} (B, h) \Leftrightarrow A = B$.

(6) $(A, g) \mathcal{D} (B, h) \Leftrightarrow a^{-1} \cdot A = B$ *for some* $a \in A$. ■

We may now prove that free inverse semigroups are, in particular, E-unitary.

Theorem 3 *The free inverse semigroup $FIS(X)$ is E-unitary, combinatorial and completely semisimple. Furthermore, each non-idempotent σ-class contains a maximum element.*

Proof We show first that $FIS(X)$ is E-unitary. Let $(A, 1) \leq (B, g)$, where $(A, 1)$ is an idempotent by Proposition 2(1). Then $g = 1$ by Proposition 2(3). Thus (B, g) is an idempotent.

To show that $FIS(X)$ is combinatorial let $(A, g) \mathcal{H} (B, h)$. Then $g^{-1} \cdot A = h^{-1} \cdot B$ and $A = B$ by Proposition 2(4) and (5). Hence $g = h$ and so $(A, g) = (B, h)$.

From Proposition 2(6), we see immediately that the $\mathcal{D}$-classes are finite. Hence the free inverse semigroup $FIS(X)$ is completely semisimple.

To prove the final assertion we first show that $(A, g) \sigma (B, h)$ if, and only if, $g = h$. By definition, $(A, g) \sigma (B, h)$ if, and only if, there is an element (C, k) such that $(C, k) \leq (A, g), (B, h)$. Thus $k = g = h$ and $A, B \subseteq C$ by Proposition 2(3). But then $A \cup B \subseteq C$. Thus $g = h$ is a necessary and sufficient condition for $(A, g) \sigma (B, h)$. Hence the σ-classes have the form

$$\{(A, g) \in \mathcal{Y} \times FG(X) \colon g \in A\}$$

where g is fixed. If $g \neq 1$ then $([g]^{\downarrow}, g)$ is the maximum element in this set. ■

We have defined the free inverse semigroup $FIS(X)$ in terms of the specific set X. In reality, it is only the cardinality of X which is important.

Proposition 4 *Let X and Y be two non-empty sets. Then $FIS(X)$ is isomorphic to $FIS(Y)$ if, and only if, X and Y have the same cardinality.*

Proof Observe that the set of maximal idempotents in $FIS(X)$ is

$$\{x^{-1}x \colon x \in X\} \cup \{xx^{-1} \colon x \in X\}.$$

The result is now clear. ■

Because of Proposition 4, we may define FIS_n to be *the* free inverse semigroup on n generators for each integer $n \geq 1$.

6.4 Munn normal form

The solution of the word problem presented in Section 6.2 is just the kind of solution to be expected on the basis of the solution of the word problem for free groups: the normal form is a special kind of string. However, there is a radically different approach to the solution of the word problem due to Douglas Munn, which is important in combinatorial inverse semigroup theory. Munn's central insight was that the elements of the free inverse semigroup are most naturally described as *trees* with two distinguished vertices. Before we can describe Munn's approach we need some elementary ideas from graph-theory.

Some graph theory

A *graph* is a digraph equipped with an involution on the set of edges, $e \mapsto e^{-1}$, such that

$$(e^{-1})^{-1} = e, \quad e^{-1} \neq e, \quad \partial_0(e^{-1}) = \partial_1(e), \quad \partial_1(e^{-1}) = \partial_0(e)$$

for all edges e.

In describing a graph we often make use of the following idea. An *orientation* for a graph is a subset E^+ of E such that E is the disjoint union of E^+ and $(E^+)^{-1}$. A graph is determined if we know its set of vertices, an orientation E^+ and the restrictions of ∂_0 and ∂_1 to E^+; thus every graph is determined by an associated digraph. In what follows graphs will always be presented in this way.

A *subgraph* of a graph $\mathcal{G} = (V, E)$ is determined by a subset of the set of vertices together with a subset of the set of edges together with the additional condition that if e is an edge in the subgraph so is e^{-1}. We shall say that a graph is *finite* if it has finitely many vertices and edges.

A *path of length n* in a graph $\mathcal{G}$ is an ordered n-tuple $(e_1, \ldots, e_n)$ of edges such that $\partial_1(e_i) = \partial_0(e_{i+1})$ for $i = 1, \ldots, n-1$. The vertices $\partial_0(e_1)$ and $\partial_1(e_n)$ are called the *initial* and *terminal* vertices of the path. For $n = 0$ and any vertex v there is a unique path of length zero with initial and terminal vertex equal to v. A path is called a *circuit* if its initial and terminal vertices coincide. A graph is said to be *connected* if any two vertices are joined by a path. A *round-trip* is a path of the form ee^{-1}. A path is said to be *reduced* if it contains no round-trips. A connected graph is said to be a *tree* if the only reduced circuits have length 0.

Cayley graphs of free groups

The starting point for Munn's approach is the observation that the free group has a natural tree structure. The *Cayley graph* of the free group $FG(X)$ is defined as follows: the vertices are the set of reduced words of the free group, and the orientation, called the *standard orientation*, is given by $FG(X) \times X$, where for each $(u, x) \in FG(X) \times X$ we have

$$\partial_0(u, x) = u \text{ and } \partial_1(u, x) = \text{red}(ux).$$

It follows that $(u, x)^{-1} = (\text{red}(ux), x^{-1})$. This graph is a tree, because the group is free: the existence of reduced circuits with non-zero length would imply the existence of non-trivial relations in the group.

The Cayley graph of $FG(X)$ and the partially ordered set $(FG(X), \leq_p)$, which we used in Section 6.2, are closely related. We can construct a digraph from $(FG(X), \leq_p)$, called the *Hasse digraph* as follows. The vertices are the elements of $FG(X)$; if $u, v \in FG(X)$ then there is an edge from u to v when $v = ux$ where $x \in X \cup X^{-1}$ and ux is reduced as it stands. Thus $u \leq_p v$ and 'v covers u'. The Hasse digraph is essentially the Hasse diagram of the partially ordered set $(FG(X), \leq_p)$.

Proposition 1 *The Hasse digraph of $(FG(X), \leq_p)$ is an orientation for the Cayley graph of $FG(X)$.*

Proof By definition, the vertices of the Hasse digraph are the same as the vertices of the Cayley graph. Suppose that $u \leq_p v$ where $v = ux$ for some $x \in X \cup X^{-1}$. Then (u, x) is an element of the Cayley graph with $\partial_0(u, x) = u$ and $\partial_1(u, x) = v$. Thus every edge of the Hasse digraph is an edge of the Cayley graph. Now let (u, x) be any edge of the Cayley graph. There are two possible cases. Firstly, if $\text{red}(ux) = ux$ then (u, x) is an edge of the Hasse digraph. Secondly, suppose that $\text{red}(ux) \neq ux$. Then $u = u'x^{-1}$ and $\text{red}(ux) = u'$. The edge $(u, x)^{-1}$ has source u' and target u where $u = u'x^{-1}$ is reduced as it stands. Thus $(u, x)^{-1}$ belongs to the Hasse digraph. ■

Munn trees

In view of Proposition 1, every non-empty, finite order ideal of the partially ordered set $(FG(X), \leq_p)$ determines and is determined by a finite, connected subgraph of the Cayley graph of $FG(X)$ containing the vertex 1. Thus by Theorem 6.2.12, the idempotents of the free inverse semigroup are determined by such subgraphs which contain at least two vertices. More generally, every element of the free inverse semigroup, once converted into Scheiblich normal form, determines a finite connected subgraph containing the vertex 1 and with one vertex marked as the root. We call this the *Cayley graph* of the element.

It is clear that Cayley graphs of elements and their Scheiblich normal forms contain exactly the same information. However, there is a certain redundancy in the Cayley graph, in that both edges and vertices are labelled. This is unnecessary, because Cayley graphs are always trees. If we relabel the vertex 1 by α and label the root vertex by β and then erase all other vertex labels, then we arrive at what are called *Munn trees*. Munn trees and Cayley graphs are in bijective correspondence: given a Munn tree we can easily construct a the Cayley graph: every vertex v is labelled by the shortest string in $FG(X)$ which labels a path from α to v; then α is re-labelled 1 and β is marked as the root. Thus two strings in $FIS(X)$ are equivalent precisely when they have the same Munn tree. We denote the Munn tree of $u \in FIS(X)$ by $MT(u)$.

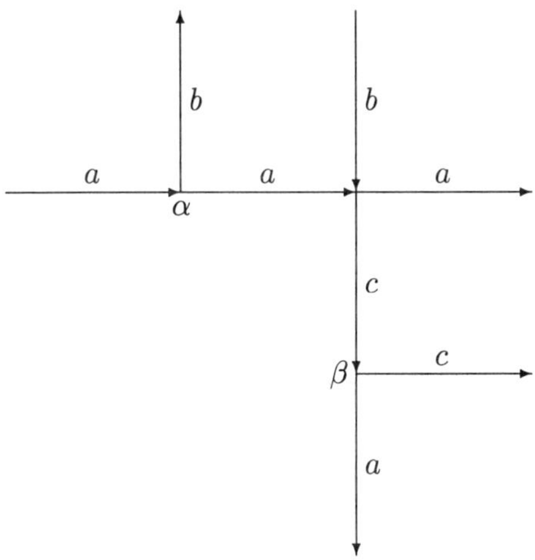

Figure 6.1: The Munn tree of $u = a^2a^{-3}abb^{-1}ab^{-1}bcaa^{-1}cc^{-1}$

The formal definition runs as follows. A *Munn tree* over $X \cup X^{-1}$ is a connected finite tree with at least one edge; the edges are labelled by elements of $X \cup X^{-1}$ such that if the edge e is labelled x then the edge e^{-1} is labelled x^{-1}; if e and e' are edges such that $\partial_0(e) = \partial_0(e')$ and the labels on e and e' are the same then $e = e'$; finally, there are two distinguished vertices labelled α and β. In representing Munn trees we choose the standard orientation.

Munn's algorithm

At the moment, the only way we have of constructing the Munn tree of a string is to convert it into Scheiblich normal form, construct the corresponding Cayley graph, and then transform the Cayley graph into a Munn tree. However, Munn discovered an extraordinarily elegant way of constructing the Munn tree directly from the string. We describe the algorithm first and then we shall justify it.

Let $u \in FIS(X)$ be any non-empty string. Then u can be used to trace out a path in the Cayley graph of the free group $FG(X)$: the path begins at the vertex of the Cayley graph labelled 1 and follows those edges which are labelled using the successive letters of u. The initial vertex of this path is labelled α and the terminal vertex is labelled β. Now erase all other vertex labellings and all edges not in the path we have traced out. This is the Munn tree of u. Of course, the Munn tree can be grown directly without passing through the Cayley graph.

Once the Munn tree of a string has been constructed we can easily write down the components and the root of the string: the components are the elements of the free group which label the leaves of the tree (those vertices which lie on exactly one edge) and the root is the reduced word labelling the β-vertex.

We now justify the above algorithm.

Proposition 2 *Let $u, v \in FSI(X)$.*

(1) $MT(u) = MT(uu^{-1}u)$.

(2) $MT(uu^{-1}vv^{-1}) = MT(vv^{-1}uu^{-1})$.

(3) *If $u \equiv v$ then $MT(u) = MT(v)$.*

Proof (1) To construct the Munn tree of $uu^{-1}u$ we first construct the Munn tree of u. Then, beginning at the vertex labelled by β, we trace out the reverse path to end up at the vertex labelled α. We then once again trace out the Munn tree of u. Thus we merely trace out the Munn tree of u, twice forward and once back.
(2) The Munn tree of $uu^{-1}vv^{-1}$ is constructed by first tracing out the circuit uu^{-1} and then tracing out the circuit vv^{-1}. The Munn tree of $vv^{-1}uu^{-1}$ is constructed in exactly the same way except that the order in which the two circuits are traced is reversed. In both cases, the same Munn tree is constructed.
(3) From Proposition 2.3.3, the fact that $u \equiv v$ implies that there is a sequence of strings $u_1, \ldots, u_n$ such that $u = u_1$, $u_n = v$ and for each pair (u_i, u_{i+1}), where $i = 1, \ldots, n-1$, u_{i+1} is obtained from u_i by using one of the relations $a \equiv aa^{-1}a$ and $aa^{-1}bb^{-1} \equiv bb^{-1}aa^{-1}$ applied to a substring of u_i. By (1) and (2), $MT(u_i) = MT(u_{i+1})$. Hence $MT(u) = MT(v)$. ∎

The Munn tree representation of the elements of a free inverse semigroup also provides an extremely elegant geometric interpretation of the various operations in the semigroup. The underlying graph of a Munn tree will be denoted by a letter such as T. The vertex labelled by α will be denoted by $\alpha(T)$ and the vertex labelled by β will be denoted by $\beta(T)$. The triple (v_1, T, v_2) represents a Munn tree where $\alpha(T) = v_1$ and $\beta(T) = v_2$. We now define some operations and a relation on the set of all Munn trees over X.

Let (v_1, T, v_2) and (v_1', T', v_2') be Munn trees. Define

$$(v_1, T, v_2)^{-1} = (v_2, T, v_1);$$

thus the labels α and β are interchanged.

Define the relation

$$(v_1, T, v_2) \subseteq (v_1', T', v_2')$$

if T is a subgraph of T' in such a way that the vertices $\alpha(T)$ and $\alpha(T')$ coincide, and the vertices $\beta(T)$ and $\beta(T')$ coincide.

Finally the product, $(v_1, T, v_2)(v_1', T', v_2')$ of two Munn trees is defined as follows: identify the vertex $\beta(T)$ with the vertex $\alpha(T')$ and call the common vertex v; erase the labels on $\beta(T)$ and $\alpha(T')$; identify every path in T begining at v with every path in T' beginning at v *which has the same edge labelling.* When this process is completed the resulting structure will be a Munn tree.

The proofs of the following are straightforward.

Proposition 3 *Let* $u, v \in FIS(X)$.

(1) $MT(u^{-1}) = MT(u)^{-1}$.

(2) $u \leq v \Leftrightarrow MT(v) \subseteq MT(u)$.

(3) $MT(uv) = MT(u)MT(v)$. ■

We now give an example of how to calculate the product of two Munn trees. Figure 6.2 consists of the Munn trees of two strings u and v. To calculate the Munn tree of uv, we translate the Munn tree of u so that its β-vertex coincides with the α-vertex of the Munn tree of v. Call this common vertex w. The process of identifying common paths beginning at w is equivalent to finding the maximum overlap in each of the two trees starting at w and then glueing the trees together on the overlaps. The overlaps have been rendered in bold in the two Munn trees. The resulting tree, Figure 6.3, is the Munn tree of the product string uv.

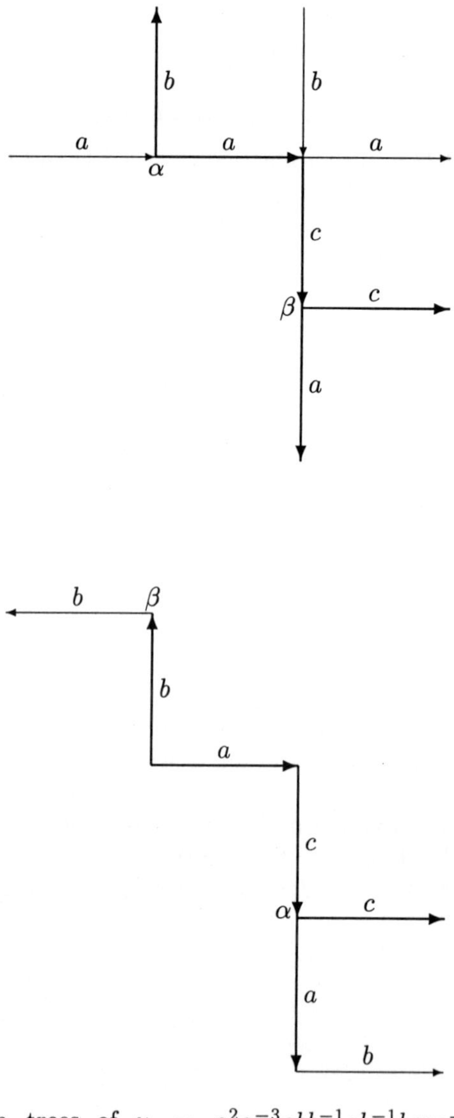

Figure 6.2: Munn trees of $u = a^2a^{-3}abb^{-1}ab^{-1}bcaa^{-1}cc^{-1}$ and $v = cc^{-1}aa^{-1}bb^{-1}a^{-1}c^{-1}a^{-1}bbb^{-1}$ with overlaps highlighted

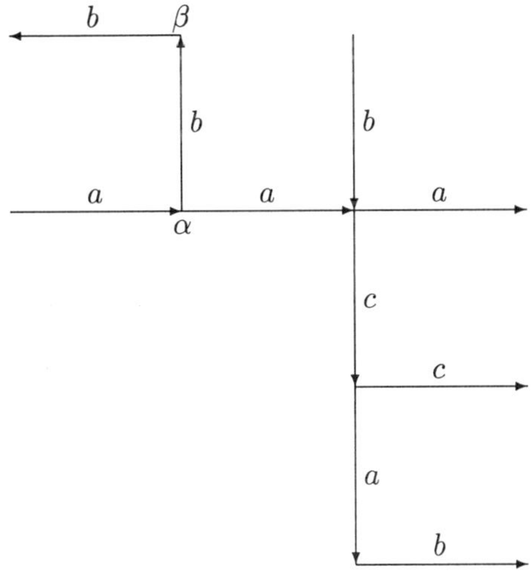

Figure 6.3: The Munn tree of uv

The free monogenic inverse semigroup

Munn trees may be used to determine the structure of the free monogenic inverse semigroup FIS_1. The Munn tree of an element of FIS_1 has no branches. Consequently, every such tree is characterised by its length and the positions of α and β. Thus the elements of the free monogenic inverse semigroup can be described by means of triples of integers. There are many ways to do this; we give one here. Let p be the number of edges from the leftmost vertex to the vertex labelled α. Clearly $p \geq 0$. The vertex labelled β may be to the left or right of the vertex labelled α. Let q be the number of edges from the vertex labelled α to the vertex labelled β, defined to be negative if β is to the left of α and positive if it is to the right. Finally, let r be the number of edges from the vertex labelled β to the rightmost vertex. Thus every Munn tree in FIS_1 determines and is determined by an element from

$$GB = \{(p, q, r) \in \mathbb{Z}^3 \colon p \geq 0,\, p + q \geq 0,\, q + r \geq 0,\, r \geq 0,\, p + q + r > 0\}.$$

The correct multiplication of such triples is obtained by considering the possible cases which arise when multiplying Munn trees.

Theorem 4 *Define a product on GB by*

$$(p, q, r)(p', q', r') = (\max(p, p' - q), q + q', \max(r', r - q')).$$

Then with this product GB is isomorphic to the free monogenic inverse semigroup. ■

We shall return to the free monogenic inverse semigroup in Chapter 9.

6.5 Notes on Chapter 6

Section 6.1

Theorem 6.1.2 is due to Wagner [421]. Monogenic inverse semigroups are discussed by Preston [331] and by Conway, Duncan and Paterson [49] (a gap in a proof contained in this paper is corrected in [119]).

Section 6.2

The historical development of a subject must always be distingushed from its logical development and this is certainly true of the development of the theory of free inverse semigroups. The first detailed description of the structure of free inverse semigroup, due to Scheiblich, was achieved without an explicit solution of the word problem. Instead, he obtained a structural description of free inverse semigroups essentially the same as the one contained in Section 6.3. This work was announced in 1972 [366] with details appearing in 1973 [368].

Subsequently, a number of alternative descriptions of free inverse semigroups were obtained by McAlister and McFadden [245], [270], Preston [330], and Schein [379]. The normal form I describe emerges from this work; one sees it in both Preston's and Schein's papers. The term Scheiblich normal form is both a tribute and a convenient designation. A useful reference for this first wave of work up to 1976 is Reilly's survey article [343].

My discussion of the word problem for free inverse semigroups is based on Schein's paper [379] combined with the heuristic arguments of Scheiblich [366]. The normal form is slightly different from the one used by Schein, but is the most convenient for carrying out calculations. McAlister's perceptive observations on the structure of free inverse semigroups described in his survey paper [241] influenced my approach considerably.

Interestingly, both Howie [148] and Petrich [312] prefer to follow Scheiblich in presenting the structural description of free inverse semigroups first; Howie then subsequently solves the word problem. In both cases, they use the structure theory of E-unitary inverse semigroups which, historically, was actually motivated by the theory of free inverse semigroups. Such things are, of course, a matter of taste; but it is evident from my approach that structural features of the free inverse semigroup are an important aspect of solving the word problem.

The second wave of work on free inverse semigroups has been heavily influenced by Munn's approach. Margolis and Meakin [215] showed that the

Cayley graph of any group presentation could be used to construct an E-unitary inverse semigroup in a manner generalising the construction of free inverse semigroups from the Cayley graphs of free groups. Margolis and Pin [224] obtained another description of free inverse semigroups within the framework of formal language theory. The theory of free inverse semigroups has now been subsumed by the flourishing subject of combinatorial inverse semigroup theory which has very close connections with combinatorial group theory (see Section 2.1). Meakin's survey article [250] is a good introduction to this important field.

Section 6.3

The semigroup $P(FG(X), \mathcal{X}, \mathcal{Y})$ is essentially the form of the free inverse semigroup obtained by Scheiblich. Analysing this construction, McAlister and McFadden [245] and later McAlister [231] introduced P-semigroups which are discussed in Chapter 7.

Section 6.4

Munn trees originated in Munn's paper [270] where they are called *birooted word trees*. The rôle of the Cayley graph of the free group in the theory of free inverse semigroups was discovered McAlister [241]. Free monogenic inverse semigroups were first described by Gluskin [98]. Munn's work forms the foundation for Stephen's approach to presentations of inverse semigroups [397] which in turn forms the basis of all current work in this area [217], [30].

To conclude this chapter, let me briefly summarise the main areas of combinatorial inverse semigroup.

The coproduct in the category of inverse semigroups is the free product. A structure theorem for free products of inverse semigroups was first obtained by Jones [156] using Scheiblich's description of free inverse semigroups. A description which is based on the Munn–Stephen approach using digraphs is presented in [161]. A useful survey of this area is [159].

A refinement of the theory of free products is the theory of free products with amalgamation. Such products are closely connected with the study of semigroup amalgams: in general, an amalgam strongly embeds into some semigroup if, and only if, it strongly embeds into its free product with amalgamation. We refer the reader to Howie [148] for definitions and proofs. Tom Hall proved [115] that the class of inverse semigroups has the strong amalgamation property, consequently every amalgam of inverse semigroups may be strongly embedded in its free product with amalgamation. The strong embedding property for inverse semigroups was later reproved by Pastijn [301] using inductive groupoids. Hall's result has numerous applications many of which are described in Hall's survey [116]. See also Chapter XIII of [312].

More enigmatic, as Jones himself says [159], is the structure of free products with amalgamation themselves. Recent work in this area using Stephen's approach is [22], [23], and [42], whereas the inductive groupoid approach of [282] is utilised in [121] combined with the Bass–Serre theory of graphs of groups.

Finally, there are also the beginnings of a theory of HNN-extensions of inverse semigroups; see, for example, [437].

Chapter 7

E-unitary inverse semigroups

The class of E-unitary semigroups is one of the most important in inverse semigroup theory. There are two main reasons for this: firstly, McAlister's covering theorem (Theorem 2.2.4) tells us that every inverse semigroup is an idempotent-separating homomorphic image of an E-unitary semigroup; and, secondly, many naturally occurring inverse semigroups are E-unitary, the bicyclic monoid and free inverse semigroups being particularly prominent examples. For these reasons, we now embark on a detailed investigation of the structure of these semigroups.

The simplest examples of E-unitary inverse semigroups are the semidirect products of semilattices by groups. Although inverse subsemigroups of such semigroups are E-unitary, they need not themselves be semidirect products. This leads to our first major result: every E-unitary inverse semigroup can be embedded in a semidirect product of a semilattice by a group. Our proof of this result relies on the properties of the completion, $C(S)$, of an E-unitary inverse semigroup S (Section 1.4) and the properties of Billhardt congruences (Section 5.3). Combining this result with the McAlister covering theorem, we deduce that every (finite) inverse semigroup divides a (finite) semidirect product of a semilattice by a group.

If we look at the homomorphic images of semidirect products of semilattices by groups, rather than their inverse subsemigroups, we obtain a class of inverse semigroups very closely related to factorisable inverse monoids. This gives us further insight into the relationship between E-unitary semigroups and factorisable inverse monoids which we first observed in Section 2.2.

To obtain a structure theory for E-unitary inverse semigroups, we generalise the semidirect product construction. This leads to the introduction of what are termed P-semigroups; such semigroups were in fact tacitly introduced in Section 6.3, when we described the structure of free inverse semigroups. The P-theorem then asserts that every E-unitary inverse semigroup is isomorphic to a P-semigroup. The proof of this result makes essential use of the idea of extending partial bijections which we introduced in Chapter 2. We illustrate this theory by using it to determine the structure of the inverse monoid generated by the Möbius transformations.

We then turn to a special class of E-unitary semigroups called F-inverse semigroups. These are intermediate between semidirect products of semilattices by groups and arbitrary E-unitary inverse semigroups; free inverse semigroups, which we know to be E-unitary, belong to this class. An examination of the structure of such semigroups leads to the concept of an enlargement; we prove that an inverse semigroup is F-inverse precisely when it has an enlargement which is a semidirect product of a semilattice by a group. Finally, we refine the McAlister covering theorem by proving that every inverse semigroup has an F-inverse cover.

7.1 Division theorems

The results of this section are centred on the properties of semidirect products of semilattices by groups. We characterise both the inverse subsemigroups and homomorphic images of such semigroups; the former are precisely the E-unitary inverse semigroups and the latter are the almost factorisable inverse semigroups, a class of semigroups generalising the factorisable inverse monoids. We are concerned solely with classical semidirect products, although the semidirect products of interest to us are in fact identical to the corresponding λ-semidirect products (by Proposition 5.3.2(1)).

Semidirect products of semilattices by groups

Let G be a group and Y a set. Recall that G acts on Y (on the left) if there is a function $G \times Y \to Y$ denoted by $(g, e) \mapsto g \cdot e$ satisfying $1 \cdot e = e$ for all $e \in Y$ and $g \cdot (h \cdot e) = (gh) \cdot e$ for all $g, h \in G$ and $e \in Y$. If Y is a partially ordered set, then we say that G acts on Y by *order automorphisms* if for all $e, f \in Y$ we have that

$$e \leq f \Leftrightarrow g \cdot e \leq g \cdot f.$$

Observe that in the case of a group action, it is enough to assume that $e \leq f$ implies $g \cdot e \leq g \cdot f$, because if $g \cdot e \leq g \cdot f$ then $g^{-1} \cdot (g \cdot e) \leq g^{-1} \cdot (g \cdot f)$ and so $1 \cdot e \leq 1 \cdot f$, which gives $e \leq f$. If Y is a meet semilattice on which G acts by order automorphisms, then it is easy to show that

$$g \cdot (e \wedge f) = g \cdot e \wedge g \cdot f$$

for all $g \in G$ and $e, f \in Y$.

Let $P(G,Y)$ be the set $Y \times G$ equipped with the multiplication

$$(e,g)(f,h) = (e \wedge g \cdot f, gh).$$

Theorem 1 *$P(G,Y)$ is an E-unitary inverse semigroup in which the semilattice of idempotents is isomorphic to $(Y, \leq)$ and $P(G,Y)/\sigma$ is isomorphic to G. The semigroup $P(G,Y)$ is a monoid precisely when Y has an identity.*

Proof In Section 5.3, we proved that $P(G,Y)$ is an inverse semigroup; the inverse of (e,g) is the element $(g^{-1} \cdot e, g^{-1})$, and the idempotents of $P(G,Y)$ are the elements of the form $(e,1)$. It is clear from the definition of the multiplication in $P(G,Y)$ that the function $(e,1) \mapsto e$ is an isomorphism of semilattices. The natural partial order is given by

$$(e,g) \leq (f,h) \Leftrightarrow e \leq f \text{ and } g = h.$$

If $(e,1) \leq (f,g)$ then $g = 1$ and so $P(G,Y)$ is E-unitary. It also follows from the description of the natural partial order that

$$(e,g) \, \sigma \, (f,h) \Leftrightarrow g = h.$$

Consequently, $P(G,Y)/\sigma$ is isomorphic to G.

We now show that $P(G,Y)$ is a monoid precisely when Y has an identity. Suppose that Y has an identity i. Then i is the maximum element of Y, consequently $(i,1)$ is an idempotent in $P(G,Y)$ satisfying

$$(i,1)(e,g) = (i \wedge 1 \cdot e, g) = (i \wedge e, g) = (e,g).$$

Similarly, $(e,g)(i,1) = (e,g)$. Thus $(i,1)$ is the identity of $P(G,Y)$. Conversely, suppose that $(e,1)$ is the identity of $P(G,Y)$. Then, in particular, $(e,1)(f,1) = (f,1)$ for all idempotents $(f,1) \in P(G,Y)$. Thus $e \wedge f = f$ for all $f \in Y$. Consequently, e is the identity of Y. ■

There are a number of abstract characterisations of semidirect products of semilattices by groups; they are based on the properties of the minimum group congruence.

Theorem 2 *Let S be an inverse semigroup. Then the following are equivalent:*

(1) *The semigroup S is isomorphic to a semidirect product of a semilattice by a group.*

(2) *S is E-unitary and for each $a \in S$ and $e \in E(S)$ there exists $b \in S$ such that $b \sim a$ and $b^{-1}b = e$.*

(3) *$\sigma^\natural\colon S \to S/\sigma$ is $\mathcal{L}$-bijective.*

(4) *There is an $\mathcal{L}$-bijective homomorphism from S to a group.*

(5) *The function $\theta\colon S \to E(S) \times S/\sigma$ defined by $\theta(a) = (a^{-1}a, \sigma(a))$ is a bijection.*

(6) *The function $\phi\colon S \to E(S) \times S/\sigma$ defined by $\phi(a) = (aa^{-1}, \sigma(a))$ is a bijection.*

Proof (1) $\Rightarrow$ (2). Without loss of generality, we may assume that S is a semidirect product of a meet semilattice Y by a group G. The semigroup S is E-unitary by Theorem 1. Let $(e,g) \in S$ and $(f,1) \in E(S)$. Then the element $(g \cdot f, g)$ of S satisfies

$$(g \cdot f, g) \sim (e,g) \text{ and } (g \cdot f, g)^{-1}(g \cdot f, g) = (f,1)$$

as required.
(2) $\Rightarrow$ (3). Since S is E-unitary, the homomorphism $\sigma^\natural\colon S \to S/\sigma$ is idempotent pure, or $\mathcal{L}$-injective, by Theorem 2.4.6. Let $e \in E(S)$ and $\sigma(a) \in S/\sigma$. By assumption there exists $b \in S$ such that $b^{-1}b = e$ and $b \sim a$. But $b \sim a$ implies $\sigma(b) = \sigma(a)$. Thus $\sigma^\natural$ is also $\mathcal{L}$-surjective.
(3) $\Rightarrow$ (4). Immediate.
(4) $\Rightarrow$ (3). Let $\theta\colon S \to G$ be an $\mathcal{L}$-bijective homomorphism to a group G. Since σ is the minimum group congruence, $\sigma \subseteq \ker\theta$ by Theorem 2.4.1. But θ is idempotent pure by assumption, and so $\sigma^\natural$ is idempotent pure by Proposition 2.4.5. In particular, S is E-unitary by Theorem 2.4.6.

To show that $\sigma^\natural$ is $\mathcal{L}$-surjective, let $s \in S$ and $e \in E(S)$. There exists $t \in S$ such that $t^{-1}t = e$ and $\theta(t) = \theta(s)$, since θ is $\mathcal{L}$-surjective. Now $\theta(s^{-1}t)$ is the identity of G, and so $s^{-1}t$ is an idempotent of S since θ is $\mathcal{L}$-injective. Similarly, st^{-1} is an idempotent. Hence $s \sim t$ and so $(s,t) \in \sigma$. Thus for each $e \in E(S)$ and $\sigma(s) \in S/\sigma$, there exists $t \in S$ such that $t^{-1}t = e$ and $\sigma(t) = \sigma(s)$. Thus $\sigma^\natural$ is $\mathcal{L}$-surjective.
(3) $\Rightarrow$ (5). Straightforward.
(5) $\Rightarrow$ (6). Suppose that $\phi(a) = \phi(b)$. Then $aa^{-1} = bb^{-1}$ and $\sigma(a) = \sigma(b)$. But $\sigma(a^{-1}) = \sigma(b^{-1})$ and so $\theta(a^{-1}) = \theta(b^{-1})$. By assumption θ is bijective and so $a^{-1} = b^{-1}$, giving $a = b$. Hence ϕ is injective.

Now let $(e, \sigma(s)) \in E \times S/\sigma$. Since θ is surjective there exists $t \in S$ such that $\theta(t) = (e, \sigma(s^{-1}))$. Thus $t^{-1}t = e$ and $t \,\sigma\, s^{-1}$. Hence t^{-1} is such that $t^{-1} \,\sigma\, s$ and $t^{-1}(t^{-1})^{-1} = e$. Thus $\phi(t^{-1}) = (e, \sigma(s))$, and so ϕ is surjective.
(6) $\Rightarrow$ (5). A similar argument to (5) $\Rightarrow$ (6).
(6) $\Rightarrow$ (1). We shall use the fact that both the functions ϕ and θ defined above are bijections.

First of all S is E-unitary. For suppose that $e \leq a$ where e is an idempotent. Then $\sigma(e) = \sigma(a)$, and $\sigma(e) = \sigma(a^{-1}a)$, so that $\sigma(a) = \sigma(a^{-1}a)$. Thus $\theta(a) = \theta(a^{-1}a)$, and so $a = a^{-1}a$, since θ is a bijection.

We shall define an action of S/σ on $E(S)$ using θ, and then show that ϕ defines an isomorphism from the semidirect product of $E(S)$ by S/σ to S.

Define $\sigma(s) \cdot e = tt^{-1}$ where $\theta(t) = (e, \sigma(s))$. This is well-defined because θ is a bijection. The two defining properties of an action hold. Firstly, if $\sigma(e)$ is the identity of S/σ then $\theta(e) = (e, \sigma(e))$ and so $\sigma(e) \cdot e = e$; secondly, $\sigma(u) \cdot (\sigma(v) \cdot e) = \sigma(u) \cdot aa^{-1}$ where $\theta(a) = (e, \sigma(v))$, and $\sigma(u) \cdot aa^{-1} = bb^{-1}$ where $\theta(b) = (aa^{-1}, \sigma(u))$. Now $a \,\sigma\, v$ and $b \,\sigma\, u$ so that $ba \,\sigma\, uv$. Also $a^{-1}a = e$ and $b^{-1}b = aa^{-1}$ so that $(ba)^{-1}ba = a^{-1}a$. Hence $\theta(ba) = (e, \sigma(uv))$. Thus

$$\sigma(uv) \cdot e = (ba)(ba)^{-1} = bb^{-1} = \sigma(u) \cdot (\sigma(v) \cdot e).$$

Next, we show that S/σ acts on $E(S)$ by means of order automorphisms. Suppose that $e \leq f$. Then $\sigma(a) \cdot e = uu^{-1}$ and $\sigma(a) \cdot f = vv^{-1}$ where

$$\theta(u) = (e, \sigma(a)) \text{ and } \theta(v) = (f, \sigma(a)).$$

Consequently, $e = u^{-1}u$ and $f = v^{-1}v$ and $u \,\sigma\, v$. But S is E-unitary, and so σ is equal to the compatibility relation by Theorem 2.4.6. From $u^{-1}u \leq v^{-1}v$ and $u \sim v$ we obtain $u \leq v$ by Lemma 1.4.14. Hence $uu^{-1} \leq vv^{-1}$ and so $\sigma(a) \cdot e \leq \sigma(a) \cdot f$.

It only remains to prove that ϕ is a homomorphism. By definition

$$\phi(a)\phi(b) = (aa^{-1}, \sigma(a))(bb^{-1}, \sigma(b)) = (aa^{-1} \wedge \sigma(a) \cdot bb^{-1}, \sigma(ab)).$$

But $\sigma(a) \cdot bb^{-1} = tt^{-1}$ where $\theta(t) = (bb^{-1}, \sigma(a))$. Thus

$$\phi(a)\phi(b) = (aa^{-1}tt^{-1}, \sigma(ab))$$

whereas

$$\phi(ab) = (ab(ab)^{-1}, \sigma(ab)).$$

It remains to show that $aa^{-1}tt^{-1} = ab(ab)^{-1}$. We know that $t^{-1}t = bb^{-1}$ and $t \,\sigma\, a$. But $t \sim a$ since S is E-unitary. Thus $tt^{-1}a = at^{-1}t = abb^{-1}$ by Lemma 1.4.12. Hence $tt^{-1}aa^{-1} = abb^{-1}a^{-1} = ab(ab)^{-1}$. ■

Property (3) shows that an inverse semigroup is a semidirect product of a semilattice by a group precisely when $\sigma^{\natural}$ is $\mathcal{L}$-bijective. Arbitrary E-unitary inverse semigroups are characterised by the property that $\sigma^{\natural}$ is $\mathcal{L}$-injective (Theorem 2.4.6). The relationship between $\mathcal{L}$-injective and $\mathcal{L}$-bijective maps is the theme of Chapter 8.

An embedding theorem

In Section 5.3, we defined a Billhardt congruence to be a congruence ρ on an inverse semigroup S such that for every $s \in S$ the set $\{t^{-1}t \colon t \in \rho(s)\}$ contains a maximum element. A special class of such congruences are those in which each ρ-class contains a maximum element.

Proposition 3 *Let S be an inverse semigroup in which each σ-class contains a maximum element. Then S is an E-unitary inverse monoid.*

Proof The identity of the group S/σ is of the form $\sigma(e)$ where e is any idempotent. Let i be the maximum element of $\sigma(e)$. Since $E(S) \subseteq \sigma(e)$, the element i is larger than every idempotent. We prove that i is an idempotent. Clearly, $i^{-1}i \in \sigma(e)$, and so $i^{-1}i \leq i$. Hence $i \leq i^2$. But then $i^2 \in \sigma(e)$, and so $i^2 \leq i$. Hence $i = i^2$. Now for any $s \in S$, we have that

$$is = i(ss^{-1})s = ss^{-1}s = s,$$

and similarly $si = s$. Thus i is the identity of S, and so S is a monoid. Finally, since the maximum element of $\sigma(e)$ is an idempotent, every element of $\sigma(e)$ is an idempotent. Thus S is E-unitary by Theorem 2.4.6. ■

Inverse semigroups in which each σ-class contains a maximum element are called *F-inverse monoids.*

Proposition 4 *Every E-unitary inverse semigroup S can be embedded in an F-inverse monoid T such that T/σ is isomorphic to S/σ.*

Proof We shall use the completion, $C(S)$ of S, described in Theorem 1.4.23. We proved there that S is embedded in $C(S)$, so our result will be proved if we show that $C(S)$ is an F-inverse monoid.

Recall that the natural partial order in $C(S)$ is subset inclusion and that the idempotents are the order ideals of $E(S)$. We start by showing that $C(S)$ is E-unitary. Let $E \leq A$ where E is an idempotent in $C(S)$ and $A \in C(S)$. Then $E \subseteq A$ and $E \subseteq E(S)$. Let $a \in A$ and let $e \in E$. Then $a \sim e$, since A is a permissible subset. But S is E-unitary and so $a \, \sigma \, e$ by Theorem 2.4.6(2). Thus a is an idempotent by Theorem 2.4.6(4). Hence every element of A is an idempotent, and so A is an idempotent of $C(S)$.

We now prove that $C(S)$ is F-inverse. Denote the minimum group congruence on $C(S)$ by $\sigma_{C(S)}$. Let $s \in S$. Then $\sigma(s)$ is a permissible subset of S. We prove that $\sigma(s)$ is the maximum element of its $\sigma_{C(S)}$-class. Let A be $\sigma_{C(S)}$-related to $\sigma(s)$. Then there exists a permissible subset B such that $B \subseteq A$ and $B \subseteq \sigma(s)$. Thus, in particular, there exists $a \in A$ such that $a \in \sigma(s)$. Every element of A is compatible with a, and a is compatible with every element in $\sigma(s)$, and so every element of A is compatible with every element in $\sigma(s)$, because S is E-unitary. Thus $A \subseteq \sigma(s)$. It follows that the maximum elements of the $\sigma_{C(S)}$-classes are the permissible subsets of the form $\sigma(s)$, where $s \in S$.

It remains to show that S/σ is isomorphic to $C(S)/\sigma_{C(S)}$. Define θ: $S/\sigma \to C(S)/\sigma_{C(S)}$ by $\theta(\sigma(s)) = \sigma_{C(S)}(\sigma(s))$. By our calculations above, θ is a bijection, and it is easy to check that it is a homomorphism. ■

We may now prove our first structural characterisation of E-unitary inverse semigroups.

Theorem 5 *Let S be an inverse semigroup. Then S is E-unitary if, and only if, S can be embedded in a semidirect product of a semilattice by a group.*

Proof Let S be an E-unitary inverse semigroup. By Proposition 4, the semigroup S can be embedded in an F-inverse monoid T, in such a way that S/σ and T/σ are isomorphic. The minimum group congruence σ on T is a Billhardt congruence. Thus by Theorem 5.3.5, T can be embedded in $\operatorname{Ker}\sigma \operatorname{Wr}^{\lambda} T/\sigma$. Since T is F-inverse, the congruence σ is idempotent pure and so $\operatorname{Ker}\sigma$ is a semilattice, and T/σ is isomorphic to S/σ. Thus by Proposition 5.3.3, $\operatorname{Ker}\sigma \operatorname{Wr}^{\lambda} T/\sigma$ is a λ-semidirect product of a semilattice by the group S/σ. The action of the group S/σ on the semilattice of functions from S/σ to $\operatorname{Ker}\sigma$ satisfies axiom (AE3) by Proposition 5.3.4. Thus by Proposition 5.3.2, the λ-semidirect product is actually a (classical) semidirect product of a semilattice by the group S/σ.

The converse holds because every semidirect product of a semilattice by a group is E-unitary (Theorem 1), and inverse subsemigroups of E-unitary inverse semigroups are themselves E-unitary. ■

If the above theorem is combined with the McAlister covering theorem (Theorem 2.2.4), we obtain the following description of all inverse semigroups.

Theorem 6 *Every (finite) inverse semigroup divides a (finite) semidirect product of a semilattice by a group.* ■

A covering theorem

Having characterised the inverse subsemigroups of semidirect products of semilattices by groups, we now characterise their homomorphic images. To do this, we shall again make use of the completion, $C(S)$, of the inverse semigroup S. The inverse semigroup $C(S)$ is a monoid, with identity $E(S)$, and so $C(S)$ has a group of units which consists of those permissible subsets A of S such that $A^{-1}A = E(S) = AA^{-1}$.

An inverse semigroup S is said to be *almost factorisable* if for each $s \in S$ there exists a unit A of $C(S)$ such that $s \in A$. First we justify our choice of terminology.

Proposition 7 *Let S be an inverse monoid. Then S is almost factorisable if, and only if, it is factorisable.*

Proof We show first that the units of $C(S)$ are precisely the subsets of the form $[g]$ where $g \in U(S)$. Let A be a unit in $C(S)$. Then $A^{-1}A = E(S)$. By Lemma 1.4.22, there is a unique element $a \in A$ such that $a^{-1}a = 1$. We show that $A = [a]$. Let $b \in A$. Then $b \sim a$ and $b^{-1}b \leq a^{-1}a = 1$. Thus $b \leq a$ by Lemma 1.4.14. Hence $A \subseteq [a]$. But A is an order ideal and so $A = [a]$. By assumption, $AA^{-1} = E(S)$. Thus $[aa^{-1}] = E(S)$. Hence $aa^{-1} = 1$, and so $a \in U(S)$. Conversely, if $g \in U(S)$, then it is straightforward to show that $[g]$ is a unit of $C(S)$.

Now suppose that S is almost factorisable. By definition, for each $s \in S$ there exists a unit A of $C(S)$ such that $s \in A$. But then $A = [g]$ for some $g \in U(S)$ by the result above. Thus $s \leq g$, and so S is factorisable. The converse is immediate by the result above. ■

Almost factorisable semigroups are the *semigroup* analogues of factorisable inverse monoids. They provide the right algebraic idea for succinctly describing semidirect products of semilattices by groups.

Theorem 8 *An inverse semigroup is a semidirect product of a semilattice by a group if, and only if, it is E-unitary and almost factorisable.*

Proof Let S be isomorphic to a semidirect product of a semilattice by a group. Without loss of generality, we may assume that S is of the form $P(G, Y)$ for some group G and semilattice Y. By Theorem 1 such a semigroup is E-unitary, so it remains to show that it is almost factorisable. Observe that every subset A of $P(G, Y)$ of the form $Y \times \{g\}$, where $g \in G$, is a permissible subset of S, and that $A^{-1}A = E(S) = AA^{-1}$. Thus every such A is a unit of $C(S)$. Clearly, every element of S belongs to one of these subsets. Hence $P(G, Y)$ is almost factorisable.

To prove the converse, let S be an E-unitary, almost factorisable inverse semigroup. We show that condition (2) of Theorem 2 holds. Let $s \in S$ and $e \in E(S)$. Since S is almost factorisable there exists a unit A of $C(S)$ such that $s \in A$. But $A^{-1}A = E(S)$ so that there exists $a \in A$ such that $e = a^{-1}a$. But $s, a \in A$ implies that $s \sim a$. ■

Let S and T be inverse semigroups. Then a homomorphism θ: $S \to T$ is said to be *full* if $E(T) \subseteq \operatorname{im}\theta$. Surjective homomorphisms are always full.

Proposition 9 *Let θ: $S \to T$ be a full homomorphism between inverse semigroups. Then θ induces a homomorphism from the group of units of $C(S)$ to the group of units of $C(T)$.*

Proof Let A be a unit in $C(S)$. Consider the image $\theta(A)$ in T. Firstly, $\theta(A)$ is a compatible subset of T because A is a compatible subset of S. To show that it is an order ideal of T let $t \leq \theta(a)$, where $a \in A$. Then $t = \theta(a)t^{-1}t$. But $t^{-1}t$ is an idempotent, and so, since θ is a full homomorphism, there exists $e \in S$ such that $\theta(e) = t^{-1}t$. Thus $t = \theta(ae)$. But A is an order ideal of S and so $ae \in A$. Hence $t \in \theta(A)$. It follows that $\theta(A)$ is a permissible subset of T. We now prove that $\theta(A)$ is a unit in $C(T)$. Let $e' \in E(T)$. Then there exists an idempotent $e \in E(S)$ such that $\theta(e) = e'$. But A is a unit in $C(S)$ and so $A^{-1}A = E(S)$. Consequently, there exists $a \in A$ such that $e = a^{-1}a$. Thus $e' = \theta(a)^{-1}\theta(a)$. We may similarly show that there exists $b \in A$ such that $e' = \theta(b)\theta(b)^{-1}$. Hence $\theta(A)$ is a unit of $C(T)$.

Define ψ: $U(C(S)) \to U(C(T))$ by $\psi(A) = \theta(A)$. We have shown that this is a well-defined function, and it is easy to check that it is a homomorphism. ■

Full homomorphisms between inverse semigroups form a category, and the construction above is the object part of a functor from this category to the category of groups.

The main result of this section is the following.

Theorem 10 *An inverse semigroup is almost factorisable if, and only if, it is a homomorphic image of a semidirect product of a semilattice by a group.*

Proof Let F be an almost factorisable inverse semigroup. Denote the group of units of $C(F)$ by G. Put

$$P = \{(s, A) \in F \times G\colon s \in A\}$$

and define θ: $P \to F$ by $\theta(s, A) = s$. It is easy to check that P is an inverse subsemigroup of the direct product $F \times G$. The function θ is a homomorphism, which is surjective because F is almost factorisable. The idempotents of P may be found as follows. If $(s, A)^2 = (s, A)$, then $s^2 = s$ and $A^2 = A$. Thus A is an idempotent in the group of units of $C(S)$. Hence $A = E(F)$ and s is an idempotent. Conversely, every element of the form $(e, E(F))$, where e is an idempotent of F, is an idempotent of P. Thus

$$E(P) = \{(e, E(F))\colon e \in E(F)\}.$$

It is worth observing, that this result implies that θ is idempotent-separating. To prove that P is a semidirect product of a semilattice by a group, we use

Theorem 2(2). Firstly, P is E-unitary, for suppose that $(e, E(F)) \leq (s, A)$. Then $(e, E(F)) = (e, E(F))(s, A)$, so that $e = es$ and $E(F) = E(F)A$. Hence $A = E(F)$, and so s is an idempotent, which implies that (s, A) is an idempotent. To finish off this part of the proof, let $(s, A) \in P$ and $(e, E(F)) \in E(P)$. Since F is almost factorisable there exists $a \in A$ such that $a^{-1}a = e$. The pair (a, A) is a well-defined element of P, $(a, A)^{-1}(a, A) = (e, E(F))$, and $(s, A)^{-1}(a, A)$ and $(s, A)(a, A)^{-1}$ are both idempotents because $s, a \in A$ and so $s \sim a$. Thus $(s, A) \sim (a, A)$. Hence the almost factorisable inverse semigroup F is covered by P, which is a semidirect product of a semilattice by a group.

To prove the converse we actually prove that homomorphic images of almost factorisable inverse semigroups are again almost factorisable; the result then follows because we proved in Theorem 8 that semidirect products of semilattices by groups are almost factorisable. Let θ: $S \to T$ be a surjective homomorphism from an almost factorisable semigroup S to an inverse semigroup T. Let $t \in T$ and let $s \in S$ be such that $\theta(s) = t$. Since S is almost factorisable there exists a unit A of $C(S)$ such that $s \in A$. By Proposition 9, $\theta(A)$ is a unit of $C(T)$, and $t \in \theta(A)$. Consequently, T is almost factorisable. ■

If F is a factorisable inverse monoid then the semigroup P constructed above will also be a monoid. The proof of the following is now immediate by Proposition 7 and Theorem 10.

Corollary 11 *An inverse monoid is factorisable if, and only if, it is a monoid homomorphic image of a monoid semidirect product of a semilattice by a group.* ■

We may now account for the duality between E-unitary inverse semigroups and almost factorisable inverse semigroups: the former arise as inverse subsemigroups of semidirect products of semilattices by groups, whereas the latter arise as homomorphic images of such semigroups.

The structure of almost factorisable semigroups

The structure of almost factorisable inverse semigroups can easily be described in terms of factorisable inverse monoids. We begin by showing how to construct examples of almost factorisable inverse semigroups.

Proposition 12 *Let F be a factorisable inverse monoid. Put $S = F \setminus U(F)$. Then S is an almost factorisable inverse semigroup.*

Proof Observe first that if $a \in F$ then $a^{-1}a = 1$ precisely when $aa^{-1} = 1$. To see why, let $a^{-1}a = 1$. Since F is factorisable there exists a unit $g \in U(F)$

such that $a < g$. Thus $a = ga^{-1}a = g$. Hence a is a unit and so $aa^{-1} = 1$. The reverse implication is proved similarly. This result implies that if $a \in S$ then $a^{-1}a, aa^{-1} \in S$. In particular, S is closed under inverses.

The set S is closed under composition, for suppose that $a, b \in S$ and $ab \in U(F)$. Then $1 = (ab)^{-1}ab = b^{-1}a^{-1}ab \leq b^{-1}b$. Hence $b \in U(F)$, which is a contradiction. Thus $ab \in S$. Consequently, S is a well-defined inverse semigroup.

It remains to show that S is almost factorisable. Let $a \in S$ and suppose that $a \leq g$ where $g \in U(F)$. Put $A = [g] \cap S$. Clearly, A is a permissible subset of S containing a. We show that it is a unit of $C(S)$. Let $e \in E(S)$ and put $b = ge$. Then $b \leq g$ and $b^{-1}b = e$. Thus $b \in S$ and $b \in [g]$, which implies $b \in A$. Hence $A^{-1}A = E(S)$. We may similarly show that $AA^{-1} = E(S)$. ■

We may now prove that every almost factorisable inverse semigroup arises in this way.

Theorem 13 *Let S be an almost factorisable inverse semigroup. Then there is a factorisable inverse monoid F such that S is isomorphic to $F \setminus U(F)$.*

Proof Let $\iota\colon S \to C(S)$ be the embedding of S in its completion. Let F be the subset of $C(S)$ consisting of $\iota(S)$ and the group of units of $C(S)$. Clearly, F is closed under inversion.

To show that F is an inverse subsemigroup of $C(S)$, it is enough to prove that for every $s \in S$ and unit A of $C(S)$ the product $[s]A \in \iota(S)$. Since $AA^{-1} = E(S)$ there exists $a \in A$ such that $s^{-1}s = aa^{-1}$. We show that $[sa] = [s]A$. Let $x \in [sa]$. Then $x = sa(x^{-1}x) = s(ax^{-1}x)$. But $ax^{-1}x \leq a$ and $a \in A$, an order ideal, so that $ax^{-1}x \in A$. Thus $x \in [s]A$. Conversely, let $x \in [s]A$. Then $x = tb$ where $t \leq s$ and $b \in A$. Clearly, $x = t(t^{-1}tb)$ and $t^{-1}tb \in A$, since $b \in A$ and A is an order ideal. Now

$$t^{-1}tb(t^{-1}tb)^{-1} = bb^{-1}t^{-1}t \leq t^{-1}t \leq s^{-1}s = aa^{-1},$$

and $t^{-1}tb, a \in A$ so that $t^{-1}tb \sim a$. Thus $t^{-1}tb \leq a$ by Lemma 1.4.14. Hence

$$x = tb = t(t^{-1}tb) \leq ta \leq sa,$$

and so $x \in [sa]$. Hence $[s]A = [sa]$ as required.

The inverse semigroup F has identity $E(S)$ and its group of units is the same as the group of units of $C(S)$. The natural partial order on F is just the restriction of the natural partial order on $C(S)$. Thus $[s] \leq A$ precisely when $s \in A$. We can now show that F is a factorisable inverse monoid. Let $[s] \in F$. Then $s \in S$ and so $s \in A$ for some unit $A \in C(S)$. But A is a unit of F and $[s] \leq A$ in F. Hence F is a factorisable inverse monoid, and by construction S is isomorphic to $F \setminus U(F)$. ■

7.2 The P-theorem

In this section, we shall obtain a more precise description of E-unitary inverse semigroups by generalising the semidirect product construction to what are called P-semigroups: we prove that every E-unitary semigroup is isomorphic to a P-semigroup. The proof of the P-theorem is divided into two parts: the first starts with an E-unitary inverse semigroup of partial bijections on a set and leads up to the definition of P-semigroups; the second begins with an abstract E-unitary inverse semigroup and concludes with a proof of its representability as a P-semigroup.

Motivation

Let S be an E-unitary inverse subsemigroup of the symmetric inverse monoid $I(A)$ on the set A. In an E-unitary semigroup the compatibility relation equals the minimum group congruence by Theorem 2.4.6. We shall use this fact to analyse S. We shall further use the fact that $I(A)$ has joins of all non-empty compatible subsets and that $I(A)$ is infinitely distributive (Proposition 1.2.1).

Let $f \in S$. Then the set $\sigma(f)$ consists of pairwise compatible elements in S. Thus the union of the set $\sigma(f)$, which we shall denote by α_f, is a well-defined element of $I(A)$; of course, there is no reason for it to belong to S. Put $G' = \{\alpha_f \colon f \in S\}$. The set G' has some interesting properties, which the following lemma describes.

Lemma 1 *With the notation above:*

(1) *For each pair α and β in G' there exists a unique element γ in G' such that $\alpha\beta \subseteq \gamma$.*

(2) *α_e is an idempotent when e is an idempotent.*

(3) *If $\alpha \in G'$ then $\alpha^{-1} \in G'$.*

Proof (1) Firstly, we show that $\alpha\beta$ is contained in some element of G'. By assumption, $\alpha = \alpha_f$ and $\beta = \alpha_g$ for some $f, g \in S$. Put $\gamma = \alpha_{fg}$. We show that $\alpha\beta \subseteq \gamma$. Let $x \in \operatorname{dom}(\alpha_f\alpha_g)$. Then there exist $g' \in \sigma(g)$ and $f' \in \sigma(f)$ such that $g'(x)$ and $f'(g'(x))$ exist. Thus $(f'g')(x)$ exists. But $fg \,\sigma\, f'g'$, so that $f'g' \subseteq \gamma$. Thus $x \in \operatorname{dom}\gamma$ and $(\alpha_f\alpha_g)(x) = \gamma(x)$. Hence $\alpha\beta \subseteq \gamma$.

To show that γ is unique, suppose that there is a $\delta \in G'$ such that $\alpha\beta \subseteq \delta$ where $\delta = \alpha_h$. Then $fg \subseteq \delta$ and $h \subseteq \delta$, so that fg and h are compatible by Lemma 1.4.14. Hence $fg \,\sigma\, h$. Consequently $\delta = \gamma$.

(2) If $e \in S$ is an idempotent of S, then $\sigma(e) = E(S)$ by Theorem 2.4.6. Thus α_e is an idempotent.

(3) Observe that $(\alpha_f)^{-1} = \alpha_{f^{-1}}$ for each $f \in S$. ■

Define a binary operation $\odot$ on G' by $\alpha \odot \beta = \gamma$ where γ is the unique element of G' containing $\alpha\beta$ guaranteed by Lemma 1(1).

Lemma 2 *$(G', \odot)$ is a group isomorphic to S/σ; the identity is α_e, where e is any idempotent of S, and the inverse of $\alpha \in G'$ is the partial bijection α^{-1}.*

Proof By Lemma 1(1), $\alpha_f \odot \alpha_g = \alpha_{fg}$. To prove that $(G', \odot)$ is a group it is enough to show that there is a bijection from S/σ to G' preserving binary operations. This is easily achieved by mapping $\sigma(f)$ to α_f. Next, observe that $\alpha_e\alpha_e = \alpha_e$ by Lemma 1(2), and so $\alpha_e \odot \alpha_e = \alpha_e$. But the only idempotent in the group $(G', \odot)$ is its identity. Finally, we have to calculate the group inverse of $\alpha \in G'$. If $\alpha \in G'$ then $\alpha^{-1} \in G'$ by Lemma 1(3). Observe that $\alpha^{-1}\alpha \subseteq \alpha_e$. Thus $\alpha^{-1} \odot \alpha = \alpha_e$. It follows that the group inverse of α is α^{-1}. ■

Although G' is a group under the binary operation $\odot$, it is represented in a rather unusual way; we would expect it to consist of bijections, rather than partial bijections, and for the binary operation to be composition of functions rather than $\odot$. A representation of the group G' by means of bijections under composition can be achieved using the central idea of Chapter 2: we embed the set A into a larger set B in such a way that every element of G' can be extended to a bijection on B. We shall further require that the set of bijections on B we obtain should form a group under composition isomorphic to G'. With this in mind, we make the following assumptions, which will be fully justified in the next section.

Let S be an E-unitary inverse subsemigroup of $I(A)$ and let G' be constructed as above. We shall assume that there is a set B containing A and a set G of bijections of B such that the following three conditions hold:

(A1) Every element of G' is contained in a unique element of G, and every element of G contains a unique element of G'.

(A2) The set G forms a group under composition isomorphic to $(G', \odot)$, where the identity of G is 1_B.

(A3) Let $\beta \in G$ and let $1_C, 1_D \in S$ be idempotents. Put $\alpha = 1_C\beta$ and suppose that $\alpha^{-1}\alpha \subseteq 1_D$. Then $\alpha \in S$.

We shall now look at the consequences of these assumptions for the structure of the inverse semigroup S.

Lemma 3 *Each element of S is contained in a unique element of G.*

Proof Let $f \in S$, and suppose that $f \subseteq \beta, \gamma$ where $\beta, \gamma \in G$. Then $\alpha_g \subseteq \beta$ and $\alpha_h \subseteq \gamma$ for some $g, h \in S$ by assumption (A1). Thus $g, f \subseteq \beta$ and $f, h \subseteq \gamma$. By Lemma 1.4.14, $g \sim f$ and $f \sim h$ and so $g \, \sigma \, f$ and $f \, \sigma \, h$. Hence $g \, \sigma \, h$ and so $\alpha_g = \alpha_h$. But then $\beta = \gamma$ by assumption (A1). ■

For each $f \in S$, we shall denote by β_f the unique element of G in which it is contained. In what follows, we shall denote by Y the set of idempotents of S.

If $f \in S$ then $ff^{-1} \in Y$ and $f \subseteq \beta_f$. Thus $f = (ff^{-1})\beta_f$. Observe that $f^{-1}f = \beta_f^{-1}(ff^{-1})\beta_f$ and $f^{-1}f \in Y$. These observations motivate the following definition. Put

$$P = \{(e, \beta) \in Y \times G \colon \beta^{-1} e \beta \in Y\}.$$

Lemma 4 *There is a bijection $\Theta \colon S \to P$ given by $\Theta(f) = (ff^{-1}, \beta_f)$.*

Proof Our comments above show that Θ is well-defined. To show that Θ is surjective, let $(e, \beta) \in P$. Put $g = e\beta$. Then $g^{-1}g = \beta^{-1}e\beta \in Y$, by assumption. Thus $g \in S$ by assumption (A3). Now $gg^{-1} = e$ and $g \subseteq \beta$, so that $\Theta(g) = (e, \beta)$.

To show that Θ is injective, suppose that $\Theta(f) = \Theta(g)$. By definition $ff^{-1} = gg^{-1}$ and $\beta_f = \beta_g$. Hence $f, g \subseteq \beta_f$ and so $f \, \sigma \, g$ by Lemma 1.4.14. But S is E-unitary and so σ is idempotent pure by Theorem 2.4.6. Thus $f = g$. ■

We have coordinatised S, but now we need to take care of the multiplication. Let $f, g \in S$. The product

$$(ff^{-1}\beta_f)(gg^{-1}\beta_g)$$

is equal to

$$(ff^{-1})(\beta_f gg^{-1}\beta_f^{-1})\beta_f\beta_g$$

using the proof of Lemma 1.4.2. The idempotent gg^{-1} is of course an element of Y, but there is no reason to suppose that $\beta_f gg^{-1}\beta_f^{-1}$ is. We are thus led to define an auxiliary set

$$X = \{\beta^{-1}e\beta \colon e \in Y \text{and } \beta \in G\},$$

which we regard as being partially ordered by functional-inclusion. The group G acts on the set X if we define

$$\beta \cdot e = \beta e \beta^{-1}.$$

We now list some properties of the triple (G, X, Y).

Lemma 5 *With the notation above, G acts on X by order automorphisms and in addition:*

(1) *Y is an order ideal of X.*

(2) *$G \cdot Y = X$.*

(3) *$\beta \cdot Y \cap Y \neq \emptyset$ for all $\beta \in G$.*

Proof It is clear that G acts by order automorphisms.
(1) Let $i \subseteq e$ where $i \in X$ and $e \in Y$. By assumption, $i = \beta^{-1} j \beta$ for some $\beta \in G$ and $j \in Y$. Put $g = j\beta$. Then $g^{-1}g = \beta^{-1}j\beta = i \subseteq e$. Thus $g \in S$ by assumption (A3), and so $i = g^{-1}g \in Y$.
(2) Immediate from the definition of X.
(3) Let $\beta \in G$. Then $\beta^{-1} \in G$ and so by assumption (A1) it contains a unique element of G'. If this element is α_f, where $f \in S$, then $f \subseteq \beta^{-1}$. Thus $f = ff^{-1}\beta^{-1}$. Hence $f^{-1}f = \beta f f^{-1}\beta^{-1} = \beta \cdot (ff^{-1}) \in Y$, and so $ff^{-1}, \beta \cdot (ff^{-1}) \in Y$. ■

Define a binary operation on P by

$$(e, \beta)(f, \gamma) = (e(\beta \cdot f), \beta\gamma).$$

Lemma 6 *The above operation is well-defined.*

Proof Let $(e, \beta), (f, \gamma) \in P$. Firstly, $e(\beta \cdot f) \in Y$ since Y is an order ideal of X by Lemma 5(1). It remains to show that $(\beta\gamma)^{-1} \cdot (e(\beta \cdot f))$ belongs to Y. By definition,

$$(\beta\gamma)^{-1} \cdot (e(\beta \cdot f)) = \gamma^{-1}((\beta^{-1} \cdot e)f)\gamma.$$

Now $(\beta^{-1} \cdot e)f \subseteq f$ and so

$$\gamma^{-1}((\beta^{-1} \cdot e)f)\gamma \subseteq \gamma^{-1}f\gamma = \gamma^{-1} \cdot f \in Y,$$

since $(f, \gamma) \in P$. Thus $\gamma^{-1}((\beta^{-1} \cdot e)f)\gamma \in Y$ since Y is an order ideal of X by Lemma 5(1). Hence the operation is well-defined. ■

By Lemma 4 and the definitions above, the following is now immediate.

Theorem 7 *The inverse semigroup S is isomorphic to P under the function Θ.* ■

Proof of the P-theorem

Everything we have done so far was modulo the assumptions (A1), (A2) and (A3). We shall justify these assumptions in this section. We begin by abstracting the construction of the semigroup P from triples of the form (G, X, Y).

Let G be a group and X a partially ordered set. We shall suppose that G acts on X on the left by order automorphisms. We denote the action of $g \in G$ on $x \in X$ by $g \cdot x$. Let Y be a subset of X partially ordered by the induced ordering. We say that (G, X, Y) is a *McAlister triple* if the following three axioms hold:

(MT1) Y is an order ideal of X and a meet semilattice under the induced ordering.

(MT2) $G \cdot Y = X$.

(MT3) $g \cdot Y \cap Y \neq \emptyset$ for every $g \in G$.

Let (G, X, Y) be a McAlister triple. Put

$$P(G, X, Y) = \{(y, g) \in Y \times G \colon g^{-1} \cdot y \in Y\}.$$

Lemma 8 *Let $(e, g), (f, h) \in P(G, X, Y)$. Then $e \wedge g \cdot f$ exists in the partially ordered set X and $(e \wedge g \cdot f, gh) \in P(G, X, Y)$.*

Proof By assumption, $g^{-1} \cdot e \in Y$. Thus $g^{-1} \cdot e \wedge f$ exists since Y is a semilattice. Put $i = g^{-1} \cdot e \wedge f$. Then $i \leq g^{-1} \cdot e$ and $i \leq f$. Thus $g \cdot i \leq e$ and $g \cdot i \leq g \cdot f$. Now let $j \leq e, g \cdot f$. Then $g^{-1} \cdot j \leq g^{-1} \cdot e$ and $g^{-1} \cdot j \leq f$. Hence $g^{-1} \cdot j \leq i$, and so $j \leq g \cdot i$. We have therefore shown that the meet $e \wedge g \cdot f$ exists.

To show that $(e \wedge g \cdot f, gh) \in P(G, X, Y)$, we have to show that the element $(gh)^{-1} \cdot (e \wedge g \cdot f)$ belongs to Y. Now

$$(gh)^{-1} \cdot (e \wedge g \cdot f) = h^{-1} \cdot (g^{-1} \cdot (e \wedge g \cdot f))$$

and

$$h^{-1} \cdot (g^{-1} \cdot (e \wedge g \cdot f)) \leq h^{-1} \cdot (g^{-1} \cdot (g \cdot f)) \leq h^{-1} \cdot f \in Y.$$

But Y is an order ideal of X and so $(gh)^{-1} \cdot (e \wedge g \cdot f) \in Y$. ∎

Define a product on $P(G, X, Y)$ by

$$(e, g)(f, h) = (e \wedge g \cdot f, gh).$$

It is well-defined by Lemma 8.

Theorem 9 *$P(G,X,Y)$ is an E-unitary inverse semigroup, with semilattice of idempotents isomorphic to Y and maximum group homomorphic image isomorphic to G.*

Proof Put $P = P(G,X,Y)$. It is straightforward to show that the product is associative. Let $(e,g) \in P$. Then it is straightforward to check that the pair $(g^{-1}\cdot e, g^{-1})$ belongs to P, and that it is an inverse of (e,g). Thus P is regular. The idempotents of P are the elements of the form $(e,1)$. The product of the idempotents $(e,1)$ and $(f,1)$ is $(e \wedge f, 1)$, so that not only do the idempotents commute, making P inverse, but the semilattice of idempotents is isomorphic to Y under the map which takes $(e,1)$ to e. Direct calculation shows that the natural partial order is given by

$$(e,g) \leq (f,h) \Leftrightarrow e \leq f \text{ and } g = h.$$

It is immediate from this that P is E-unitary. A simple calculation shows that

$$(e,g)\,\sigma\,(f,h) \Leftrightarrow g = h.$$

Define a function from P to G by $(e,g) \mapsto g$. It is clearly a homomorphism. To show that it is surjective, let $g \in G$. By axiom (MT3), $g^{-1} \cdot Y \cap Y \neq \emptyset$. Let $e \in Y$ such that $g^{-1} \cdot e \in Y$. Then $(e,g) \in P$ and $(e,g) \mapsto g$. It is now immediate that G is isomorphic to P/σ. ■

Semigroups of the form $P(G,X,Y)$ are called *P-semigroups*. They are generalisations of semidirect products of groups and semilattices: for if $X = Y$ is a semilattice, then $P(G,X,X) = P(G,X)$. In Section 6.3, we proved that every free inverse semigroup is isomorphic to a P-semigroup.

We can now justify the assumptions made in the previous section, and so prove that every E-unitary inverse semigroup is isomorphic to a P-semigroup. Let S be an E-unitary inverse semigroup. By the Wagner–Preston representation theorem, Theorem 1.5.1, S can be embedded in $I(S)$ by means of the homomorphism θ: $S \to I(S)$ given by

$$\theta(a) = \theta_a \colon a^{-1}aS \to aa^{-1}S$$

where $\theta_a(x) = ax$. Notice that S plays two rôles: on the left-hand side it is regarded as an inverse semigroup, whereas on the right-hand side it is regarded as a set. Our first step is to replace the *set* S by a set in bijective correspondence with it; essentially, we shall re-coordinatise S so that the action of S on itself determined by the Wagner–Preston representation will have a simple form.

Put $G = S/\sigma$. Define κ: $S \to E(S) \times G$ by $\kappa(s) = (s^{-1}s, \sigma(s))$. Since S is E-unitary, the congruence σ is idempotent pure and so κ is injective. Put $A = \kappa(S)$ and $B = E(S) \times G$. We shall regard κ as a bijection from S to

A. Define an injective homomorphism ϕ: $S \to I(A)$ by $\phi(a) = \kappa\theta(a)\kappa^{-1}$. Of course, $I(A)$ is an inverse subsemigroup of $I(B)$. We now look at the effect of $\phi(a)$ on A.

Lemma 10 *With the above notation, the function*

$$\phi(a)\colon \kappa(a^{-1}aS) \to \kappa(aa^{-1}S)$$

has the form

$$\phi(a)(x^{-1}x, \sigma(x)) = (x^{-1}x, \sigma(a)\sigma(x)).$$

Proof Let $x \in a^{-1}aS$. Then $\kappa(x) = (x^{-1}x, \sigma(x))$. Thus $\phi(a)(x^{-1}x, \sigma(x)) = \phi(a)(\kappa(x))$. But $\phi(a)(\kappa(x)) = \kappa(\theta(a)(x)) = \kappa(ax)$. However

$$\kappa(ax) = ((ax)^{-1}(ax), \sigma(ax)) = (x^{-1}a^{-1}ax, \sigma(a)\sigma(x)) = (x^{-1}x, \sigma(a)\sigma(x)),$$

since $a^{-1}ax = x$. ∎

The above result shows that the form of the action of $\phi(a)$ on A is remarkably straightforward, and it motivates the following definition. Define an action $G \times B \to B$ by $g \cdot (e, h) = (e, gh)$.

Lemma 11 *With the above notation, G acts without fixed-points on B. In particular, G is embedded in $I(B)$.*

Proof Suppose that $g \cdot (e, h) = (e, h)$. Then $(e, gh) = (e, h)$, and so $gh = h$ thus $g = 1$, the identity.

Let β_g be the permutation of B determined by the action of g on B. If $\beta_g = \beta_h$ then $g \cdot (e, k) = h \cdot (e, k)$ for all $(e, k) \in B$. But then $gk = hk$ and so $g = h$. Thus G is embedded in $I(B)$. ∎

Let $\hat{G}$ be the image of G in $I(B)$; the elements of $\hat{G}$ are of the form β_g where β_g is the permutation of B determined by the action of g. Define, as before, $\alpha_{\phi(a)} = \bigcup \sigma(\phi(a))$ and put $G' = \{\alpha_{\phi(a)}\colon a \in S\}$. To summarise: $\phi(S)$ is an E-unitary inverse subsemigroup of $I(A)$; G' consists of the partial bijections of $I(A)$ obtained from the σ-classes of $\phi(S)$; B is a set extending A; and $\hat{G}$ is a group of permutations of B.

Lemma 12 *The assumptions* (A1) *and* (A2) *hold for G' and $\hat{G}$.*

Proof For every $a \in S$, we have that $\phi(a) \subseteq \beta_{\sigma(a)}$ by Lemma 10. Thus $\alpha_{\phi(a)} \subseteq \beta_{\sigma(a)}$, and so each element of G' is contained in an element of $\hat{G}$, and vice-versa. Uniqueness, in both cases, is guaranteed by Lemma 11. Thus assumption (A1) holds. By definition, $\hat{G}$ is isomorphic to G, and 1_B is the identity of $\hat{G}$ and so assumption (A2) holds. ■

It remains to show that assumption (A3) holds. Extend the group action to arbitrary subsets of B: if $C \subseteq B$ then $g \cdot C = \{(e, gh) \colon (e, h) \in C\}$.

Lemma 13 *If $g \cdot \kappa(eS) \subseteq \kappa(fS)$ then there exists $u \in fS$ such that $e = u^{-1}u$ and $g = \sigma(u)$ and $g \cdot \kappa(eS) = \kappa(uS)$.*

Proof Since $e \in eS$ we have that $\kappa(e) = (e, 1) \in \kappa(eS)$ and so $g \cdot \kappa(e) = (e, g) \in \kappa(fS)$. Thus there exists $u \in fS$ such that $\kappa(u) = (e, g)$ where $u^{-1}u = e$ and $g = \sigma(u)$. It remains to establish the final claim. Let $(x^{-1}x, g\sigma(x)) \in g \cdot \kappa(eS)$, where $ex = x$. Put $y = ux$. Then

$$y^{-1}y = (ux)^{-1}ux = x^{-1}u^{-1}ux = x^{-1}ex = x^{-1}x$$

and $\sigma(y) = \sigma(ux) = g\sigma(x)$. Thus

$$(x^{-1}x, g\sigma(x)) = (y^{-1}y, \sigma(y)) \in \kappa(uS).$$

Conversely, let $(y^{-1}y, \sigma(y)) \in \kappa(uS)$, where $y \in uS$. Then $y = ux$ for some $x \in eS$, since $e = u^{-1}u$. It follows that $y^{-1}y = x^{-1}x$ and $\sigma(y) = g\sigma(x)$. Thus

$$(y^{-1}y, \sigma(y)) = (x^{-1}x, g\sigma(x)) \in g \cdot \kappa(eS).$$

Hence $g \cdot \kappa(eS) = \kappa(uS)$. ■

Proposition 14 *With the notation above, assumption* (A3) *holds.*

Proof Let $\phi(e)$ and $\phi(f)$ be idempotents in $\phi(S)$ and let $\beta_g \in \hat{G}$. Put $\alpha = \phi(e)\beta_g$ and suppose that $\alpha^{-1}\alpha \subseteq \phi(f)$. We shall prove that $\alpha \in \phi(S)$. By assumption,

$$\alpha^{-1}\alpha = \beta_g^{-1}\phi(e)\beta_g \subseteq \phi(f).$$

Thus

$$\operatorname{dom}(\beta_g^{-1}\phi(e)\beta_g) \subseteq \operatorname{dom}\phi(f).$$

But $\operatorname{dom}(\beta_g^{-1}\phi(e)\beta_g) = g^{-1} \cdot \kappa(eS)$ and $\operatorname{dom}\phi(f) = \kappa(fS)$. Thus $g^{-1} \cdot \kappa(eS) \subseteq \kappa(fS)$. By Lemma 13, there exists $u \in fS$ such that $g^{-1} = \sigma(u)$, $u^{-1}u = e$ and $g^{-1} \cdot \kappa(eS) = \kappa(uS)$. We prove that $\phi(u^{-1}) = \alpha$. Firstly,

$$\operatorname{dom}\alpha = \beta_g^{-1}(\operatorname{dom}\phi(e)) = g^{-1} \cdot \kappa(eS) = \kappa(uS) = \kappa(uu^{-1}S),$$

and

$$\operatorname{im}\alpha = \operatorname{im}\phi(e) = \kappa(u^{-1}uS).$$

Secondly, by construction α is a restriction of β_g, and $g = \sigma(u^{-1})$. Thus by Lemma 10, $\alpha = \phi(u^{-1})$. ■

We can now prove the P-theorem.

Theorem 15 *Every E-unitary inverse semigroup is isomorphic to a P-semigroup.*

Proof Let S be an E-unitary inverse semigroup, and let θ: $S \to I(S)$ be the Wagner–Preston representation of S. Define κ: $S \to E(S) \times G$, where $G = S/\sigma$, where $\kappa(s) = (s^{-1}s, \sigma(s))$. This is injective because S is E-unitary. Put $A = \kappa(S)$ and $B = E(S) \times G$. Thus, in particular, $A \subseteq B$. The function ϕ: $S \to I(A)$, given by $\phi(a) = \kappa\theta(a)\kappa^{-1}$, is an injective homomorphism, and $I(A)$ is an inverse subsemigroup of $I(B)$. Define G' and $\hat{G}$ as above. Then by Lemma 12 and Proposition 14, the assumptions (A1), (A2) and (A3) hold. Let Y be the set of idempotents of $\phi(S)$, and $X = \{\beta^{-1}e\beta^{-1}: e \in Y \text{and } \beta \in \hat{G}\}$. Then $(\hat{G}, X, Y)$ is a McAlister triple by Lemma 5, and so $P(\hat{G}, X, Y)$ is isomorphic to $\phi(S)$ by Theorem 7. But G is isomorphic to $\hat{G}$ and S is isomorphic to $\phi(S)$, and so S is isomorphic to the P-semigroup $P(G, X, Y)$. ■

In Chapter 8, we prove that every E-unitary inverse semigroup is isomorphic to an essentially unique P-semigroup.

7.3 The Möbius inverse monoid

The theory developed in the previous section will now be illustrated by means of a concrete example.

Möbius transformations

A *Möbius transformation* is a partial function α of the complex plane having the form: $\alpha(z) = \frac{az+b}{cz+d}$ where a, b, c, d are complex numbers and $ad - bc \neq 0$. Of course, there are many ways to write the function α because multiplying a, b, c and d by a non-zero complex number results in a different form but the same function. This must always be borne in mind.

Möbius transformations are *partial* functions rather than functions because of their denominators; if the coefficient c is non-zero then there is one value of z satisfying $cz + d = 0$. Thus the transformations are of two types: the functions in which $c = 0$, and the partial functions in which $c \neq 0$. Of course, the partial transformations are 'only just' partial; the domain of the function omits the point $\frac{-d}{c}$ and the image omits the point $\frac{a}{c}$. The condition $ad - bc \neq 0$ ensures that in both cases the transformations are injective. Thus Möbius transformations are either bijections or partial bijections of $\mathbb{C}$. The inverse transformation is also a Möbius transformation, namely $\alpha^{-1}(z) = \frac{dz-b}{-cz+a}$.

The composition of two Möbius transformations is not in general a Möbius transformation. To see this, let

$$\alpha\colon \mathbb{C}\setminus\{1\}\to\mathbb{C}\setminus\{2\} \text{ given by } \alpha(z)=\frac{2z+1}{z-1}$$

and

$$\beta\colon \mathbb{C}\setminus\{-1\}\to\mathbb{C}\setminus\{2\} \text{ given by } \beta(z)=\frac{2z+1}{z+1}$$

be Möbius transformations. Then their composite $\alpha\beta$ is such that

$$\mathrm{dom}(\alpha\beta)=\mathbb{C}\setminus\{0,-1\} \text{ and } \mathrm{im}(\alpha\beta)=\mathbb{C}\setminus\{2,5\}.$$

Thus the partial function composition of α and β is not a Möbius transformation, because two points are omitted from the domain of the resulting function.

Möbius transformations can be 'multiplied' by an operation which, as we have seen above, cannot be composition of partial functions. If $\alpha(z)=\frac{az+b}{cz+d}$ and $\beta(z)=\frac{Az+B}{Cz+D}$ are Möbius transformations then their product, which we shall denote by $\alpha\odot\beta$, is defined to be $\frac{(aA+bC)z+(aB+bD)}{(cA+dC)z+(cB+dD)}$. The result is certainly a Möbius transformation, as can easily be checked. Consequently, $\odot$ is a well-defined binary operation on the set of Möbius transformations. With respect to this operation the Möbius transformations form a group, called the *Möbius group*.

The Möbius group and the Möbius inverse monoid

Since Möbius transformations are partial bijections of $\mathbb{C}$ we can regard them as elements of the symmetric inverse monoid $I(\mathbb{C})$. The inverse submonoid of $I(\mathbb{C})$ generated by the Möbius transformations we call the *Möbius inverse monoid*, denoted by $\mathcal{M}$. In this section, we shall how the Möbius group can be obtained from the Möbius inverse monoid.

The following inverse subsemigroup of $I(X)$ will be useful in our analysis of $\mathcal{M}$.

Lemma 1 *Let X be an infinite set. The set of bijections between the cofinite subsets of X forms an inverse subsemigroup of $I(X)$.*

Proof Let A and B be finite subsets of X. Then $(X\setminus A)\cap(X\setminus B)=X\setminus(A\cup B)$. Thus the intersection of two cofinite subsets of X is again cofinite. It is now straightforward to prove that the bijections between the cofinite subsets form an inverse subsemigroup of $I(X)$. ■

We shall refer to the elements of the above inverse semigroup as *cofinite partial bijections*.

Lemma 2 *The idempotents of the Möbius inverse monoid are precisely the cofinite idempotents of $I(\mathbb{C})$.*

Proof Every element of $\mathcal{M}$ is a finite composition of Möbius transformations each of which is cofinite. Thus every element of $\mathcal{M}$ is cofinite; in particular, every idempotent of $\mathcal{M}$ is cofinite.

To prove the converse, it is enough to prove that every subset of $\mathbb{C}$ of the form $\mathbb{C} \setminus \{w\}$, where $w \in \mathbb{C}$, is the domain of a Möbius transformation, because the identity defined on the subset $\mathbb{C} \setminus \{w_1, \ldots w_n\}$ can be obtained by composing the identities defined on the subsets $\mathbb{C} \setminus \{w_1\}, \ldots, \mathbb{C} \setminus \{w_n\}$. Let w be a complex number. When $w \neq 0$ the domain of the Möbius transformation $\frac{z}{z-w}$ is $\mathbb{C} \setminus \{w\}$, and when $w = 0$ the domain of the Möbius transformation $\frac{z+1}{z}$ is $\mathbb{C} \setminus \{0\}$. ■

The proof of the following is immediate from the definitions.

Lemma 3 $\alpha\beta \subseteq \alpha \odot \beta$. ■

The following is a fundamental property of Möbius transformations.

Theorem 4 *A Möbius transformation which fixes three or more points is the identity. Hence two Möbius transformations which agree on three or more points must be equal.* ■

The above result has the following important consequence for the structure of $\mathcal{M}$.

Lemma 5 *If $\alpha \in \mathcal{M}$ and $\alpha \subseteq \beta, \gamma$ where β and γ are Möbius transformations, then $\beta = \gamma$.*

Proof The functions β and γ agree on the set $\operatorname{dom} \alpha$. By Lemma 2, this set is cofinite and so always contains at least three elements. The result now follows from Theorem 4. ■

Our first important result is the following.

Theorem 6 *In the Möbius inverse monoid $\mathcal{M}$, the following hold:*

(1) *The Möbius transformations form the maximal elements of $\mathcal{M}$, and every element is contained in a unique Möbius transformation.*

(2) *Two elements are σ-related if, and only if, they are bounded above by the same Möbius transformation.*

(3) *The Möbius inverse monoid is F-inverse.*

(4) *$\mathcal{M}/\sigma$ is isomorphic to the Möbius group.*

Proof (1) By Lemma 3, the composition of any finite number of Möbius transformations is contained in a Möbius transformation, and so every element of $\mathcal{M}$ is contained in a Möbius transformation. Now suppose that $\alpha \subseteq \beta$ where α is a Möbius transformation and β an arbitrary element of $\mathcal{M}$. Then $\beta \subseteq \gamma$ for some Möbius transformation γ. Thus $\alpha \subseteq \gamma$. But then $\alpha = \gamma$ by Theorem 4, and so $\alpha = \beta$. Thus the Möbius transformations are all maximal. Every element of $\mathcal{M}$ is contained in a unique Möbius transformation by Lemma 5.
(2) Let α and β be two elements of $\mathcal{M}$ which are σ-related. By definition, there exists $\gamma \in \mathcal{M}$ such that $\gamma \subseteq \alpha$ and $\gamma \subseteq \beta$. Now α and β are contained in maximal elements α' and β'. Thus γ is contained in both α' and β'. Hence $\alpha' = \beta'$ by (1) above. It follows that $\alpha, \beta \subseteq \alpha' = \beta'$. Conversely, two elements which are bounded above by the same element must be σ-related.
(3) By (1) and (2), each σ-class contains a unique Möbius transformation which is the maximum element of its σ-class. Thus $\mathcal{M}$ is an F-inverse monoid.
(4) Consider now the group $\mathcal{M}/\sigma$. The elements of this group are in bijective correspondence with the maximum elements of the σ-classes and so with the Möbius transformations. Let $\sigma(\alpha)$ and $\sigma(\beta)$ be two σ-classes containing the maximum elements α and β. Their product in $\mathcal{M}/\sigma$ will be a σ-class $\sigma(\gamma)$ containing the maximum element γ. Clearly, $\alpha\beta \subseteq \gamma$. But $\alpha\beta \subseteq \alpha \odot \beta$ by Lemma 3, $\alpha \odot \beta$ is a Möbius transformation, and $\alpha\beta$ is contained in a unique Möbius transformation. Thus $\gamma = \alpha \odot \beta$. ■

The structure of the Möbius inverse monoid

It is tempting to think that the structure of the Möbius inverse monoid must be rather simple, perhaps that it is a semidirect product of a semilattice by a group; this is not the case.

Proposition 7 *The Möbius inverse monoid is not isomorphic to a semidirect product of a semilattice by a group.*

Proof Let α be any Möbius transformation which omits a point w_1 from its domain, and let ι be the identity transformation. If $\mathcal{M}$ were a semidirect product, there would exist by Theorem 7.1.2(2) an element $\beta \in \mathcal{M}$ such that $\beta \sim \alpha$ and $\beta^{-1}\beta = \iota$. By Theorem 6, β is contained in some Möbius transformation γ. But then $\beta = \gamma\beta^{-1}\beta = \gamma$. Thus β is a Möbius transformation. Since $\beta^{-1}\beta = \iota$, it omits no complex number from its domain and so β is a bijection on $\mathbb{C}$. From $\beta \sim \alpha$ and $\alpha^{-1}\alpha \subseteq \beta^{-1}\beta = \iota$ we obtain $\alpha \subseteq \beta$ by Lemma 1.4.14. But α and β are both Möbius transformations and so must be equal by Theorem 4. However, this contradicts the fact that β is a bijection and α omits a point. ■

We shall now obtain a McAlister triple which describes the Möbius inverse monoid.

Adjoin a new symbol ∞ to the set $\mathbb{C}$ and denote the resulting set by $\mathbb{C}^*$. This new symbol is required to satisfy the usual arithmetic properties of infinity: $z + \infty = \infty = \infty + z$, for all $z \in \mathbb{C}$; $z\infty = \infty = \infty z$ for all $z \in \mathbb{C} \setminus \{0\}$; $z/\infty = 0$ and $z/0 = \infty$ for all $z \in \mathbb{C} \setminus \{0\}$, and $\infty\infty = \infty$.

The inverse semigroup $I(\mathbb{C})$ will be regarded as an inverse subsemigroup of $I(\mathbb{C}^*)$. Each Möbius transformation $\alpha(z) = \frac{az+b}{cz+d}$ can be extended to a bijection α^* of $\mathbb{C}^*$, called an *extended Möbius transformation*, in the following way: if $c \neq 0$, then $\alpha^*(\frac{-d}{c}) = \infty$ and $\alpha^*(\infty) = \frac{a}{c}$ and elsewhere agrees with α, whereas if $c = 0$ then $\alpha^*(\infty) = \infty$ and elsewhere agrees with α.

Lemma 8 *With the terminology above.*

(1) *For every Möbius transformation α, we have that $\alpha = 1_{\mathbb{C}}\alpha^* 1_{\mathbb{C}}$.*

(2) *If α and β are Möbius transformations then $(\alpha \odot \beta)^* = \alpha^*\beta^*$.*

Proof (1) Let α be the Möbius transformation $\alpha(z) = \frac{az+b}{cz+d}$. If $c = 0$ then the result is clear since $\alpha^*(\infty) = \infty$. Suppose that $c \neq 0$. Then a simple calculation shows that $\mathrm{dom}(1_{\mathbb{C}}\alpha^*1_{\mathbb{C}}) = \mathbb{C} \setminus \{\frac{-d}{c}\}$ and $\mathrm{im}(1_{\mathbb{C}}\alpha^*1_{\mathbb{C}}) = \mathbb{C} \setminus \{\frac{a}{c}\}$.
(2) Straightforward calculation. ■

The extended Möbius transformations form a subgroup H of the group of units of $I(\mathbb{C}^*)$, which is isomorphic to the Möbius group.

Define $Y = E(\mathcal{M})$ and

$$X = \{g^{-1}eg\colon g \in H \text{ and } e \in Y\}.$$

The group H acts on the set X when we define $g \cdot e = geg^{-1}$.

Lemma 9 *The set X consists of just the non-identity cofinite idempotents of $I(\mathbb{C}^*)$.*

Proof Since every element of Y is a cofinite idempotent of $I(\mathbb{C})$, it is immediate from the definition that every element of X is a cofinite idempotent of $I(\mathbb{C}^*)$. Furthermore, because every cofinite idempotent of $I(\mathbb{C})$ omits the point ∞ from its domain, every element of X must omit at least one element of $\mathbb{C}^*$ from its domain. Thus, in particular, the identity on $\mathbb{C}^*$ is not an element of X. It remains to show that every non-identity cofinite idempotent of $I(\mathbb{C}^*)$ belongs to X. Let $1_{\mathbb{C}\setminus A}$ be a non-identity cofinite idempotent in $I(\mathbb{C}^*)$, where $A = \{w_1, \ldots, w_n\}$. If any of the $w_i = \infty$ then $1_{\mathbb{C}\setminus A}$ is a cofinite idempotent of $I(\mathbb{C})$ and consequently an element of Y, and so of X. We may therefore suppose that none of the w_i is equal to ∞. Consider the extended Möbius transformation $\alpha^* = \frac{1}{z-w_1}$. Then $\alpha^*(\infty) = 0$ and $\alpha^*(w_1) = \infty$. Thus the image of $\alpha^* 1_{\mathbb{C}\setminus A}$ is a cofinite idempotent of $I(\mathbb{C})$. Hence $1_{\mathbb{C}\setminus A} \in X$. ■

We now have enough information to describe the structure of the Möbius inverse monoid.

Theorem 10 *The Möbius inverse monoid is isomorphic to the P-semigroup $P(H, X, Y)$, where H is the Möbius group, X is the set of non-identity cofinite idempotents in $I(\mathbb{C}^*)$, and Y is the set of cofinite idempotents in $I(\mathbb{C})$.*

Proof We use the arguments of the motivational part of Section 7.2. The Möbius inverse monoid is an inverse subsemigroup of $I(\mathbb{C})$. It is F-inverse, and each σ-class contains a maximum element which is a Möbius transformation. Thus the union of each σ-class is the Möbius transformation it contains. The set $\mathbb{C}$ can be extended to a larger set $\mathbb{C}^*$ with one extra element. Each Möbius transformation extends to a unique extended Möbius transformation on $\mathbb{C}^*$. These extended Möbius transformations are bijections and form a group isomorphic to the Möbius group. It is clear that the assumptions (A1) and (A2) hold. The fact that assumption (A3) holds follows from Lemma 8(1). Thus (H, X, Y) is a McAlister triple and $\mathcal{M}$ is isomorphic to $P(H, X, Y)$. ■

7.4 F-inverse semigroups

We defined a semigroup to be an F-inverse monoid when each of its σ-classes has a maximum element (Section 7.1). We shall now introduce the semigroup analogue of these monoids.

Definition and basic properties

Let S be an inverse semigroup and let e and f be any idempotents. Put $\sigma_{e,f} = \sigma \cap (eSf \times eSf)$. The set eSf is a subsemigroup of S, not necessarily inverse.

Lemma 1 *Let S be an inverse semigroup and let $e, f \in E(S)$. Then $eSf/\sigma_{e,f}$ is isomorphic to S/σ, and $\sigma_{e,f}$ is the minimum group congruence on eSf.*

Proof Define θ: $eSf/\sigma_{e,f} \to S/\sigma$ by $\theta(\sigma_{e,f}(s)) = \sigma(s)$. This is a well-defined embedding. To show it is surjective, let $\sigma(s) \in S/\sigma$. Then $\theta(\sigma_{e,f}(esf)) = \sigma(esf) = \sigma(s)$.

Now let ρ be any group congruence on eSf, and let $(a, b) \in \sigma_{e,f}$. In particular, $a, b \in eSf$. By definition, there exists an idempotent u in S such that $au = bu$. But then $a(fue) = b(fue)$ where $fue \in E(eSf)$. Thus $\rho(a) = \rho(b)$. ■

Proposition 2 *Let S be an inverse semigroup. Then every σ-class of S contains a maximum element if, and only if, S is a monoid and every $\sigma_{e,f}$-class has a maximum element.*

Proof Suppose that every σ-class of S contains a maximum element. By Proposition 7.1.3, S has an identity. Let $e, f \in E(S)$. We show that every $\sigma_{e,f}$-class contains a maximum element. Consider the class $\sigma_{e,f}(esf)$. Let a be the maximum element of $\sigma(s)$. Now $\sigma_{e,f}(esf) \subseteq \sigma(s)$ and so every element of $\sigma_{e,f}(esf)$ lies beneath a. Thus every element of $\sigma_{e,f}(esf)$ lies beneath eaf. But $(a, s) \in \sigma$ implies that $(eaf, esf) \in \sigma_{e,f}$. Thus eaf is the maximum element of $\sigma_{e,f}(esf)$.

The converse follows from the fact that $\sigma = \sigma_{1,1}$. ■

It is not true that an inverse semigroup in which each $\sigma_{e,f}$-class contains a maximum element is E-unitary: the semigroup B_2 is a counterexample.

In Theorem 3.2.7, we proved that inverse semigroups not containing a copy of B_2 were E-reflexive.

Proposition 3 *Let S be an inverse semigroup in which each $\sigma_{e,f}$-class contains a maximum element. Then S is E-reflexive if, and only if, it is E-unitary.*

Proof Let S be E-reflexive, and suppose that $i \leq s$ where i is an idempotent. Let $e = ss^{-1}$ and $f = s^{-1}s$. Clearly, $i, s \in eSf$. Also, since $i \leq s$, the element $\sigma_{e,f}(s)$ is an idempotent of $eSf/\sigma_{e,f}$. Thus $(s^2, s) \in \sigma_{e,f}$. Suppose now that a is the maximum element of $\sigma_{e,f}(s)$. Then $aa^{-1} \leq e$ and $a^{-1}a \leq f$. But also $e \leq aa^{-1}$ and $f \leq a^{-1}a$. Thus $e = aa^{-1}$ and $f = a^{-1}a$. Now, $s \leq a$ and so $s = ss^{-1}a = aa^{-1}a = a$. Thus s is the maximum element of $\sigma_{e,f}$. But then, $s^2 \leq s$. However, S is E-reflexive and so $s^2 = s$ by Theorem 3.2.7.

Conversely, let S be E-unitary. Suppose that $s^2 \leq s$. Then $(s^2, s) \in \sigma$. But this implies that $\sigma(s)$ is an idempotent. Thus by Theorem 2.4.6, the element s is an idempotent. The fact that S is E-reflexive now follows from Theorem 3.2.7. ■

An inverse semigroup S is said to be F-*inverse* if it is E-unitary and every $\sigma_{e,f}$-class contains a maximum element. The F-inverse monoids are, by Proposition 2 and Proposition 7.1.3, precisely the inverse semigroups in which every σ-class contains a maximum element.

Lemma 4 *An E-unitary inverse semigroup S is F-inverse if, and only if, for every pair of idempotents e and f and for every element s in S the subset $eSf \cap \sigma(s)$ has a maximum element.*

Proof Observe first that subsets of the form $eSf \cap \sigma(s)$ are always non-empty: for the element $esf \leq s$, so that $esf \in \sigma(s)$ and clearly, $esf \in eSf$. Thus

$$eSf \cap \sigma(s) = eSf \cap \sigma(esf).$$

But $eSf \cap \sigma(esf) = \sigma_{e,f}(esf)$. ∎

The F-inverse semigroups are a special class of E-unitary inverse semigroups, so it is natural to characterise them via their McAlister triples.

Theorem 5 *The P-semigroup $P(G, X, Y)$ is F-inverse if, and only if, X is a meet semilattice.*

Proof We begin with a general observation. To show that X is a meet semilattice, it is enough to prove that $x \wedge y$ exists for all $x \in X$ and all $y \in Y$; to see why, let x and x' be arbitrary elements of X. Since $GY = X$ there exists $g \in G$ and $y \in Y$ such that $g \cdot y = x'$. By assumption, $g^{-1} \cdot x \wedge y$ exists, but it is easy to check that $g \cdot (g^{-1} \cdot x \wedge y) = x \wedge x'$.

Suppose that $P = P(G, X, Y)$ is F-inverse. Let $x \in X$ and $y \in Y$. Then $g \cdot z = x$ for some $z \in Y$ by axiom (MT2). Clearly, $(y, 1), (z, 1) \in E(P)$. The set

$$\{(e, g) \in P\colon (e, g) \in (y, 1)P(z, 1)\}$$

has a maximum element, (i, g) say, by Lemma 4. From $(i, g)(z, 1) = (i, g)$ and $(y, 1)(i, g) = (i, g)$ we obtain $g^{-1} \cdot i \leq z$ and $i \leq y$. Thus $i \leq x, y$. Now suppose that $j \leq x, y$. Then $g^{-1} \cdot j \leq z$ and $j \leq y$ and so

$$(j, g) \in \{(e, g)\colon (e, g) \in (y, 1)P(z, 1)\}.$$

Thus $(j, g) \leq (i, g)$, which implies $j \leq i$. Hence $i = x \wedge y$.

Now suppose that X is a semilattice. Let $(e, 1)$ and $(f, 1)$ be idempotents of P and let $g \in G$. We show that the set

$$\{(i, g)\colon (i, g) \in (e, 1)P(f, 1)\}$$

has a maximum element. Since X is a semilattice, the meet $e \wedge g \cdot f$ exists. Consider the ordered pair $(e \wedge g \cdot f, g)$. Then $e \wedge g \cdot f \leq e$ and so $e \wedge g \cdot f \in Y$ since Y is an order ideal of X. Also $g^{-1} \cdot (e \wedge g \cdot f) \leq f \in Y$. Thus $(e \wedge g \cdot f, g) \in P$. Clearly,

$$(e \wedge g \cdot f, g) \in \{(i, g)\colon (i, g) \in (e, 1)P(f, 1)\}.$$

It is easy to check that it is the largest element of this set. ∎

Enlargements

We proved in Section 7.1, that E-unitary inverse semigroups are essentially the inverse subsemigroups of semidirect products of semilattices by groups. This raises the question of just which inverse subsemigroups of such semidirect products are also F-inverse. The point is that if X is a meet semilattice, we can form the semidirect product inverse semigroup $P(G, X)$, which contains $P(G, X, Y)$ as an inverse subsemigroup. In the following proposition, we shall describe the *abstract* relationship between $P(G, X, Y)$ and $P(G, X)$.

Proposition 6 *Let (G, X, Y) be a McAlister triple, where X is a meet semilattice.*

(1) *The idempotents of $P(G, X, Y)$ form an order ideal of $P(G, X)$.*

(2) *If $(x, g) \in P(G, X)$ is such that*

$$(x, g)^{-1}(x, g), (x, g)(x, g)^{-1} \in P(G, X, Y)$$

then $(x, g) \in P(G, X, Y)$.

(3) *For each idempotent $(x, 1) \in P(G, X)$ there exists an idempotent $(y, 1) \in P(G, X, Y)$ such that $(x, 1) \mathcal{D} (y, 1)$.*

Proof (1) Let $(y, 1)$ be an idempotent in $P(G, X, Y)$ and $(x, 1) \in P(G, X)$ such that $(x, 1) \leq (y, 1)$. Then $x \leq y$ and so $x \in Y$ by axiom (MT1).
(2) By assumption

$$(x, g)^{-1}(x, g) = (g^{-1} \cdot x, 1) \text{ and } (x, g)(x, g)^{-1} = (x, 1)$$

are both idempotents in $P(G, X, Y)$, and so both $g^{-1} \cdot x$ and x belong to Y. Hence $(x, g) \in P(G, X, Y)$ from the definition of a P-semigroup.
(3) By axiom (MT2), there exists $g^{-1} \in G$ and $y \in Y$ such that $g^{-1} \cdot y = x$. Consider the element $(y, g) \in P(G, X)$. Then

$$(y, g)(y, g)^{-1} = (y, 1) \text{ and } (y, g)^{-1}(y, g) = (x, 1).$$

Thus $(x, 1) \mathcal{D} (y, 1)$. ■

On the basis of the above proposition, we make the following definition. Let S be an inverse subsemigroup of an inverse semigroup T. We say that T is an *enlargement* of S if the following three axioms hold:

(E1) $E(S)$ is an order ideal of $E(T)$.

(E2) If $t \in T$ and $t^{-1}t, tt^{-1} \in S$ then $t \in S$.

(E3) For every idempotent $e \in T$ there exists an idempotent $f \in S$ such that $e \mathcal{D} f$.

The following is easy to prove.

Lemma 7 *Let S be an inverse subsemigroup of T. Then axiom* (E1) *holds if, and only if, S is an order ideal of T.* ∎

In the following result, σ_T and σ_S denote the minimum group congruences on T and S respectively.

Lemma 8 *Let T be an enlargement of the inverse subsemigroup S.*

(1) $\sigma_T \cap (S \times S) = \sigma_S$.

(2) *If T is F-inverse then S is F-inverse.*

Proof (1) Clearly $\sigma_S \subseteq \sigma_T \cap (S \times S)$. Conversely, let $(a,b) \in \sigma_T \cap (S \times S)$. Then there exists $c \leq a, b$ where $c \in T$. But S is an order ideal of T by Lemma 7, and $a, b \in S$ so that $c \in S$. Hence $(a,b) \in \sigma_S$.
(2) By assumption T is F-inverse and so it is E-unitary. Thus the inverse subsemigroup S is E-unitary. Let $s \in S$ and $e, f \in E(S)$. Consider the subset $\sigma_S(s) \cap eSf$ of S. Clearly, $eSf \subseteq eTf$; the reverse inclusion also holds, for if $t \in eTf$ then $et = t$ and so $tt^{-1} \leq e$. But $e \in E(S)$, so that by axiom (E1), $tt^{-1} \in E(S)$. Similarly, $t^{-1}t \in E(S)$. Thus by axiom (E2), $t \in S$. Hence $eSf = eTf$. Thus $\sigma_S(s) \cap eSf = \sigma_T(s) \cap eTf$. By assumption, this set has a maximum element. Thus S is F-inverse. ∎

We now have the following characterisation of F-inverse semigroups in terms of semidirect products of semilattices by groups.

Theorem 9 *An inverse semigroup S is F-inverse if, and only if, it has an enlargement which is a semidirect product of a semilattice by a group.*

Proof Let S be F-inverse. Then S is isomorphic to a P-semigroup $P(G, X, Y)$ where X is a meet semilattice by Theorem 5. But $P(G, X)$ is an enlargement of $P(G, X, Y)$ by Proposition 6. Consequently, S has an enlargement which is isomorphic to a semidirect product of a semilattice by a group.

Conversely, let $P(G, X)$ be a semidirect product of a semilattice by a group. Suppose that $P(G, X)$ is an enlargement of the inverse subsemigroup S. Then S is F-inverse by Lemma 8. ∎

The above theorem is equivalent to finding a P-representation of an F-inverse semigroup as we now show.

Theorem 10 *Let G be a group and X a semilattice, and let S be an inverse subsemigroup of the inverse semigroup $P(G,X)$. Suppose that $P(G,X)$ is an enlargement of S. Let $Y \times \{1\} = E(S)$. Then (G,X,Y) is a McAlister triple and $S = P(G,X,Y)$.*

Proof We show first that the defining conditions for a McAlister triple hold for (G,X,Y).

Axiom (MT1) holds: by assumption Y is a semilattice. Let $y \in Y$ and suppose that $x \leq y$. Then $(x,1) \leq (y,1)$ in P. But $(y,1) \in S$. Thus $(x,1) \in S$ by axiom (E1). Thus $x \in Y$.

Axiom (MT2) holds: let $x \in X$. Then $(x,1) \in E(P(G,X))$. By axiom (E3), $(x,1)\,\mathcal{D}\,(y,1)$ for some $(y,1) \in S$. Thus there exists $(z,g) \in P$ such that $(z,g)(z,g)^{-1} = (x,1)$ and $(z,g)^{-1}(z,g) = (y,1)$. Thus $x = z$ and $y = g^{-1}z$. Hence $x = gy$ where $y \in Y$.

Axiom (MT3) holds: let $g \in G$. Let $(x,g) \in P(G,X)$ and let $y' \in Y$. Then $(y',1)(x,g)(y',1) = (y,g)$, say, is an element of S by axioms (E1) and (E2). But then $y \in Y$ and $g \cdot y = y' \in Y$. Thus $g^{-1} \cdot Y \cap Y \neq \emptyset$.

Finally we show that $S = P(G,X,Y)$. By definition, $(x,g) \in P(G,X,Y)$ if, and only if, $(x,g) \in Y \times G$ and $g^{-1} \cdot x \in Y$. But this is equivalent to $(x,g)^{-1}(x,g), (x,g)(x,g)^{-1} \in E(S)$. By axiom (E2) these are precisely the elements of S. ■

The above result suggests that we can regard Theorem 9 as being a coordinate-free version of the P-theorem for F-inverse semigroups.

7.5 F-inverse covers

We say that an inverse semigroup S has an *F-inverse cover* if it has an E-unitary cover which is also F-inverse. The aim of this section is to prove that every inverse semigroup has an F-inverse cover.

The homomorphism factorisation theorem

Let ρ be a congruence on an inverse semigroup S. Define a relation $\rho_{\min}$ by

$$(a,b) \in \rho_{\min} \Leftrightarrow \exists e \in E(S) \text{ such that } ae = be \text{ and } (e, a^{-1}a), (e, b^{-1}b) \in \rho.$$

Proposition 1 *Let S be an inverse semigroup and let ρ be a congruence on S.*

(1) *$\rho_{\min}$ is a congruence on S such that* $\operatorname{tr}\rho = \operatorname{tr}\rho_{\min}$.

(2) *If τ is any congruence on S with the same trace as ρ then $\rho_{\min} \subseteq \tau$.*

(3) $\rho_{\min} \subseteq \sigma$.

Proof (1) We begin by showing that $\rho_{\min}$ is an equivalence relation. It is easy to check that $\rho_{\min}$ is reflexive, and immediate from the definition that it is symmetric, so it remains to show that it is transitive. Let $(a,b),(b,c) \in \rho_{\min}$. By definition

$$ae = be \text{ and } (e, a^{-1}a), (e, b^{-1}b) \in \rho$$

for some idempotent e, and

$$bf = cf \text{ and } (f, b^{-1}b), (f, c^{-1}c) \in \rho$$

for some idempotent f. Clearly, $aef = cef$. From $(e, b^{-1}b), (f, b^{-1}b) \in \rho$ we obtain $(e,f) \in \rho$. Thus $(e, ef) \in \rho$. But $(e, a^{-1}a) \in \rho$ so that $(ef, a^{-1}a) \in \rho$. Also $(f, c^{-1}c) \in \rho$ and $(ef, f) \in \rho$ imply $(ef, c^{-1}c) \in \rho$. Hence $(a,c) \in \rho_{\min}$.

Next we prove that $\rho_{\min}$ is a congruence. Let $(a,b),(c,d) \in \rho_{\min}$. By definition

$$ae = be \text{ and } (e, a^{-1}a), (e, b^{-1}b) \in \rho$$

for some idempotent e, and

$$cf = df \text{ and } (f, c^{-1}c), (f, d^{-1}d) \in \rho$$

for some idempotent f. Now $aecf = bedf$ and so

$$ac(c^{-1}ecf) = bd(d^{-1}edf).$$

Put $i = (c^{-1}ec)f(d^{-1}ed)$. Then $aci = bdi$. Now

$$(e, a^{-1}a), (cf, c) \in \rho$$

and $cf = df$. Thus $(df, c) \in \rho$. Hence, writing $\equiv$ for ρ, we obtain

$$i = (c^{-1}ec)f(d^{-1}ed) = c^{-1}ec(df)^{-1}e(df) \equiv c^{-1}ec \equiv (ac)^{-1}(ac).$$

Thus

$$(i, (ac)^{-1}ac) \in \rho.$$

We may similarly show that

$$(i, (bd)^{-1}bd) \in \rho.$$

Hence $(ac, bd) \in \rho_{\min}$.

Let e and f be idempotents. If $(e,f) \in \rho_{\min}$, then $ei = fi$ and $(i,e),(i,f) \in \rho$ for some idempotent $i \in S$. Thus $(e,f) \in \rho$. Conversely, if $(e,f) \in \rho$, then e(ef) = f(ef) and $(ef, e) \in \rho$ and $(ef, f) \in \rho$. Thus $(e,f) \in \rho_{\min}$.

(2) Let τ be any congruence with the same trace as ρ. Let $(a,b) \in \rho_{\min}$. Then

$ae = be$ and $(e, a^{-1}a), (e, b^{-1}b) \in \rho$ for some idempotent e. Then by assumption $(e, a^{-1}a), (e, b^{-1}b) \in \tau$. Thus $(ae, a), (be, b) \in \tau$. But $ae = be$ so that $(a, b) \in \tau$.
(3) Immediate from the definition. ■

Thus $\rho_{\min}$ is the smallest congruence having the same trace as ρ.

Lemma 2 *Let θ: $S \to T$ be a surjective homomorphism between inverse semigroups, and suppose that* $\ker\theta \subseteq \sigma$. *Then S/σ is isomorphic to T/σ.*

Proof Denote the minimum group congruence on S by σ_S and the minimum group congruence on T by σ_T. By Theorem 2.4.2, there is a homomorphism θ^*: $S/\sigma_S \to T/\sigma_T$ defined by $\theta^*(\sigma_S(s)) = \sigma_T(\theta(s))$. Clearly, θ^* is surjective, so it remains to prove that it is injective. Suppose that $\theta^*(\sigma_S(s)) = \theta^*(\sigma_S(s'))$. Then there exists $t \in T$ such that $t \leq \theta(s), \theta(s')$. By Proposition 1.4.21, there exists $a \in S$ such that $\theta(a) = t$ and $a \leq s$, and there exists $b \in S$ such that $\theta(b) = t$ and $b \leq s'$. But $\theta(a) = \theta(b)$ and so, by assumption, $(a, b) \in \sigma_S$. Thus $\sigma_S(s) = \sigma_S(s')$, and so θ^* is injective. ■

Because of Lemma 2, we say that a congruence *preserves maximal group homomorphic images* if it is contained in the minimum group congruence.

The following is the homomorphism factorisation theorem for inverse semigroup homomorphisms.

Theorem 3 *Let θ: $S \to T$ be a surjective homomorphism between inverse semigroups. Then there is an inverse semigroup U and surjective homomorphisms θ_1: $S \to U$ and θ_2: $U \to T$ such that $\theta = \theta_2\theta_1$, where θ_1 preserves maximal group homomorphic images, and θ_2 is idempotent-separating.*

Proof Since θ is a surjective homomorphism we may factorise it as $\theta = \psi\rho^\natural$ where $\rho = \ker\theta$ and ψ is an isomorphism from S/ρ to T, by Theorem 2.3.1. Since $\rho_{\min} \subseteq \rho$ there is by Theorem 2.3.2 a unique surjective homomorphism ϕ: $S/\rho_{\min} \to S/\rho$ such that $\phi\rho_{\min}^\natural = \rho^\natural$. It is straightforward to check that ϕ is idempotent-separating. Put $\theta_1 = \rho_{\min}^\natural$ and $\theta_2 = \psi\phi$. Then θ_1 preserves maximal group homomorphic images and θ_2 is idempotent-separating. ■

F-inverse covers

We shall use Theorem 3 to prove that every inverse semigroup has an F-inverse cover. First we prove that idempotent pure homomorphisms preserve certain properties.

Proposition 4 *Let $\theta: S \to T$ be a surjective, idempotent pure homomorphism.*

(1) *If S is E-unitary, then T is E-unitary.*

(2) *If S is F-inverse, then T is F-inverse.*

Proof (1) Let $e \leq t$ in T where e is an idempotent. By Proposition 1.4.21, there exist $s, f \in S$ such that $\theta(s) = t$, $\theta(f) = e$ and $f \leq s$. But θ is idempotent pure and so f is an idempotent. By assumption S is E-unitary and so s is an idempotent. Hence t is an idempotent.
(2) To prove that T is F-inverse, it remains to check that the conditions of Lemma 7.4.4 hold. Thus we have to show that for all idempotents $e, f \in T$ and for every element $t \in eTf$ the set $eTf \cap \sigma(t)$ has a maximum element. By Lallement's lemma, there exist idempotents i and j in S such that $\theta(i) = e$ and $\theta(j) = f$. Let $s \in S$ such that $\theta(s) = t$. Observe that $\theta(isj) = etf = t$, so we can assume that $s \in iSj$. Since S is F-inverse, the set $\sigma(s) \cap iSj$ has a maximum element a, say. By assumption, $a \in iSj$ and so $\theta(a) = \theta(iaj) = \theta(i)\theta(a)\theta(j) = e\theta(a)f$. Thus $\theta(a) \in eTf$. Also $a \,\sigma\, s$, so that there exists $b \in S$ such that $b \leq a, s$. Hence $\theta(b) \leq \theta(a), \theta(s)$, and so $\theta(a) \,\sigma\, t$. It follows that $\theta(a) \in eTf \cap \sigma(t)$.

We now show that $\theta(a)$ is the maximum element of $eTf \cap \sigma(t)$. Let $t' \in eTf \cap \sigma(t)$. Then there exists $s' \in iSj$ such that $\theta(s') = t'$. Also $t' \,\sigma\, t$ and so there exists $u \in T$ such that $u \leq t', t$. By Proposition 1.4.21(7), from $\theta(s') = t'$ and $u \leq t'$, we obtain an element $b \in S$ such that $b \leq s'$ and $\theta(b) = u$. Similarly, there exists an element $c \in S$ such that $c \leq s$ and $\theta(c) = u$. But $\theta(b) = u = \theta(c)$, and θ is idempotent pure and so $b \sim c$ by Proposition 2.4.5. Thus $s' \,\sigma\, s$ in S. Hence $s' \in iSj \cap \sigma(s)$. But by assumption $s' \leq a$ and so $t' = \theta(s') \leq \theta(a)$. ■

Using Proposition 4, we may specialise Theorem 3 as follows.

Theorem 5 *Let θ: $S \to T$ be a surjective homomorphism between inverse semigroups where S is E-unitary. Then there is an E-unitary inverse semigroup U and surjective homomorphisms θ_1: $S \to U$ and θ_2: $U \to T$ such that $\theta = \theta_2\theta_1$, where θ_1 is idempotent pure, and θ_2 is idempotent-separating. If S is F-inverse then so is U*

Proof Let S be E-unitary. Then by Theorem 2.4.6 the compatibility relation is equal to σ. Thus by Proposition 2.4.5 every congruence on S which preserves maximal group homomorphic images is idempotent pure. By Proposition 4 idempotent pure homomorphic images of E-unitary semigroups are again E-unitary. The result now follows by Theorem 3. ■

The above result tells us that to find E-unitary or F-inverse covers, we do not need to assume that the covering homomorphism is idempotent-separating.

Proposition 6 *The free inverse semigroup $FIS(X)$ is F-inverse.*

Proof We proved in Theorem 6.3.3, that every σ-class of $FIS(X)$ except the idempotent σ-class contains a maximum element. Thus the free inverse monoid $FIM(X)$ is an F-inverse monoid. It is now easy to verify that if S is any inverse semigroup for which S^1 is F-inverse then S is F-inverse. ■

Every inverse semigroup is a homomorphic image of a free inverse semigroup. Thus by Proposition 6 and Theorem 5, we arrive at the following refinement of the McAlister covering theorem.

Theorem 7 *Every inverse semigroup has an F-inverse cover.* ■

7.6 Notes on Chapter 7

Section 7.1

The embeddability of E-unitary inverse semigroups in semidirect products of semilattices by groups was first established by O'Carroll [293]. A direct proof was given by Wilkinson [434]. Margolis and Pin [223] derive the same result from their more general theory. Pedro Silva obtains a strengthened form of the result in terms of normal-convex embeddings [395]: he proves that every E-unitary inverse semigroup admits a normal-convex embedding into a semidirect product of a semilattice by a group.

The class of F-inverse monoids was introduced by McFadden and O'Carroll [247]. The 'F' comes from the first letter of the French word *fermé*; the origin of this class of semigroups lies in the theory of partially ordered semigroups. The fact that an inverse semigroup is E-unitary precisely when it can be embedded in an F-inverse monoid was proved by O'Carroll [290]. Szendrei [406] shows that F-inverse monoids arise naturally from the Birget–Rhodes expansion of groups.

The proof I have given of the embeddability theorem is a special case of a more general theory of Billhardt [27]. He proved that given an inverse semigroup S and idempotent pure congruence ρ on S then S may be embedded in an inverse semigroup T equipped with a Billhardt congruence τ such that S/ρ is isomorphic to T/τ.

Almost factorisable inverse semigroups were introduced and studied in [235]. McAlister termed them covering semigroups. The term almost factorisable was introduced in my paper [188]. Pedro Silva [395] proves that every inverse semigroup can be embedded in an idempotent-separating homomorphic image of a semidirect product of a semilattice by a free group. Thus, in particular, every inverse semigroup admits a normal-convex embedding into an almost factorisable inverse semigroup.

Section 7.2

The P-theorem was first proved by McAlister [232] in 1974 using a Rees-type coordinatisation. A number of authors subsequently produced alternative proofs. Schein's proof [378] is the shortest; I took this as the basis for my proof expanding on the motivational comments which form the conclusion to Schein's paper. Douglas Munn [273] also produced a very elegant proof. A different approach to the structure of E-unitary semigroups is developed by Petrich and Reilly [313] using dual prehomomorphisms (see Section 3.1).

Two of the ingredients, the group and the semilattice, in the P-representation of an E-unitary inverse semigroup are easy to determine but the third, the partially ordered set X, is something of a puzzle. Three other proofs of the P-theorem use category theory to explain its rôle. Loganathan [206] obtained a proof from his cohomology theory of inverse semigroups. Because homological methods have yet to play a major rôle in inverse semigroup theory his work has yet to be developed. Margolis and Pin [223] prove the P-theorem using the Grothendieck construction (although they refer to it as the derived covering). Their proof has been enormously influential because it can be generalised to much broader classes of semigroups; see the important paper [84] for details. The third proof, due to myself [180], is discussed in Chapter 8.

All these proofs indicate that the P-theorem is a multifaceted result. It is therefore perhaps not surprising, that there are precursors to it in the literature. Schein [384] argues that an *ur*-P-theorem was known by Gołąb in 1939 [101]. A result equivalent to the existence of E-unitary covers was proved by Joubert in 1966 [162] (see [184] for a description), and used to study topological foliations.

Section 7.3

This is due to the author [192].

Section 7.4

My use of the term F-inverse semigroup is non-standard but extremely convenient. These semigroups were introduced by McAlister [237] who called them 'E-unitary inverse semigroups over semilattices'. In the literature, F-inverse semigroups are what I have called F-inverse monoids.

The characterisation of F-inverse semigroups in terms of enlargements is due to myself [188]; but see the notes to Chapter 8 for the origins of this notion.

Section 7.5

The factorisation theorem for homomorphisms with E-unitary inverse semigroups as domains is due to Reilly and Munn [346]. The general factorisation theorem for homomorphisms is given by Petrich [312] in Proposition III.5.10.

In addition to the congruence $\rho_{\min}$, there is also the congruence $\rho_{\max}$ which is the *largest* congruence with a trace equal to that of ρ [347]. With these concepts, the minimum group congruence is $\rho_{\min}$ where $\rho = E(S) \times E(S)$, whereas the maximum idempotent-separating congruence is $\rho_{\max}$ where ρ is the equality congruence on $E(S)$. Dual notions of $\rho^{\min}$ and $\rho^{\max}$ arising from Kernels were introduced by Green [108].

The existence of F-inverse covers was first established in [346], and refined in [237]. It is important to note that if an inverse semigroup is finite, then the F-inverse cover constructed in this section is still infinite. It appears to be an open question whether finite inverse semigroups have finite F-inverse covers; it was first posed by Henckell and Rhodes [126] (note that they may have misread Szendrei's paper [406] in formulating an equivalent version of this question).

Chapter 8

Enlargements

In the last chapter, we introduced the concept of an enlargement of an inverse semigroup, and used it to obtain a coordinate-free version of the P-theorem for F-inverse semigroups. In this chapter, we extend the definition of enlargements to arbitrary ordered groupoids. We use this notion to study the P-theorem, the theory of E-unitary covers, and the theory of idempotent pure (pre)homomorphisms.

Our first application of enlargements will be a coordinate-free version of the P-theorem for arbitrary E-unitary semigroups (Theorem 8.1.4). Then we prove that every E-unitary cover of an inverse semigroup S arises from an embedding of S into a factorisable inverse monoid (Theorem 8.2.10); this formalises the relationship between E-unitary covers and the property of extending partial bijections introduced in Chapter 2. The key result of this chapter is a proof of Ehresmann's maximum enlargement theorem (Theorem 8.3.3). This theorem describes the structure of arbitrary star injective ordered functors in terms of ordered covering functors; it was originally conceived as a fundamental ingredient in Ehresmann's pseudogroup theory of local structures. We use Ehresmann's theorem to determine the structure of idempotent pure extensions and so prove results first obtained by O'Carroll. This work is summarised in Theorem 8.4.8.

In this chapter, we make essential use of Chapter 4.

8.1 A coordinate-free P-theorem

In Section 7.4, we showed that the P-theorem for F-inverse semigroups could be expressed in a coordinate-free way using enlargements. In this section, we shall prove an analogous result for arbitrary E-unitary semigroups.

Ordered groupoid enlargements

Let G be an ordered subgroupoid of the ordered groupoid H. We say that H is an *enlargement* of G if the following three axioms hold:

(GE1) G_o is an order ideal of H_o.

(GE2) If $x \in H$ and $\mathbf{d}(x), \mathbf{r}(x) \in G$ then $x \in G$.

(GE3) If $e \in H_o$ then there exists $x \in H$ such that $\mathbf{r}(x) = e$ and $\mathbf{d}(x) \in G$.

When we restrict the above definition to inductive groupoids, we obtain exactly the same notion of enlargement as that introduced in Section 7.4.

Observe that if H is an enlargement of G then G is an order ideal of H; to see this, let $h \leq g$ where $g \in G$ and $h \in H$. Then $\mathbf{d}(h) \leq \mathbf{d}(g)$ and $\mathbf{r}(h) \leq \mathbf{r}(g)$. Thus $\mathbf{d}(h), \mathbf{r}(h) \in G_o$ by axiom (GE1). But then $h \in G$ by axiom (GE2), as required.

Groups acting on posets

A McAlister triple (G, X, Y) (Section 7.2) is defined in terms of a group G acting on a partially ordered set X by order automorphisms. Such actions can be completely described in terms of ordered groupoids. Define a partial multiplication on $X \times G$ by

$$(x,g)(y,h) = \begin{cases} (x, gh) & \text{if } x = g \cdot y \\ \text{undefined} & \text{otherwise,} \end{cases}$$

and define a partial order on $X \times G$ by

$$(x,g) \leq (y,h) \Leftrightarrow x \leq y \text{ and } g = h.$$

The set $X \times G$ equipped with this partial multiplication and partial order is denoted by $P(G, X)$. In what follows, we regard the group G as being ordered by equality.

Theorem 1 *With the above definitions, $P(G,X)$ is an ordered groupoid, and the function π_2: $P(G,X) \rightarrow G$ defined by $\pi_2(x,g) = g$ is a surjective, ordered covering functor. In addition, if X is a meet semilattice then the ordered groupoid $P(G,X)$ is inductive and the corresponding inverse semigroup is the usual semidirect product of a semilattice by a group.*

Proof Let $P = P(G, X)$. It is straightforward to check that P is a groupoid in which the identities are the elements of the form $(x, 1)$ where $x \in X$, inverses are given by $(x,g)^{-1} = (g^{-1} \cdot x, g^{-1})$ and

$$\mathbf{d}(x,g) = (g^{-1} \cdot x, 1) \text{ and } \mathbf{r}(x,g) = (x,1).$$

It is also clear that $\leq$ is a partial order. To show that $P(G,X)$ is an ordered groupoid we verify that the axioms of Proposition 4.1.4 hold.

Axiom (OG1) holds: let $(x,g) \leq (y,g)$. Then $x \leq y$. Thus $g^{-1} \cdot x \leq g^{-1} \cdot y$ since G acts by order automorphisms. Now

$$(x,g)^{-1} = (g^{-1} \cdot x, g^{-1}) \text{ and } (y,g)^{-1} = (g^{-1} \cdot y, g^{-1}).$$

Hence $(x,g)^{-1} \leq (y,g)^{-1}$ as required.

Axiom (OG2) holds: let $(x,g) \leq (y,g)$ and $(u,h) \leq (v,h)$ and suppose that the products $(x,g)(u,h)$ and $(y,g)(v,h)$ are defined. Then $(x,g)(u,h) = (x,gh)$ and $(y,g)(v,h) = (y,gh)$. Hence $(x,gh) \leq (y,gh)$.

Axiom (OI) holds: let $(x,1)$ be an identity of $P(G,X)$ and let $(y,g) \leq (x,1)$. Then $g = 1$, so that (y,g) is an identity.

Axiom (OG4) holds: let $(x,g) \in P(G,X)$ and let $(y,1) \leq \mathbf{d}(x,g)$. Since $\mathbf{d}(x,g) = (g^{-1} \cdot x, 1)$, we have that $y \leq g^{-1} \cdot x$. Consider the element $(g \cdot y, g)$ of $P(G,X)$. Then $\mathbf{d}(g \cdot y, g) = (y,1)$, and from $y \leq g^{-1} \cdot x$ we obtain $g \cdot y \leq x$, so that $(g \cdot y, g) \leq (x,g)$. It follows that we may define

$$((x,g) \mid (y,1)) = (g \cdot y, g).$$

Similarly, if $(y,1) \leq \mathbf{r}(x,g)$ we may define

$$((y,1) \mid (x,g)) = (y,g).$$

Hence $P(G,X)$ is an ordered groupoid.

We now turn to the function $\pi_2\colon P \to G$. The set of identities of P is just $X \times \{1\}$. Thus π_2 maps every identity of P to the identity of G. Let (x,g) and (y,h) be elements of P such that $(x,g)(y,h)$ exists. Then $(x,g)(y,h) = (x,gh)$. Now $\pi_2(x,g) = g$, $\pi_2(y,h) = h$, and $\pi_2(x,gh) = gh$. Thus π_2 is a functor. If $(x,g) \leq (y,h)$ in P then $g = h$. Thus $\pi_2(x,g) \leq \pi_2(y,h)$ in G, and so π_2 is an ordered functor. To show that π_2 is a covering functor we have to show that it is star injective and star surjective. Suppose that $\mathbf{d}(x,g) = \mathbf{d}(y,h)$ and $\pi_2(x,g) = \pi_2(y,h)$. Then $(g^{-1} \cdot x, 1) = (h^{-1} \cdot y, 1)$ and $g = h$. Hence $(x,g) = (y,h)$, and π_2 is star injective. Finally, let $g \in G$ and let $(x,1)$ be an identity of P. The pair $(g \cdot x, g)$ belongs to P, and $\mathbf{d}(g \cdot x, g) = (x,1)$ and $\pi_2(g \cdot x, g) = g$. Thus π_2 is a star surjective.

We may now describe the pseudoproduct. Let $(x,g),(y,h) \in P(G,X)$. Then the pseudoproduct $(x,g) \otimes (y,h)$ exists if, and only if, $g^{-1} \cdot x \wedge y$ exists in X. To see this, observe that $\mathbf{d}(x,g) = (g^{-1} \cdot x, 1)$ and $\mathbf{r}(y,h) = (y,1)$. Thus the pseudoproduct of (x,g) and (y,h) exists precisely when the meet of $(g^{-1} \cdot x, 1)$ and $(y,1)$ exists. If this meet does exist it is equal to $(g^{-1}x \wedge y, 1)$. We claim that the pseudoproduct is given by

$$(x,g) \otimes (y,h) = (x \wedge g \cdot y, gh).$$

This follows by observing that

$$((x,g)\,|\,(g^{-1}\cdot x\wedge y,1)) = (g\cdot(g^{-1}\cdot x\wedge y),g)$$

and

$$((g^{-1}\cdot x\wedge y,1)\,|\,(y,h)) = (g^{-1}\cdot x\wedge y,h).$$

But $g\cdot(g^{-1}\cdot x\wedge y) = x\wedge g\cdot y$. Thus $(x,g)\otimes(y,h) = (x\wedge g\cdot y, gh)$.

Clearly, $P(G,X)$ is inductive precisely when X is a meet semilattice, in which case the pseudoproduct is everywhere defined and coincides with the usual product in the semidirect product of the semilattice X by the group G. ■

We say that the ordered groupoid $P(G,X)$ is a *semidirect product of a poset by a group*. The result above guarantees that when X is a meet semilattice $(P(G,X),\otimes)$ is the usual semidirect product of a semilattice by a group.

Ordered star injective functors are the main focus of this section, and the following property of such functors will be used repeatedly.

Lemma 2 *Let $\theta\colon G\to H$ be an ordered star injective functor between ordered groupoids and let $x,y\in G$. Then $x\le y$ if, and only if, $\mathbf{d}(x)\le\mathbf{d}(y)$ and $\theta(x)\le\theta(y)$.*

Proof Let $\mathbf{d}(x)\le\mathbf{d}(y)$ and $\theta(x)\le\theta(y)$. The restriction $(y\,|\,\mathbf{d}(x))$ exists and $\mathbf{d}(y\,|\,\mathbf{d}(x)) = \mathbf{d}(x)$. By Proposition 4.1.2

$$\theta(y\,|\,\mathbf{d}(x)) = (\theta(y)\,|\,\mathbf{d}(\theta(x))).$$

Now

$$(\theta(y)\,|\,\mathbf{d}(\theta(x)))\le\theta(y)\le\theta(x) \text{ and } \mathbf{d}(\theta(y)\,|\,\mathbf{d}(\theta(x))) = \mathbf{d}(\theta(x)).$$

Thus $(\theta(y)\,|\,\mathbf{d}(\theta(x))) = \theta(x)$, and so $\theta(x) = \theta((y\,|\,\mathbf{d}(x)))$. Since θ is star injective $x = (y\,|\,\mathbf{d}(x))$ and so $x\le y$. The converse is immediate. ■

We may now obtain an abstract characterisation of ordered groupoids of the form $P(G,X)$.

Theorem 3 *Let $\pi\colon\Pi\to G$ be an ordered covering functor from the ordered groupoid Π onto a group G. Then G acts on the poset $X=\Pi_o$ by order automorphisms, and there is an isomorphism of ordered groupoids $\theta\colon\Pi\to P(G,X)$ such that $\pi_2\theta=\pi$.*

Proof We begin by showing that G acts on $X = \Pi_o$. Let $e \in X$ and $g \in G$. Define

$$g \cdot e = \mathbf{r}(x)$$

where

$$x \in \Pi,\ e = \mathbf{d}(x) \text{ and } g = \pi(x);$$

the element x is unique with these properties since π is a covering functor. We now check that this defines an action. Let $e \in X$. Then $e \in \Pi_o$ is the unique element such that $\mathbf{d}(e) = e$ and $\pi(e) = 1$. Thus $1 \cdot e = e$. Now let $g, h \in G$ and $e \in X$. Let

$$g \cdot e = \mathbf{r}(x) \text{ where } \pi(x) = g \text{ and } \mathbf{d}(x) = e,$$

and let

$$h \cdot \mathbf{r}(x) = \mathbf{r}(y) \text{ where } \pi(y) = h \text{ and } \mathbf{d}(y) = \mathbf{r}(x).$$

The product yx is defined in Π since $\mathbf{d}(y) = \mathbf{r}(x)$. But then $\pi(yx) = hg$ and $\mathbf{d}(yx) = \mathbf{d}(x)$, so that $(hg) \cdot e$ is defined and equals $\mathbf{r}(y)$, as required.

To show that G acts on X by order automorphisms, let $e \leq f$ in X and $g \in G$. Then

$$g \cdot e = \mathbf{r}(x) \text{ where } e = \mathbf{d}(x) \text{ and } g = \pi(x)$$

and

$$g \cdot f = \mathbf{r}(y) \text{ where } f = \mathbf{d}(y) \text{ and } g = \pi(y).$$

From $\pi(x) = \pi(y)$ and $\mathbf{d}(x) \leq \mathbf{d}(y)$, we obtain $x \leq y$ by Lemma 2. Hence $g \cdot e \leq g \cdot f$. We may now construct the ordered groupoid $P(G, X)$.

Define a function θ: $\Pi \to P(G, X)$ by

$$\theta(x) = (\mathbf{r}(x), \pi(x)).$$

Clearly, θ maps identities to identities. Let $x, y \in \Pi$ such that xy is defined. Now $\mathbf{d}(\theta(x)) = (\pi(x)^{-1} \cdot \mathbf{r}(x), 1)$ and

$$\pi(x)^{-1} \cdot \mathbf{r}(x) = \pi(x^{-1}) \cdot \mathbf{d}(x^{-1}) = \mathbf{r}(x^{-1}) = \mathbf{d}(x).$$

Thus $\mathbf{d}(\theta(x)) = (\mathbf{d}(x), 1)$. Also $\mathbf{r}(\theta(y)) = (\mathbf{r}(y), 1)$, so that the product $\theta(x)\theta(y)$ is defined. It is easy to check that $\theta(x)\theta(y) = \theta(xy)$. Hence θ is a functor. It is clear that θ preserves the order, so let $\theta(x) \leq \theta(y)$. Then $(\mathbf{r}(x), \pi(x)) \leq (\mathbf{r}(y), \pi(y))$. Thus

$$\mathbf{d}(x^{-1}) \leq \mathbf{d}(y^{-1}) \text{ and } \pi(x^{-1}) = \pi(y^{-1}).$$

Hence $x^{-1} \leq y^{-1}$ by Lemma 2, giving $x \leq y$. It follows also that θ is injective. It remains to prove that θ is surjective. Let $(e, g) \in \Pi(G, X)$. Since π is star surjective, there exists an element $x \in \Pi$ such that $\mathbf{d}(x) = e$ and $\pi(x) = g^{-1}$. Hence $\theta(x^{-1}) = (e, g)$. ■

The structure of E-unitary inverse semigroups

Theorem 4 *Let S be an inverse semigroup with associated inductive groupoid $\mathbf{G}(S)$. Then S is E-unitary if, and only if, there is an ordered embedding $\iota: \mathbf{G}(S) \to P(G,X)$ into some semidirect product of a poset by a group such that $P(G,X)$ is an enlargement of $\iota(\mathbf{G}(S))$ in such a way that the function $\pi_2: P(G,X) \to G$ restricted to $\iota(\mathbf{G}(S))$ is surjective.*

Proof If an inverse semigroup S has an ordered groupoid enlargement which is a semidirect product of a poset by a group then it is straightforward to check that S is E-unitary.

To prove the converse, let S be an E-unitary inverse semigroup. Without loss of generality we may assume that S is a P-semigroup $P = P(G,X,Y)$, where (G,X,Y) is a McAlister triple. Form the ordered groupoid $P(G,X)$. Recall that

$$P(G,X,Y) = \{(y,g) \in Y \times G: g^{-1} \cdot y \in Y\}.$$

For the remainder of the proof, we shall regard P as an inductive groupoid. The order on P is the restriction of the ordering on $P(G,X)$; inversion in P is just the restriction of the inversion of $P(G,X)$, and

$$\exists (e,g) \cdot (f,h) \Leftrightarrow g \cdot f = e$$

in which case $(e,g) \cdot (f,h) = (e,gh)$. Thus the restricted product in P is the restriction of the partial product of $P(G,X)$. Hence P is an ordered subgroupoid of $P(G,X)$. We now prove that $P(G,X)$ is an enlargement of $\mathbf{G}(P)$.

Axiom (GE1) holds: to show that P_o is an order ideal of $P(G,X)$ let $(x,1) \leq (y,1)$ where $(y,1) \in P_o$. Then $x \leq y$. But $y \in Y$ and $Y \subseteq X$ is an order ideal. Thus $x \in Y$. Hence $(x,1) \in P_o$.

Axiom (GE2) holds: let $(x,g) \in P(G,X)$ such that $\mathbf{d}(x,g), \mathbf{r}(x,g) \in P_o$. Then $\mathbf{d}(x,g) = (g^{-1} \cdot x, 1) \in P_o$ and $\mathbf{r}(x,g) = (x,1) \in P_o$. Thus $x \in Y$ and $g^{-1} \cdot x \in Y$. Hence $(x,g) \in P$.

Axiom (GE3) holds: let $(x,1) \in P(G,X)_o$. Since $G \cdot Y = X$ there exists an element $y \in Y$ and $g \in G$ such that $g \cdot y = x$. Consider now the element $(g \cdot y, g)$. Then $\mathbf{r}(g \cdot y, g) = (g \cdot y, 1) = (x,1)$ and $\mathbf{d}(g \cdot y, g) = (y,1) \in P$.

Finally, we prove that π_2 restricted to P is surjective. Let $g \in G$. Because (G,X,Y) is a McAlister triple, the set $g^{-1} \cdot Y \cap Y$ is non-empty. Let $y \in Y$ such that $g^{-1} \cdot y \in Y$. Then $(y,g) \in P$ and $\pi_2(y,g) = g$. ■

The above result is equivalent to finding a P-representation for S as we now show.

Theorem 5 *Let $P(G,X)$ be a semidirect product of a poset by a group, and let $\pi_2: P(G,X) \to G$ be the associated surjective ordered covering functor. Let*

H be an ordered subgroupoid of $P(G,X)$, which is an inductive groupoid with respect to the induced order. Suppose that $P(G,X)$ is an enlargement of H and that π_2 restricted to H is surjective. Put $Y \times \{1\} = H_o$. Then (G,X,Y) is a McAlister triple and $H = P(G,X,Y)$.

Proof We show first that (G,X,Y) is a McAlister triple.

Axiom (MT1) holds: H_o is an order ideal of $P(G,X)_o$ by axiom (GE1). Thus $Y \times \{1\}$ is an order ideal of $X \times \{1\}$. Hence Y is an order ideal of X. By assumption, H is an inductive groupoid and so Y is a meet semilattice under the induced order.

Axiom (MT2) holds: let $x \in X$. Then $(x,1) \in P(G,X)_o$. By axiom (GE3), there exists an element $(e,g) \in P(G,X)$ such that $\mathbf{r}(e,g) = (x,1)$ and $\mathbf{d}(e,g) \in H_o$. Now $\mathbf{r}(e,g) = (e,1)$ and $\mathbf{d}(e,g) = (g^{-1} \cdot e, 1)$. Thus $x = e$ and $g^{-1} \cdot e = y \in Y$. Hence $x = g^{-1} \cdot y$.

Axiom (MT3) holds: let $g \in G$. Then $g^{-1} \in G$. Since π_2 is surjective when restricted to H, there exists $(y, g^{-1}) \in H$, for some $y \in Y$. Thus $y, g \cdot y \in Y$ and so $g \cdot Y \cap Y$ is non-empty.

Finally,

$$H = \{(x,g) \in P(G,X) \colon \mathbf{d}(x,g), \mathbf{r}(x,g) \in H_o\}$$

by axiom (GE2). This implies that $H = P(G,X,Y)$, as an inductive groupoid. ■

Theorems 4 and 5 together constitute a coordinate-free version of the P-theorem.

8.2 E-unitary covers

In Section 2.2, we showed that every embedding of an inverse semigroup S into a factorisable inverse monoid gives rise to an E-unitary cover of S. In Theorem 10, we shall prove that every E-unitary cover arises in this way. There are three basic steps needed to prove this theorem. Firstly, we prove an extension theorem which tells us that any identity-separating ordered congruence extends to an enlargement (Theorem 4). Secondly, factorisable ordered groupoids are introduced as generalisations of factorisable inverse monoids: Lemmas 6 and 7 tell us how to construct certain factorisable ordered groupoids. Thirdly, we show how to translate results from the category of ordered groupoids into the category of inverse semigroups (Lemma 8).

A congruence extension theorem

We prove that certain ordered congruences may be extended from an ordered groupoid to an enlargement of that groupoid. The ordered congruences in question are generalisations of idempotent-separating congruences.

Let G be an ordered groupoid and let ρ be an ordered congruence on G. Then ρ is said to be *identity-separating* if the following axiom holds:

(IS) If $e, f \in G_o$ and $(e, f) \in \rho$ then $e = f$.

Proposition 1 *Let ρ be an equivalence relation on an ordered groupoid G. Then ρ is an identity-separating ordered congruence if, and only if, axiom* (IS) *holds together with axioms* (OCON1), (OCON2), *and* (OCON4) *of* Section 4.3.

Proof By definition, an identity-separating congruence satisfies the given conditions, so we need only prove the converse. Let ρ be an equivalence relation which satisfies (OCON1), (OCON2), (OCON4) and (IS). The axioms (OCON3), (OCON5) and (OCON6) all follow from (IS). It remains to show that (AC) holds. The product of two equivalence classes $\rho(x)\rho(y)$ is defined precisely when there exist $x' \in \rho(x)$ and $y' \in \rho(y)$ such that $x'y'$ exists. But $x'y'$ exists precisely when $\mathbf{d}(x') = \mathbf{r}(y')$. But by axioms (OCON2) and (IS), $x \,\rho\, x'$ implies $\mathbf{d}(x') = \mathbf{d}(x)$, and $y \,\rho\, y'$ implies $\mathbf{r}(y') = \mathbf{r}(y)$. Thus $\rho(x)\rho(y)$ exists precisely when xy exists. It is now straightforward to check that (AC) holds. ■

The following result simply assures us that when G is an inductive groupoid, the identity-separating ordered congruences on G are precisely the idempotent-separating congruences on the inverse semigroup $(G, \otimes)$.

Proposition 2 *Let G be an inductive groupoid. Then ρ is an identity-separating ordered congruence on G precisely when ρ is an idempotent-separating congruence on the inverse semigroup $(G, \otimes)$.*

Proof Let ρ be an identity-separating ordered congruence on G. We prove that ρ is a congruence on $(G, \otimes)$. Suppose that $x \,\rho\, y$. Then $\mathbf{r}(x) = \mathbf{r}(y)$ since ρ is identity-separating. Let $a \in G$ and put $e = \mathbf{d}(a) \wedge \mathbf{r}(x)$. Then $a \otimes x = (a \,|\, e)(e \,|\, x)$ and $a \otimes y = (a \,|\, e)(e \,|\, y)$. From $x \,\rho\, y$ and $(e \,|\, x) \leq x$ there exists an element y' such that $y' \leq y$ and $y' \,\rho\, (e \,|\, x)$ by axiom (OCON4). In particular, $\mathbf{r}(y') = e$, since ρ is identity-separating. Thus $y' = (e \,|\, y)$. Hence $(e \,|\, x) \,\rho\, (e \,|\, y)$. But then

$$(a \,|\, e)(e \,|\, x) \,\rho\, (a \,|\, e)(e \,|\, y)$$

by axiom (OCON1), and so $a \otimes x \,\rho\, a \otimes y$. Thus ρ is a left congruence. We may similarly show that ρ is a right congruence. The converse is clear. ■

If H is an enlargement of the ordered subgroupoid G, we may define a functor from H to G. To construct such a functor we need the following notion.

A *choice function* for the enlargement is any function ϕ: $H_0 \to H$ satisfying the following axiom:

(CF) For each $e \in H_0$, $\mathbf{d}(\phi(e)) \in G_0$ and $\mathbf{r}(\phi(e)) = e$.

Such a function always exists by axiom (GE3). Given a choice function ϕ, define a function Φ: $H \to G$ by $\Phi(x) = \phi(\mathbf{r}(x))^{-1}x\phi(\mathbf{d}(x))$. It is well-defined by axiom (GE2). It is straightforward to check that Φ is a functor. The ordered pair (Φ, ϕ) is called a *reflection of H in G*; this terminology is consistent with the notion of reflection in category theory.

Lemma 3 *Let H be an enlargement of the ordered subgroupoid G.*

(1) *Let (Φ, ϕ) and (Φ', ϕ') be any two reflections of H in G. Define a function γ: $H_o \to G$ by $\gamma(e) = \phi(e)^{-1}\phi'(e)$. Then*

$$\Phi'(x) = \gamma(\mathbf{r}(x))^{-1}\Phi(x)\gamma(\mathbf{d}(x))$$

for all $x \in H$.

(2) *For each pair of elements $x, y \in H$ where $x \leq y$, there exists a reflection (Φ, ϕ) of H in G such that $\Phi(x) \leq \Phi(y)$.*

Proof (1) Straightforward calculation.
(2) Let ϕ: $H_o \setminus \{\mathbf{d}(x), \mathbf{r}(x)\} \to H$ be any function satisfying axiom (CF). Since

$$\mathbf{d}(x) \leq \mathbf{d}(y) = \mathbf{r}(\phi(\mathbf{d}(y))) \text{ and } \mathbf{r}(x) \leq \mathbf{r}(y) = \mathbf{r}(\phi(\mathbf{r}(y)))$$

the elements

$$\phi(\mathbf{d}(x)) = (\mathbf{d}(x) \mid \phi(\mathbf{d}(y))) \text{ and } \phi(\mathbf{r}(x)) = (\mathbf{r}(x) \mid \phi(\mathbf{r}(y)))$$

are well-defined. From

$$\mathbf{d}(\phi(\mathbf{d}(x))) = \mathbf{d}(\mathbf{d}(x) \mid \phi(\mathbf{d}(y))) \leq \mathbf{d}(\phi(\mathbf{d}(y)))$$

and $\mathbf{d}(\phi(\mathbf{d}(y))) \in G$, we have that $\mathbf{d}(\phi(\mathbf{d}(x))) \in G$ since G is an order ideal of H. Similarly, $\mathbf{d}(\phi(\mathbf{r}(x))) \in G$. Thus ϕ is a choice function for the enlargement. Finally, since

$$\phi(\mathbf{d}(x)) \leq \phi(\mathbf{d}(y)) \text{ and } \phi(\mathbf{r}(x)) \leq \phi(\mathbf{r}(y)),$$

we have that

$$\Phi(x) = \phi(\mathbf{r}(x))^{-1}x\phi(\mathbf{d}(x)) \leq \phi(\mathbf{r}(y))^{-1}y\phi(\mathbf{d}(y)) = \Phi(y)$$

as required. ■

Let H be an enlargement of the ordered subgroupoid G. Let α be an ordered congruence on G, and let β be an ordered congruence on H. We say that β *extends* α if $\alpha = \beta \cap (G \times G)$.

Theorem 4 *Let H be an enlargement of the ordered subgroupoid G, and let α be an identity-separating, ordered congruence on G. Let $(\Phi, \phi)\colon H \to G$ be a reflection, and define a relation β on H by*

$$x\,\beta\,y \Leftrightarrow \mathbf{d}(x) = \mathbf{d}(y),\ \mathbf{r}(x) = \mathbf{r}(y) \ \textit{and}\ \Phi(x)\,\alpha\,\Phi(y).$$

Then β is independent of the choice of reflection (Φ, ϕ), and is the unique, identity-separating ordered congruence extending α to H.

Proof We begin by showing that the definition of β is independent of the choice of reflection. Let $(\Phi', \phi')\colon H \to G$ be any other reflection. Let β' be the relation defined by

$$x\,\beta'\,y \Leftrightarrow \mathbf{d}(x) = \mathbf{d}(y),\ \mathbf{r}(x) = \mathbf{r}(y) \text{ and } \Phi'(x)\,\alpha\,\Phi'(y).$$

Define $\gamma\colon H_0 \to G$ by $\gamma(e) = \phi(e)^{-1}\phi'(e)$. Then

$$\Phi'(x) = \gamma(\mathbf{r}(x))^{-1}\Phi(x)\gamma(\mathbf{d}(x)) \text{ and } \Phi'(y) = \gamma(\mathbf{r}(y))^{-1}\Phi(y)\gamma(\mathbf{d}(y))$$

by Lemma 3. Thus

$$\Phi'(x)\,\alpha\,\Phi'(y) \Leftrightarrow \Phi(x)\,\alpha\,\Phi(y)$$

by (OCON1). It follows that $x\,\beta'\,y$ if, and only if, $x\,\beta\,y$.

It is always possible to choose Φ so that $\Phi(x) = x$ for each $x \in G$: we simply pick a choice function ϕ which is the identity on G_o. Thus $\beta \cap (G \times G) = \alpha$.

To show that β is an identity-separating ordered congruence on H, we have to show that the axioms of Proposition 1 hold. The only axiom which causes any difficulty is (OCON4). Let $x \leq y \in H$ and $y\,\beta\,y'$. Choose $\Phi\colon H \to G$ so that $\Phi(x) \leq \Phi(y)$ by Lemma 3(2). Then $\Phi(y)\,\alpha\,\Phi(y')$ and $\Phi(x) \leq \Phi(y)$. Since α is an ordered congruence, there exists an element $x'' \in G$ such that $x''\,\alpha\,\Phi(x)$ and $x'' \leq \Phi(y')$. Define $x' = \phi(\mathbf{r}(x))x''\phi(\mathbf{d}(x))^{-1}$. Then $x'\,\beta\,x$ and, since both $\phi(\mathbf{d}(x)) \leq \phi(\mathbf{d}(y))$ and $\phi(\mathbf{r}(x)) \leq \phi(\mathbf{r}(y))$, we have that $x' \leq y'$ as required.

We now finish off by showing that the extension is unique. Let β' be any identity-separating ordered congruence on H extending α. Let $x\,\beta'\,y$. Then

$$\phi(\mathbf{r}(x))^{-1}x\phi(\mathbf{d}(x))\,\beta'\,\phi(\mathbf{r}(x))^{-1}y\phi(\mathbf{d}(x))$$

since β' is an ordered congruence. Thus $\Phi(x)\,\beta'\,\Phi(y)$. But $\Phi(x)$ and $\Phi(y)$ are elements of G, so that $\Phi(x)\,\alpha\,\Phi(y)$ since β' is an extension of α. Thus $x\,\beta\,y$, and so $\beta' \subseteq \beta$. The reverse inclusion holds by symmetry. Hence $\beta = \beta'$. ■

The proof of the following corollary is immediate by Theorem 4 and Proposition 2.

Corollary 5 *Let T be an enlargement of the inverse subsemigroup S. Then for every idempotent-separating congruence α on S there exists a unique idempotent-separating congruence β on T such that $\alpha = \beta \cap (T \times T)$.* ■

Factorisable ordered groupoids

We say that an ordered groupoid has a *maximum identity* if its set of identities has a maximum element. We shall usually denote any maximum identity by 1. Clearly, an inductive groupoid with a maximum identity is just the inductive groupoid of an inverse monoid.

Let F be an ordered groupoid with a maximum identity. Define the *group of units* of F to be the group $U(F) = \text{end}(1)$. If F has the property that for each $x \in F$ there exists $g \in U(F)$ such that $x \leq g$, then we say that F is *factorisable.*

An ordered embedding $\iota: G \to F$ of an ordered groupoid G in a factorisable ordered groupoid F is said to be *strict* if for each $g \in U(F)$ there exists $x \in G$ such that $\iota(x) \leq g$.

Let $(X, \leq)$ be a poset. Denote by $(X^1, \leq)$ the poset X with a maximum element 1 adjoined. The proof of the following is straightforward.

Lemma 6 *Let $P(G, X)$ be a semidirect product of a poset by a group, and suppose that X does not have a maximum element. Extend the action of G on X to an action of G on the poset X^1 by defining $g \cdot 1 = 1$ for all $g \in G$. The ordered groupoid $P(G, X^1)$ is factorisable with group of units $\{1\} \times G$.* ■

The following construction is of fundamental importance.

Lemma 7 *Let $P(G, X)$ be a semidirect product of a poset by a group, and let $\Theta: P(G, X) \to H$ be a surjective identity-separating special ordered functor to an ordered groupoid H.*

(1) *Let $F = H \cup G$, regarded as a groupoid. Define an order $\leq$ on F as follows: it agrees with the order on H; it is equality on G; and for $g \in G$ and $h \in H$ satisfies*

$$h \leq g \Leftrightarrow h = \Theta(x, g)$$

for some $x \in X$. Then F is a factorisable ordered groupoid.

(2) *There is a surjective, identity-separating special ordered functor*

$$\Theta': P(G, X^1) \to F$$

which extends Θ.

Proof (1) We have first to show that we really have defined a partial order on F. The relation is reflexive because it agrees with the given orders on G and H. It is also easy to see that the relation is antisymmetric. The only property which needs to be checked is one case of transitivity. Suppose that $h' \leq h$ in H and $h \leq g$ where $g \in G$. By definition $\Theta(x, g) = h$. But Θ

is a special ordered functor and so there exists $(x', g') \in P(G, X)$ such that $(x', g') \leq (x, g)$ and $\Theta(x', g') = h'$. Thus $g' = g$ and so $h' \leq g$. To show that F with this partial order is an ordered groupoid, we check that axioms (OG1), (OG2), (OI), (OG4) hold by Proposition 4.1.4.

Axiom (OG1) holds: we have to prove that if $a \leq b$ in F then $a^{-1} \leq b^{-1}$. The only case that needs any checking is when $h \leq g$ where $h \in H$ and $g \in G$. By definition $h = \Theta(x, g)$, so that $h^{-1} = \Theta(x, g)^{-1} = (g^{-1} \cdot x, g^{-1})$. Hence $h^{-1} \leq g^{-1}$.

Axiom (OG2) holds: we have to prove that if $a \leq b$ and $c \leq d$ and $\exists ac$ and $\exists bd$ then $ac \leq bd$. The only case we need to check is where $h \leq g$ and $h' \leq g'$ and $\exists hh'$ and $\exists gg'$ and $h, h' \in H$ and $g, g' \in G$. By definition $h = \Theta(x, g)$ and $h' = \Theta(x', g')$ for some $(x, g), (x', g') \in P(G, X)$. Since the product hh' exists, so too does the product $(x, g)(x', g')$ because Θ is identity-separating. But $(x, g)(x', g') = (x, gg')$. Thus $hh' = \Theta(x, gg')$ so that $hh' \leq gg'$.

Axiom (OI) holds: the identities of F are the identities of H together with the single identity, 1, of G. To show that (OI) holds it is enough to show that if $h \leq 1$ where $h \in H$ then h is an identity. By definition, $h = \Theta(x, 1)$, but $(x, 1)$ is an identity and so h is an identity.

Axiom (OG4) holds: the one case which needs to be checked is where $g \in G$ and $e \in H_o$. Let $\Theta(x, 1) = e$. Then $\Theta(g \cdot x, g) \leq g$ and $\mathbf{d}(\Theta(g \cdot x, g)) = \Theta(x, 1) = e$.

It remains to show that F is factorisable. Firstly, 1 is the maximum identity. To see this let $e \in H$ be any identity. Then $e = \Theta(x, 1)$ for some identity $(x, 1) \in P(G, X)$. Thus $e \leq 1$. The group of units of F is evidently G. Finally, if $h \in H$ then $h = \Theta(x, g)$ for some $(x, g) \in P(G, X)$ since Θ is surjective. Hence $h \leq g$, and so F is factorisable.

(2) Extend the function Θ to a function $\Theta'\colon P(G, X^1) \to F$ by defining $\Theta'(1, g) = g$ and $\Theta'(x, g) = \Theta(x, g)$ for $x \in X$. It is clear that Θ is a surjective functor. Suppose that $(x, g) \leq (1, g)$ in $P(G, X^1)$. Then $\Theta'(x, g) = \Theta(x, g)$ and $\Theta'(1, g) = g$. Thus $\Theta'(x, g) \leq \Theta'(1, g)$. It follows that Θ' is an ordered functor. It is immediate from the definition that Θ' is identity-separating. Let $h \in H$ and $g \in G$ and suppose that $h \leq g$ and $\Theta'(1, g) = g$. By definition, $h = \Theta(x, g)$ for some $(x, g) \in P(G, X)$. Thus $h = \Theta'(x, g)$ and $(x, g) \leq (1, g)$. It follows that Θ' reflects partial orders. Consequently, Θ' is a special ordered functor. ∎

The next result provides us with a route back to the category of inverse semigroups from the category of ordered groupoids in the one case that matters.

Lemma 8 *Let $\iota\colon S \to F$ be a strict ordered embedding of the inductive groupoid S in the factorisable ordered groupoid F which preserves pseudoproducts. Then there exists a strict embedding $\kappa\colon S \to F'$ of the inverse semigroup S in a* *factorisable inverse monoid F' such that $U(F') = U(F)$, and for each $s \in S$ and $g \subset U(F)$,*

$$\iota(s) \leq g \text{ if, and only if, } \kappa(s) \leq g.$$

Proof By Theorem 4.1.9, there is an ordered embedding ϕ: $F \to I(F)$ which is pseudoproduct preserving. Define

$$F' = \{\alpha \in I(F) \colon \alpha \leq \phi(g) \text{ for some } g \in U(F)\}.$$

Then F' is a factorisable inductive groupoid with a group of units isomorphic to $U(F)$; we identify $U(F)$ and $U(F')$, and regard ϕ as an ordered embedding of F in F' which is the identity on $U(F)$. Define κ: $S \to F'$ by $\kappa = \phi\iota$; this is pseudoproduct preserving because both ι and ϕ are pseudoproduct preserving. But S and F' are inductive groupoids, and so κ is an inductive functor. Let $g \in U(F') = U(F)$. Since ι is a strict embedding of S in F there exists $s \in S$ such that $\iota(s) \leq g$. Hence $\kappa(s) \leq g$, and so κ is a strict embedding of S in F'. If $\iota(s) \leq g$, then $\kappa(s) \leq g$. Conversely, suppose that $\kappa(s) \leq g$. Then $\phi(\iota(s)) \leq \phi(g) = g$. But this implies $\iota(s) \leq g$, because ϕ is an ordered embedding. ■

E-unitary covers and factorisable embeddings

We can now prove our first main result.

Theorem 9 *Let θ: $P \to S$ be an E-unitary cover of the inverse semigroup S over the group $G = P/\sigma$. Then there is a strict embedding κ: $S \to F'$ into a factorisable inverse monoid F' with group of units G such that for each $s \in S$ and $g \in G$,*

$$\kappa(s) \leq g \Leftrightarrow \theta(a) = s$$

for some $a \in P$ with $\sigma(a) = g$

Proof By the P-theorem, we may assume that $P = P(G,X,Y)$ is a P-semigroup. We therefore have to prove that there is a strict embedding κ: $S \to F'$ into a factorisable inverse monoid F' with group of units G such that for $s \in S$ and $g \in G$,

$$\kappa(s) \leq g \Leftrightarrow \theta(x,g) = s$$

for some $(x,g) \in P(G,X,Y)$.

Until the conclusion of the proof we shall work in the category of ordered groupoids and ordered functors.

From the proof of Theorem 8.1.4, the ordered groupoid $P(G,X)$ is an enlargement of the inductive groupoid $P(G,X,Y)$. From Section 4.3, $\ker\theta$ is an identity-separating ordered congruence on $P(G,X,Y)$. Thus by Theorem 4, there exists a unique, identity-separating ordered congruence ρ on $P(G, X)$ which extends $\ker\theta$. Put $H = P(G, X)/\rho$, and let Θ: $P(G, X) \to H$ be the associated natural ordered functor. Define ι: $S \to H$ by $\iota(\theta(x, g)) = \Theta(x, g)$; this function is well-defined because Θ extends θ. It is straightforward to check that ι is an ordered embedding. Because ρ is an ordered congruence, the functor Θ is a special ordered functor. Thus $\iota(S)$ is an order ideal of H because $P(G, X, Y)$ is an order ideal of $P(G, X)$. It follows that ι preserves pseudoproducts.

Extend the action of G on X to an action of G on the poset X^1, and construct the ordered groupoid $P(G, X^1)$ by Lemma 6. Construct the ordered groupoid $F = H \cup G$ as in Lemma 7, and extend Θ to $P(G, X^1)$ by defining $\Theta(1, g) = g$. The ordered embedding ι: $S \to H$ will be regarded as an ordered embedding of S in F; it still preserves pseudoproducts. Observe that $\iota(s) \leq g$ in F if, and only if, $\theta(y, g) = s$ for some $(y, g) \in P(G, X, Y)$. Let $g \in U(F)$. Then by Theorem 8.1.4, there exists $(y, g) \in P(G, X, Y)$. Thus $\iota(\theta(y, g)) \leq g$ in F. Hence ι: $S \to F$ is a strict ordered embedding of the inductive groupoid S into the factorisable ordered groupoid F which preserves pseudoproducts. We now use Lemma 8 to conclude the proof. ■

The above result shows us how to pass from an E-unitary cover to a closely related strict factorisable embedding. However, from Theorem 2.4.7 we know that such an embedding itself gives rise to an E-unitary cover. The relationship between this new cover and the old one can now be described.

Theorem 10 *Every E-unitary cover of an inverse semigroup S over a group G is isomorphic to one constructed from a strict embedding of S into a factorisable inverse monoid with group of units isomorphic to G.*

Proof Let θ: $P(G, X, Y) \to S$ be an E-unitary cover of S. By Theorem 9, there is a strict embedding ι: $S \to F$ into a factorisable inverse monoid F, where $U(F) = G$, such that

$$\iota(s) \leq g \Leftrightarrow \theta(y, g) = s$$

for some $(y, g) \in P(G, X, Y)$.

Let π_1: $P \to S$ be the E-unitary cover constructed from the strict embedding ι by means of Theorem 2.2.4. Define ψ: $P(G, X, Y) \to P$ by $\psi(y, g) = (\theta(y, g), g)$. This is a well-defined map, because $\iota(\theta(y, g)) \leq g$. It is clearly a homomorphism, so it remains to prove that ψ is a bijection. It is clear that ψ is injective. To prove that ψ is surjective let $(s, g) \in P$ be any element. Then $\iota(s) \leq g$. Thus there exists $(y, g) \in P(G, X, Y)$ such that $\theta(y, g) = s$. Hence $\psi(y, g) = (s, g)$. ■

F-inverse covers and almost factorisable embeddings

The theory of E-unitary covers can be specialised to the case of F-inverse covers.

Proposition 11 *Let θ: $T \to W$ be a surjective, idempotent-separating homomorphism, and let S be an inverse subsemigroup of T such that T is an enlargement of S. Then W is an enlargement of $\theta(S)$.*

Proof We show that axioms (E1), (E2) and (E3) hold.

Axiom (E1) holds: let f be an idempotent of $\theta(S)$ and let $e \leq f$ in W. By Theorem 1.5.1, there is an idempotent $f' \in S$ such that $\theta(f') = f$ and an idempotent $e' \in T$ such that $e' \leq f'$ and $\theta(e') = e$. But T is an enlargement of S. Thus $e' \in S$ and so $e \in \theta(S)$.

Axiom (E2) holds: let $w \in W$ such that $w^{-1}w, ww^{-1} \in \theta(S)$. Let $t \in T$ such that $\theta(t) = w$. Then $\theta(t^{-1}t), \theta(tt^{-1}) \in \theta(S)$. Thus there are idempotents $e', f' \in S$ such that $\theta(t^{-1}t) = \theta(e')$ and $\theta(tt^{-1}) = \theta(f')$. But θ is idempotent-separating, and so $t^{-1}t, tt^{-1} \in S$. Hence $t \in S$, since T is an enlargement of S, and so $w \in \theta(S)$.

Axiom (E3) holds: let e be an idempotent of W. Then there is an idempotent $e' \in T$ such that $\theta(e') = e$. Since T is an enlargement of S there exists $t \in T$ such that $tt^{-1} = e'$ and $t^{-1}t \in S$. Put $w = \theta(t)$. Then $ww^{-1} = e$ and $w^{-1}w \in \theta(S)$. ■

To make the statement of our two main results more succinct we make the following definition. A strict embedding ι: $S \to F$ of an inverse semigroup S into a factorisable inverse monoid F such that $F \setminus U(F)$ is an enlargement of $\iota(S)$ is called a *factorisable envelope* of S.

Theorem 12 *Let θ: $P \to S$ be an F-inverse cover of the inverse semigroup S over the group $G = P/\sigma$. Then there is a factorisable envelope ι: $S \to F$, where F has group of units G, such that for each $s \in S$ and $g \in G$,*

$$\iota(s) \leq g \Leftrightarrow \theta(a) = s$$

for some $a \in P$ with $\sigma(a) = g$.

Proof The proof uses the same argument as the proof of Theorem 9 with the exception that we do not need to apply Lemma 8. Also we work entirely with inverse semigroups. By the P-theorem, we may assume that $P = P(G, X, Y)$ is a P-semigroup where X is a meet semilattice, since P is F-inverse.

From the proof of Proposition 7.4.6, the inverse semigroup $P(G, X)$ is an enlargement of the inverse semigroup $P(G, X, Y)$. From Section 4.3, $\ker\theta$ is an identity-separating congruence on $P(G, X, Y)$. Thus by Corollary 5, there exists a unique, idempotent-separating congruence ρ on $P(G,X)$ which extends $\ker\theta$. Put $H = P(G,X)/\rho$, and let $\Theta: P(G,X) \to H$ be the associated natural homomorphism. Define $\iota: S \to H$ by $\iota(\theta(x,g)) = \Theta(x,g)$; this function is well-defined because Θ extends θ. It is straightforward to check that ι is an embedding. Since $P(G,X)$ is an enlargement of $P(G,X,Y)$, it follows that H is an enlargement of $\iota(S)$ by Proposition 11.

Extend the action of G on X to an action of G on the poset X^1, and construct the factorisable inverse semigroup $P(G,X^1)$ by Lemma 6. Construct the ordered groupoid $F = H \cup G$ as in Lemma 7. It is clearly inductive and so can be treated as an inverse semigroup. Extend Θ to $P(G,X^1)$ by defining $\Theta(1,g) = g$. The embedding $\iota: S \to H$ will be regarded as an embedding of S in F. Observe that $\iota(s) \leq g$ in F if, and only if, $\theta(y,g) = s$ for some $(y,g) \in P(G,X,Y)$. Finally, $F \setminus U(F) = H$, which is an enlargement of $\iota(S)$. ■

The final result of this section is the appropriate F-inverse analogue of Theorem 10. It is immediate from Theorem 12 and Theorem 10.

Theorem 13 *Every F-inverse cover of an inverse semigroup S over a group G is isomorphic to one constructed from a factorisable envelope $\iota: S \to F$ where $U(F)$ is isomorphic to G.* ■

By Theorem 7.1.13, inverse semigroups of the form $F \setminus U(F)$, where F is factorisable, are almost factorisable. Thus the theory of F-inverse covers of an inverse semigroup S is intimately connected with the theory of almost factorisable enlargements of S.

8.3 The maximum enlargement theorem

In this section, we prove a theorem which provides information about the structure of ordered star injective functors. To motivate this result, we return to the theory of E-unitary inverse semigroups, and consider it from a different point of view.

The essential difference between arbitrary E-unitary semigroups and semidirect products of semilattices by groups resides in the behaviour of the minimum group congruence. When S is an E-unitary semigroup the natural homomorphism $\sigma^{\natural}: S \to S/\sigma$ is $\mathcal{L}$-injective, but when S is a semidirect product of a semilattice by a group it is $\mathcal{L}$-bijective. The P-theorem can be interpreted in these terms. To see how, consider first the simpler case of an F-inverse semigroup S. The P-theorem for F-inverse semigroups is equivalent to the existence of an enlargement T of S which is a semidirect product of a semilattice by a group; thus the $\mathcal{L}$-injective natural homomorphism $\sigma^\natural: S \to S/\sigma$ is extended to the $\mathcal{L}$-bijective natural homomorphism $\sigma^\natural: T \to S/\sigma$. For arbitrary E-unitary semigroups the same result holds good except that we have to work with ordered groupoids; thus $\mathcal{L}$-bijective homomorphisms are replaced by star bijective ordered functors. When viewed in this way, the P-theorem says that an ordered star injective functor from an E-unitary inductive groupoid to a group can be extended to a star bijective ordered functor.

The maximum enlargement theorem generalises the P-theorem: it says that every ordered star injective functor can be extended to an ordered covering functor. The precise statement is as follows:

The maximum enlargement theorem
Let p: $H \to K$ be an ordered, star injective functor between ordered groupoids. Then there is an ordered groupoid G, an ordered embedding i: $H \to G$, and an ordered covering functor p': $G \to K$ such that $p'i = p$ where G is an enlargement of $i(H)$. In addition, p' is the 'best possible' such covering functor.

Our proof of the maximum enlargement theorem will be independent of our proof of the P-theorem. By using Theorem 8.1.7, we will then have a new proof of the P-theorem. More generally, the theorem can be used to study idempotent pure prehomomorphisms and to set up a theory of idempotent pure extensions. Thus it provides a unifying framework for many of the results we have proved in this chapter.

The enlargement theorem and local structures

The proof of the maximum enlargement theorem, which we describe in the next section, is based on some technical constructions in category theory. These can best be motivated by looking into the meaning of the maximum enlargement theorem within the theory of local structures.

In Section 1.2, we described how a complete pseudogroup, which we here denote by H, could be used to construct local structures on topological spaces by means of atlases. In the 1950s, Ehresmann axiomatised this construction using category theory. His starting point was the forgetful functor p: $H \to K$ from H to the ordered groupoid of sets and partial bijections K. The aim of the axiomatisation was to show how the corresponding category of local structures and isomorphisms between them could be constructed from this functor.[1] It is important to notice that the topological substrate, which we assumed pre-existing in our informal discussion in Section 1.2, is not assumed here, instead the construction of this substrate is to be part of the axiomati-

[1] Arbitrary morphisms between local structures require the use of ordered categories rather than ordered groupoids.

sation. Ehresmann discovered that there were two basic steps involved in the axiomatisation:

(1) The construction of an enlargement.

(2) The construction of a completion.

We shall discuss the second construction briefly at the end of this section, but first we deal with the construction of the enlargement which is the basis of the maximum enlargement theorem.

The construction of an enlargement relies on the notion of transportability. Informally, transportability means that if a set X is equipped with a structure, and Y is a set in bijective correspondence with X, then a structure can be defined on Y in such a way that the bijection becomes an isomorphism. Transportability is defined categorically as follows. Let C be a category and θ: $C \to$ **Sets** a (forgetful) functor to the category of sets. We say that θ is *transportable* if for every object $X \in C$ and every bijection α: $\theta(X) \to Y$ in **Sets** there exists a unique morphism $f \in C$ with domain X such that $\theta(f) = \alpha$. If C is a groupoid, and we restrict attention to the groupoid of bijections of **Sets**, then transportability says precisely that θ is star bijective.

The relevance of transportability to local structures is explained as follows. The forgetful functor p: $H \to K$ is clearly star injective, whereas the forgetful functor from the category of local structures is star bijective. Thus the first step in constructing the category of local structures from p is to extend p so that it is star bijective.

Now that we know why we wish to construct covering extensions of star injective functors, we can now turn to the method for constructing such extensions. We begin with a concrete example.

Let (X, τ) be a topological space with open sets τ. Let α: $X \to Y$ be a bijection from the *set* X to the set Y. Define τ' to be the set of all subsets of Y of the form $\alpha(U)$ where $U \in \tau$. Then (Y, τ') is a topological space and α is a homeomorphism. It is evident that τ' is the only topology on Y which makes α into a homeomorphism. We say that the topology τ on X has been *transported* to the set Y by means of the bijection α. Observe that the topological space (Y, τ') is completely determined by the pair $(\alpha, (X, \tau))$ consisting of the function α: $X \to Y$ and the topological space (X, τ). However, this may not be the only such ordered pair describing (Y, τ'). Let (Z, σ) be a topological space and let β: $Z \to Y$ be a bijection. Suppose that τ' is also the topology obtained when σ is transported to Y by β. Then $\beta^{-1}\alpha$ is a homeomorphism from (X, τ) to (Z, σ). The converse also holds: let τ' be the topology transported to Y by α and let σ' be the topology transported to Y by β. Let γ: $(X, \tau) \to (Z, \sigma)$ be a homeomorphism such that $\beta\gamma = \alpha$. Then $\sigma' = \tau'$. We conclude that the pair $(\alpha, (X, \tau))$ and the pair $(\beta, (Z, \sigma))$ denote the same topological space precisely when α and β have the same image and there is a homeomorphism γ: $X \to Z$ such that $\gamma\beta = \alpha$.

The above example motivates the construction of an ordered covering functor extension of an arbitrary ordered star injective functor p: $H \to K$. Let $e \in H_o$ and let y: $p(e) \to \mathbf{r}(y)$ in K. In our example, e was a topological space, $p(e)$ the underlying set and y a bijection. Thus $(\mathbf{r}(y), y, e)$ denotes the structure on e transported to $\mathbf{r}(y)$ by y. In the first instance, these triples form the identities of a new category $N = N(H, K, p)$ whose morphisms are simply the triples of the form (x, y, e) where $\mathbf{d}(x) = \mathbf{r}(y)$; the domain of (x, y, e) is $(\mathbf{r}(y), y, e)$. However, the category N is too big: we have to deal with the problem that different triples may denote the same structure. This is solved in a way suggested by our topological example. Define an eqivalence relation on the set of triples as follows: we say that $(\mathbf{r}(y), y, e)\, \rho\, (\mathbf{r}(y'), y', e')$ if $\mathbf{r}(y) = \mathbf{r}(y')$ and there exists $z \in H$ such that z: $e' \to e$ and $y' = yp(z)$. The relation ρ can be extended to all morphisms in N by defining $(x, y, e)\, \rho\, (x', y', e')$ if, and only if, $x = x'$ and $(\mathbf{r}(y), y, e)\, \rho\, (\mathbf{r}(y'), y', e')$. This equivalence relation is used to construct a new category $G = N/\rho$. It is this category which proves the maximum enlargement theorem, because G is an enlargement of H, and there is an ordered covering functor p': $G \to K$ which extends p. This completes the motivation of the proof of the maximum enlargement theorem.

It only remains to say a few words about the second part of Ehresmann's axiomatisation of the construction of local structures. We need some definitions first. A compatibility relation can be defined in ordered groupoids extending the definition we have already given for inverse semigroups. Our starting point is the characterisation of the inverse semigroup compatibility relation provided by Lemma 1.4.11. Let G be an ordered groupoid. We define $x \sim y$ if the meet $x \wedge y$ exists in the partially ordered set $(G, \leq)$ and both $\mathbf{d}(x) \wedge \mathbf{d}(y)$ and $\mathbf{r}(x) \wedge \mathbf{r}(y)$ exist and satisfy

$$\mathbf{d}(x \wedge y) = \mathbf{d}(x) \wedge \mathbf{d}(y) \text{ and } \mathbf{r}(x \wedge y) = \mathbf{r}(x) \wedge \mathbf{r}(y).$$

Let θ: $G \to H$ be an ordered functor. A non-empty subset A of G will be said to be *θ-compatible* if it is a compatible subset of G, if its image $\theta(A)$ is a compatible subset of H and if $\bigvee \theta(A)$ exists in $(H, \leq)$. We say that G is *θ-complete* if every θ-compatible subset of G has a join. This notion of completeness generalises the usual notion: if the functor θ is chosen to be the identity functor on G then θ-complete means that every non-empty compatible subset has a join.

We now return to our example; thus p': $G \to K$ is the ordered covering functor extending p: $H \to K$ constructed above. Let A be a p'-compatible subset of G_o. If we write down what A is in terms of our original data, we discover that the set A can be regarded as a class of charts (and so it is an atlas) with changes of co-ordinates (that is transition functions) in the groupoid H.

Thus the p'-completion of G will be a category whose objects can be identified with the complete atlases compatible with H.

In conclusion, the classical procedure for constructing local structures, described in Section 1.2, can be characterised in terms of properties of ordered groupoids. In particular, the idea of extending star injective functors to star bijective functors plays a pivotal rôle. We have already noted that the same idea lies behind the structure theory of E-unitary inverse semigroups. Thus the maximum enlargement theorem provides an important point of contact between Ehresmann's work, motivated by pseudogroup theory, and the algebraic theory of inverse semigroups.

Proof of the maximum enlargement theorem

Let $p\colon H \to K$ be an ordered, star injective functor between ordered groupoids. In accordance with the arguments described in the previous section, the first step in extending p to an ordered covering functor is to adjoin new elements to H. To do this we shall use a special case of a comma category.

Let $e' \in K_o$ and define

$$(e' \downarrow K) = \{(x,y) \in K \times K \times\colon e' = \mathbf{d}(y),\, \mathbf{d}(x) = \mathbf{r}(y)\}$$

equipped with the partial product given by

$$(x',y')(x,y) = \begin{cases} (x'x,y) & \text{if } y' = xy \\ \text{undefined} & \text{otherwise.} \end{cases}$$

It is easy to check that $(e' \downarrow K)$ is a groupoid in which the identities are the elements of the form $(\mathbf{r}(x),x)$, inverses are given by $(x,y)^{-1} = (x^{-1},xy)$ and

$$\mathbf{d}(x,y) = (\mathbf{r}(y),y) \text{ and } \mathbf{r}(x,y) = (\mathbf{r}(xy),xy).$$

The category $(e' \downarrow K)$ is called the *category of objects under* e'.

Proposition 1 *Let* $p\colon H \to K$ *be an ordered star injective functor between ordered groupoids. Define* $N = N(H,K,p)$ *by*

$$N = \{(x,y,e) \in K \times K \times H_o\colon p(e) = \mathbf{d}(y), \mathbf{d}(x) = \mathbf{r}(y)\}.$$

Define a partial product on N *by*

$$(x',y',e')(x,y,e) = \begin{cases} (x'x,y,e) \text{ if } e' = e \text{ and } y' = xy \\ \text{undefined otherwise} \end{cases}$$

and equip N *with the direct product order.*

(1) N *is an ordered groupoid.*

(2) *π_1: $N \to K$ is an ordered covering functor.*

(3) *The function ζ: $H \to N$ given by $\zeta(h) = (p(h), p(\mathbf{d}(h)), \mathbf{d}(h))$ is a well-defined injective function, such that $h \leq k$ if, and only if, $\zeta(h) \leq \zeta(k)$.*

(4) $\pi_1\zeta = p$.

Proof (1) Observe that N is the disjoint union of the groupoids $(p(e) \downarrow K)$ as e varies over H_o Thus N is a groupoid in which

$$\mathbf{d}(x, y, e) = (\mathbf{r}(y), y, e), \mathbf{r}(x, y, e) = (\mathbf{r}(xy), xy, e) \text{ and } (x, y, e)^{-1} = (x^{-1}, xy, e).$$

To show that N is an ordered groupoid, we verify that the axioms (OG1), (OG2), (OI), and (OG4) of Proposition 4.1.4 hold. Of these, only axiom (OG4) needs any special comment. Let $(u, v, f) \in N$ and $(\mathbf{r}(y), y, e) \in N_o$ such that

$$(\mathbf{r}(y), y, e) \leq \mathbf{d}(u, v, f) = (\mathbf{r}(v), v, f).$$

Define

$$((u, v, f) \mid (\mathbf{r}(y), y, e)) = ((u \mid \mathbf{r}(y)), y, e).$$

It is easy to check that this is a well-defined element of N with the appropriate properties.
(2) It is routine to check that π_1 is an ordered functor. It remains to show that π_1 is star bijective. We begin by showing that it is star injective. Let $(x, y, e), (x', y', e') \in N$ such that

$$\mathbf{d}(x, y, e) = \mathbf{d}(x', y', e') \text{ and } \pi_1(x, y, e) = \pi_1(x', y', e').$$

Then $x = x'$ and $(\mathbf{r}(x), y, e) = (\mathbf{r}(x'), y', e')$, so that $(x, y, e) = (x', y', e')$. Finally, we show that π_1 is star surjective. Let $(\mathbf{r}(y), y, e)$ be an identity of N and let x be any element of K such that $\mathbf{d}(x) = \pi_1(\mathbf{r}(y), y, e)$. Then (x, y, e) is a well-defined element of N such that

$$\pi_1(x, y, e) = x \text{ and } \mathbf{d}(x, y, e) = (\mathbf{r}(y), y, e).$$

(3) It is clear that ζ is well-defined, and that $h \leq k$ implies $\zeta(h) \leq \zeta(k)$. Conversely, suppose that $\zeta(h) \leq \zeta(k)$. Then $p(h) \leq p(k)$ and $\mathbf{d}(h) \leq \mathbf{d}(k)$. But then $h \leq k$ by Lemma 8.1.2, since p is star injective. It is now clear that ζ is injective.
(4) Immediate from the definitions. ■

The ordered covering functor π_1: $N \to K$ was essentially obtained by adjoining all possible 'missing' identities, so that the functor becomes star bijective, but we have not taken account of the structure of the ordered groupoid

H. This is reflected in the fact that ζ, although an order-preserving function, is not a functor. However, as we shall show below, we can construct a quotient ordered groupoid N/ρ of N in such a way that the composition of the function ζ with $\rho^{\natural}$ yields an embedding of H in N/ρ. Furthermore, there is an ordered covering functor from N/ρ to K. It remains therefore to define the congruence ρ.

Define a relation ρ on N as follows: $(x, y, e)\, \rho\, (x', y', e')$ if, and only if, $x = x'$ and there exists $z \in H$ such that $\mathbf{r}(z) = e$ and $\mathbf{d}(z) = e'$ and $y' = yp(z)$.

Proposition 2 *With the above definitions, we have the following:*

(1) *If $(x, y, e)\, \rho\, (x', y', e')$ and $(x, y, e) \leq (x', y', e')$ then $(x, y, e) = (x', y', e')$.*

(2) *The relation ρ is an ordered congruence on the ordered groupoid N.*

(3) *The composite $i = \rho^{\natural}\zeta\colon H \to N/\rho$ is an ordered embedding.*

Proof (1) By definition $x = x'$, $y \leq y'$ and $e \leq e'$, and there exists $z \in H$ such that $\mathbf{r}(z) = e$, $\mathbf{d}(z) = e'$, and $y' = yp(z)$. Now $\mathbf{d}(x) = \mathbf{d}(x')$, so that $\mathbf{r}(y) = \mathbf{r}(y')$. But $y \leq y'$ so that $y = y'$. Also $y' = yp(z)$ so that $p(z)$ is an identity. But p is star injective and so z is an identity. Hence $e = e'$.
(2) It is easy to check that ρ is an equivalence relation. We now verify that the axioms for an ordered congruence hold replacing axiom (OCON3) by (OCON3)* (Lemma 4.3.7) so that axiom (AC) does not have to be checked. The verifications of (OCON1) and (OCON2) are routine and omitted.

Axiom (OCON3)* holds: let $(x, y, e), (u, v, f) \in N$ such that

$$\mathbf{d}(x, y, e)\, \rho\, \mathbf{r}(u, v, f).$$

Then $\mathbf{r}(y) = \mathbf{r}(u)$ and there exists an element z such that

$$\mathbf{r}(z) = e,\ \mathbf{d}(z) = f \text{ and } uv = yp(z).$$

Consider the triple (x, uv, f). Then

$$\mathbf{d}(x) = \mathbf{r}(y) = \mathbf{r}(u) = \mathbf{r}(uv) \text{ and } \mathbf{d}(uv) = \mathbf{d}(v) = p(f).$$

Thus $(x, uv, f) \in N$. But

$$\mathbf{d}(x, uv, f) = (\mathbf{r}(uv), uv, f) = (\mathbf{r}(u), uv, f) = \mathbf{r}(u, v, f)$$

and $(x, y, e)\, \rho\, (x, uv, f)$.

Axiom (OCON4) holds: let

$$(k, f, e) \leq (l, g, i)\, \rho\, (l', g', i').$$

Then

$$k \leq l,\ f \leq g \text{ and } e \leq i,$$

and $l = l'$ and there exists an element $x \in H$ such that

$$\mathbf{r}(x) = i,\ \mathbf{d}(x) = i' \text{ and } g' = gp(x).$$

Consider the triple $(k, fp(e \,|\, x), \mathbf{d}(e \,|\, x))$. We have first to check that the components of this triple are well-defined. The element $(e \,|\, x)$ is defined in H since $e \leq i = \mathbf{r}(x)$; next observe that

$$\mathbf{d}(f) = p(e) \text{ and } \mathbf{r}(p(e \,|\, x)) = p(\mathbf{r}(e \,|\, x)) = p(e)$$

so that the element $fp(e \,|\, x)$ is defined. Now we have to show that the triple is a well-defined element of N. We have that

$$\mathbf{r}(fp(e \,|\, x)) = \mathbf{r}(f) \text{ and } \mathbf{d}(k) = \mathbf{r}(f)$$

so that $\mathbf{d}(k) = \mathbf{r}(fp(e \,|\, x))$. Finally

$$\mathbf{d}(fp(e \,|\, x)) = \mathbf{d}(p(e \,|\, x)) = p(\mathbf{d}(e \,|\, x)).$$

It now follows that $(k, fp(e \,|\, x), \mathbf{d}(e \,|\, x)) \in N$. Observe that

$$(k, fp(e \,|\, x), \mathbf{d}(e \,|\, x)) \leq (l', g', i')$$

since $k \leq l = l'$, $fp(e \,|\, x) \leq gp(x) = g'$ and $\mathbf{d}(e \,|\, x) \leq \mathbf{d}(x) = i'$. In conclusion, we show that

$$(k, f, e)\, \rho\, (k, fp(e \,|\, x), \mathbf{d}(e \,|\, x)).$$

To this end, observe that $\mathbf{r}(e \,|\, x) = e$ and $\mathbf{d}(e \,|\, x) = \mathbf{d}(e \,|\, x)$, also $fp(e \,|\, x) = fp(e \,|\, x)$, and the result is proved.

Axiom (OCON5) holds: let $(x', y', e') \leq (x, y, e)$, and let $(\mathbf{r}(z), z, f)$ be an identity such that $(\mathbf{r}(z), z, f) \leq \mathbf{d}(x, y, e)$ and $(\mathbf{r}(z), z, f)\, \rho\, \mathbf{d}(x', y', e')$. Then

$$x' \leq x,\ y' \leq y \text{ and } e' \leq e,$$

and

$$\mathbf{r}(z) \leq \mathbf{r}(y),\ z \leq y \text{ and } f \leq e,$$

and $\mathbf{r}(z) = \mathbf{r}(y')$ and there exists an element w such that

$$e' = \mathbf{r}(w),\ f = \mathbf{d}(w) \text{ and } z = y'p(w).$$

Consider the triple (x', z, f). Then

$$\mathbf{d}(x') = \mathbf{r}(y') = \mathbf{r}(z) \text{ and } \mathbf{d}(z) = p(f).$$

Thus $(x', z, f) \in N$. It is immediate that $(x', z, f) \leq (x, y, e)$, and $\mathbf{d}(x', z, f) = (\mathbf{r}(z), z, f)$. It remains to show that $(x', y', e')\, \rho\, (x', z, f)$ but this follows from the fact that $\mathbf{r}(w) = e'$, $\mathbf{d}(w) = f$ and $z = y'p(w)$.

Axiom (OCON6) holds: immediate from (1).

(2) Suppose that $h \leq k$ in H. Then $\zeta(h) \leq \zeta(k)$ by Proposition 1, and so $\rho^{\natural}(\zeta(h)) \leq \rho^{\natural}(\zeta(k))$, since ρ is an ordered congruence. Thus $i(h) \leq i(k)$. Conversely, suppose that $i(h) \leq i(k)$. Then $\rho^{\natural}(\zeta(h)) \leq \rho^{\natural}(\zeta(k))$. Thus there exists $(x, y, e)\, \rho\, \zeta(h)$ such that $(x, y, e) \leq \zeta(k)$. By definition,

$$(x, y, e)\, \rho\, (p(h), p(\mathbf{d}(h)), \mathbf{d}(h)) \text{ and } (x, y, e) \leq (p(k), p(\mathbf{d}(k)), \mathbf{d}(k)).$$

Thus y is an identity, $x = p(h)$, and there exists $z \in H$ such that $\mathbf{r}(z) = e$, $\mathbf{d}(z) = \mathbf{d}(h)$ and $p(\mathbf{d}(h)) = yp(z)$. Hence $p(z) = p(\mathbf{d}(h))$ and so z is an identity since p is star injective. Thus $(x, y, e) = (p(h), p(\mathbf{d}(h)), \mathbf{d}(h))$ and so $p(h) \leq p(k)$ and $\mathbf{d}(h) \leq \mathbf{d}(k)$. Hence $h \leq k$ by Lemma 8.1.2.

It remains to show that i is a functor. It is clear that i maps identities to identities. Let hk be defined in H, so that $\mathbf{d}(h) = \mathbf{r}(k)$. It is easy to check that $\mathbf{d}(\zeta(h))\, \rho\, \mathbf{r}(\zeta(k))$. Thus the product $i(h)i(k)$ is defined in N/ρ. To calculate what this product is, observe that $(p(h), p(k), \mathbf{d}(k))$ belongs to N, $\zeta(h)\, \rho\, (p(h), p(k), \mathbf{d}(k))$, and $\mathbf{d}(p(h), p(k), \mathbf{d}(k)) = \mathbf{r}(\zeta(k))$. Thus

$$i(h)i(k) = \rho((p(h), p(k), \mathbf{d}(k))(p(k), p(\mathbf{d}(k)), \mathbf{d}(k))) = i(hk).$$

■

The following result is Ehresmann's maximum enlargement theorem.

Theorem 3 *Let $p\colon H \to K$ be an ordered star injective functor. Then there is an ordered groupoid G, an ordered embedding $i\colon H \to G$ and an ordered covering functor $p'\colon G \to K$ such that G is an enlargement of $i(H)$ and $p'i = p$.*

Proof Let $N = N(H, K, p)$ and $\zeta\colon H \to N$ by Proposition 1. Put $G = N(H, K, p)/\rho$, and $i = \rho^{\natural}\zeta\colon H \to G$ by Proposition 2. Define $p'\colon G \to K$ by $p'(\rho(x, y, e)) = x$. This is well-defined from the definition of ρ and it is immediate that $p'i = p$. It is easy to see that $p'\colon G \to K$ is an ordered functor, we prove that it is star bijective. Let $x \in K$ and let $\rho(\mathbf{d}(x), y', e') \in G$ be an identity such that $p'(\rho(\mathbf{d}(x), y', e')) = \mathbf{d}(x)$. Then $\rho(x, y', e')$ is the unique element in G such that $p'(\rho(x, y', e')) = x$ and $\mathbf{d}(\rho(x, y', e')) = \rho(\mathbf{d}(x), y', e')$. It remains to prove that G is an enlargement of $i(H)$.

Axiom (GE1) holds: let $\rho(x, y, e) \leq i(h)$ where h is an identity of H. Then by definition, there exists an element $(x', y', e') \in N$ such that

$$(x', y', e')\, \rho\, (x, y, e) \text{ and } (x', y', e') \leq (p(h), p(h), h).$$

Thus x' and y' are identities of K, so that $x' = y' = p(e')$, where $e' \in H_o$. Thus $(x', y', e') = (p(e'), p(e'), e') = i(e')$. Hence $\rho(x, y, e) \subset i(H)$.

Axiom (GE2) holds: let $\rho(x, y, e) \in G$ such that

$$\mathbf{d}(\rho(x, y, e)), \mathbf{r}(\rho(x, y, e)) \in i(H).$$

Then there exist identities $h, h' \in H_o$ such that

$$\mathbf{d}(\rho(x, y, e)) = \rho(\mathbf{r}(y), y, e) = i(h)$$

and

$$\mathbf{r}(\rho(x, y, e)) = \rho(\mathbf{r}(xy), xy, e) = i(h').$$

Hence there exists $z \in H$ such that

$$\mathbf{r}(y) = p(h),\ \mathbf{r}(z) = e,\ \mathbf{d}(z) = h \text{ and } p(h) = yp(z),$$

and there exists $w \in H$ such that

$$\mathbf{r}(x) = p(h'),\ \mathbf{r}(w) = e,\ \mathbf{d}(w) = h' \text{ and } p(h') = xyp(w).$$

Now $\mathbf{d}(w^{-1}) = \mathbf{r}(w) = e = \mathbf{r}(z)$ and so the product $w^{-1}z$ is defined in H. Since $\mathbf{r}(x) = p(h')$, we have $\mathbf{r}(x) = xyp(w)$ and so $p(w) = (xy)^{-1}$. Also since $\mathbf{r}(y) = p(h)$, we have that $\mathbf{r}(y) = yp(z)$ and so $p(z) = y^{-1}$. It follows that $p(w^{-1}z) = x$. Hence

$$i(w^{-1}z) = \rho(x, \mathbf{r}(y), h).$$

But it is easy to see that $(x, \mathbf{r}(y), h)\, \rho\, (x, y, e)$. Thus $\rho(x, y, e) \in i(H)$.

Axiom (GE3) holds: let $\rho(\mathbf{r}(y), y, e) \in G_o$. Then $(y, p(e), e) \in N$ and so $\rho(y, p(e), e) \in G$. But

$$\mathbf{d}(\rho(y, p(e), e)) = \rho(p(e), p(e), e) = i(e)$$

and $\mathbf{r}(\rho(y, p(e), e)) = \rho(\mathbf{r}(yp(e)), yp(e), e) = \rho(\mathbf{r}(y), y, e)$. ■

The next result will enable us to prove that enlargements are essentially unique.

Theorem 4 *Let p: $H \to K$ be an ordered star injective functor, i: $H \to G$ an ordered embedding such that G is an enlargement of $i(H)$, and let p': $G \to K$ be an ordered covering functor such that $p'i = p$. Let j: $H \to G'$ be an ordered embedding and let p'': $G' \to K$ be an ordered covering functor such that $p''j = p$. Then there is a unique ordered functor θ: $G \to G'$ such that $\theta i = j$ and $p''\theta = p'$.*

Proof Without loss of generality we may assume that i is an inclusion of ordered groupoids. We shall define a function from G to G' in stages.

Let $x \in G$ such that $\mathbf{d}(x) \in H_o$. Then since p'' is star bijective there is a unique element in G', which we denote by $\theta(x)$, such that

$$\mathbf{d}(\theta(x)) = j(\mathbf{d}(x)) \text{ and } p''(\theta(x)) = p'(x).$$

We shall extend θ to the whole of G below. But first we prove a result which we shall use later.

With x as above, let $y \in G$ such that $\mathbf{r}(y) = \mathbf{r}(x)$ and $\mathbf{d}(y) \in H_o$. Then since G is an enlargement of H, we have that $x^{-1}y \in H$. Now $\mathbf{d}(\theta(x)) = j(\mathbf{d}(x)) = \mathbf{r}(j(x^{-1}y))$. Thus $\theta(x)j(x^{-1}y)$ is defined in G'. But $p''(\theta(x)j(x^{-1}y)) = p'(y)$ and $\mathbf{d}(\theta(x)j(x^{-1}y)) = j(\mathbf{d}(y))$. Thus $\theta(y) = \theta(x)j(x^{-1}y)$.

Let (Φ, ϕ) be a reflection of G in H (Section 8.2). Then for each $x \in G$ we can write

$$x = \phi(\mathbf{r}(x))\Phi(x)\phi(\mathbf{d}(x))^{-1}.$$

Extend θ to arbitrary elements of G by defining

$$\theta(x) = \theta(\phi(\mathbf{r}(x)))j(\Phi(x))\theta(\phi(\mathbf{d}(x))^{-1}).$$

To see that the definition of θ is independent of the choice of reflection we argue as follows using the fact that G is an enlargement of H. Let $a, b \in G$ such that

$$\mathbf{r}(a) = \mathbf{r}(x),\ \mathbf{r}(b) = \mathbf{d}(x),\ \text{ and } h = a^{-1}xb \in H.$$

Similarly let $c, d \in G$ such that

$$\mathbf{r}(c) = \mathbf{r}(x),\ \mathbf{r}(d) = \mathbf{d}(x),\ \text{ and } k = c^{-1}xd \in H.$$

Now $a = c(c^{-1}a)$ where $c^{-1}a \in H$ and $b = d(d^{-1}b)$ where $d^{-1}b \in H$. Thus by the result above $\theta(a) = \theta(c)j(c^{-1}a)$ and $\theta(b) = \theta(d)j(d^{-1}b)$. Hence

$$\theta(a)j(h)\theta(b)^{-1} = \theta(c)j(k)\theta(d)^{-1}.$$

Thus θ is well-defined and clearly a functor. Observe that $p''(\theta(x)) = p'(x)$, and $\theta i = j$ by choosing the reflection so that Φ is the identity when restricted to H.

It remains to show that θ is order-preserving. Let $x, y \in G$ such that $\mathbf{d}(x), \mathbf{d}(y) \in H$ and $x \leq y$. Now $p'(x) \leq p'(y)$ and $j(\mathbf{d}(x)) \leq j(\mathbf{d}(y))$. But then $p''(\theta(x)) \leq p''(\theta(y))$ and $\mathbf{d}(\theta(x)) \leq \mathbf{d}(\theta(y))$. Thus $\theta(x) \leq \theta(y)$ by Lemma 8.1.2. Now let $x, y \in G$ be arbitrary. We may choose a reflection (Φ, ϕ) such that $\Phi(x) \leq \Phi(y)$ by Lemma 8.2.3. It is now straightforward to check that $\theta(x) \leq \theta(y)$. Hence we have proved that θ is an ordered functor such that $\theta i = j$ and $p''\theta = p'$.

Now let $\theta'\colon G \to G'$ be any ordered functor such that $\theta' i = j$ and $p''\theta' = p'$. Let $x \in G$. Then we can write $x = ahb^{-1}$ where $h \in H$ and $a, b \in G$. Thus $\theta'(x) = \theta'(a)\theta'(h)\theta'(b^{-1})$. But $h \in H$ and so $\theta'(h) = j(h)$ since $\theta' i = j$. Also,

$$\mathbf{d}(\theta'(a)) = j(\mathbf{d}(a)) \text{ and } p''(\theta'(a)) = p'(a)$$

since $\theta' i = j$ and $p''\theta' = p'$. Thus $\theta'(a) = \theta(a)$. Similarly, $\theta(b') = \theta(b)$. Hence $\theta'(x) = \theta(x)$, as required. ∎

The proof of the following is immediate by Theorem 4.

Theorem 5 *Let $p\colon H \to K$ be an ordered star injective functor. Let $i\colon H \to G$ be an ordered embedding such that G is an enlargement of $i(H)$, and let $p'\colon G \to K$ be an ordered covering functor such that $p'i = p$; now let $j\colon H \to G'$ be any other ordered embedding such that G is an enlargement of $j(H)$, and let $p''\colon G' \to K$ be an ordered covering functor such that $p = p''j$. Then there is an isomorphism of ordered groupoids $\theta\colon G \to G'$ such that $\theta i = j$ and $p''\theta = p'$.* ∎

It is worth spelling out what Theorems 3, 4 and 5 tell us about star injective ordered functors $p\colon H \to K$. We say that p *extends to an ordered covering functor* if there is an ordered embedding $j_1\colon H \to G_1$ and an ordered covering functor $p_1\colon G_1 \to K$ such that $p_1 j_1 = p$. The ordered pair (j_1, p_1) is called an *extension* of p. If (j_1, p_1) and (j_2, p_2) are extensions of p then a *morphism* from (j_1, p_1) to (j_2, p_2) is an ordered functor $\theta\colon G_1 \to G_2$ such that $\theta j_1 = j_2$ and $p_2\theta = p_1$. Thus with every star injective ordered functor we can associate a category of extensions. Theorems 3, 4 and 5 tell us that this category has an initial object (i, p') which is characterised by the fact that i embeds H in an ordered groupoid which is an enlargement.

8.4 Applications

The maximum enlargement theorem can be used to obtain information about arbitrary idempotent pure prehomomorphisms. But we shall concentrate on two applications: the first will deal with P-semigroups and the second with idempotent pure homomorphisms.

Applications to P-semigroups

The P-theorem can be proved easily using the maximum enlargement theorem.

Theorem 1 *Each E-unitary inverse semigroup is isomorphic to a P-semigroup.*

Proof Let S be an E-unitary inverse semigroup. Then $\sigma^\natural: S \to S/\sigma$ is a surjective idempotent pure homomorphism by Theorem 2.4.6. We now work inside the category of ordered groupoids and ordered functors using the translation provided by the Ehresmann–Schein–Nambooripad theorem. Thus the function $\sigma^\natural$ is a surjective, star injective ordered functor. Consequently by the maximum enlargement theorem there is an enlargement T of S which is equipped with a surjective star bijective ordered functor $\theta: T \to G$ which extends $\sigma^\natural$. However T is a semidirect product of a poset by the group G by Theorem 8.1.3, and so S is isomorphic to a P-semigroup by Theorem 8.1.5. ■

We now take up a point we raised earlier: the uniqueness of the P-representation of an E-unitary inverse semigroup. Let (G, X) and (G', X') be two actions of groups on partially ordered sets by order automorphisms. A *homomorphism* from (G, X) to (G', X') is a pair (θ, ω) consisting of a group homomorphism $\theta: G \to G'$ and an order-preserving function $\omega: X \to X'$ such that $\omega(g \cdot x) = \theta(g) \cdot \omega(x)$. Such a homomorphism can be used to construct an ordered functor $\alpha: P(G, X) \to P(G', X')$ by defining $\alpha(x, g) = (\omega(x), \theta(g))$. This functor has the additional property that $\theta\pi_2 = \pi_2\alpha$. In fact every ordered functor from $P(G, X)$ to $P(G', X')$ which satisfies this condition arises in this way as we now show.

Proposition 2 *Let $\alpha: P(G, X) \to P(G', X')$ be an ordered functor, and let $\theta: G \to G'$ be a group homomorphism such that $\pi_2\alpha = \theta\pi_2$. Then $\alpha = (\omega, \theta)$ where (ω, θ) is a homomorphism from (G, X) to (G', X').*

Proof Since α is a functor, identities of $P(G, X)$ are mapped to identities of $P(G', X')$. Define $\omega: X \to X'$ by $\alpha(x, 1) = (\omega(x), 1)$. Then $\omega: X \to X'$ is order-preserving. Since $\pi_2\alpha = \theta\pi_2$, the second component of $\alpha(x, g)$ is $\theta(g)$. Let $\alpha(x, g) = (x', \theta(g))$, where x' is to be determined. Then

$$\mathbf{r}(\alpha(x, g)) = \alpha(\mathbf{r}(x, g)) = \alpha(x, 1) = (\omega(x), 1).$$

But $\mathbf{r}(\alpha(x, g)) = \mathbf{r}(x', g) = (x', 1)$ Hence $\alpha(x, g) = (\omega(x), \theta(g))$. We now check that (ω, θ) is a homomorphism from (G, X) to (G', X'). Let $x \in X$ and $g \in G$. Then $(x, g^{-1}) \in P(G, X)$ and $(x, g^{-1})^{-1} = (g \cdot x, g)$. Thus

$$\alpha(x, g^{-1})^{-1} = \alpha(g \cdot x, g) = (\omega(g \cdot x), \theta(g)).$$

But $\alpha(x, g^{-1})^{-1} = (\alpha(x, g^{-1}))^{-1}$ and so

$$\alpha(x, g^{-1})^{-1} = (\omega(x), \theta(g^{-1}))^{-1} = (\theta(g) \cdot \omega(x), \theta(g)).$$

Hence $\omega(g \cdot x) = \theta(g) \cdot \omega(x)$. ■

Let (G, X, Y) and (G', X', Y') be two McAlister triples. A *morphism* from (G, X, Y) to (G', X', Y') is a homomorphism (θ, ω) from (G, X) to (G', X') such that ω maps Y to Y' and is a semilattice map. We say that (θ, ω) is an isomorphism if θ is a group isomorphism and ω is an order isomorphism.

We can now prove the uniqueness of the P-representation of an E-unitary inverse semigroup.

Theorem 3 *Let (G, X, Y) and (G', X', Y') be two McAlister triples. Then the inverse semigroups $P(G, X, Y)$ and $P(G', X', Y')$ are isomorphic if, and only if, the McAlister triples (G, X, Y) and (G', X', Y') are isomorphic.*

Proof Let (ω, θ) be an isomorphism from (G, X, Y) to (G', X', Y'). Define β: $P(G, X, Y) \to P(G', X', Y')$ by $\beta(y, g) = (\omega(y), \theta(g))$. It is routine to check that β is an isomorphism of semigroups such that $\theta\pi_2 = \pi_2\beta$.

Conversely, let β: $P(G, X, Y) \to P(G', X', Y')$ be an isomorphism of semigroups. By the remarks following Theorem 2.4.2, β induces an isomorphism from $P(G, X, Y)/\sigma$ to $P(G', X', Y')/\sigma$. But the natural maps from $P(G, X, Y)$ to $P(G, X, Y)/\sigma$ and from $P(G', X', Y')$ to $P(G', X', Y')/\sigma$ are isomorphic to the functions π_2: $P(G, X, Y) \to G$ and π_2: $P(G', X', Y') \to G'$ respectively. Thus there is an isomorphism θ: $G \to G'$ such that $\theta\pi_2 = \pi_2\beta$.

Let ι: $P(G, X, Y) \to P(G, X)$ and κ: $P(G', X', Y') \to P(G', X')$ be the obvious inclusions. Then $\kappa\beta$ is an ordered embedding of $P(G, X, Y)$ in $P(G', X')$ and $P(G', X')$ is an enlargement of the image of $P(G, X, Y)$. Furthermore, $\theta^{-1}\pi_2$ is an ordered covering functor from $P(G', X')$ to G, and $(\theta^{-1}\pi_2)(\kappa\beta)$ is the projection π_2 from $P(G, X, Y)$ to G. Thus by Theorem 8.3.5, there is an isomorphism α: $P(G, X) \to P(G', X')$ such that $\pi_2\alpha = \pi_2\theta$ and $\alpha\iota = \kappa\beta$. By Proposition 2, α is induced by an isomorphism from (G, X) to (G', X') which maps Y to Y' since $\alpha\iota = \kappa\beta$. The restriction of α to $\iota(Y)$ is a semilattice map because β is an inverse semigroup homomorphism. Hence there is an isomorphism from the McAlister triple (G, X, Y) to (G', X', Y'). ■

Idempotent pure extensions

Let θ: $S \to T$ be a surjective idempotent pure homomorphism. In the category of ordered groupoids and ordered functors, the maximum enlargement theorem tells us that there is an ordered groupoid enlargement G of S and a surjective ordered covering functor Θ: $G \to T$ extending θ. In this section, we shall use this result to obtain a description of S in terms which generalise the P-representation of E-unitary semigroups. There are two basic ingredients in realising this idea:

(1) An *ordered representation* of an inverse semigroup T generalises the idea of a group acting on a partially ordered set by order automorphisms. We

prove that ordered representations of T correspond to ordered covering functors onto T (Theorem 4); this generalises the relationship we found between a group G acting on a partially ordered set and surjective ordered covering functors onto G (Theorems 8.1.1 and 8.1.3).

(2) Ordered representations form part of the definition of *McAlister–O'Carroll triples*, which generalise McAlister triples. Every McAlister–O'Carroll triple can be used to construct an *L-semigroup*, which is a generalisation of a P-semigroup. The properties of McAlister–O'Carroll triples and L-semigroups are described in Theorems 5 and 6.

We begin by introducing ordered representations of inverse semigroups. Let $(X, \leq)$ be a partially ordered set. The inverse semigroup $\Sigma(X, \leq)$ consists of all order isomorphisms between order ideals of X. Let T be an inverse semigroup, $(X, \leq)$ a partially ordered set, $\gamma\colon T \to \Sigma(X, \leq)$ a homomorphism, and $p\colon X \to E(T)$ a surjective order-preserving function. Then (γ, p) is said to be an *ordered representation* of T if the following two axioms hold:

(OR1) $\operatorname{dom}\gamma(e) = p^{-1}[e]$ for all $e \in E(T)$.

(OR2) If $t^{-1}t = p(x)$ then $tt^{-1} = p(\gamma(t)(x))$.

An ordered representation is essentially a *partial action* of T on the partially ordered set X. For this reason, we often write $t \cdot x$ instead of $\gamma(t)(x)$.

The group of units of $\Sigma(X, \leq)$ consists of the order automorphisms of X. Thus an action of a group G on X by order automorphisms corresponds to a homomorphism from G to the group of units of $\Sigma(X, \leq)$. The only unexpected feature of the general case is the presence of the function $p\colon X \to E(T)$. This function is invisible when T is a group because then T has only one idempotent.

Let $\theta\colon G \to T$ and $\theta'\colon G' \to T$ be surjective ordered covering functors onto the inductive groupoid T. We say that θ is *isomorphic* to θ' if there is an isomorphism of ordered groupoids $\phi\colon G \to G'$ such that $\theta'\phi = \theta$.

The following theorem is a generalisation of Theorems 8.1.1 and 8.1.3; it tells us that ordered representations of an inverse semigroup T are equivalent to ordered covering functors onto the inductive groupoid T.

Theorem 4 *Let T be an inverse semigroup, and (γ, p) an ordered representation where $\gamma\colon T \to \Sigma(X, \leq)$. Put*

$$\Pi(\gamma, p) = \{(x, t) \in X \times T\colon p(x) = tt^{-1}\}$$

and define a partial product on $\Pi(\gamma, p)$ by $(x, s)(y, t) = (x, st)$ if, and only if, st is a restricted product and $x = s \cdot y$. Endowed with the direct product ordering, $\Pi(\gamma, p)$ is an ordered groupoid and $\pi_2\colon \Pi(\gamma, p) \to T$ is a surjective ordered covering functor.

Every surjective ordered covering functor to T is isomorphic to one constructed in this way.

Proof Throughout the proof we shall denote the product in T by concatenation. The first part of the theorem is a straightforward verification; we simply note the key properties. In the groupoid $\Pi(\gamma, p)$, the identities are the elements of the form $(x, p(x))$ where $x \in X$ and $(x, s)^{-1} = (s^{-1} \cdot x, s^{-1})$. Thus

$$\mathbf{d}(x, s) = (s^{-1} \cdot x, s^{-1}s) \text{ and } \mathbf{r}(x, s) = (x, ss^{-1}).$$

If $(y, p(y)) \leq \mathbf{d}(x, s)$ then

$$((x, s) \mid (y, p(y))) = ((sp(y)) \cdot y, sp(y)),$$

and if $(y, p(y)) \leq \mathbf{r}(x, s)$ then

$$((y, p(y)) \mid (x, s)) = (y, p(y)s).$$

To prove the converse let Θ: $G \to T$ be a surjective ordered covering functor. We first show how to construct an ordered representation (γ, p) from this covering. Then we show that the covering π_2: $\Pi(\gamma, p) \to T$ constructed from this ordered representation is isomorphic to the covering Θ: $G \to T$.

Put $X = G_o$, with the induced order, and let $p = (\Theta \mid X)$ so that p: $X \to E(T)$. Clearly, p is a surjective, order-preserving function.

Let $s \in T$. Define $\gamma(s) \in \Sigma(X, \leq)$ as follows: $\operatorname{dom} \gamma(s) = p^{-1}[s^{-1}s]$ and if $x \in \operatorname{dom} \gamma(s)$ then

$$\gamma(s)(x) = \mathbf{r}(g) \text{ where } \mathbf{d}(g) = x \text{ and } \Theta(g) = sp(x).$$

Then $\gamma(s)$ is well-defined as a function because Θ is star bijective. It remains to show that $\gamma(s)$ really is an element of $\Sigma(X, \leq)$.

First of all, we claim that $\operatorname{im} \gamma(s) = p^{-1}[ss^{-1}]$. Let $y \in \operatorname{im} \gamma(s)$. Then there exists $x \in \operatorname{dom} \gamma(s)$ such that $\gamma(s)(x) = y$. Thus by definition there exists $g \in G$ such that $\mathbf{d}(g) = x$, $\Theta(g) = sp(x)$, and $\gamma(s)(x) = \mathbf{r}(g) = y$. But $p(y) = \Theta(\mathbf{r}(g)) = \mathbf{r}(sp(x)) \leq ss^{-1}$. Hence $\operatorname{im} \gamma(s) \subseteq p^{-1}[ss^{-1}]$. To prove the reverse inclusion, let $p(y) \leq ss^{-1}$. Then $y \in \operatorname{dom} \gamma(s^{-1})$. Put $x = \gamma(s^{-1})(y)$. Then there exists $g \in G$ such that $\mathbf{d}(g) = y$, $\Theta(g) = s^{-1}p(y)$ and $\mathbf{r}(g) = x$. Observe that $\Theta(g^{-1}) = p(y)s \leq s$ and $\mathbf{d}(\Theta(g^{-1})) = \Theta(\mathbf{r}(g)) = \Theta(x) = p(x)$. Hence $\Theta(g^{-1}) = sp(x)$. But $\mathbf{d}(g^{-1}) = x$ and $\mathbf{r}(g^{-1}) = y$, so that $y = \gamma(s)(x)$. Thus

$$\gamma(s)\colon p^{-1}[s^{-1}s] \to p^{-1}[ss^{-1}].$$

It is also clear from these calculations that $\gamma(s^{-1}) = \gamma(s)^{-1}$.

We now show that $\gamma(s)$ is an order isomorphism. Let $x \leq y$ where $x, y \in \operatorname{dom}\gamma(s)$. Then

$$\gamma(s)(x) = \mathbf{r}(g) \text{ where } \mathbf{d}(g) = x \text{ and } \Theta(g) = sp(x),$$

and

$$\gamma(s)(y) = \mathbf{r}(h) \text{ where } \mathbf{d}(h) = y \text{ and } \Theta(h) = sp(y).$$

Hence $\mathbf{d}(g) \leq \mathbf{d}(h)$ and $\Theta(g) \leq \Theta(h)$. Thus $g \leq h$ by Lemma 8.1.2, and so $\mathbf{r}(g) \leq \mathbf{r}(h)$ giving $\gamma(x) \leq \gamma(y)$. We have already proved that $\gamma(s)$ is a bijection and its inverse is $\gamma(s^{-1})$, which is also order-preserving. Thus $\gamma(s)$ is an order isomorphism.

Define γ: $T \to \Sigma(X, \leq)$. We showed above that this was a well-defined function. To show that it is a semigroup homomorphism let $s, t \in T$. We show first that

$$\gamma(t)^{-1}(p^{-1}[s^{-1}s] \cap p^{-1}[tt^{-1}]) = p^{-1}[(st)^{-1}st].$$

Let $\gamma(t^{-1})(x)$ be an element of the left-hand side of this equation. Then $\gamma(t^{-1})(x) = \mathbf{r}(g)$ where $\mathbf{d}(g) = x$ and $\Theta(g) = t^{-1}p(x)$. Hence

$$p(\gamma(t^{-1})(x)) = \Theta(\mathbf{r}(g)) = \mathbf{r}(\Theta(g)) = \mathbf{r}(t^{-1}p(x)) \leq (st)^{-1}st.$$

Now let x be an element of the right-hand side of this equation. Then $p(x) \leq t^{-1}t$ and so $x \in \operatorname{dom}\gamma(t)$. Let $\gamma(t)(x) = y$, say. Thus there exists $g \in G$ such that $\mathbf{d}(g) = x$, $\Theta(g) = tp(x)$ and $\gamma(t)(x) = \mathbf{r}(g) = y$. But

$$p(y) = \Theta(\mathbf{r}(g)) = \mathbf{r}(\Theta(g)) = \mathbf{r}(tp(x)) = tp(x)t^{-1} \leq tt^{-1}s^{-1}s$$

and so $y \in p^{-1}[s^{-1}s] \cap p^{-1}[tt^{-1}]$ and $x = \gamma(t^{-1})(y)$. Hence $\operatorname{dom}(\gamma(s)\gamma(t)) = \operatorname{dom}\gamma(st)$. We can now complete the proof that γ is a homomorphism. Let $x \in \operatorname{dom}\gamma(st)$. Then

$$\gamma(st)(x) = \mathbf{r}(g) \text{ where } \mathbf{d}(g) = x \text{ and } \Theta(g) = stp(x).$$

Also

$$\gamma(s)(\gamma(t)(x)) = \gamma(s)(\mathbf{r}(k)) \text{ where } \mathbf{d}(k) = x \text{ and } \Theta(k) = tp(x),$$

and

$$\gamma(s)(\mathbf{r}(k)) = \mathbf{r}(h) \text{ where } \mathbf{d}(h) = \mathbf{r}(k) \text{ and } \Theta(h) = sp(\mathbf{r}(k)).$$

Now $\mathbf{d}(h) = \mathbf{r}(k)$ and so hk exists in G. But $\mathbf{d}(hk) = \mathbf{d}(k) = x$ and $\Theta(hk) \leq st$ and $\mathbf{d}(\Theta(hk)) = p(x)$. Thus $\Theta(hk) = \Theta(g)$. Hence by Lemma 8.1.2 we have that $hk = g$. Thus $\gamma(s)\gamma(t) = \gamma(st)$. To prove that (γ, p) is an ordered representation, observe that axiom (OR1) holds by definition, and axiom (OR2) is a simple deduction from the definition.

We now show that Θ is isomorphic to the covering we construct from the associated ordered representation. Define $\phi: G \to \Pi(\gamma, p)$ by $\phi(g) = (\mathbf{r}(g), \Theta(g))$. This is clearly a well-defined function and $\pi_2\phi = \Theta$. We show that ϕ is an isomorphism of ordered groupoids. First of all ϕ is a functor. It is clear that ϕ maps identities to identities. Suppose that gh is defined in G. Then $\phi(g)\phi(h) = \phi(gh)$. By definition, $\phi(g) = (\mathbf{r}(g), \Theta(g))$ and $\phi(h) = (\mathbf{r}(h), \Theta(h))$. Clearly, $\Theta(g)\Theta(h)$ is a restricted product in the inverse semigroup T and $\mathbf{r}(g) = \gamma(\Theta(g))(\mathbf{r}(h))$. Thus the product $(\mathbf{r}(g), \Theta(g))(\mathbf{r}(h), \Theta(h))$ is defined in $\Pi(\gamma, p)$ and equals $(\mathbf{r}(g), \Theta(gh))$ which is equal to $\phi(gh)$.

It is immediate that ϕ is an ordered functor; also if $\phi(g) \leq \phi(h)$ then $g \leq h$ by Lemma 8.1.2. In particular, ϕ is injective. It remains to prove that ϕ is surjective. Let $(x, t) \in \Pi(\gamma, p)$. Then $t \in T$ and $x \in G_o$ such that $p(x) = \Theta(x) = tt^{-1}$. Since Θ is a covering functor, there exists $g \in G$ such that $\mathbf{r}(g) = x$ and $\Theta(g) = t$. But then $\phi(g) = (x, t)$. ∎

Actions of groups on partially ordered sets are an essential ingredient in the definition of McAlister triples. We have generalised such actions to ordered representations. We now generalise McAlister triples.

Let T be an inverse semigroup, X a partially ordered set, and Y an order ideal of X which is a semilattice under the induced order. Let $\gamma: T \to \Sigma(X, \leq)$ be a homomorphism, and $q: Y \to E(T)$ a surjective semilattice map. We call (γ, q, Y) a *McAlister–O'Carroll triple* if the following axioms hold:

(MOT1) $\gamma(T)(Y) = X$.

(MOT2) For each $y \in Y$ and $e \in E(T)$ we have that $y \in \operatorname{dom}\gamma(e)$ if, and only if, $q(y) \leq e$.

(MOT3) For each $s \in T$ there exists $y \in Y$ such that $q(y) = ss^{-1}$and $\gamma(s^{-1}) \in Y$.

The above definition contains a subtlety: we have not assumed that γ is an ordered representation of T. In particular, notice that the domain of the function q is Y and not the whole of X. Our first task, therefore, is to show that every McAlister–O'Carroll triple (γ, q, Y) determines a unique ordered representation of T.

Theorem 5 *Let (γ, q, Y) be a McAlister–O'Carroll triple, where*

$$\gamma: T \to \Sigma(X, \leq),\ Y \subseteq X \text{ and } q: Y \to E(T).$$

Then there exists a unique order-preserving surjection $p: X \to E(T)$ extending q such that (γ, p) is an ordered representation.

Proof We prove this result in a number of steps.

For each $x \in X$ there exists $s \in T$ such that

$$\gamma(s^{-1})(x) \in Y \text{ and } q(\gamma(s^{-1})(x)) = s^{-1}s.$$

To prove this, let $x \in X$. By axiom (MOT1), there is $t \in T$ and $y \in Y$ such that $\gamma(t)(y) = x$. By axiom (MOT2), $y \in \operatorname{dom}\gamma(t) = \operatorname{dom}\gamma(t^{-1}t)$ implies that $q(y) \leq t^{-1}t$. Put $s = tq(y)$. Clearly, $q(y) = s^{-1}s$. Now $s \leq t$ implies that $\gamma(s) \leq \gamma(t)$, since γ is a homomorphism, and $y \in \operatorname{dom}\gamma(s)$ from $q(y) = s^{-1}s$ and axiom (MOT2). Hence $\gamma(s)(y) = \gamma(t)(y) = x$. Thus $y = \gamma(s^{-1})(x)$ and $q(\gamma(s^{-1})(x)) = q(y) = s^{-1}s$, which completes the proof of our claim.

Define p: $X \to E(T)$ by $p(x) = ss^{-1}$ if

$$s \in T,\ \gamma(s^{-1})(x) \in Y \text{ and } q(\gamma(s^{-1})(x)) = s^{-1}s.$$

We prove that p is a well-defined function. By our result above, p is defined for every $x \in X$. Let $x \in X$. Suppose that $s, t \in T$ are such that

$$\gamma(s^{-1})(x) = y \in Y \text{ and } q(y) = s^{-1}s$$

and

$$\gamma(t^{-1})(x) = y' \in Y \text{ and } q(y') = t^{-1}t.$$

We have to show that $ss^{-1} = tt^{-1}$. We have that

$$x = \gamma(s)(y) = \gamma(t)(y').$$

Thus

$$y = (\gamma(s)^{-1}\gamma(t))(y') \text{ and } y' = (\gamma(t)^{-1}\gamma(s))(y).$$

But γ is a homomorphism, so that

$$y = \gamma(s^{-1}t)(y') \text{ and } y' = \gamma(t^{-1}s)(y).$$

Now, $y' \in Y$ and $y' \in \operatorname{dom}\gamma(s^{-1}t)$ imply that

$$q(y') \leq (s^{-1}t)^{-1}(s^{-1}t)$$

by axiom (MOT2). Similarly, $q(y) \leq (t^{-1}s)^{-1}t^{-1}s$. But $q(y') = t^{-1}t$ and $q(y) = s^{-1}s$. Thus

$$t^{-1}t \leq (s^{-1}t)^{-1}s^{-1}t \text{ and } s^{-1}s \leq (t^{-1}s)^{-1}t^{-1}s,$$

and so $tt^{-1} = ss^{-1}$ as required.

To show that p extends q let $y \in Y$. By axiom (MOT2), $y \in \operatorname{dom}\gamma(q(y))$ and $\gamma(q(y))(y) = y$, since $\gamma(q(y))$ is an idempotent. Thus $\gamma(q(y))(y) \in Y$ and

$$q(\gamma(q(y))(y)) = q(y) = q(y)^{-1}q(y).$$

Hence $p(y) = q(y)q(y)^{-1} = q(y)$ by the definition of p. Thus p extends q.

We now show that p is order-preserving. Let $w \leq x$ in X and let $s, t \in T$ such that

$$\gamma(s^{-1})(w) = y \in Y \text{ and } q(y) = s^{-1}s$$

and

$$\gamma(t^{-1})(x) = z \in Y \text{ and } q(z) = t^{-1}t.$$

Thus $p(w) = ss^{-1}$ and $p(x) = tt^{-1}$. We prove that $ss^{-1} \leq tt^{-1}$. Now, $x \in \operatorname{dom}\gamma(t^{-1})$ and $\operatorname{dom}\gamma(t^{-1})$ is an order ideal. Thus from $w \leq x$ we have that $w \in \operatorname{dom}\gamma(t^{-1})$. Since $\gamma(t^{-1})$ is order-preserving we have that

$$\gamma(t^{-1})(w) \leq \gamma(t^{-1})(x) = z \in Y.$$

But Y is an order ideal of X, and so $\gamma(t^{-1})(w) = y' \in Y$. We now have that $\gamma(t)(y') = \gamma(s)(y)$, so that $y' = \gamma(t^{-1}s)(y)$. By axiom (MOT2), $q(y) \leq (t^{-1}s)^{-1}t^{-1}s$. But $q(y) = s^{-1}s$. Thus $s^{-1}s \leq (t^{-1}s)^{-1}t^{-1}s$. Hence $ss^{-1} \leq tt^{-1}$, and so $p(w) \leq p(x)$.

We have shown that p is a surjective, order-preserving function from X to $E(T)$ which extends q. We now show that (γ, p) is an ordered representation.

To show that axiom (OR1) holds, let $e \in E(T)$. We prove that $\operatorname{dom}\gamma(e) = p^{-1}[e]$. Let $x \in p^{-1}[e]$. By the definition of p there exists $s \in T$ such that $ss^{-1} = p(x)$ and $\gamma(s^{-1})(x) \in Y$. Thus $x \in \operatorname{dom}\gamma(ss^{-1}) \subseteq \operatorname{dom}\gamma(e)$. Hence $x \in \operatorname{dom}\gamma(e)$. To prove the reverse inclusion, let $x \in \operatorname{dom}\gamma(e)$. Thus $x = \gamma(e)(x)$. There exists $s \in T$ such that $\gamma(s^{-1})(x) = y \in Y$, $q(y) = s^{-1}s$, and $p(x) = ss^{-1}$. Now $x = \gamma(es)(y)$, where $y \in Y$. Thus by axiom (MOT2), $q(y) \leq (es)^{-1}es$. Hence $s^{-1}s \leq (es)^{-1}es$. Thus $ss^{-1} \leq e$, and so $p(x) \leq e$ as required.

To prove that axiom (OR2) holds, let $t^{-1}t = p(x)$. By definition, there exists $s \in T$ such that

$$p(x) = ss^{-1},\ \gamma(s^{-1})(x) = y \in Y, \text{ and } q(y) = s^{-1}s.$$

Put $u = ts$. This is a restricted product and so $u^{-1}u = s^{-1}s$. Let $x' = \gamma(t)(x)$, which is defined by axiom (OR1). We calculate $p(x')$. Now $x = \gamma(s)(y)$ and so $x' = \gamma(ts)(y)$. Hence $x' = \gamma(u)(y)$, and so $\gamma(u^{-1})(x') = y \in Y$. But

$$q(\gamma(u^{-1})(x')) = q(y) = s^{-1}s = u^{-1}u.$$

Thus by the definition of p, we have that $p(\gamma(t)(x)) = p(x') = uu^{-1} = tt^{-1}$. Hence $p(x) = t^{-1}t$ implies $p(\gamma(t)(x)) = tt^{-1}$. We have therefore proved that (γ, p) is an ordered representation.

It only remains to prove uniqueness. Let p': $X \rightarrow E(T)$ be a surjective order-preserving function extending q such that (γ, p') is an ordered representation. Let $x \in X$ and $s \in T$ such that

$$\gamma(s^{-1})(x) = y \in Y \text{ and } q(y) = s^{-1}s.$$

Thus $p(x) = ss^{-1}$. Now $p'(y) = s^{-1}s$, since p' extends q, so that by axiom (OR2), we have that $ss^{-1} = p'(\gamma(s)(y))$. But $ss^{-1} = p(x)$ and $\gamma(s)(y) = x$. Hence $p(x) = p(x')$, as required. ■

We call (γ, p) the ordered representation *associated* with the McAlister–O'Carroll triple (γ, q, Y). It follows that such triples are defined from ordered representations in the same way as McAlister triples are defined from actions of groups on posets. We now show how to construct inverse semigroups from McAlister–O'Carroll triples.

Let (γ, q, Y) be a McAlister–O'Carroll triple. Define

$$L(\gamma, q, Y) = \{(y, s) \in Y \times S\colon q(y) = ss^{-1}, \gamma(s)^{-1}(y) \in Y\},$$

and define a product on $L(\gamma, q, Y)$ by

$$(x, s)(y, t) = (s \cdot (s^{-1} \cdot x \wedge y), st).$$

That this product is well-defined is a consequence of the following result.

Theorem 6 *Let (γ, q, Y) be a McAlister–O'Carroll triple. Then with the above product $L = L(\gamma, q, Y)$ is an inverse semigroup whose semilattice of idempotents is isomorphic to Y; furthermore, $\pi_2\colon L \to T$ is a surjective idempotent pure homomorphism.*

Let (γ, p) be the associated ordered representation. Then $\Pi = \Pi(\gamma, p)$ is an enlargement of $L(\gamma, q, Y)$.

Proof We shall use the associated ordered representation (γ, p) throughout the proof. Since p extends q we have that $L \subseteq \Pi$.

We shall first prove that L is an ordered subgroupoid of Π which is inductive; the pseudoproduct on L will be the binary operation defined above.

We begin by showing that L is a subgroupoid of Π. Let $(x, s), (y, t) \in L$ and suppose that $\exists (x, s)(y, t)$ in Π. Then $(x, s)(y, t) = (x, st)$, where st is a restricted product in T and $x = s \cdot y$. We need to prove that $q(x) = \mathbf{r}(st)$ and $(st)^{-1} \cdot x \in Y$. Now $(x, s) \in L$ implies $q(x) = \mathbf{r}(s) = \mathbf{r}(st)$ since st is a restricted product. Also $s^{-1} \cdot x = y$ and $t^{-1} \cdot y \in Y$, so that $(st)^{-1} \cdot x \in Y$. Hence L is closed under partial products. To show that L is closed under inverses, let $(x, s) \in L$. Consider the ordered pair $(x, s)^{-1} = (s^{-1} \cdot x, s^{-1})$ which belongs to Π. Since $(x, s) \in L$, we have by definition that $s^{-1} \cdot x \in Y$. It only remains to compute $q(s^{-1} \cdot x)$. Now $q(x) = \mathbf{r}(s) = \mathbf{d}(s^{-1})$. Thus by axiom (OR2), $q(s^{-1} \cdot x) = \mathbf{r}(s^{-1})$. Hence $(x, s)^{-1} \in \Pi$. It follows that L is a subgroupoid of Π. It is clear that

$$L_o = \{(y, q(y))\colon y \in Y\}.$$

To show that L is an ordered subgroupoid of Π, let $(x,s) \in L$ and $(y,q(y)) \in L$ and $(y,q(y)) \leq \mathbf{d}(x,s)$. Then in Π we have that

$$((x,s) \mid (y,q(y))) = ((sq(y)) \cdot y, sq(y)).$$

To show that this element belongs to L, it is enough to prove that $(sq(y)) \cdot y \in Y$. Now $\gamma(sq(y)) \subseteq \gamma(s)$. Thus $\gamma(sq(y))(y) = \gamma(s)(y) \leq x \in Y$. But Y is an order ideal of X. Hence the result. Thus L is an ordered subgroupoid of Π.

To show that L is an inductive groupoid, let $(x,q(x)),(y,q(y)) \in L_o$. Since $x,y \in Y$ and Y is a meet semilattice, the meet $x \wedge y$ exists in Y. But q is a semilattice map so that $q(x \wedge y) = q(x) \wedge q(y)$. Thus $(x \wedge y, q(x \wedge y))$ is the meet of $(x,q(x))$ and $(y,q(y))$ in L_o.

We now compute the form of the pseudoproduct in L. Let $(x,s),(y,t) \in L$. By definition

$$\mathbf{d}(x,s) \wedge \mathbf{r}(y,t) = (s^{-1} \cdot x \wedge y, q(s^{-1} \cdot x \wedge y)) = \mathbf{e}.$$

Thus

$$((x,s) \mid \mathbf{e}) = (s \cdot (s^{-1} \cdot x \wedge y), stt^{-1})$$

and

$$(\mathbf{e} \mid (y,t)) = (s^{-1} \cdot x \wedge y, s^{-1}st).$$

Hence

$$(x,s) \otimes (y,t) = (s \cdot (s^{-1} \cdot x \wedge y), st).$$

This is none other the binary operation we defined in L. We have therefore proved that L is an inverse semigroup with semilattice of idempotents isomorphic to Y.

The function $\pi_2 \colon L \to T$ is surjective by axiom (MOT3), because if $t \in T$ then there exists $y \in Y$ such that $q(y) = tt^{-1}$ and $\gamma(t^{-1})(y) \in Y$. Thus $(y,t) \in L$ and $\pi_2(y,t) = t$. It is easy to check that π_2 is an idempotent pure homomorphism.

We now have to show that $\Pi(\gamma,p)$ is an enlargement of $L(\gamma,q,Y)$. Axiom (GE1) holds because Y is an order ideal of X and so L_o is an order ideal of Π_o. Axiom (GE2) holds simply from the definition of L: if $(x,t) \in \Pi$ then $(x,t) \in L$ if, and only if, $\mathbf{d}(x,t), \mathbf{r}(x,t) \in L$. Finally, to show that axiom (GE3) holds, let $(x,p(x)) \in \Pi_o$. Since $x \in X$ there exists by axiom (MOT1) elements $y \in Y$ and $t' \in T$ such that $t' \cdot y = x$. Put $t = t'q(y)$. Then $q(y) = t^{-1}t$ since $q(y) \leq \mathbf{d}(t')$ by axiom (MOT2), and $t \cdot y = x$. Thus by axiom (OR2), we have that $p(t \cdot y) = tt^{-1}$. Hence $(x,t) \in \Pi$. Furthermore, $\mathbf{r}(x,t) = (x,p(x))$ and $\mathbf{d}(x,t) = (y,q(y))$. ∎

Let (γ,q,Y) be a McAlister–O'Carroll triple. Then the inverse semigroup $L(\gamma,q,Y)$ is called an *L-semigroup*. As we have seen, it comes equipped with a surjective, idempotent pure homomorphism to the inverse semigroup T.

Theorem 7 *Let (γ, p) be an ordered representation where $\gamma\colon T \to \Sigma(X, \leq)$. Let S be an inductive subgroupoid of $\Pi(\gamma, p)$ such that $\Pi(\gamma, p)$ is an enlargement of S, and the restriction of π_2 to S is surjective and preserves the meet operation on the identities of S. Put $Y = \{x \in X\colon (x, p(x)) \in S_o\}$ and let q be the restriction of p to Y. Then (γ, q, Y) is a McAlister–O'Carroll triple, and $S = L(\gamma, p, Y)$.*

Proof We first prove that (γ, q, Y) is a McAlister–O'Carroll triple. Observe that Y is an order ideal of X because S_o is an order ideal of Π_o by axiom (GE1). Since S is an inductive groupoid, Y is a semilattice. By assumption, q maps Y onto $E(T)$ and is a semilattice map. Axiom (MOT1) holds by axiom (GE3), axiom (MOT2) holds from the definition of an ordered representation and the fact that q is a restriction of p, and axiom (MOT3) holds since π_2 maps S onto T.

Finally, observe that $S = L(\gamma, q, Y)$. To see this let $(y, s) \in S$. Then $y \in Y$, and $\mathbf{d}(y, s) \in S$ so that $\gamma(s^{-1})(y) \in Y$. Thus $(y, s) \in L(\gamma, q, Y)$. The converse follows by axiom (GE2). ■

We can now determine the structure of an idempotent pure extension.

Theorem 8 *Let $\theta\colon S \to T$ be a surjective idempotent pure homomorphism. Then there is a McAlister–O'Carroll triple (γ, q, Y) such that S is isomorphic to $L(\gamma, q, Y)$ by an isomorphism ϕ which satisfies $\pi_2\phi = \theta$.*

Proof We shall work in the category of ordered groupoids and ordered functors until the conclusion of the proof. Thus $\theta\colon S \to T$ is a surjective star injective ordered functor. By the maximum enlargement theorem (Theorem 8.3.3) there exists an ordered groupoid G, an ordered embedding $\iota\colon S \to G$, and a surjective ordered covering functor $\Theta\colon G \to T$ such that $\theta = \Theta\iota$ and G is an enlargement of $\iota(S)$. By Theorem 4, there is an ordered representation (γ, p) where $\gamma\colon T \to \Sigma(X, \leq)$ such that $\Theta\colon G \to T$ is isomorphic to $\pi_2\colon \Pi(\gamma, p) \to T$. Thus, without loss of generality, we can regard ι as being the inclusion of S in $\Pi(\gamma, p)$. The conditions of Theorem 7 are satisfied and so there is a McAlister–O'Carroll triple (γ, q, Y) such that $S = L(\gamma, q, Y)$ as inverse semigroups. The remainder of the theorem is evident from the construction and the identification we have made between S and its image in $\Pi(\gamma, p)$. ■

8.5 Notes on Chapter 8

Section 8.1

The fact that a group acting on a set can be regarded as a groupoid is a classical construction of groupoid theory [133]. It is often the case that when the group, set and action possess additional structure, then this structure is inherited by the groupoid.

The remaining results of this section are due to myself [180], [190]. However, the central concept of an enlargement derives from both Ehresmann and McAlister. The definition of an enlargement for ordered groupoids is due to Ehresmann, although axiom (GE1) is not explicitly mentioned. The semigroup definition of an enlargement is essentially due to McAlister [235], where axioms (E1) and (E2) are the basis for what McAlister calls 'heavy' subsemigroups. Axiom (E3) is mentioned in passing at the end of this paper. My contribution in [190] really lies in recognising the importance of this notion in inverse semigroup theory. This paper contains many more results on the theory of enlargements. For instance, I show how to construct all inverse enlargements of an inverse monoid using Rees matrix semigroups.

The notion of enlargement is clearly important, as the results of this chapter demonstrate, but I suspect that it is only an elementary aspect of a more general theory. In particular, enlargements appear to be involved in the Morita theory of semigroups developed by Talwar [408].

Section 8.2

The two-way relationship between E-unitary covers and factorisable embeddings was discovered by McAlister and Reilly [246]. My proof is identical to theirs when viewed purely *set-theoretically*. The extra machinery which I use explains why their proof works.

McAlister and Reilly [246] discovered a number of different, but equivalent, ways in which E-unitary covers can arise. We mention two here. The existence of an E-unitary cover of S through the group G is equivalent to:

(1) The existence of an idempotent pure prehomomorphism $\phi: S \to K(G)$ such that for each $g \in G$ there exists $s \in S$ such that $g \in \theta(s)$.

(2) The existence of a dual prehomomorphism $\phi: G \to C(S)$ such that for each $s \in S$ there exists $g \in G$ such that $s \leq \phi(g)$.

Section 8.3

The maximum enlargement theorem is due to Ehresmann. Further information about this result may be found in [71]. I found the description of the basic construction given by Joubert in Proposition II.11 of [162] very useful. Variants of this result were proved independently by Eilenberg [72] in some unpublished lecture notes.

Section 8.4

The structure theory of E-unitary semigroups was the basis for the theory of arbitrary idempotent pure homomorphisms. The elements of this theory were first described by O'Carroll [291], [292] who generalised Schein's approach [378] to P-semigroups. A good description of O'Carroll's work, which formed the basis of my work, may be found in [241]. Loganathan [206] deduced this work from his theory of inverse semigroup cohomology.

O'Carroll applies his theory to the study of strongly E-reflexive semigroups in [294], [295]. These semigroups may be characterised as those whose minimum Clifford congruence is idempotent pure.

My approach to the theory of idempotent pure homomorphisms is based on the maximum enlargement theorem and was developed in a sequence of papers [180], [182], [189]. It can be regarded as a coordinate-free version of O'Carroll's theory, but has the added advantage that it also encompasses idempotent pure prehomomorphisms.

Gomes and Szendrei [103] have extended Margolis and Pin's work on E-unitary semigroups [223] to idempotent pure homomorphisms. Once again, O'Carroll's work can be deduced from their approach. Furthermore, they establish a connection with Billhardt's λ-semidirect products [27].

Chapter 9

0-E-unitary inverse semigroups

One of the themes of this book has been the importance of E-unitary inverse semigroups, but this is only half the story. It is easy to see that an E-unitary inverse semigroup with zero must be a semilattice, which from our point of view is a rather degenerate inverse semigroup. We have met similar situations to this before: a definition reduces to something relatively trivial when applied to inverse semigroups with zero. As usual, we get around this difficulty by modifying the definition. In this case, the following is more useful for inverse semigroups with zero. An inverse semigroup with zero is said to be *0-E-unitary* if $0 \neq e \leq s$, where e is an idempotent, implies that s is an idempotent.

The class of 0-E-unitary inverse semigroups was only explicitly recognised in the late 1980s, but there are at least two reasons for regarding it as both natural and important. The first is mathematical and is based on a construction due to Johannes Kellendonk. He showed how to construct a topological groupoid from any inverse semigroup. In general, the topology is T_1 but in the case of 0-E-unitary semigroups the topology is hausdorff, which is the hallmark of a 'nice' structure. This work is discussed in Section 9.2. The second reason for believing the class of 0-E-unitary inverse semigroups to be important is the range of interesting examples belonging to it; in this chapter, we discuss polycyclic monoids, McAlister semigroups and tiling semigroups. Polycyclic monoids (Section 9.3) are generalisations of the bicyclic monoid and arise naturally in connection with self-similarity; in this capacity, these monoids are a good example of the rôle we ascribed to inverse semigroup theory in Section 1.3. We discuss applications of polycyclic monoids to rings and touch on the early stages of Girard's 'geometry of interaction programme' for linear logic. McAlister semigroups (Section 9.4) are two-sided analogues of polycyclic monoids which are closely related to our third, and most remarkable, class of examples: tiling semigroups (Section 9.5). Kellendonk showed how every tiling in $\mathbb{R}^n$ can be used to construct a 0-E-unitary inverse semigroup which we term a tiling semigroup. These semigroups have properties which are analogous to those of free inverse semigroups, and have proven applications to solid-state physics.

The theory of 0-E-unitary inverse semigroups is currently in its infancy. Those semigroups which are categorical at zero can be described using the theory of idempotent pure congruences (Section 9.1); such congruences were described in Chapter 8. However, none of the examples of 0-E-unitary inverse semigroups mentioned above possesses this property. The right way to describe more general classes of 0-E-unitary inverse semigroups is discussed in the notes at the end of the chapter.

9.1 The structure of a class of 0-E-unitary inverse semigroups

We show how the theory we have developed for E-unitary semigroups can be generalised to those 0-E-unitary semigroups which are categorical at zero.

The compatibility relation plays a central rôle in the theory of E-unitary semigroups. For 0-E-unitary semigroups a new relation is more important. Let S be an arbitrary inverse semigroup with zero. A pair of elements $s, t \in S$ are said to be *strongly compatible* if either $s = t = 0$ or $s, t \neq 0$ and $s^{-1}t, st^{-1} \in E(S) \setminus \{0\}$. We shall write $s \approx t$ if s and t are strongly compatible. Clearly, the strong compatibility relation is reflexive and symmetric but not necessarily transitive.

In an inverse semigroup with zero any two elements are bounded below by zero, for this reason we shall only be interested in the existence of *non-zero* lower bounds.

Lemma 1 *Let S be an inverse semigroup with zero.*

(1) *If s and t are non-zero and $s \approx t$ then s and t have a non-zero lower bound.*

(2) *Let S be 0-E-unitary. Then a pair of non-zero elements s and t have a non-zero lower bound if, and only if, $s \approx t$.*

Proof (1) Put $z = ss^{-1}t$. Then clearly $z \leq t$. Also $z \leq s$ since by assumption $s^{-1}t$ is an idempotent. Suppose that $z = 0$. Then $ss^{-1}t = 0$ and so $s^{-1}t = 0$, which contradicts $s^{-1}t \neq 0$. Thus $z \neq 0$, and so s and t have a non-zero lower bound.
(2) Let $z \leq s, t$ be a non-zero lower bound of s and t. Then $z^{-1}z \leq s^{-1}t$ and

$zz^{-1} \leq st^{-1}$. But $z^{-1}z$ and zz^{-1} are non-zero idempotents and the semigroup is 0-E-unitary. Thus $s^{-1}t$ and st^{-1} are non-zero idempotents. Hence $s \approx t$. The converse follows from (1). ∎

E-unitary semigroups were characterised as those inverse semigroups having a transitive compatibilty relation. It is natural, therefore, to investigate those semigroups in which $\approx$ is transitive. We shall use the term *directed* here to mean that any two elements in a poset have a common lower bound.

Theorem 2 *Let S be an inverse semigroup with zero. Then $\approx$ is transitive if, and only if, S is 0-E-unitary and for every non-zero element s the ordered set $[s] \setminus \{0\}$ is directed.*

Proof Suppose that $\approx$ is transitive. To show that S is 0-E-unitary let $e \leq s$ where e is a non-zero idempotent. Then $s^{-1}e \leq s^{-1}s$. Thus $s^{-1}e$ is an idempotent. Suppose that $s^{-1}e = 0$. Then $es = (s^{-1}e)^{-1} = 0$. But $e \leq s$ implies $e = es = se$. Hence $0 = es = e$, which is a contradiction. Thus $s^{-1}e$ is a non-zero idempotent, and, since $e = se$, it follows that se is a non-zero idempotent. Hence $e \approx s$. Clearly, $e \approx s^{-1}s$. Thus $s \approx s^{-1}s$ by transitivity. But then $s(s^{-1}s)$ is a non-zero idempotent. Hence s is an idempotent, and so S is 0-E-unitary.

It remains to show that if s and t are non-zero and $s, t \leq u$, then s and t have a non-zero lower bound. From $s \leq u$ we obtain $s^{-1}s \leq s^{-1}u$. The idempotent $s^{-1}s$ is non-zero since s is non-zero. Thus $s^{-1}u$ is an idempotent since S is 0-E-unitary. From $s \leq u$ we also have that $su^{-1} \leq uu^{-1}$. Thus su^{-1} is an idempotent. Suppose that $su^{-1} = 0$. Then $us^{-1} = (su^{-1})^{-1} = 0$ and so $s = us^{-1}s = 0$, which is a contradiction. Thus su^{-1} is a non-zero idempotent. We have therefore shown that $s \approx u$. Similarly, $t \approx u$. But then $s \approx t$ by transitivity. Consequently s and t have a non-zero lower bound by Lemma 1.

We now prove the converse. Let S be 0-E-unitary and suppose that for every non-zero element s the ordered set $[s] \setminus \{0\}$ is directed. Let $s \approx t$ and $t \approx u$ where s, t, u are all non-zero. Then by Lemma 1(2), there are non-zero elements $x, y \in S$ such that $x \leq s, t$ and $y \leq t, u$. Thus $x, y \leq t$. By assumption there is a non-zero element z such that $z \leq x, y$. But then $z \leq s, u$ and so $s \approx u$ by Lemma 1(2). ∎

Although the strong compatibility relation is an equivalence relation in 0-E-unitary semigroups, it need not be a congruence. To guarantee that it is, we need an additional condition.

A semigroup with zero is said to be *categorical at zero* if $abc = 0$ implies $ab = 0$ or $bc = 0$. The following result provides a link with Theorem 2.

Lemma 3 *If S is categorical at zero then the posets $[s] \setminus \{0\}$ are directed for every non-zero element s.*

Proof Let $a, b \leq s$ be non-zero elements. Then $a = aa^{-1}s$ and $b = sb^{-1}b$. Put $u = aa^{-1}sb^{-1}b$. Then $u \leq a, b$. Suppose that $u = aa^{-1}sb^{-1}b = 0$. Then either $0 = aa^{-1}s = a$ or $0 = sb^{-1}b = b$, since S is categorical at zero. But neither a nor b is zero by assumption. Thus u is non-zero, and so the poset $[s] \setminus \{0\}$ is directed. ■

Let S be an inverse semigroup with zero. Define the relation β on S by $(s,t) \in \beta$ if either $s = t = 0$ or $s, t \neq 0$ and there exists a non-zero element $u \leq s, t$. In 0-E-unitary semigroups β coincides with the strong compatibility relation by Lemma 1(2).

A congruence is *0-restricted* if the congruence class containing the zero contains only zero. A congruence is *primitive* if the quotient by that congruence is a primitive inverse semigroup.

Proposition 4 *Let S be an inverse semigroup with zero which is categorical at zero. Then β is the minimum 0-restricted primitive congruence on S.*

Proof The relation β is clearly reflexive and symmetric, and it is transitive because the sets $[s] \setminus \{0\}$ are directed by Lemma 3. Thus β is an equivalence relation.

To show that β is a congruence let $(a, b), (c, d) \in \beta$. Without loss of generality we may assume that none of a, b, c, d is zero. From the definition of β there are non-zero elements $u, v \in S$ such that $u \leq a, b$ and $v \leq c, d$. Suppose that $ac = 0$. Then $uu^{-1}ac = 0$. However $u = uu^{-1}a = uu^{-1}b$ since $u \leq a, b$, and so $uu^{-1}bc = 0$. But the semigroup is categorical at zero and so $uu^{-1}b = 0$ or $bc = 0$. The product $uu^{-1}b = u$, which is non-zero. Thus $bc = 0$. In a similar way, we can show that $bc = 0$ implies $bd = 0$. Thus

$$ac = 0 \Rightarrow bc = 0 \Rightarrow bd = 0.$$

The converses are proved similarly. Hence

$$ac = 0 \Leftrightarrow bc = 0 \Leftrightarrow bd = 0.$$

We may therefore assume without loss of generality that $ac, bc, bd \neq 0$.

We have that $uc \leq ac, bc$ and $bv \leq bc, bd$. Suppose $uc = 0$. Since $u = uu^{-1}a$, we have that $0 = uc = uu^{-1}ac$. Thus either $uu^{-1}a = 0$ or $ac = 0$. But both $uu^{-1}a$ $(= u)$ and ac are non-zero by assumption. Thus $uc \neq 0$. Similarly $bv \neq 0$. But $uc, bv \leq bc$. Hence by Lemma 3, uc and bv have a non-zero lower bound and so $(ac, bd) \in \beta$. Thus β is a congruence.

By definition, β is 0-restricted, and S/β is an inverse semigroup with zero. To see that it is primitive let $\beta(s)$ and $\beta(t)$ be non-zero elements of S/β and let $\beta(t) \leq \beta(s)$. Then by Proposition 1.4.21(7) there is an element t' such that

$t' \leq s$ and $\beta(t') = \beta(t)$. Now $\beta(t)$ non-zero implies t' is non-zero, and $\beta(s)$ non-zero implies s is non-zero. Thus $(s, t') \in \beta$ and so $\beta(s) = \beta(t)$. Hence S/β is a primitive inverse semigroup.

Let γ be any 0-restricted primitive congruence on S. Let $(a, b) \in \beta$ where $a, b \neq 0$. Then there exists $u \leq a, b$ where u is non-zero. Thus $\gamma(u) \leq \gamma(a), \gamma(b)$ in S/γ. Now γ is 0-restricted and so none of $\gamma(u)$, $\gamma(a)$ and $\gamma(b)$ is zero. But S/γ is a primitive inverse semigroup and so $\gamma(u) = \gamma(a) = \gamma(b)$. Hence $(a, b) \in \gamma$. ■

In the next result we characterise those inverse semigroups S which are categorical at zero for which S/β is a Brandt semigroup.

Lemma 5 *Let S be an inverse semigroup with zero which is categorical at zero. Then S/β is a Brandt semigroup if, and only if, for any two non-zero ideals I, J in S the ideal $I \cap J$ is non-zero.*

Proof Let S/β be a Brandt semigroup and let I, J be any two nonzero ideals in S. Then $\beta(I)$ and $\beta(J)$ are ideals of S/β and neither is the zero ideal since β is 0-restricted. Thus $\beta(I) = \beta(J) = S/\beta$ since S/β is 0-simple. Now

$$(S/\beta)^2 = \beta(IJ) \subseteq \beta(I \cap J).$$

But S/β is 0-simple and so $\beta(IJ)$ is non-zero. Thus $\beta(I \cap J)$ contains a non-zero element, and so $I \cap J$ contains a non-zero element.

Conversely, suppose that S satisfies the condition that any two non-zero ideals have non-zero intersection. Let $\beta(s)$ and $\beta(t)$ be any two non-zero elements of S/β. By assumption, the two principal ideals SsS and StS intersect in a non-zero element. Thus there exist elements a, b, c, d such that $asb = ctd \neq 0$. By Proposition 3.1.3, there exist elements

$$a' \leq a,\ s' \leq s,\ b' \leq b \text{ and } c' \leq c,\ t' \leq t,\ d' \leq d$$

such that

$$a' \cdot s' \cdot b' = c' \cdot t' \cdot d' \neq 0.$$

Thus $(s', t') \in \mathcal{D}$ by Proposition 3.2.5(3). But $\beta(s) = \beta(s')$ and $\beta(t) = \beta(t')$, so that $\beta(s)\,\mathcal{D}\,\beta(t)$. Hence S/β is a 0-bisimple primitive inverse semigroup, and so is a Brandt semigroup. ■

Lemma 6 *Let S be an inverse semigroup with zero which is categorical at zero. Then β is idempotent pure if, and only if, S is 0-E-unitary.*

Proof Suppose that β is idempotent pure. Let $e \leq a$ where e is non-zero. Then $\beta(e) = \beta(a)$. But by assumption this implies that a is an idempotent.

Conversely, let S be 0-E-unitary and suppose first that $\beta(e) = \beta(a)$ where e is non-zero. Then there exists $f \leq e, a$ where f is non-zero. But then $f \leq a$ and so, since S is 0-E-unitary, a is an idempotent. Now suppose that $\beta(0) = \beta(a)$. Then $a = 0$ since β is 0-restricted. Thus β is idempotent pure. ■

We now summarise the above results.

Theorem 7 *Let S be a 0-E-unitary inverse semigroup which is categorical at zero.*

(1) *$\beta^{\natural}$: $S \to S/\beta$ is an idempotent pure homomorphism onto a primitive inverse semigroup.*

(2) *If the intersection of any two non-zero ideals of S is again a non-zero ideal then S/β is a Brandt semigroup.* ■

The structure of 0-E-unitary inverse semigroups which are categorical at zero can now be determined using the theory described in Section 8.4.

9.2 Kellendonk's topological groupoid

In this section, we shall describe a method for constructing a topological groupoid from an inverse semigroup. The topological groupoid we obtain will always be T_1, but when the inverse semigroup is 0-E-unitary it will be hausdorff. This provides additional evidence for the importance of 0-E-unitary inverse semigroups. The topological groupoid is constructed using two basic constructions: Prim(S) and $\tilde{S}$ which are described in turn below.

Let S be an inverse semigroup with zero. Recall that a non-zero element $a \in S$ is said to be *primitive* if $b \leq a$ implies $b = 0$ or $b = a$. The set Prim(S) consists of zero together with all the primitive elements of S.

Proposition 1 *Let S be an inverse semigroup with zero. If $a, b \in$* Prim(S) *are non-zero then their product ab is non-zero precisely when it is a restricted product. Furthermore,* Prim(S) *is a primitive inverse subsemigroup of S.*

Proof Let $a, b \in$ Prim(S) be non-zero elements and suppose ab is non-zero. Then $ab = a' \cdot b'$ where $a' \leq a$ and $b' \leq b$ by Theorem 3.1.2. But a and b are primitive. Thus $a' = 0$ or $a' = a$, and $b' = 0$ or $b' = b$. But ab is non-zero, and so $a' = a$ and $b' = b$. Hence ab is a restricted product. Conversely, let $ab = a \cdot b$ be a restricted product where a and b are non-zero. Suppose that $ab = 0$. Then $a^{-1}ab = 0$. But by assumption $a^{-1}a = bb^{-1}$ and so $b = 0$, which is a contradiction. Thus ab is non-zero.

Let $a, b \in \mathrm{Prim}(S)$ and suppose that $ab \neq 0$. Let $c \leq ab$ where $c \neq 0$. Then by Theorem 3.1.2(3), $c = a' \cdot b'$ where $a' \leq a$ and $b' \leq b$. Thus $a' = 0$ or $a' = a$, and $b' = 0$ or $b' = b$. But $c \neq 0$, and so $a' = a$ and $b' = b$, which gives $c = ab$. Hence ab is a primitive element of S. It is straightforward to check that $\mathrm{Prim}(S)$ is closed under inverses. Thus $\mathrm{Prim}(S)$ is an inverse subsemigroup of S, which is clearly primitive. ∎

We now turn to the second construction. The inverse semigroup $S^{\mathbb{N}}$ is just the direct product of countably many copies of S with pointwise operations. The subset S' of $S^{\mathbb{N}}$ consists of all decreasing sequences (a_n) (with respect to the natural partial order). In fact S' is an inverse subsemigroup of $S^{\mathbb{N}}$; this is a consequence of the compatibility of the multiplication with the natural partial order and the fact that the order is preserved by inverses. Clearly, S can be embedded in S' via the map $\iota \colon S \to S'$ given by $\iota(s) = (s)$, the constant sequence with every component equal to s.

Define the relation $\preceq$ on S' by $(a_n) \preceq (b_n)$ if, and only if, for every i there exists a j such that $a_j \leq b_i$.

Proposition 2 *The relation $\preceq$ is a preorder on S' and the corresponding equivalence relation $\cong$ is a congruence on S'.*

Proof The relation $\preceq$ is clearly reflexive and easily seen to be transitive. Thus $\cong$ is an equivalence relation on S'. To prove that $\cong$ is a congruence let $(a_n) \cong (b_n)$ and $(c_n) \cong (d_n)$. We shall show that $(a_n c_n) \cong (b_n d_n)$. Since, in particular, $(a_n) \preceq (b_n)$ for every $i \in \mathbb{N}$ there exists j such that $a_j \leq b_i$. Similarly, since $(c_n) \preceq (d_n)$ there exists k such that $c_k \leq d_i$. Let $l = \max\{j, k\}$. Then $a_l \leq a_j$ and $c_l \leq c_k$, since the sequences are decreasing. Thus $a_l c_l \leq b_i d_i$. Hence $(a_n c_n) \preceq (b_n d_n)$. We may similarly show that $(b_n d_n) \preceq (a_n c_n)$. Thus $(a_n c_n) \cong (b_n d_n)$. ∎

Perhaps a few motivational comments are in order here. The intention is that elements of S' are to be regarded as limits of decreasing sequences in S. However, distinct elements of S' may have the same limit. For this reason, we need to define a congruence on S' which identifies those elements which have the same limit. This is precisely the function of the congruence $\cong$.

We shall denote the $\cong$-congruence class containing (a_n) by $[a_n]$. Observe that this is the second meaning we have assigned to square brackets (the other being order ideal). There should be no confusion. For each element $a \in S$, the constant sequence (a) gives rise to the congruence class $[a]$. Put $\tilde{S} = S'/\cong$. Observe that if (a) and (b) are two constant sequences then $(a) \cong (b)$ precisely when $a = b$. Thus there is an embedding $\iota \colon S \to \tilde{S}$ which takes a to $[a]$. Informally, we may regard the inverse semigroup $\tilde{S}$ as a kind of completion of

S with respect to decreasing sequences, although this does not imply that $\tilde{S}$ is itself closed under decreasing sequences.

Lemma 3 *Let S be an inverse semigroup.*

(1) *If $[a_n]$ is an idempotent in $\tilde{S}$ then the class $[a_n]$ contains an element of the form (e_n) where all the e_n are idempotents.*

(2) *Let $(a_i) \in S'$. Let $m \in \mathbb{N}$ be a fixed natural number and define $c_i = a_{i+m}$. Then $(c_i) \cong (a_i)$.*

(3) *Let $(a_i) \in S'$. Let $j\colon \mathbb{N} \to \mathbb{N}$ be any unbounded function such that $j(n+1) \geq j(n)$. Then $(a_{j(i)}) \cong (a_i)$.*

(4) *$[a_n] \leq [b_n]$ in $\tilde{S}$ if, and only if, $(a_n) \preceq (b_n)$.*

(5) *Let (a_n) be a decreasing sequence of elements in S. Then*

$$[a_n] = \bigwedge\{[a_i]\colon i \in \mathbb{N}\}$$

in $\tilde{S}$.

(6) *$[0]$ is the zero element of $\tilde{S}$ and $(a_i) \cong (0)$ precisely when $a_n = 0$ for some n; clearly $a_{n+i} = 0$ for all $i \in \mathbb{N}$.*

Proof (1) The idempotents in S' are the elements of the form (e_n) where the e_n are idempotents. Thus the result follows from Proposition 1.4.21(3).
(2) By definition

$$a_0 \geq a_1 \geq \ldots a_{m-1} \geq a_m = c_0$$

and $c_i = a_{i+m}$ for $i \in \mathbb{N}$. Thus $(c_i) \preceq (a_i)$. The fact that $(a_i) \preceq (c_i)$ is immediate from the definition of (c_i).
(3) Clearly, $(a_i) \preceq (a_{j(i)})$. To prove that $(a_{j(i)}) \preceq (a_i)$ we use the fact that j is an unbounded function; thus given any i there is an i' such that $j(i') \geq i$. But then $a_{j(i')} \leq a_i$.
(4) Suppose that $[a_n] \leq [b_n]$ in $\tilde{S}$. Then by the definition of the natural partial order and by (1) above, there exists an idempotent $[e_n]$ where each e_n is an idempotent of S such that $[a_n] = [b_n][e_n] = [b_n e_n]$. Thus for each i there exists j such that $a_j \leq b_i e_i$, so that $a_j \leq b_i$. Hence $(a_n) \preceq (b_n)$.

To prove the converse, suppose that $(a_n) \preceq (b_n)$. We shall prove that $[a_n] \leq [b_n]$. There are two cases to consider.

Suppose first that there is an m such that $a_m \leq b_{m+i}$ for all $i \in \mathbb{N}$. Consider the decreasing sequence (c_i) where $c_i = a_{m+i}$. Then, $[c_i] \leq [b_i]$ since $c_i \leq b_i$ and $[a_i] = [c_i]$ by (2) above. Thus $[a_n] \leq [b_n]$.

For the second case, we suppose that for every m there exists i such that $a_m \not\leq b_{m+i}$. Define a new sequence $(a_{j(i)})$ as follows: for every i let $j(i)$ be the smallest suffix such that $a_{j(i)} \leq b_i$. Observe that $j(i+1) \geq j(i)$ by construction, so that the sequence $a_{j(0)}, a_{j(1)}, a_{j(2)}, \ldots$ is decreasing. Thus $[a_{j(i)}] \leq [b_i]$. Next, observe that the subscripts $j(i)$ are unbounded, for if they were bounded there would be a value of i such that $j(i) = j(i+1) = \ldots$, which would contradict our assumption. Thus $[a_{j(i)}] = [a_i]$ by (3) above and so $[a_i] \leq [b_i]$
(5) Clearly $[a_0] \geq [a_1] \geq [a_2] \geq \ldots$ is a decreasing sequence of elements in $\iota(S)$. It is also clear that for each $i \in \mathbb{N}$ we have that $[a_n] \leq [a_i]$. Now suppose that $[b_n] \leq [a_i]$ for every $i \in \mathbb{N}$. Then for each i there exists j such that $b_j \leq a_i$. Hence $[b_n] \leq [a_n]$.
(6) It is clear that $[0]$ is the zero. If $(a_i) \cong (0)$ then there exists n such that $a_n \leq 0$. Thus $a_n = 0$. The converse is clear. ∎

We are now ready to construct the groupoid which will underlie our topological groupoid. Let S be an inverse semigroup with zero. Define

$$\mathcal{G}(S) = \text{Prim}(\tilde{S}) \setminus \{0\}.$$

This is a groupoid under the restricted product by Proposition 1 and our work on the associated groupoid. A decreasing sequence (a_n) of S is said to be *approximating* if $[a_n]$ is a primitive element of $\tilde{S}$.

Before defining a topology on $\mathcal{G}(S)$, we have to address the following: it is possible for $\mathcal{G}(S)$ to be empty. However, if S is finite then $\tilde{S}$ will be finite and so will have primitive elements. The existence of primitive elements of $\tilde{S}$ when S is countable is guaranteed by the following result.

Proposition 4 *Let S be a countable inverse semigroup with zero. Then for each non-zero element $x \in S$ there exists a primitive element $[y_i] \in \tilde{S}$ such that $[y_i] \leq [x]$.*

Proof Let $\gamma\colon \mathbb{N} \to S \setminus \{0\}$ be any bijection such that $\gamma(0) = x$. Define a new function $\gamma'\colon \mathbb{N} \to S$ as follows: $\gamma'(0) = x$, and for each $i \geq 1$

$$\gamma'(i) = \begin{cases} \gamma'(i-1)\mathbf{d}(\gamma(i)) & \text{if the product is non-zero} \\ \gamma'(i-1) & \text{otherwise.} \end{cases}$$

By definition, $(\gamma'(i))$ is a decreasing sequence, none of whose terms is zero, and $[\gamma'(i)] \leq [x]$. We shall prove that $[\gamma'(i)]$ is a primitive element of $\tilde{S}$.

Let $[z_i]$ be a non-zero element of $\tilde{S}$ such that $[z_i] \leq [\gamma'(i)]$. We need to show that $[\gamma'(i)] = [z_i]$. To do this, it is enough to show that $(\gamma'(i)) \preceq (z_i)$.

First observe that for any $m, n \in \mathbb{N}$ the elements z_n and $\gamma'(m)$ are bounded below by a non-zero element, since for each m there exists p such that $z_p \leq$

$\gamma'(m)$. Either $z_n \leq z_p$ or $z_p \leq z_n$. Thus z_n and $\gamma'(m)$ are bounded below by either z_n or z_p. This observation has the following consequence. If $z \leq \gamma'(m), z_n$ is a non-zero lower bound then $z = zz^{-1}z \leq \gamma'(m)\mathbf{d}(z_n)$, and so the product $\gamma'(m)\mathbf{d}(z_n)$ is non-zero.

Since $(z_i) \preceq (\gamma'(i))$ and $\gamma'(0) = x$, there is a smallest subscript m such that $z_m \leq x$. Let $n \geq m$. Then

$$z_n = z_m\mathbf{d}(z_n) = x\mathbf{d}(z_n)$$

since (z_i) is a decreasing sequence. Define $q = \gamma^{-1}(z_n)$, so that $\gamma(q) = z_n$. Suppose that $q = 0$. Then $\gamma'(q) = \gamma(q) = \gamma(0) = x = z_n$. Thus trivially $\gamma'(q) \leq z_n$. Suppose that $q \geq 1$. We have already shown that $\gamma'(q-1)\mathbf{d}(z_n)$ is non-zero. But $\mathbf{d}(z_n) = \mathbf{d}(\gamma(q))$ so that

$$\gamma'(q-1)\mathbf{d}(\gamma(q)) = \gamma'(q-1)\mathbf{d}(z_n)$$

is non-zero. Thus $\gamma'(q) = \gamma'(q-1)\mathbf{d}(\gamma(q))$ by definition. Now $\gamma'(q-1) \leq x$ and $z_n = x\mathbf{d}(z_n)$. Thus

$$\gamma'(q) = \gamma'(q-1)\mathbf{d}(\gamma(q)) = \gamma'(q-1)\mathbf{d}(z_n) \leq x\mathbf{d}(z_n) = z_n.$$

Hence for each $n \geq m$ there exists q such that $\gamma'(q) \leq z_n$. It follows easily that $(\gamma'(i)) \preceq (z_i)$ as required. ■

We now recall some definitions from topology. Let (X, τ) be a topological space. The *product topology* on $X \times X$ consists of all subsets of $X \times X$ which are unions of sets of the form $U \times V$ where $U, V \in \tau$. If Y is a subset of $X \times X$ then the *subspace topology* on Y consists of all sets of the form $Y \cap W$ where W is an open set of $X \times X$.

Let G be a groupoid and put

$$G * G = \{(a, b) \in G \times G\colon \mathbf{d}(a) = \mathbf{r}(b)\}.$$

The partial multiplication in G is a function μ from $G * G$ to G. If τ is a topology on G then $G \times G$ is equipped with the product topology and $G * G$ is equipped with the subspace topology from $G \times G$. A *topological groupoid* is a groupoid G equipped with a topology τ such that inversion and the partial product $G * G \rightarrow G$ are continuous functions. A topological groupoid having one identity is just a topological group.

The topology τ on our topological groupoid will be described by means of a *subbasis*; this is a set τ' such that every element of τ is a union of finite intersections of elements of τ'.

If we are given a groupoid G and a set of subsets τ' of G which is a subbasis for a topology τ, then to show that (G, τ) is a topological groupoid, it is enough to prove the following two statements: (1) $U^{-1} \in \tau'$ for every $U \in \tau'$; and (2) for every $U \in \tau'$ the set

$$\mu^{-1}(U) = \{(a, b) \in G * G \colon ab \in U\}$$

is a union of sets of the form $(V \times V') \cap (G * G)$ where $V, V' \in \tau'$.

We shall define a topology on the groupoid $\mathcal{G}(S)$ by means of a subbasis with respect to which $\mathcal{G}(S)$ is a topological groupoid.

For each $a \in S$, put

$$\mathcal{U}_a = \{[a_n] \in \mathcal{G}(S) \colon [a_n] \leq [a]\}.$$

Lemma 5 *Let S be an inverse semigroup with zero.*

(1) *Let $a, b \in S$. Then $\mathcal{U}_a\mathcal{U}_b = \mathcal{U}_{ab}$ in the groupoid $\mathcal{G}(S)$.*

(2) $(\mathcal{U}_a)^{-1} = \mathcal{U}_{a^{-1}}$.

(3) $\mathcal{G}(S)_o = \bigcup_{e \in E(S)} \mathcal{U}_e$.

Proof (1) The result is clear if either a or b is zero. Thus in what follows, we shall assume that both a and b are non-zero. Let $[a_n] \in \mathcal{U}_a$ and $[b_n] \in \mathcal{U}_b$. By definition $[a_n] \leq [a]$ and $[b_n] \leq [b]$ and both $[a_n]$ and $[b_n]$ are primitive elements of $\tilde{S}$. Suppose $[a_n][b_n] \neq 0$. Then $[a_n][b_n] \in \mathcal{G}(S)$ and $[a_n][b_n] \leq [a][b] = [ab]$. Thus $[a_n][b_n] \in \mathcal{U}_{ab}$ Hence $\mathcal{U}_a\mathcal{U}_b \subseteq \mathcal{U}_{ab}$.

To prove the reverse inclusion let $[c_n] \in \mathcal{U}_{ab}$. Then $[c_n]$ is a primitive element of $\tilde{S}$ and $[c_n] \leq [ab]$. Thus $[c_n] = [c_n][c_n]^{-1}[a][a]^{-1}[c_n]$. Define

$$[a_n] = [c_n c_n^{-1} a] \text{ and } [b_n] = [a^{-1} c_n].$$

Then $[c_n] = [a_n][b_n]$, and $[a_n] \leq [a]$ and $[b_n] \leq [b]$. The result will be proved if we show that both $[a_n]$ and $[b_n]$ are primitive elements of $\tilde{S}$.

We begin with $[a_n]$. Let $[x_n]$ be a non-zero element of $\tilde{S}$ such that $[x_n] \leq [a_n]$. Then $[x_n a^{-1}] \leq [c_n c_n^{-1} a a^{-1}] \leq [c_n c_n^{-1}]$. Observe that $[x_n][a^{-1}a] = [x_n]$ so that $[x_n a^{-1}]$ is non-zero. Thus $[x_n a^{-1}] = [c_n c_n^{-1}]$ because $[c_n c_n^{-1}]$ is primitive since $[c_n]$ is primitive. Hence $[x_n a^{-1} a] = [c_n c_n^{-1} a]$ and so $[x_n] = [a_n]$, as required.

We now show that $[b_n]$ is primitive. Let $[y_n]$ be a non-zero element of $\tilde{S}$ such that $[y_n] \leq [b_n]$. Then $[a y_n] \leq [a a^{-1} c_n] \leq [c_n]$. Observe that $[a^{-1}a][y_n] = [y_n]$ so that $[a y_n]$ is non-zero. Thus $[a y_n] = [c_n]$ because $[c_n]$ is primitive. Hence $[a^{-1} a y_n] = [a^{-1} c_n] = [b_n]$, as required.

(2) Straightforward.

(3) If e is a non-zero idempotent of S, then every element of $\mathcal{U}_e$ is an idempotent of $\mathcal{G}(S)$. Conversely, if $[a_n]$ is an idempotent of $\mathcal{G}(S)$, then we can assume without loss of generality that a_0 is an idempotent by Lemma 3(1). Clearly

$[a_n] \leq [a_0]$. Thus $[a_n] \in \mathcal{U}_{a_0}$. ■

Let τ be the topology on $\mathcal{G}(S)$ having the set $\{\mathcal{U}_a\colon a \in S\}$ as a subbasis.

Theorem 6 *Let S be an inverse semigroup with zero. Then the pair $(\mathcal{G}(S), \tau)$ is a T_1 topological groupoid in which the set of identities forms an open subset. Moreover, if S is 0-E-unitary then $\mathcal{G}(S)$ is hausdorff.*

Proof Put $\mathcal{G} = \mathcal{G}(S)$. To show that $\mathcal{G}$ is a topological groupoid we have to prove that the product and inversion functions are continuous. The inversion function is continuous because $\mathcal{U}_a^{-1} = \mathcal{U}_{a^{-1}}$ by Lemma 5(2), and the identities form an open subset by Lemma 5(3). Thus it remains to prove that the product is continuous. For each $c \in S \setminus \{0\}$, we shall show that

$$\mu^{-1}(\mathcal{U}_c) = \bigcup_{0 \neq ab \leq c} (\mathcal{U}_a \times \mathcal{U}_b) \cap (\mathcal{G} * \mathcal{G}).$$

Let $a, b \in S$ such that $0 \neq ab \leq c$. Then $\mathcal{U}_{ab} \subseteq \mathcal{U}_c$, and $\mathcal{U}_a\mathcal{U}_b = \mathcal{U}_{ab}$ by Lemma 5(1). It is easy to check that

$$(\mathcal{U}_a \times \mathcal{U}_b) \cap (\mathcal{G} * \mathcal{G}) \subseteq \mu^{-1}(\mathcal{U}_c).$$

Conversely, let $([a_i], [b_i]) \in \mu^{-1}(\mathcal{U}_c)$. Then $([a_i], [b_i]) \in \mathcal{G} * \mathcal{G}$, and $[a_ib_i] \leq [c]$. Thus there exists an n such that $a_nb_n \leq c$. Now $[a_i] \in \mathcal{U}_{a_n}$ and $[b_i] \in \mathcal{U}_{b_n}$, so that

$$([a_i], [b_i]) \in (\mathcal{U}_{a_n} \times \mathcal{U}_{b_n}) \cap (\mathcal{G} * \mathcal{G}).$$

Hence $\mathcal{G}$ is a topological groupoid.

We now show that the topology is T_1. Let $[a_i]$ and $[b_i]$ be two distinct elements of $\mathcal{G}(S)$. Since both are primitive elements of $\tilde{S}$ we have that

$$[a_i] \not\leq [b_i] \text{ and } [b_i] \not\leq [a_i]$$

and so there exist m and n such that

$$a_i \not\leq b_m \text{ and } b_j \not\leq a_n$$

for all i and j. Let $p = \max\{m, n\}$. Then for all i and j we have that

$$a_i \not\leq b_p \text{ and } b_j \not\leq a_p.$$

Thus

$$[a_i] \in \mathcal{U}_{a_p} \text{ but } [b_i] \notin \mathcal{U}_{a_p},$$

and

$$[a_i] \notin \mathcal{U}_{b_p} \text{ but } [b_i] \in \mathcal{U}_{b_p}.$$

Hence the topology is T_1.

Suppose now that S is 0-E-unitary. We continue with the notation and definitions above. Assume that for some $i = s$ the elements a_p and b_s are not strongly compatible. Then they have no non-zero lower bounds. Consequently $\mathcal{U}_{a_p} \cap \mathcal{U}_{b_s}$ is empty. But $[a_i] \in \mathcal{U}_{a_p}$ and $[b_i] \in \mathcal{U}_{b_s}$. Thus the topology is hausdorff. It therefore only remains to verify our assumption.

Suppose to the contrary, therefore, that a_p and b_i are strongly compatible for all i. Then $a_p^{-1}b_i$ and $a_p b_i^{-1}$ are non-zero idempotents for all i. Thus $[b_i\mathbf{d}(a_p)] \neq [0]$. But $[b_i\mathbf{d}(a_p)] \leq [b_i]$ and $[b_i]$ is primitive, so that $[b_i\mathbf{d}(a_p)] = [b_i]$. Hence for some q and r we have that $b_r \leq b_q\mathbf{d}(a_p)$. But $b_q a_p^{-1}$ is an idempotent and so $b_r \leq a_p$, which is a contradiction. Hence a_p and b_i are not strongly compatible for all i. ■

9.3 Polycyclic monoids

In this section, we investigate the polycyclic monoids; these were introduced by Nivat and Perrot [286] as generalistions of the bicyclic monoid (Section 3.4). Whereas the bicyclic monoid is E-unitary the polycyclic monoids are 0-E-unitary.

Definition

Recall that the bicyclic monoid can be represented in terms of the successor function on the natural numbers. More generally, if X is any set which is isomorphic to a proper subset Y of itself by means of an isomorphism $\theta: X \to Y$, then the partial bijections θ and θ^{-1} in $I(X)$ generate an inverse submonoid isomorphic to the bicyclic monoid. Thus the bicyclic monoid arises in contexts where a structure is isomorphic to a proper substructure. To see how this situation can be generalised, we consider the analogous properties of the Cantor set.

The Cantor set $\mathcal{C}$ is a subset of the real line obtained in the following way. Let $F_1 = [0,1]$, the closed unit interval. Next, let $F_2 = F_1 \setminus [\frac{1}{3}, \frac{2}{3}]$. Then let $F_3 = F_2 \setminus ([\frac{1}{9}, \frac{2}{9}] \cup [\frac{7}{9}, \frac{8}{9}])$; the process continues by removing at each stage the middle third of each interval remaining. Then $\mathcal{C} = \bigcap_{i=0}^{\infty} F_i$.

An important property of the Cantor set is that it is isomorphic to a disjoint union of two copies of itself. Observe first that

$$\mathcal{C} = ([0, \frac{1}{3}] \cap \mathcal{C}) \cup ([\frac{2}{3}, 1] \cap \mathcal{C})$$

is a disjoint union. Define two elements of $I(\mathcal{C})$ as follows:

$$p\colon [0, \frac{1}{3}] \cap \mathcal{C} \to \mathcal{C} \text{ by } x \mapsto 3x$$

and

$$q\colon [\frac{2}{3}, 1] \cap \mathcal{C} \to \mathcal{C} \text{ by } x \mapsto 3x - 2.$$

The elements p and q generate an inverse submonoid of $I(\mathcal{C})$ which encodes the self-similarity properties of the Cantor set. In the inverse monoid $I(\mathcal{C})$ the equations $pp^{-1} = 1$ and $qq^{-1} = 1$ hold, since the image of each map is the whole Cantor set. Also $p^{-1}pq^{-1}q = 0$ since the functions have disjoint domains. This equation is equivalent to $pq^{-1} = 0$.

This example motivates the following definition. For $n \geq 2$, the *polycyclic monoid on n generators* P_n is the semigroup with zero given by the presentation:

$$P_n = \langle p_1, \ldots, p_n, p_1^{-1}, \ldots, p_n^{-1}\colon p_i p_i^{-1} = 1,\, p_i p_j^{-1} = 0 \text{ for } i \neq j\rangle.$$

We denote the polycyclic monoid on a countably infinite number of generators by P_∞. We shall prove later that the partial symmetries of the Cantor set really do form an inverse monoid isomorphic to P_2.

Properties of polycyclic monoids

The bicyclic monoid, as we recalled above, can be represented by partial bijections on the natural numbers. A similar representation can be obtained for polycyclic monoids by means of partial bijections on free monoids. Let $\Sigma = \{x_1, \ldots, x_n\}$. Define the function $\alpha_i\colon \Sigma^* \to \Sigma^*$ by $\alpha_i(u) = x_i u$. This has domain Σ^* and image $x_i\Sigma^*$. We may regard α_i as an element of $I(\Sigma^*)$. Denote by B_n the inverse submonoid of $I(\Sigma^*)$ generated by the α_i. Observe that $\alpha_i^{-1}\alpha_i = 1$, the identity function on Σ^*, whereas if $i \neq j$ then $\alpha_i^{-1}\alpha_j = 0$, the empty function on Σ^*. Define the function $\theta\colon P_n \to B_n$ by $\theta(p_i^{-1}) = \alpha_i$ and $\theta(p_i) = \alpha_i^{-1}$. Then by Proposition 2.3.5, B_n is a homomorphic image of P_n.

In the following result, we shall say that two elements of the free semigroup with zero are 'equivalent' if they are equal in P_n; we write '$u \equiv v$'. If $p_{i_1} \ldots p_{i_m}$ is a string then define $(p_{i_1} \ldots p_{i_m})^{-1}$ to be $p_{i_m}^{-1} \ldots p_{i_1}^{-1}$.

Proposition 1 *P_n is isomorphic to B_n.*

Proof It is straightforward to check that every non-zero element of P_n is equivalent to a string of the form $u^{-1}v$ where $u, v \in \{p_1, \ldots, p_n\}^*$.

To prove that this representation is unique we shall need the following result: in the semigroup B_n, if

$$(\alpha_{i_1} \ldots \alpha_{i_m})(\alpha_{j_1} \ldots \alpha_{j_n})^{-1} = (\alpha_{k_1} \ldots \alpha_{k_p})(\alpha_{l_1} \ldots \alpha_{l_q})^{-1}$$

then $m = p$, $n = q$,

$$\alpha_{i_1} = \alpha_{k_1}, \ldots, \alpha_{i_m} = \alpha_{k_p} \text{ and } \alpha_{j_1} = \alpha_{l_1}, \ldots, \alpha_{j_n} = \alpha_{l_q}.$$

To see this, observe that the element

$$\alpha = \alpha_{i_1} \ldots \alpha_{i_m}$$

of B_n has domain Σ^* and image $x_{i_1} \ldots x_{i_m} \Sigma^*$, whereas the element

$$\beta = (\alpha_{j_1} \ldots \alpha_{j_n})^{-1}$$

has image Σ^* and domain $x_{j_n} \ldots x_{j_1} \Sigma^*$. Thus $\alpha\beta$ has domain $x_{j_n} \ldots x_{j_1} \Sigma^*$ and image $x_{i_1} \ldots x_{i_m} \Sigma^*$. Hence elements in B_n of the form

$$(\alpha_{i_1} \ldots \alpha_{i_m})(\alpha_{j_1} \ldots \alpha_{j_n})^{-1}$$

are uniquely determined by their domains and images.

We can now prove that every non-zero element of P_n is equivalent to a unique string of the form $u^{-1}v$ where $u, v \in \{p_1, \ldots, p_n\}^*$. Suppose that $u^{-1}v \equiv x^{-1}y$. Then $\theta(u^{-1}v) = \theta(x^{-1}y)$ in B_n. Thus $\theta(u)^{-1}\theta(v) = \theta(x)^{-1}\theta(y)$. The result now follows by our result above.

It is now clear that the surjective homomorphism from P_n to B_n is an injection. ∎

To describe the form taken by the product in P_n we shall generalise the monus operation on the natural numbers introduced in Section 3.4. Let $x, y \in \{x_1, \ldots, x_n\}^*$. Define

$$yx^{-1} = \begin{cases} u & \text{if } y = ux \\ 1 & \text{otherwise} \end{cases} \quad \text{and} \quad x^{-1}y = \begin{cases} v & \text{if } y = xv \\ 1 & \text{otherwise.} \end{cases}$$

This usage of the superscript '-1' we call *string cancellation* and should be carefully distinguished from its usage in inverse semigroups. Although potentially confusing the context will usually make clear which meaning is intended; we shall warn the reader in other cases. The basic properties of string cancellation are given below.

Lemma 2 *Let $x, y, u, v \in \{x_1, \ldots, x_n\}^*$.*

(1) *$x(x^{-1}y) = y$ if, and only if, x is a prefix of y.*

(2) *$(yx^{-1})x = y$ if, and only if, x is a suffix of y.*

(3) *The set $\Sigma^*u \cap \Sigma^*v$ is non-empty precisely when either u is a suffix of v or v is a suffix of u. When it is non-empty then $(vu^{-1})u = (uv^{-1})v$, and*

$$\Sigma^*u \cap \Sigma^*v = \Sigma^*(vu^{-1})u = \Sigma^*(uv^{-1})v.$$

(4) *The set* $u\Sigma^* \cap v\Sigma^*$ *is non-empty precisely when either* u *is a prefix of* v *or* v *is a prefix of* u. *When it is non-empty then* $v(v^{-1}u) = u(u^{-1}v)$, *and*

$$u\Sigma^* \cap v\Sigma^* = v(v^{-1}u)\Sigma^* = u(u^{-1}v)\Sigma^*.$$

Proof (1) If $x(x^{-1}y) = y$, then clearly x is a prefix of y. Conversely, suppose that x is a prefix of y then $y = xu$ for some u. By definition $u = x^{-1}y$. Thus $x(x^{-1}y) = y$.
(2) Similar proof to (1).
(3) From the properties of free monoids, it is easy to see that $\Sigma^*u \cap \Sigma^*v$ is non-empty precisely when u is a suffix of v or vice-versa. Suppose u is a suffix of v. Then $v = (vu^{-1})u$ by (2), and $(uv^{-1})v = 1v = v$. Clearly, $\Sigma^*v \subseteq \Sigma^*u$ and so $\Sigma^*u \cap \Sigma^*v = \Sigma^*v$. The case where v is a suffix of u is dealt with similarly.
(4) Similar to (3). ■

With this notation we may now obtain a succinct description of the multiplication in polycyclic monoids.

Lemma 3 *Let* $x^{-1}y$ *and* $u^{-1}v$ *be two elements of* P_n. *Their product is non-zero precisely when either* y *is a suffix of* u *or* u *is a suffix of* y; *that is, when the intersection* $\Sigma^*y \cap \Sigma^*u$ *is non-empty.*

If $(x^{-1}y)(u^{-1}v) \neq 0$ *then*

$$(x^{-1}y)(u^{-1}v) = ((uy^{-1})x)((yu^{-1})v).$$

Proof We first consider all products of the form yu^{-1} where $y, u \in \Sigma^*$. Suppose that y is a suffix of u. Then $u = wy$ for some string $w \in \Sigma^*$ and so $yu^{-1} = yy^{-1}w^{-1} = w^{-1}$. If on the other hand u is a suffix of y, then $y = wu$ for some $w \in \Sigma^*$ and $yu^{-1} = wuu^{-1} = w$. In both cases the product is non-zero. Now suppose that neither y is a suffix of u nor u is a suffix of y. Then there is a string v in Σ^* such that $y = y_1v$ and $u = u_1v$ and the last letters of y_1 and u_1 are different. It is clear that $yu^{-1} = 0$.

Now consider the product $(x^{-1}y)(u^{-1}v)$. From the above calculations it is enough to consider the case where y is a suffix of u or vice-versa. Suppose that y is a suffix of u then

$$(x^{-1}y)(u^{-1}v) = x^{-1}w^{-1}v = (wx)^{-1}v$$

which is non-zero. On the other hand, suppose that u is a suffix of y then

$$(x^{-1}y)(u^{-1}v) = x^{-1}wv$$

which is nonzero.

To finish off, we obtain an explicit description of the product $(x^{-1}y)(u^{-1}v)$ when it is non-zero. Using the calculations above, we find that if $u = wy$ then

$$(x^{-1}y)(u^{-1}v) = (wx)^{-1}v$$

and if $y = wu$ then

$$(x^{-1}y)(u^{-1}v) = x^{-1}wv.$$

It is easy to check that the product given in the statement of the lemma incorporates both of these possibilities and no others. ■

The above description of the product leads to a natural representation of the polycyclic monoids. For $n \geq 2$, the monoid P_n is isomorphic to the set $\Sigma^* \times \Sigma^* \cup \{0\}$ with the following multiplication:

$$(x,y)(u,v) = \begin{cases} ((uy^{-1})x, (yu^{-1})v) & \text{if } \Sigma^*y \cap \Sigma^*v \neq \emptyset \\ 0 & \text{else} \end{cases}$$

The element 0 is defined to act as a zero.

The monoid P_1 is isomorphic to $\Sigma^* \times \Sigma^*$ where Σ contains exactly one element and with the multiplication above, except that the zero is omitted. In view of the isomorphism between the free monoid on a one-letter alphabet and the integers under addition, the bicyclic monoid is then isomorphic to the set $\mathbb{N} \times \mathbb{N}$ with the product:

$$(a,b)(c,d) = ((c \dot{-} b) + a, (b \dot{-} c) + d)$$

which is the product described in Section 3.4.

We now determine some important properties of polycyclic monoids using the ordered pair representation.

Theorem 4 *Polycyclic monoids are combinatorial, 0-bisimple, 0-E-unitary inverse semigroups.*

Proof We begin by showing that the polycyclic monoids are inverse. To do this, we have to characterise the idempotents. Suppose that $(u,v)^2 = (u,v)$. Then $\Sigma^*v \cap \Sigma^*u \neq \emptyset$ and $u = (uv^{-1})u$ and $v = (vu^{-1})v$. Thus $u = v$. It is easy to check that every element of the form (u,u) is an idempotent.

Next we show that the idempotents commute. Clearly, $(u,u)(v,v) = 0$ precisely when $(v,v)(u,u) = 0$. So suppose that $(u,u)(v,v)$ is non-zero. Then u is a suffix of v, or vice-versa. Thus

$$(u,u)(v,v) = ((vu^{-1})u, (uv^{-1})v) \text{ and } (v,v)(u,u) = ((uv^{-1})v, (vu^{-1})u).$$

It is immediate from Lemma 2(3) that these two products are equal and are idempotents.

Observe that $(u,v)(v,w) = (u,w)$, and so $(u,v)(v,u)(u,v) = (u,v)$. Thus polycyclic monoids are inverse and $(u,v)^{-1} = (u,v)$. The natural partial order has a simple characterisation. Suppose that $(u,v) \leq (x,y)$. Then $(u,v) = (x,y)(v,v)$. Thus $\Sigma^* y \cap \Sigma^* v$ is non-empty and $u = (vy^{-1})x$ and $v = (yv^{-1})v$. Thus y is a suffix of v and x is a suffix of u. In particular, $(u,v) = (vy^{-1})(x,y)$. Thus $(u,v) = p(x,y) = (px, py)$ for some $p \in \Sigma^*$. Conversely, if $(u,v) = p(x,y)$ for some $p \in \Sigma^*$ then $(u,v) \leq (x,y)$, as can easily be verified.

To show that the semigroup is combinatorial suppose that $(u,v)\,\mathcal{H}\,(x,y)$. Then $u = x$ and $v = y$ from our results above.

To show that the semigroup is 0-bisimple let (u,u) and (v,v) be any two non-zero idempotents. Then

$$(v,v) = (u,v)^{-1}(u,v) \text{ and } (u,u) = (u,v)(u,v)^{-1}$$

so that $(u,u)\,\mathcal{D}\,(v,v)$.

Finally, to show that the semigroup is 0-E-unitary suppose that $(u,u) \leq (x,y)$. Then $(u,u) = p(x,y)$ for some $p \in \Sigma^*$. Thus $x = y$, and so (x,y) is an idempotent. ■

A fundamental property of polycyclic monoids for $n \geq 2$ is the following.

Theorem 5 *For every $n \geq 2$, the polycyclic monoid on n generators is congruence-free.*

Proof By Theorem 4, P_n is 0-bisimple and so by Section 3.2 it is 0-simple. Thus P_n has no non-trivial ideals. Let ρ be any non-universal congruence on P_n. As we noted in Section 2.3, $\rho(0)$ is an ideal of P_n. Hence $\rho(0) = \{0\}$ or $\rho(0) = P_n$. Thus if ρ is a non-universal congruence $\rho(0) = \{0\}$. Suppose that ρ is not the diagonal congruence. Then we can find a pair of distinct ρ-related elements (x,y) and (x',y') such that the length

$$|\,x\,| + |\,y\,| + |\,x'\,| + |\,y'\,|$$

is a minimum. We can assume, without loss of generality, that $x \neq 1$; this is because at least one of the strings x, y, x', y' must be non-empty, otherwise $(x,y) = (x',y')$, which contradicts our choice of elements. But then this string can be assumed to be x, either by interchanging (x,y) and (x',y') or by observing that $(y,x)\,\rho\,(y',x')$ also holds. Thus $x \in \Sigma^* a$ for some letter a and so $x = ua$ for some string u.

We shall now prove that $x' \notin \Sigma^* a$. Suppose to the contrary that $x' = va$ for some string v. Then

$$(1,a)(x,y)\,\rho\,(1,a)(x',y')$$

and so $(u,y)\,\rho\,(v,y')$. Clearly

$$|u|+|y|+|v|+|y'|<|x|+|y|+|x'|+|y'|.$$

By mimimality this implies that $(u,y)=(v,y')$, from which we deduce that $(x,y)=(x',y')$, a contradiction. Thus x' cannot end with an a.

There are now two possibilities: either $x'\in\Sigma^*b$ where $b\neq a$, possible since $n\geq 2$, or $x'=1$. Suppose first that $x'=vb$. Then

$$(1,b)(x,y)\,\rho\,(1,b)(x',y').$$

Now $(1,b)(x,y)=0$, whereas $(1,b)(x',y')=(v,y')\neq 0$, which contradicts the fact that $\rho(0)=\{0\}$. Thus $x'=1$ must hold. Let b be any letter such that $b\neq a$. Then

$$(1,b)(x,y)\,\rho\,(1,b)(1,y').$$

However, $(1,b)(x,y)=0$ whereas $(1,b)(1,y')=(1,by')\neq 0$. Again, we have a contradiction. At fault is our original assumption that ρ is not the diagonal congruence. ■

A consequence of the above result is that the inverse semigroup obtained from the Cantor set at the beginning of this section really is isomorphic to P_2.

In the following result, we prove that P_2 contains copies of all polycyclic monoids on countable sets. It is convenient to let

$$P_2=\langle p,q,p^{-1},q^{-1}\colon pp^{-1}=1=qq^{-1},\,pq^{-1}=0=qp^{-1}\rangle.$$

Proposition 6 *P_n is embedded in P_2 for every $n\geq 2$, including the case $n=\infty$.*

Proof We begin with some preliminary results. Put $p_i=pq^i$ for $i\in\mathbb{N}$. Then

$$p_ip_i^{-1}=(pq^i)(pq^i)^{-1}=pq^iq^{-i}p^{-1}=1.$$

Now consider $p_ip_j^{-1}$ where $i\neq j$. Clearly, $p_ip_j^{-1}=pq^iq^{-j}p^{-1}$. Suppose that $j>i$. Then $j=i+k$ for some $k>0$, and so $q^iq^{-j}=q^{-k}$. Thus $p_ip_j^{-1}=0$. Similarly, when $j<i$ then $p_ip_j^{-1}=0$.

Let $n\geq 2$ be a finite natural number. We construct an inverse subsemigroup of P_2 isomorphic to P_n. Define a subset $\{p_{n,i}\colon 1\leq i\leq n\}$ of P_2 as follows:

$$p_{n,i}=\begin{cases} pq^{i-1} & \text{for } i=1,\ldots,n-1\\ q^{n-1} & \text{for } i=n.\end{cases}$$

From our preliminary calculations, it is clear that $p_{n,i}p_{n,i}^{-1}=1$ and $p_{n,i}p_{n,j}^{-1}=0$ if $i\neq j$ and $i\neq n$. It is straightforward to check that $p_{n,n}p_{n,n}^{-1}=1$ and that for $i \neq n$ we have that $p_{n,i}p_{n,n}^{-1} = 0$. Thus the set $\{p_{n,i}\colon 1 \leq i \leq n\}$ generates an inverse submonoid P_n' of P_2 which is a non-trivial homomorphic image of P_n. However, for $n \geq 2$ the monoid P_n is congruence-free by Theorem 5. Thus P_n is isomorphic to P_n'.

Now suppose that $n = \infty$. Then it is proved in a similar way that the set $\{p_i\colon i \in \mathbb{N}\}$ generates an inverse submonoid isomorphic to P_∞. ■

Embeddings of polycyclic monoids into rings

In Section 1.3, we argued that the relationship between inverse semigroups and groups mirrored the relationship between partial and global symmetries. From this point of view, polycyclic monoids appear to be closely connected with self-similarity phenomena. We have already noted that the polycyclic monoid on two generators can be represented by those partial symmetries of the Cantor set which express its self-similarity properties. In this, and the next, section we shall examine two other examples where the presence of P_2 is indicative of self-similarity phenomena. In this section, we prove that that a ring containing a copy of the polycyclic monoid on two generators in a sufficiently nice way must be isomorphic to all finite matrix rings over itself.

Let R be a ring with identity. We say that P_n is *embedded in R* if there is a monoid homomorphism θ from P_n to the multiplicative monoid of R which maps the zero of P_n to the zero of R. We say that P_n is *strongly embedded* in R if it satisfies the further condition

$$\sum_{i=1}^{n} \theta(p_i^{-1}p_i) = 1.$$

In what follows we shall assume that P_n is a submonoid of the multiplicative monoid of the ring to simplify notation.

Proposition 7 *Let R be a ring.*

(1) *If P_n is embedded in R, then $e = \sum_{i=1}^{n} p_i^{-1}p_i$ is an idempotent, and eRe is isomorphic to $M_n(R)$.*

(2) *If P_n is strongly embedded in R, then R is isomorphic to $M_n(R)$.*

Proof (1) The element e is an idempotent since it is a sum of pairwise orthogonal idempotents. Define

$$\mathbf{h}_n = \begin{pmatrix} p_1^{-1} \\ p_2^{-1} \\ \cdots \\ p_n^{-1} \end{pmatrix} \text{ and } \mathbf{v}_n = \begin{pmatrix} p_1 \\ p_2 \\ \cdots \\ p_n \end{pmatrix}$$

Observe that $\mathbf{v}_n\mathbf{h}_n^t = I_n$, the $n \times n$-identity matrix, and $\mathbf{h}_n^t\mathbf{v}_n = e$.

Define $\Phi_n: M_n(R) \to eRe$ by $\Phi_n(A) = \mathbf{h}_n^t A\mathbf{v}_n$. This is well defined for

$$e\Phi_n(A)e = (\mathbf{h}_n^t\mathbf{v}_n)(\mathbf{h}_n^t A\mathbf{v}_n)(\mathbf{h}_n^t\mathbf{v}_n) = \mathbf{h}_n^t A\mathbf{v}_n = \Phi_n(A).$$

It is also a ring homomorphism since for any matrices $A, B \in M_n(R)$ we have that

$$\Phi_n(A)\Phi_n(B) = (\mathbf{h}_n^t A\mathbf{v}_n)(\mathbf{h}_n^t B\mathbf{v}_n) = \mathbf{h}_n^t A(\mathbf{v}_n\mathbf{h}_n^t)B\mathbf{v}_n = \mathbf{h}_n^t AI_nB\mathbf{v}_n$$

which is just $\Phi_n(AB)$. It is clear that addition is preserved, and that the identity I_n of $M_n(R)$ is mapped to e, the identity of eRe.

Define $\Psi_n: eRe \to M_n(R)$ by $\Psi_n(r) = \mathbf{v}_n r\mathbf{h}_n^t$. This is a ring homomorphism since for any elements $r, s \in eRe$ we have that

$$\Psi_n(r)\Psi_n(s) = (\mathbf{v}_n r\mathbf{h}_n^t)(\mathbf{v}_n s\mathbf{h}_n^t) = \mathbf{v}_n(r(\mathbf{h}_n^t\mathbf{v}_n)s)\mathbf{h}_n^t = \mathbf{v}_n(res)\mathbf{h}_n^t = \mathbf{v}_n rs\mathbf{h}_n^t$$

which is just $\Psi_n(rs)$. It is clear that addition is preserved and that e is mapped to I_n.

Finally, for all $r \in eRe$,

$$\Phi_n(\Psi_n(r)) = \mathbf{h}_n^t\mathbf{v}_n r\mathbf{h}_n^t\mathbf{v}_n = ere = r,$$

and for all $A \in M_n(R)$,

$$\Psi_n(\Phi_n(A)) = \mathbf{v}_n\mathbf{h}_n^t A\mathbf{v}_n\mathbf{h}_n^t = I_nAI_n = A.$$

Hence, Φ and Ψ are mutually inverse ring isomorphisms.
(2) This is immediate from (1), since in this case $e = 1$. ■

Proposition 8 *If P_2 is strongly embedded in R, then so is P_n, for $n \geq 2$.*

Proof We use the proof of Proposition 6. The result holds trivially for $n = 2$. Observe that

$$\sum_{j=1}^{n+1} p_{n+1,j}^{-1}p_{n+1,j} = p^{-1}p + q^{-1}\left(\sum_{i=1}^{n} p_{n,i}^{-1}p_{n,i}\right)q.$$

It follows that if P_n is strongly embedded in R, then P_{n+1} is too. Hence the result follows by induction. ■

The following is now immediate from Propositions 7 and 8.

Theorem 9 *If P_2 is strongly embedded in a ring R, then $M_n(R)$ is isomorphic to R for all n.* ■

An example of a ring satisfying the condition of Theorem 9 is the ring $B(l^2)$ of all bounded linear operators on infinite square-summable sequences.

Embeddings of polycyclic monoids into $I(\mathbb{N})$

We continue with our investigation of the relationship between polycyclic monoids and self-similarity phenomena by examining a class of representations of polycyclic monoids by partial bijections of the natural numbers; these are analogous to the ring embeddings considered above.

A monoid homomorphism $\theta: P_n \to I(\mathbb{N})$ which maps the zero of P_n to the zero of $I(\mathbb{N})$ will be called *strong* if

$$\bigcup_{i=1}^{n} \theta(p_i^{-1}p_i) = 1.$$

In this section, we shall describe ways of constructing strong embeddings of P_2 and P_∞ into $I(\mathbb{N})$. We begin with P_2.

The disjoint union $\mathbb{N} \sqcup \mathbb{N}$ is defined to be the set $(\mathbb{N} \times \{0\}) \cup (\mathbb{N} \times \{1\})$.

Proposition 10 *Every bijection from $\mathbb{N} \sqcup \mathbb{N}$ to $\mathbb{N}$ determines, and is determined by, a strong embedding of P_2 in $I(\mathbb{N})$.*

Proof Let $\phi: \mathbb{N} \sqcup \mathbb{N} \to \mathbb{N}$ be a bijection. Define

$$p^{-1}(n) = \phi(n,0) \text{ and } q^{-1}(n) = \phi(n,1).$$

Clearly, p^{-1} and q^{-1} are injections with domains equal to $\mathbb{N}$. Thus they are partial bijections of $\mathbb{N}$ with inverses denoted by p and q such that $pp^{-1} = 1 = qq^{-1}$. Since ϕ is a bijection, the images of p^{-1} and q^{-1} are disjoint, so that $p^{-1}pq^{-1}q = 0$. Hence $pq^{-1} = 0$. Finally, $p^{-1}p \cup q^{-1}q$ is the union of the images of p^{-1} and q^{-1} which is $\mathbb{N}$. Thus p and q give rise to a strong embedding of P_2 in $I(\mathbb{N})$.

Conversely, suppose that there is a strong embedding of P_2 in $\mathbb{N}$. We can assume without loss of generality that $p, q \in I(\mathbb{N})$ and so $pp^{-1} = 1 = qq^{-1}$, $p^{-1}p \cup q^{-1}q = 1$, and $p^{-1}pq^{-1}q = 0$. Define

$$\phi: (\mathbb{N} \times \{0\}) \cup (\mathbb{N} \times \{1\}) \to \mathbb{N}$$

by $\phi(n,0) = p^{-1}(n)$ and $\phi(n,1) = q^{-1}(n)$. This is a well-defined function, since $pp^{-1} = 1 = qq^{-1}$. Suppose that $\phi(m,i) = \phi(n,j)$. If $i \neq j$, then either $p^{-1}(m) = q^{-1}(n)$ or $q^{-1}(m) = p^{-1}(n)$. In each case, this would contradict the fact that $p^{-1}pq^{-1}q = 0$. Thus $i = j$. If $i = j = 0$ then $p^{-1}(m) = p^{-1}(n)$ hence $m = n$. Similarly, if $i = j = 1$ then $m = n$. Thus ϕ is injective. Finally, we prove that ϕ is surjective. Let $m \in \mathbb{N}$. By assumption, $p^{-1}p \cup q^{-1}q = 1$ so that either $(p^{-1}p)(m)$ or $(q^{-1}q)(m)$ is defined. If $(p^{-1}p)(m)$ is defined, then $\phi(p(m), 0) = m$. Similarly, if $(q^{-1}q)(m)$ is defined, then $\phi(q(m), 1) = m$. ■

We may prove an analogous result to Proposition 10 for strong embeddings of P_∞ in $I(\mathbb{N})$.

Proposition 11 *Every bijection from* $\mathbb{N} \times \mathbb{N}$ *to* $\mathbb{N}$ *determines, and is determined by, a strong embedding of* P_∞ *in* $I(\mathbb{N})$.

Proof Let $\psi\colon \mathbb{N} \times \mathbb{N} \to \mathbb{N}$ be a bijection. Define a family of partial bijections $\{p_i^{-1}\colon i \in \mathbb{N}\}$ of $\mathbb{N}$ by $p_i^{-1}(n) = \psi(i,n)$ for each $i \in \mathbb{N}$, the inverse of p_i^{-1} being denoted by p_i. By definition the domain of every p_i^{-1} is $\mathbb{N}$ so that $p_i p_i^{-1} = 1$. Let $i \neq j$. Then since ψ is a bijection, the images of p_i^{-1} and p_j^{-1} are disjoint. Thus $p_i^{-1}p_i p_j^{-1}p_j = 0$, and so $p_i p_j^{-1} = 0$. Finally $\bigcup_{i=1}^{\infty} p_i^{-1}p_i$ is the identity on the union of the images of the p_i^{-1}. Thus $\bigcup_{i=1}^{\infty} p_i^{-1}p_i = 1$ since ψ is surjective.

Conversely, suppose that P_∞ is strongly embedded in $I(\mathbb{N})$. We may assume that $p_i \in I(\mathbb{N})$ are such that $p_i p_i^{-1} = 1$, $p_i p_j^{-1} = 0$ for $i \neq j$, and $\bigcup_{i=1}^{\infty} p_i^{-1}p_i = 1$. Define $\psi\colon \mathbb{N} \times \mathbb{N} \to \mathbb{N}$ by $\psi(i,n) = p_i^{-1}(n)$. This is a well-defined function since $p_i p_i^{-1} = 1$ for every $i \in \mathbb{N}$. Suppose that $\psi(i,m) = \psi(j,n)$. Then $p_i^{-1}(m) = p_j^{-1}(n)$. If $i \neq j$ then the images of p_i^{-1} and p_j^{-1} are disjoint. Thus $i = j$ and so $m = n$. Hence ψ is injective. Finally, we prove that ψ is surjective. Let $n \in \mathbb{N}$. By assumption, $\bigcup_{i=1}^{\infty} p_i^{-1}p_i = 1$. Thus $n = p_i^{-1}p_i(n)$ is defined for some i. Then $\psi(i, p_i(n)) = p_i^{-1}(p_i(n)) = n$. ∎

Some strong embeddings of P_2 in $I(\mathbb{N})$ may be used to construct strong embeddings of P_∞ in $I(\mathbb{N})$.

Lemma 12 *Let* $p, q \in I(\mathbb{N})$ *be such that* $pp^{-1} = 1 = qq^{-1}$, $pq^{-1} = 0$ *and* $p^{-1}p \cup q^{-1}q = 1$. *Put* $p_i = pq^i$ *for* $i \in \mathbb{N}$. *Then the* p_i *generate an inverse submonoid of* $I(\mathbb{N})$ *isomorphic to* P_∞. *This submonoid is a strong embedding of* P_∞ *if, and only if,* $\bigcap_{j=1}^{\infty} q^{-j}q^j = 0$.

Proof It is immediate from the proof of Proposition 6 that the p_i generate an inverse submonoid isomorphic to P_∞. We shall prove the second assertion in stages.

Firstly,

$$\left(\bigcup_{i=0}^{n-1} p_i^{-1}p_i\right) \cup q^{-n}q^n = 1$$

for every $n \geq 1$. The result is true when $n = 1$, since $p^{-1}p \cup q^{-1}q = 1$. Assume the result is true for n; we shall then prove it for $n+1$. Observe that

$$\left(\bigcup_{i=0}^{n} p_i^{-1}p_i\right) \cup q^{-(n+1)}q^{n+1} = p^{-1}p \cup q^{-1}\left(\bigcup_{i=0}^{n-1} p_i^{-1}p_i \cup q^{-n}q^n\right) q.$$

By the induction hypothesis this reduces to $p^{-1}p \cup q^{-1}q = 1$.

Next, $\bigcup_{i=0}^{n-1} p_i^{-1}p_i$ and $q^{-n}q^n$ are disjoint for every $n \geq 1$. Observe that for all $0 \leq i \leq n-1$ we have that

$$(p_i^{-1}p_i)^{-1}q^{-n}q^n = 0 \text{ and } p_i^{-1}p_i(q^{-n}q^n)^{-1} = 0.$$

Finally, we prove that

$$\bigcup_{i=0}^{\infty} p_i^{-1}p_i = 1 \Leftrightarrow \bigcap_{j=1}^{\infty} q^{-j}q^j = 0.$$

Suppose first that

$$\bigcup_{i=0}^{\infty} p_i^{-1}p_i = 1 \text{ but } \bigcap_{j=1}^{\infty} q^{-j}q^j \neq 0.$$

Then by assumption, there is an element $x \in \mathbb{N}$ such that

$$x \in \operatorname{dom}\left(\bigcap_{j=1}^{\infty} q^{-j}q^j\right).$$

Hence $x \in \operatorname{dom}(q^{-j}q^j)$ for all $j \geq 1$. But $x \in \operatorname{dom} 1$ and so $x \in \operatorname{dom}(p_n^{-1}p_n)$ for some n. Thus

$$x \in \operatorname{dom}\left(\bigcup_{i=0}^{n} p_i^{-1}p_i\right) \text{ and } x \in \operatorname{dom}(q^{-(n+1)}q^{n+1})$$

which contradicts disjointness.

Conversely, suppose that

$$\bigcap_{j=1}^{\infty} q^{-j}q^j = 0 \text{ but } \bigcup_{i=0}^{\infty} p_i^{-1}p_i \neq 1.$$

Then there exists $x \in \mathbb{N}$ such that $x \notin \operatorname{dom}(p_i^{-1}p_i)$ for all i. But this implies that $x \in \operatorname{dom}(q^{-j}q^j)$ for all j, which contradicts our assumption. ∎

It is easy to construct an example of a pair of elements $p, q \in I(\mathbb{N})$ which satisfy the conditions of Lemma 12. Let $\mathbb{E}$ be the set of even natural numbers and $\mathbb{O}$ the set of odd natural numbers Define

$$p\colon \mathbb{E} \to \mathbb{N} \text{ by } p(n) = \frac{n}{2}$$

and

$$q\colon \mathbb{O} \to \mathbb{N} \text{ by } q(n) = \frac{n-1}{2}.$$

Observe that $\operatorname{dom} q^i = \{2^i n + (2^i - 1) \colon n \in \mathbb{N}\}$; thus the intersection of all these sets is empty.

A construction of Girard

Strong embeddings of polycyclic monoids in $I(\mathbb{N})$ can be used to define new structures on $I(\mathbb{N})$. These structures were first identified by Girard in his 'geometry of interaction programme' for linear logic [95]. To describe what these are, it is convenient to make the following (non-standard) definition.

A *Girard monoid*, $(M, e, \Box, \alpha, \gamma)$, is a monoid M with identity e and a product, denoted by concatenation, equipped with another binary operation $\Box$ and distinguished invertible elements $\alpha, \gamma \in M$ satisfying the following axioms:

(G1) $e \Box e = e$.

(G2) $(x \Box y)(x' \Box y') = (xx') \Box (yy')$.

(G3) $\alpha(x \Box (y \Box z))\alpha^{-1} = (x \Box y) \Box z$.

(G4) $\alpha^2 = (\alpha \Box e)\alpha(e \Box \alpha)$.

(G5) $\gamma^2 = e$.

(G6) $\gamma(x \Box y)\gamma^{-1} = y \Box x$.

(G7) $\alpha\gamma\alpha = (\gamma \Box e)\alpha(e \Box \gamma)$.

Girard monoids are therefore close to being one-object symmetric monoidal categories.

We shall prove below that every strong embedding of P_2 in $I(\mathbb{N})$ induces a Girard monoid structure $(I(\mathbb{N}), 1, \oplus, t, s)$, and every strong embedding of P_∞ in $I(\mathbb{N})$ induces a Girard monoid structure $(I(\mathbb{N}), 1, \otimes, \tau, \sigma)$.

Throughout the remainder of this section, we shall use the fact that $I(\mathbb{N})$ has disjoint unions of elements and that composition of partial bijections distributes over disjoint unions of partial bijections. We shall also use the orthogonality relation introduced in Section 1.4.

Fix a strong embedding of P_2 in $I(\mathbb{N})$. Without loss of generality, let $p, q \in I(\mathbb{N})$ be such that $pq^{-1} = 0 = qp^{-1}$ and $pp^{-1} = 1 = qq^{-1}$ and $p^{-1}p \cup q^{-1}q = 1$. Let $f, g \in I(\mathbb{N})$. Define

$$f \oplus g = p^{-1}fp \cup q^{-1}gq$$

and define distinguished elements $s, t \in I(\mathbb{N})$ by

$$s = q^{-1}p \cup p^{-1}q \text{ and } t = p^{-2}p \cup (qp)^{-1}pq \cup q^{-1}q^2.$$

Lemma 13 *With the notation above, we have that:*

(1) $f \oplus g \in I(\mathbb{N})$.

(2) $s \in I(\mathbb{N})$ *and is invertible; in fact,* $s^2 = 1$.

(3) $t \in I(\mathbb{N})$ *and is invertible.*

Proof (1) We prove that $p^{-1}fp \perp q^{-1}gq$. Clearly,

$$(p^{-1}fp)^{-1}q^{-1}gq = p^{-1}f^{-1}pq^{-1}gq = 0.$$

Similarly

$$p^{-1}fp(q^{-1}gq)^{-1} = 0.$$

Thus $f \oplus g \in I(\mathbb{N})$.
(2) We first show that s is a well-defined element of $I(\mathbb{N})$. To do this we show that $q^{-1}p \perp p^{-1}q$. Clearly,

$$(q^{-1}p)^{-1}p^{-1}q = p^{-1}qp^{-1}q = 0.$$

Similarly,

$$q^{-1}p(p^{-1}q)^{-1} = 0.$$

Thus $s \in I(\mathbb{N})$. Next we prove that $s^2 = 1$, which of course implies that s is invertible. We calculate,

$$s^2 = (q^{-1}p \cup p^{-1}q)(q^{-1}p \cup p^{-1}q).$$

Which on expanding gives $p^{-1}p \cup q^{-1}q = 1$.
(3) We first show that t is a well-defined element of $I(\mathbb{N})$. To do this we have to show that

$$p^{-2}p \perp (qp)^{-1}pq, \quad p^{-2}p \perp q^{-1}q^2, \text{ and } (qp)^{-1}pq \perp q^{-1}q^2.$$

These are all straightforward. To show that t is invertible we have to prove that $t^{-1}t = 1 = tt^{-1}$. This is achieved by multiplying out the expressions and reducing in each case to $p^{-1}p \cup q^{-1}q = 1$. ■

Theorem 14 $(I(\mathbb{N}), 1, \oplus, t, s)$ *is a Girard monoid.*

Proof We prove first that $\oplus$ is a monoid homomorphism from $I(\mathbb{N}) \times I(\mathbb{N})$ to $\mathbb{N}$. This will imply that axioms (G1) and (G2) hold. Let $(f, g), (h, k) \in I(\mathbb{N}) \times I(\mathbb{N})$. Then $(f, g)(h, k) = (fh, gk)$. Now

$$f \oplus g = p^{-1}fp \cup q^{-1}gq \text{ and } h \oplus k = p^{-1}hp \cup q^{-1}kq.$$

Thus

$$(f \oplus g)(h \oplus k) = (p^{-1}fp \cup q^{-1}gq)(p^{-1}hp \cup q^{-1}kq)$$

which is just

$$(p^{-1}fp)(p^{-1}hp) \cup (p^{-1}fp)(q^{-1}kq) \cup (q^{-1}gq)(p^{-1}hp) \cup (q^{-1}gq)(q^{-1}kq)$$

which in turn simplifies to

$$p^{-1}fhp \cup q^{-1}gkq = (fh) \oplus (gk).$$

The fact that $1 \oplus 1 = 1$ follows from the fact that the embedding is strong.

Notice that $p(f \oplus g)p^{-1} = f$ and $q(f \oplus g)q^{-1} = g$ for all $f, g \in I(\mathbb{N})$, so that $\oplus$ is, in fact, injective.

To prove that axiom (G3) holds, we have to show that $t(f \oplus (g \oplus h))t^{-1} = (f \oplus g) \oplus h$. By definition and using properties of disjoint unions we have that

$$t(f \oplus (g \oplus h))t^{-1} = t(p^{-1}fp)t^{-1} \cup t((pq)^{-1}g(pq))t^{-1} \cup t(q^{-2}hq^2)t^{-1}.$$

Taking each term of this union in turn we find that

$$t(p^{-1}fp)t^{-1} = p^{-2}fp^2$$

and

$$t((pq)^{-1}g(pq))t^{-1} = (qp)^{-1}gqp$$

and finally

$$t(q^{-2}hq^2)t^{-1} = q^{-1}hq.$$

Thus

$$t(f \oplus (g \oplus h))t^{-1} = p^{-2}fp^2 \cup (qp)^{-1}gqp \cup q^{-1}hq.$$

But this is easily seen to be $(f \oplus g) \oplus h$.

To prove that axiom (G4) holds, we have to prove that $t^2 = (t \oplus 1)t(t \oplus t)$. A simple calculation shows that

$$t^2 = p^{-3}p \cup p^{-2}q^{-1}pq \cup p^{-1}q^{-1}pq^2 \cup q^{-1}q^3.$$

A rather lengthier calculation shows that this is also the value of $(t \oplus 1)t(t \oplus t)$.

Axiom (G5) was established in Lemma 13(2).

To prove that axiom (G6) holds, we have to show that $s(f \oplus g)s^{-1} = g \oplus f$. This is straightforward.

Finally, to prove that axiom (G7) holds, we have to show that

$$tst = (s \oplus 1)t(1 \oplus s).$$

A direct computation shows that

$$tst = p^{-1}q^{-1}p \cup q^{-1}pq \cup p^{-2}q^2.$$

On the other hand,

$$(s \oplus 1)t(1 \oplus s) = p^{-1}sp^{-1}p \cup q^{-1}qsq \cup p^{-1}sq^{-1}psq.$$

Calculating the value of each term we obtain the desired result. ■

Fix a strong embedding of P_∞ in $I(\mathbb{N})$. Without loss of generality let $p_i \in I(\mathbb{N})$ be such that $p_i p_j^{-1} = 0 = p_j p_i^{-1}$ for $i \neq j$ and $p_i p_i^{-1} = 1$ and $\bigcup_{i=0}^{\infty} p_i^{-1} p_i = 1$. Let $f, g \in I(\mathbb{N})$. Define

$$f \otimes g = \bigcup_{i=0}^{\infty} p_{f(i)}^{-1} g p_i$$

and define $[,]\colon \mathbb{N} \times \mathbb{N} \to \mathbb{N}$ by $[i, n] = p_i^{-1}(n)$, which is a bijection by Proposition 11.

Lemma 15 *With the definitions above we have the following:*

(1) $f \otimes g \in I(\mathbb{N})$.

(2) $(f \otimes g)(a)$ *is defined if, and only if,* $a = [m, n]$ *where* $m \in \operatorname{dom} f$ *and* $n \in \operatorname{dom} g$, *in which case* $(f \otimes g)([m, n]) = [f(m), g(n)]$.

(3) $\otimes\colon I(\mathbb{N}) \times I(\mathbb{N}) \to I(\mathbb{N})$ *is a monoid homomorphism.*

Proof (1) We show that distinct terms in the union are orthogonal. Suppose $i \neq j$. Then

$$(p_{f(i)}^{-1} g p_i)^{-1} (p_{f(j)}^{-1} g p_j) = p_i^{-1} g^{-1} (p_{f(i)} p_{f(j)}^{-1}) g p_j.$$

But $f(i) \neq f(j)$ since $i \neq j$ and f is a partial bijection. Thus $p_{f(i)} p_{f(j)}^{-1} = 0$ and so

$$(p_{f(i)}^{-1} g p_i)^{-1} (p_{f(j)}^{-1} g p_j) = 0.$$

On the other hand,

$$(p_{f(i)}^{-1} g p_i)(p_{f(j)}^{-1} g p_j)^{-1} = p_{f(i)}^{-1} g (p_i p_j^{-1}) g^{-1} p_{f(j)} = 0.$$

(2) By definition

$$(f \otimes g)[m, n] = (f \otimes g)(p_m^{-1}(n)) = \left(\bigcup_{i=0}^{\infty} p_{f(i)}^{-1} g p_i \right) (p_m^{-1}(n)).$$

But the union is disjoint, and the only suffix that provides a non-zero value is $i = m$. Thus

$$(f \otimes g)[m,n] = p_{f(m)}^{-1}(g(n)) = [f(m), g(n)].$$

(3) We prove that $(u \otimes v)(f \otimes g) = uf \otimes vg$. It is straightforward to check, using (2) above, that the left-hand side has the same domain as the right-hand side. Let $n \in \mathbb{N}$. Then there exists a unique pair $(a,b) \in \mathbb{N} \times \mathbb{N}$ such that $[a,b] = n$. Thus

$$((u \otimes v)(f \otimes g))(n) = (u \otimes v)((f \otimes g)[a,b]).$$

By (2),

$$(f \otimes g)[a,b] = [f(a), g(b)] \text{ and } (u \otimes v)[f(a), g(b)] = [(uf)(a), (vg)(b)].$$

But

$$[(uf)(a), (vg)(b)] = ((uf) \otimes (vg))[a,b] = (uf \otimes vg)(n).$$

Next, observe that

$$(1 \otimes 1)(n) = (1 \otimes 1)[a,b] = [1(a), 1(b)] = [a,b] = 1(n).$$

Thus $1 \otimes 1 = 1$. ∎

Thus axioms (G1) and (G2) hold. Define

$$\tau = \bigcup_{i=0}^{\infty} (p_i^{-1} \otimes 1) p_i.$$

Lemma 16 *With the definition above we have that:*

(1) $\tau \in I(\mathbb{N})$, *and is invertible.*

(2) $\tau([a,[b,c]]) = [[a,b],c]$ *for all* $a, b, c \in \mathbb{N}$.

(3) $\tau(u \otimes (v \otimes w))\tau^{-1} = (u \otimes v) \otimes w$.

(4) $\tau^2 = (\tau \otimes 1)\tau(1 \otimes \tau)$.

Proof (1) We begin by showing that τ is well-defined. We shall use the fact that $p_i^{-1}(n) = [i,n]$. By definition and expanding the terms $p_i^{-1} \otimes 1$, we have that

$$\tau = \bigcup_{i,j=0}^{\infty} p_{[i,j]}^{-1} p_j p_i.$$

Let $p_{[i,j]}^{-1}p_jp_i$ and $p_{[k,l]}^{-1}p_lp_k$ be any two terms of this double union. First we calculate

$$(p_{[i,j]}^{-1}p_jp_i)^{-1}(p_{[k,l]}^{-1}p_lp_k);$$

if $p_{[i,j]} \neq p_{[k,l]}$ then this product is zero, whereas if $p_{[i,j]} = p_{[k,l]}$ then $i = k$ and $j = l$ and the two terms are equal. Now we calculate

$$(p_{[i,j]}p_jp_i)(p_{[k,l]}p_lp_k)^{-1};$$

if $i \neq k$ then this product is zero. If $i = k$ but $j \neq l$ then this product is zero. If $i = k$ and $j = l$ then the terms are equal. Thus distinct terms are orthogonal.

We now show that $\tau^{-1}\tau = 1 = \tau\tau^{-1}$. We begin with $\tau^{-1}\tau$. By definition

$$\tau^{-1}\tau = \left(\bigcup_{i,j=0}^{\infty} p_{[i,j]}^{-1}p_jp_i\right)^{-1}\left(\bigcup_{k,l=0}^{\infty} p_{[k,l]}^{-1}p_lp_k\right).$$

A typical term in this product is

$$p_i^{-1}p_j^{-1}p_{[i,j]}p_{[k,l]}^{-1}p_lp_k,$$

which is zero unless $i = k$ and $j = l$. Thus

$$\tau^{-1}\tau = \bigcup_{i,j=0}^{\infty} p_i^{-1}p_j^{-1}p_jp_i.$$

However this can be rewritten as

$$\bigcup_{i=0}^{\infty} p_i^{-1}\left(\bigcup_{j=0}^{\infty} p_j^{-1}p_j\right)p_i,$$

which is equal to 1. We now calculate $\tau\tau^{-1}$. By definition a typical term of $\tau\tau^{-1}$ is

$$p_{[i,j]}^{-1}p_jp_ip_k^{-1}p_l^{-1}p_{[k,l]}.$$

This is zero unless $i = k$ and $j = l$. Thus

$$\tau\tau^{-1} = \bigcup_{i,j=0}^{\infty} p_{[i,j]}^{-1}p_{[i,j]}.$$

But $[,]$ is a bijection from $\mathbb{N} \times \mathbb{N}$ to $\mathbb{N}$ and so

$$\tau\tau^{-1} = \bigcup_{r=0}^{\infty} p_r^{-1}p_r = 1.$$

(2) By definition, $[a,[b,c]] = p_a^{-1}(p_b^{-1}(c))$. A straightforward calculation shows that

$$\tau[a,[b,c]] = p_{[a,b]}^{-1}(c) = [[a,b],c].$$

(3) Let $n \in \mathbb{N}$, $n = [a,b]$ and $b = [c,d]$. By Lemma 15(2)

$$(u \otimes (v \otimes w))(n) = [u(a),[v(c),w(d)]].$$

Hence

$$\tau(u \otimes (v \otimes w))(n) = \tau[u(a),[v(c),w(d)]] = [[u(a),v(c)],w(d)]$$

by (2) above. Now

$$[[u(a),v(c)],w(d)] = ((u \otimes v) \otimes w)([[a,c],d])$$

and

$$((u \otimes v) \otimes w)([[a,c],d]) = ((u \otimes v) \otimes w)\tau[a,[c,d]] = (((u \otimes v) \otimes w)\tau)(n)$$

by (2) above. This is true for all n, and so the result follows.
(4) Let $n \in \mathbb{N}$. We shall show that $((\tau \otimes 1)\tau(1 \otimes \tau))(n)$ and $\tau^2(n)$ are equal. Let $n = [a,b]$, $\tau(b) = [c,d]$ and $c = [e,f]$. Then

$$((\tau \otimes 1)\tau(1 \otimes \tau))(n) = [[[a,e],f],d].$$

Now let $b = [u,v]$ and $v = [x,y]$. Then $b = [u,v] = [u,[x,y]]$. Thus $\tau(b) = [[u,x],y] = [c,d]$. But $[,]$ is a bijection, and so $c = [u,x] = [e,f]$ and $y = d$. Hence $u = e$, $x = f$ and $y = d$. Now

$$\tau^2(n) = \tau^2[a,b] = \tau^2[a,[e,v]] = \tau[[a,e],v].$$

But

$$\tau[[a,e],v] = \tau[[a,e],[f,d]] = [[[a,e],f],d]$$

and so

$$\tau^2(n) = [[[a,e],f],d] = ((\tau \otimes 1)\tau(1 \otimes \tau))(n).$$

■

Thus axioms (G3) and (G4) hold.

Lemma 17 (1) *There is an element σ such that $\sigma^2 = \sigma$ and $\sigma([a,b]) = [b,a]$ for all $a,b \in \mathbb{N}$.*

(2) $\sigma(f \otimes g)\sigma^{-1} = g \otimes f$.

(3) $\tau\sigma\tau = (\sigma \otimes 1)\tau(1 \otimes \sigma)$.

Proof (1) Since [,] is a bijection, we may define $\sigma(n) = [b, a]$ if $n = [a, b]$. Clearly, $\sigma^2 = 1$.
(2) Calculate $\sigma(f \otimes g)(n)$ for every $n \in \mathbb{N}$. It is just $((g \otimes f)\sigma)(n)$.
(3) For every $n \in \mathbb{N}$ we calculate $(\tau\sigma\tau)(n)$ and $(\sigma \otimes 1)\tau(1 \otimes \sigma)(n)$ and show they are equal. Let $n = [a, b]$ and $b = [c, d]$. Then

$$(\tau\sigma\tau)(n) = \tau\sigma\tau[a, [c, d]] = [[d, a], c]$$

using Lemma 15 and (1) above. Since $b = [c, d]$ we have that $\sigma(b) = [d, c]$. Thus

$$(\sigma \otimes 1)\tau(1 \otimes \sigma)(n) = (\sigma \otimes 1)\tau(1 \otimes \sigma)[a, b] = (\sigma \otimes 1)\tau[a, \sigma(b)] = (\sigma \otimes 1)\tau[a, [d, c]]$$

which reduces to $[[d, a], c]$. ■

Thus axioms (G5), (G6) and (G7) hold. We have therefore proved the following result.

Theorem 18 *$(I(\mathbb{N}), 1, \otimes, \tau, \sigma)$ is a Girard monoid.* ■

9.4 McAlister semigroups

In this section, we shall introduce and study a family M_n of two-sided analogues of the polycyclic monoids which we call McAlister semigroups. Apart from the intrinsic merit of these semigroups, we use them in Section 9.5 to help us study 1-dimensional tiling semigroups.

Definition and basic properties

For each positive integer n, the inverse semigroup M_n, called the *McAlister semigroup on n generators*, is defined by

$$M_n = \mathrm{Inv}\langle x_1, \ldots, x_n \colon x_i^{-1}x_j = 0 = x_i x_j^{-1} \text{ for } i \neq j\rangle,$$

the inverse semigroup with zero generated by the x_i subject to the given relations. In the case of $n = 1$ the zero is omitted, and consequently M_1 is the free inverse semigroup on one generator. We shall see that the semigroups M_n are, in a real sense, generalisations of the free inverse semigroup on one generator. Observe that for $n \geq 2$, the semigroup M_n is generated by orthogonal elements.

The following notation will be useful. For every string $x_1 \dots x_n$ in Σ^* we define $(x_1 \dots x_n)^r = x_n \dots x_1$. It is clear that x is a suffix of y precisely when x^r is a prefix of y^r, and consequently $(yx^{-1})^r = (x^r)^{-1}y^r$.

The first step in solving the word problem for McAlister semigroups is to obtain concrete examples of inverse semigroups satisfying the above relations. Let $\Sigma = \{x_1, \dots, x_n\}^*$ be the free monoid on n generators and let $x \in \Sigma^*$. Define a function

$$\rho_x \colon \Sigma^* x \times \Sigma^* \to \Sigma^* \times x\Sigma^*$$

by $\rho_x(ux, v) = (u, xv)$. This is clearly a partial bijection of $\Sigma^* \times \Sigma^*$.

Lemma 1 *Define ρ: $\Sigma^\dagger \to I(\Sigma^* \times \Sigma^*)$ by $\rho(x) = \rho_x$. Then ρ is an injective homomorphism.*

Proof We show first that ρ is injective. Suppose that $\rho(x) = \rho(y)$. Then $\Sigma^* x \times \Sigma^* = \Sigma^* y \times \Sigma^*$, so that $\Sigma^* x = \Sigma^* y$. Thus $x = y$.

We show that ρ is a homomorphism by showing that $\rho_x \rho_y = \rho_{xy}$. Now

$$\operatorname{dom} \rho_x \cap \operatorname{im} \rho_y = (\Sigma^* x \times \Sigma^*) \cap (\Sigma^* \times y\Sigma^*) = \Sigma^* x \times y\Sigma^*.$$

Thus

$$\operatorname{dom}(\rho_x \rho_y) = \rho_y^{-1}(\Sigma^* x \times y\Sigma^*) = \Sigma^* xy \times \Sigma^*.$$

It is now straightforward to show that $\rho_x \rho_y = \rho_{xy}$. ■

Denote by T_n the inverse subsemigroup of $I(\Sigma^* \times \Sigma^*)$ generated by $\rho(\Sigma^\dagger)$. To ease notation, we shall denote $\rho(x_i)$ by ρ_i.

Lemma 2 *The semigroup T_n is generated as an inverse semigroup by the elements $\rho_1, \dots, \rho_n$. Also, $\rho_i^{-1}\rho_j = 0 = \rho_i \rho_j^{-1}$ if $i \neq j$.*

Proof The first part is clear by Lemma 1. Now suppose that $i \neq j$. Then

$$\operatorname{dom}(\rho_i^{-1}\rho_j) = \rho_j^{-1}(\operatorname{dom} \rho_i^{-1} \cap \operatorname{im} \rho_j).$$

But the right-hand side is just $\rho_j^{-1}(\Sigma^* \times (x_i\Sigma^* \cap x_j\Sigma^*))$ and $x_i \neq x_j$. Thus $x_i\Sigma^* \cap x_j\Sigma^* = \emptyset$. Hence $\operatorname{dom}(\rho_i^{-1}\rho_j) = 0$, so that $\rho_i^{-1}\rho_j = 0$. A similar result holds for $\rho_i\rho_j^{-1}$. ■

Lemma 2 implies that T_n is a homomorphic image of M_n. We shall show that they are isomorphic. But to do this needs some preparation.

Lemma 3 *Let* $\Sigma = \{x_1, \ldots, x_n\}$.

(1) *Suppose that* $x\Sigma^* \cap y\Sigma^* \neq \emptyset$, *then*

$$\mathrm{dom}(\rho_x^{-1}\rho_y) = \Sigma^* y \times (y^{-1}x)\Sigma^*,$$

and

$$\mathrm{im}(\rho_x^{-1}\rho_y) = \Sigma^* x \times (x^{-1}y)\Sigma^*.$$

(2) *Suppose that* $\Sigma^* y \cap \Sigma^* z \neq \emptyset$ *then*

$$\mathrm{dom}(\rho_y\rho_z^{-1}) = \Sigma^*(yz^{-1}) \times z\Sigma^*,$$

and

$$\mathrm{im}(\rho_y\rho_z^{-1}) = \Sigma^*(zy^{-1}) \times y\Sigma^*.$$

Proof (1) From the definition $\mathrm{dom}(\rho_x^{-1}\rho_y) = \rho_y^{-1}(\Sigma^* \times (x\Sigma^* \cap y\Sigma^*))$. But $x\Sigma^* \cap y\Sigma^* = y(y^{-1}x)\Sigma^*$, by Lemma 9.3.2 and the result follows. The calculation of the image is entirely analogous.
(2) Similar to (1). ■

A triple $(a, b, c) \in \Sigma^* \times \Sigma^\dagger \times \Sigma^*$ is said to be *allowable* if a is a prefix of b and c is a suffix of b. This concept will be central to the solution of the word problem for M_n. The proof of the following is easily obtained by applying Lemma 3.

Lemma 4 *Let* (x, y, z) *be an allowable triple. Then*

(1) $\mathrm{dom}((\rho_x^{-1}\rho_y)\rho_z^{-1}) = \Sigma^*(yz^{-1}) \times z\Sigma^*$.

(2) $\mathrm{im}((\rho_x^{-1}\rho_y)\rho_z^{-1}) = \Sigma^* x \times (x^{-1}y)\Sigma^*$. ■

Lemma 5 *Let* (x, y, z) *and* (u, v, w) *be allowable triples. Then*

$$\rho_x^{-1}\rho_y\rho_z^{-1} = \rho_u^{-1}\rho_v\rho_w^{-1} \Leftrightarrow (x, y, z) = (u, v, w).$$

Proof By Lemma 4, we have that

$$\mathrm{dom}(\rho_x^{-1}\rho_y\rho_z^{-1}) = \Sigma^*(yz^{-1}) \times z\Sigma^*$$

and

$$\mathrm{dom}(\rho_u^{-1}\rho_v\rho_w^{-1}) = \Sigma^*(vw^{-1}) \times w\Sigma^*.$$

Similarly

$$\mathrm{im}(\rho_x^{-1}\rho_y\rho_z^{-1}) = \Sigma^* x \times (x^{-1}y)\Sigma^*$$

and

$$\mathrm{im}(\rho_u^{-1}\rho_v\rho_w^{-1}) = \Sigma^* u \times (u^{-1}v)\Sigma^*.$$

Thus

$$z = w,\ yz^{-1} = vw^{-1},\ x = u,\ \text{and } x^{-1}y = u^{-1}v.$$

But z is a suffix of y and w is a suffix of v so that

$$y = (yz^{-1})z = (vw^{-1})w = v$$

by Lemma 9.3.2. Hence $(x, y, z) = (u, v, w)$. The converse is immediate. ∎

We shall now set about solving the word problem for M_n. In the semigroup M_n we shall use equality to mean that strings are identically equal and $\equiv$ to mean that they are equal in the presentation. A *positive word* in M_n is any string in the generators of M_n. A string w in M_n will be said to be in *normal form* if $w = x^{-1}yz^{-1}$ where x, y and z are all positive strings and x is a prefix of y and z is a suffix of y. Thus (x, y, z) is an allowable triple. The next three lemmas lead to a proof that every string, not equivalent to 0, is equivalent to a string in normal form.

Lemma 6 *In the McAlister semigroup M_n:*

(1) *Let $z = u^{-1}v$ where u and v are non-empty positive strings. Then either $z \equiv 0$ or $u = a_1u_1$ and $v = a_1v_1$ for some $a_1 \in \Sigma$ and strings u_1 and v_1 and $z \equiv u_1^{-1}v_1(v^{-1}v)$.*

(2) *Let $z = uv^{-1}$ where u and v are non-empty positive strings. Then either $z \equiv 0$ or $u = u_1a_1$ and $v = v_1a_1$ for some $a_1 \in \Sigma$ and strings u_1 and v_1 and $z \equiv (uu^{-1})u_1v_1^{-1}$.*

Proof (1) Let $u = a_1 \ldots a_m$ and $v = b_1 \ldots b_n$. Then $z = a_m^{-1} \ldots a_1^{-1}b_1 \ldots b_n$. Now if z is not to equivalent 0 then $a_1 = b_1$, otherwise we would have $a_1^{-1}b_1 = 0$. Put $u_1 = a_2 \ldots a_m$ and $v_1 = b_2 \ldots b_n$. Then $z = u_1^{-1}(a_1^{-1}a_1)v_1$. Now

$$zv_1^{-1} = u_1^{-1}(a_1^{-1}a_1)v_1v_1^{-1} \equiv u_1^{-1}(v_1v_1^{-1})(a_1^{-1}a_1).$$

Thus $z \equiv u_1^{-1}(v_1v_1^{-1})(a_1^{-1}a_1)v_1$. But $a_1v_1 = v$ and $v_1^{-1}a_1^{-1} = v^{-1}$. Thus $z \equiv (u_1^{-1}v_1)(v^{-1}v)$.
(2) Similar to (1). ∎

Iterating the above result, we obtain the following.

Lemma 7 *In the McAlister semigroup M_n:*

(1) *Let $z = u^{-1}v$, where u and v are positive words. If z is not equivalent to 0, then either u is a prefix of v or v is a prefix of u.*

(2) *Let $z = uv^{-1}$, where u and v are positive words. If z is not equivalent to 0, then either u is a suffix of v or v is a suffix of u.*

Proof (1) Suppose that $z = u^{-1}v$ is not equivalent to 0. Then $z \equiv u_1^{-1}v_1(v^{-1}v)$ where $u = a_1u_1$ and $v = a_1v_1$ by Lemma 6. Since z is not equivalent to zero nor can $u_1^{-1}v_1$ be. Thus by Lemma 6 again, $u = a_1a_2u_2$ and $v = a_1a_2v_2$ for some letters a_1 and a_2. The lemma is repeatedly applied until one of two cases occurs: if $| u |\geq| v |$ then v is a prefix of u; if $| v |\geq| u |$ then u is a prefix of v.
(2) Similar to (1). ■

We can now prove what amounts to an algorithm for converting strings to normal form.

Lemma 8 *In the McAlister semigroup M_n, let $z = a^{-1}bc^{-1}$ be a word in normal form, and let u be any positive string.*

(1) *If $w = zu^{-1}$ is not equivalent to 0, then w is equivalent to a string in normal form.*

(2) *If $w = zu$ is not equivalent to 0, then w is equivalent to a string in normal form.*

Proof (1) Now

$$w = (a^{-1}bc^{-1})u^{-1} \equiv a^{-1}b(uc)^{-1}.$$

By assumption, w is not equivalent to zero and so by Lemma 7 either b is a suffix of uc or uc is a suffix of b. In the latter case, $a^{-1}b(uc)^{-1}$ is already in normal form. Thus we may assume that b is a suffix of uc, so that $uc = db$ for some string d. By assumption, c is a suffix of b, so that $b = b'c$ for some b'. Hence $uc = db = db'c$ and so $u = db'$. Observe that

$$w \equiv a^{-1}b(uc)^{-1} = a^{-1}b(db)^{-1} \equiv a^{-1}bb^{-1}d^{-1}.$$

Thus

$$wd \equiv a^{-1}(bb^{-1})(d^{-1}d) \equiv a^{-1}(d^{-1}d)(bb^{-1}) \equiv (da)^{-1}(db)b^{-1}.$$

Hence $w \equiv (da)^{-1}(db)(db)^{-1}$, which is in normal form.
(2) Now

$$w = (a^{-1}bc^{-1})u = a^{-1}b(c^{-1}u).$$

By assumption, w is not equivalent to 0 and so by Lemma 7 either c is a prefix of u or u is a prefix of c. We consider each case separately.

Suppose that c is a prefix of u. Then $u = cv$ for some string v. Also c is a suffix of b and so $b = b'c$ for some string b'. Thus $w = a^{-1}b'cc^{-1}(cv)$ and so

$$w \equiv a^{-1}b'cv = a^{-1}bv \equiv a^{-1}(bv)1$$

which is in normal form.

Now suppose that u is a prefix of c. Thus $c = ud$ for some string d. By assumption, $b = ec$ for some string e. Thus $b = eud$. Now

$$w = a^{-1}bc^{-1}u = a^{-1}(eud)(ud)^{-1}u \equiv a^{-1}eudd^{-1}u^{-1}u$$

and

$$a^{-1}eudd^{-1}u^{-1}u \equiv a^{-1}euu^{-1}udd^{-1} \equiv a^{-1}(eud)d^{-1} = a^{-1}bd^{-1}$$

which is in normal form. ■

The key result is the following.

Theorem 9 *A string in M_n is not equivalent to* 0*, if and only if, it is equivalent to a string in normal form.*

Proof We show first that there is an algorithm which will transform any string not equivalent to zero into an an equivalent string in normal form. A string w in M_n is a string over $\{x_1, \ldots, x_n\} \cup \{x_1^{-1}, \ldots, x_n^{-1}\}$. If w begins with a positive letter this may be regarded as a normal form. Read now from left to right and multiply by each successive letter whether positive or negative and apply Lemma 8 at each stage. Either at some point the string will be shown to be equivalent to 0 or a normal form will be reached. If w begins with a negative letter x^{-1} so that $w = x^{-1}y$ for some string y, then $w \equiv (x^{-1}xx^{-1})y$ where $x^{-1} \equiv x^{-1}xx^{-1}$ is in normal form. Thus the algorithm above may now be applied.

Conversely, we have already mentioned that T_n is a homomorphic image of M_n. The images in T_n of all elements of M_n in normal form are non-zero by Lemma 4. It follows that the normal forms in M_n are not equivalent to 0. ■

The product of two normal forms in M_n can be explicitly calculated.

Proposition 10 *Let $x = a^{-1}bc^{-1}$ and $y = u^{-1}wz^{-1}$ be two elements of M_n in normal form. Then their product in M_n is not equivalent to* 0 *precisely when uc is a suffix of b (or vice versa) and uc is a prefix of w (or vice versa).*

When the product is not equivalent to 0, we have that

$$(a^{-1}bc^{-1})(u^{-1}wz^{-1}) \equiv [[(uc)b^{-1}]a]^{-1}[[(uc)^{-1}b]b[(uc)^{-1}w]][w^{-1}(uc)]^{-1},$$

a normal form, where $(uc)b^{-1}$, $(uc)^{-1}w$, *and* $w^{-1}(uc)$ *are string cancellations.*

Proof We first establish the condition for not being equivalent to 0. Suppose that xy is not equivalent to 0. Now

$$xy = (a^{-1}bc^{-1})(u^{-1}wz^{-1}) = a^{-1}((bc^{-1})u^{-1})wz^{-1}.$$

Thus b is a suffix of uc or vice-versa by Lemma 7. Using Lemma 8, we can compute $(a^{-1}bc^{-1})u^{-1}$ in each case: if uc is a suffix of b then

$$(a^{-1}bc^{-1})u^{-1} \equiv a^{-1}b(uc)^{-1}$$

and if b is a suffix of uc and $uc = db$ then

$$(a^{-1}bc^{-1})u^{-1} = (da)^{-1}(db)(uc)^{-1}.$$

In either case, the last part of $(a^{-1}bc^{-1})u^{-1}$ is $(uc)^{-1}$. Thus when postmultiplying by w, either uc is a prefix of w or vice-versa by Lemma 7. Consequently the conditions are necessary for the product to be non-zero.

Given that the conditions hold we shall show that xy is equivalent to a string in normal form. The partial product $(a^{-1}bc^{-1})u^{-1}w$ will be denoted by p. There are four cases to consider:

1. *uc is a suffix of b, and uc is a prefix of w.*
Suppose that $w = (uc)v$ for some v. Thus $p \equiv a^{-1}bv$. The assumptions imply that z is a suffix of bv. Thus $xy \equiv a^{-1}(bv)z^{-1}$, which is a normal form. This may be written in terms of string cancellations as follows:

$$a^{-1}(b[(uc)^{-1}w])z^{-1}.$$

2. *uc is a suffix of b, and w is a prefix of uc.*
Suppose that $uc = wd$ for some d. Thus $p \equiv a^{-1}bd^{-1}$. The assumptions imply that zd is a suffix of b. Thus $xy \equiv a^{-1}b(zd)^{-1}$, which is a normal form. This may be written in terms of string cancellations as follows:

$$a^{-1}b(z[w^{-1}(uc)])^{-1}.$$

3. *b is a suffix of uc, and uc is a prefix of w.*
Suppose that $uc = db$ for some d. Thus $p \equiv (da)^{-1}w$. We already know that z is a suffix of w. Thus $xy \equiv (da)^{-1}wz^{-1}$, which is a normal form. This may be written in terms of string cancellations as follows:

$$([(uc)b^{-1}]a)^{-1}([(uc)b^{-1}]b[(uc)^{-1}w])z^{-1}.$$

4. *b is a suffix of uc, and w is a prefix of uc.*
Suppose that $uc = db$ and $uc = wd_1$ for some elements d and d_1. Thus $p \equiv (da)^{-1}(uc)d_1^{-1}$. The assumptions imply that zd_1 is a suffix of uc. Thus $xy \equiv (da)^{-1}(uc)(zd_1)^{-1}$, which is a normal form. This may be written in terms of string cancellations as follows:

$$([(uc)b^{-1}]a)^{-1}([(uc)b^{-1}]b)(z[w^{-1}(uc)])^{-1}.$$

■

The above description of the product leads to a natural representation of McAlister semigroups. Let $\Sigma = \{x_1, \ldots, x_n\}$ be a finite alphabet. As before a triple $(a, b, c) \in \Sigma^* \times \Sigma^\dagger \times \Sigma^*$ is said to be *allowable* if a is a prefix of b and c is a suffix of b. The set M_1 may be identified with the set of all allowable triples whereas M_n for $n \geq 2$ may be identified with the set of all allowable triples together with a zero, 0. From now on we shall make these identifications when referring to M_n. Two allowable triples (a, b, c) and (u, w, z) over the same alphabet will be said to be *compatible* if

$$\Sigma^*(uc) \cap \Sigma^* b \neq \emptyset \text{ and } (uc)\Sigma^* \cap w\Sigma^* \neq \emptyset.$$

Clearly, all elements of M_1 are compatible. Define an operation $\otimes$ on M_n as follows: if (a, b, c) and (u, w, z) are compatible then

$$(a, b, c) \otimes (u, w, z) = ([(uc)b^{-1}]a, [(uc)b^{-1}]b[(uc)^{-1}w], z[w^{-1}(uc)]).$$

Otherwise, the product is defined to be zero, and all products in which one of the terms is zero are defined to be zero.

The basic structural properties of McAlister semigroups are listed below.

Proposition 11 *The semigroup M_n has the following properties:*

(1) *The idempotents are those allowable triples (a, b, c) for which $ac = b$.*

(2) *If (a, b, c) and (u, w, z) are idempotents then*

$$(a, b, c) \leq (u, w, z) \Leftrightarrow a = (au^{-1})u \text{ and } c = z(z^{-1}c).$$

(3) *The inverse of (a, b, c) is $(bc^{-1}, b, a^{-1}b)$. Also*

$$(a, b, c)^{-1} \otimes (a, b, c) = (bc^{-1}, b, c) \text{ and } (a, b, c) \otimes (a, b, c)^{-1} = (a, b, a^{-1}b).$$

(4) *For arbitrary allowable triples*

$$(a, b, c)\, \mathcal{D}\, (u, w, z) \Leftrightarrow b = w.$$

(5) *The elements* $C_n = \{(1, x, 1)\colon x \in \Sigma^\dagger\}$ *form a subsemigroup isomorphic to* $\Sigma^\dagger$. *Furthermore, each non-zero* $\mathcal{D}$*-class of* M_n *contains exactly one element of* C_n.

(6) *For each allowable triple* (a, b, c) *we have that*

$$(a, b, c) = (1, a, 1)^{-1} \otimes (1, b, 1) \otimes (1, c, 1)^{-1},$$

and

$$(1, a, 1) \otimes (a, b, c) \otimes (1, c, 1) = (1, b, 1).$$

(7) *The elements* $(1, x_i, 1)$ *and* $(1, x_j, 1)$ *are orthogonal if* $i \neq j$.

(8) *The second component of* $(a, b, c) \otimes (u, w, z)$ *is also equal to*

$$[b(uc)^{-1}]w[w^{-1}(uc)].$$

Proof (1) Suppose that (a, b, c) is an idempotent. Then $\Sigma^*(ac) \cap \Sigma^* b$ and $(ac)\Sigma^* \cap b\Sigma^*$ are non-empty because the product $(a, b, c) \otimes (a, b, c)$ exists. Thus

$$(b(ac)^{-1})ac = ((ac)b^{-1})b \text{ and } b(b^{-1}(ac)) = (ac)((ac^{-1})b)$$

by Lemma 9.3.2. By assumption, (a, b, c) is an idempotent, consequently

$$a = [(ac)b^{-1}]a,\ b = [(ac)b^{-1}]b[(ac)^{-1}b],\ \text{and } c = c[b^{-1}(ac)].$$

Thus

$$[(ac)b^{-1}] = [(ac)^{-1}b] = [b^{-1}(ac)] = 1.$$

Hence $b = ac$. The converse is straightforward to prove.
(2) Suppose that $(a, b, c) \leq (u, w, z)$. Then $(a, b, c) = (a, b, c) \otimes (u, w, z)$. In particular, $\Sigma^* uc \cap \Sigma^* b$ and $(uc)\Sigma^* \cap w\Sigma^*$ are non-empty. Thus

$$(b(uc)^{-1})uc = ((uc)b^{-1})b \text{ and } w(w^{-1}(uc)) = (uc)((uc^{-1})w)$$

by Lemma 9.3.2. By assumption,

$$a = [(uc)b^{-1}]a, b = [(uc)b^{-1}]b[(uc)^{-1}w] \text{ and } c = z[w^{-1}(uc)].$$

Thus

$$[(uc)b^{-1}] = 1,\ [(uc)^{-1}w] = 1 \text{ and } c = z[w^{-1}(uc)].$$

Consequently, $b = (b(uc)^{-1})(uc)$ and $c = z[w^{-1}(uc)]$. Now $b = ac$, and so $a = (b(uc)^{-1})u$. Hence u is a suffix of a. Clearly, z is a prefix of c. The converse is straightfoward.
(3) These are straightforward computations.

(4) Recall that for idempotents $(a,b,c)\,\mathcal{D}\,(e,f,g)$ precisely when there is an element (u,w,z) such that

$$(u,w,z)^{-1} \otimes (u,w,z) = (a,b,c)$$

and

$$(u,w,z) \otimes (u,w,z)^{-1} = (e,f,g)$$

(Proposition 3.2.5). The result for idempotents is now straightforward to prove. The result for arbitrary elements follows from (3).
(5) The proof of the first assertion is straightforward. The proof of the second assertion follows immediately from (4).
(6) The proof of both assertions is straightforward.
(7) Straightforward.
(8) By Lemma 9.3.2(3) and (4), we have that

$$[(uc)b^{-1}]b = (b(uc)^{-1})uc \text{ and } (uc)[(uc)^{-1}w] = w(w^{-1}(uc)).$$

Thus

$$[(uc)b^{-1}]b[(uc)^{-1}w] = [b(uc)^{-1}]w[w^{-1}(uc)].$$

■

We can now describe the algebraic properties enjoyed by McAlister semigroups.

Theorem 12 *The semigroup M_n is a combinatorial, completely semisimple, 0-E-unitary inverse semigroup.*

Proof To show that M_n is combinatorial, suppose that $(a,b,c)\,\mathcal{H}\,(u,v,w)$ in M_n. Then

$$(bc^{-1},b,c) = (vu^{-1},v,w) \text{ and } (a,b,a^{-1}b) = (u,v,v^{-1}w)$$

by Proposition 11. Thus $(a,b,c) = (u,v,w)$.

To show that M_n is completely semisimple let (a,b,c) and (u,w,z) be two idempotents such that

$$(a,b,c) \leq (u,w,z) \text{ and } (a,b,c)\,\mathcal{D}\,(u,w,z).$$

Then

$$a = (au^{-1})u, \quad c = z(z^{-1}c) \text{ and } b = w$$

by Proposition 11. Also $b = ac$ and $w = uz$, hence $ac = uz$. Thus $ac = (au^{-1})uz(z^{-1}c)$, which gives $au^{-1} = 1$ and $z^{-1}c = 1$. Hence $a = u$ and $c = z$, and so $(a,b,c) = (u,w,z)$.

To show that M_n is 0-E-unitary let (a,b,c) be a non-zero idempotent and suppose that $(a,b,c) \leq (u,w,z)$. Then $(a,b,c) = (a,b,c) \otimes (u,w,z)$. The product is non-zero so that

$$(b(uc)^{-1})uc = ((uc)b^{-1})b \text{ and } w(w^{-1}(uc)) = uc((uc)^{-1}w).$$

by Lemma 9.3.2. From the form of the product we have that

$$(uc)b^{-1} = 1, \quad (uc)^{-1}w = 1 \text{ and } c = z[w^{-1}(uc)].$$

Thus

$$b = (b(uc)^{-1})uc \text{ and } uc = w(w^{-1}(uc))$$

and multiplying c on the left by u we obtain $uc = uz[w^{-1}(uc)]$. But $uc = w[w^{-1}(uc)]$, so that $w = uz$. Hence (u,w,z) is also an idempotent. ■

Proposition 11(5) and (6) suggest that M_n is akin to a semigroup of fractions of the free semigroup on n generators. However, the fractions in question have one numerator but two denominators: one on each side.

McAlister semigroups and polycyclic monoids

In the introduction to this section, we referred to McAlister semigroups as being two-sided analogues of polycyclic monoids. We shall now establish the exact relationship between McAlister semigroups and polycyclic monoids.

For $n = 1$ put $DP_1 = P_1 \times P_1$ with the direct product operation. For $n \geq 2$ put $DP_n = (P_n \setminus \{0\}) \times (P_n \setminus \{0\}) \cup \{0\}$ with the following multiplication: the product $((a,b),(c,d)) \otimes ((x,y),(u,v))$ is zero if either $(a,b) \otimes (x,y)$ or $(c,d) \otimes (u,v)$ is zero. Otherwise the product is that obtained from the direct product. The zero operates as a zero.

Lemma 13 *DP_n is a quotient of $P_n \times P_n$ for $n \geq 2$.*

Proof Clearly, both $\{0\} \times P_n$ and $P_n \times \{0\}$ are ideals of $P_n \times P_n$, consequently $I = (\{0\} \times P_n) \cup (P_n \times \{0\})$ is an ideal of $P_n \times P_n$. We can thus form the Rees quotient semigroup $(P_n \times P_n)/I$. We claim that this Rees quotient is isomorphic to DP_n. Recall that the non-zero elements of $(P_n \times P_n)/I$ may be regarded as the elements of $(P_n \times P_n) \setminus I$. But this set is just $DP_n \setminus \{0\}$. The result is now clear. ■

Put

$$M'_n = \{((a,b),(e,f)) \in DP_n \colon 1 \neq ae^r = bf^r\} \cup \{0\},$$

a subset of DP_n.

Lemma 14 *M'_n is an inverse subsemigroup of DP_n.*

Proof Let $((a,b),(c,d))$ and $((u,v),(w,x))$ be non-zero elements of M'_n whose product is non-zero. Thus

$$(a,b) \otimes (u,v) = ((ub^{-1})a, (bu^{-1})v)$$

and

$$(c,d) \otimes (w,x) = ((wd^{-1})c, (dw^{-1})x).$$

We need to show that

$$(ub^{-1})a((wd^{-1})c)^r = ((bu^{-1})v)((dw^{-1})x)^r.$$

We begin with the left-hand side of this equation. By assumption $ac^r = bd^r$ and $uw^r = vx^r$, so that the left-hand side is

$$(ub^{-1})b((wd^{-1})d)^r,$$

whereas the right-hand side is

$$(bu^{-1})u((dw^{-1})w)^r.$$

By assumption, $\Sigma^* b \cap \Sigma^* u$ and $\Sigma^* d \cap \Sigma^* w$ are non-empty. Consequently

$$(ub^{-1})b = (bu^{-1})u \text{ and } ((wd^{-1})d)^r = ((dw^{-1})w)^r$$

by Lemma 9.3.2. Hence M'_n is closed under the multiplication. The proof that the set is closed under inverses is straightforward. ∎

The proof of the following is straightforward.

Lemma 15 *If (a,b,c) and (u,w,z) are allowable triples, then:*

(1) $\Sigma^*(uc) \cap \Sigma^* b \neq \emptyset \Leftrightarrow \Sigma^*(bc^{-1}) \cap \Sigma^* u \neq \emptyset$.

(2) $(uc)\Sigma^* \cap w\Sigma^* \neq \emptyset \Leftrightarrow c\Sigma^* \cap (u^{-1}w)\Sigma^* \neq \emptyset$. ∎

The exact relationship between McAlister semigroups and polycyclic monoids is spelt out by the following theorem.

Theorem 16 *Define a function ι: $M_n \to M'_n$ by $\iota(0) = 0$ and $\iota(a,b,c) = ((a, bc^{-1}), ((a^{-1}b)^r, c^r))$. Then ι is an isomorphism. In particular, M_n may be embedded in DP_n, and so M_n divides $P_n \times P_n$.*

Proof By Lemma 9.3.2 the function ι is well-defined. To prove that ι is a bijection, suppose first that $\iota(a,b,c) = \iota(u,v,w)$. Then $a = u$, $c = w$, $bc^{-1} = vw^{-1}$, and $a^{-1}b = u^{-1}v$. But $b = (bc^{-1})c = (vw^{-1})w = v$ by Lemma 9.3.2. Hence $(a,b,c) = (u,v,w)$. Thus ι is injective. To show that ι is surjective, let $((u,v),(x,y)) \in M'_n$. Put $a = u$, $c = y^r$ and $b = ux^r = vy^r$, the latter holding by assumption. Then it can easily be checked that $\iota(a,b,c) = ((u,v),(x,y))$.

By Lemma 14, the product $(a,b,c)\otimes(u,w,z)$ is non-zero precisely when the product $\iota(a,b,c)\otimes\iota(u,w,z)$ is non-zero. We may thus assume that $\iota(a,b,c)\otimes\iota(u,w,z)$ is non-zero. We have to prove that

$$\iota((a,b,c)\otimes(u,w,z)) = \iota(a,b,c)\otimes\iota(u,w,z).$$

We do this by comparing components. We begin with the first components. By definition

$$\iota(a,b,c) = ((a,bc^{-1}),((a^{-1}b)^r,c^r))$$

and

$$\iota(u,w,z) = ((u,wz^{-1}),((u^{-1}w)^r,z^r)).$$

A simple calculation shows that

$$(a,bc^{-1})\otimes(u,wz^{-1}) = ([(uc)b^{-1}]a,[b(uc)^{-1}](wz^{-1})).$$

By Lemma 9.3.2

$$[(uc)b^{-1}]b = (b(uc)^{-1})uc \text{ and } (uc)[(uc)^{-1}w] = w(w^{-1}(uc)).$$

Thus

$$[(uc)b^{-1}]b[(uc)^{-1}w] = [b(uc)^{-1}]w[w^{-1}(uc)].$$

By Lemma 9.3.2, $w = (wz^{-1})z$ since z is a suffix of w. It follows that

$$[(b(uc)^{-1})w(w^{-1}(uc))][z(w^{-1}(uc))]^{-1} = (b(uc)^{-1})(wz^{-1}).$$

It is now easy to check that the first component of $\iota((a,b,c)\otimes(u,w,z))$ is equal to the first component of $\iota(a,b,c)\otimes\iota(u,w,z)$. Turning now to second components, a simple calculation shows that

$$((a^{-1}b)^r,c^r)\otimes((u^{-1}w)^{-1},z^r) = (((a^{-1}b)[(uc)^{-1}w])^r,(z[w^{-1}(uc)])^r).$$

By Lemma 9.3.2, $b = a(a^{-1}b)$ since a is a prefix of b, and so

$$([(uc)b^{-1}]a)^{-1}([(uc)b^{-1}]b[(uc)^{-1}w]) = (a^{-1}b)[(uc)^{-1}w].$$

It is now straightforward to check that the second component of $\iota((a,b,c)\otimes(u,w,z))$ is equal to the second component of $\iota(a,b,c)\otimes\iota(u,w,z)$. ■

The ideal structure of McAlister semigroups

Recall from Proposition 11 that $C_n = \{(1, x, 1) \colon x \in \Sigma^\dagger\}$ forms a subsemigroup of M_n isomorphic to $\Sigma^\dagger$ where Σ contains n elements. The next result should be contrasted with Theorem 9.3.5, where we proved that the polycyclic monoids are congruence-free when $n \geq 2$.

Theorem 17 *Let Σ be a finite alphabet with n elements. Then the ideal lattice of the McAlister semigroup M_n is isomorphic to the ideal lattice of the semigroup $\Sigma^\dagger$ with a zero adjoined.*

Proof Without loss of generality, we may replace $\Sigma^\dagger$ by C_n.

Define the order-preserving function $I \mapsto M_n I M_n$ from the lattice of ideals of C_n^0 to the lattice of ideals of M_n, and define the order-preserving function $J \mapsto J \cap C_n^0$ from the lattice of ideals of M_n to the lattice of ideals of C_n^0. It remains to show that these functions are mutually inverse.

We prove first that $M_n I M_n \cap C_n^0 = I$. The key step is to prove that

$$M_n I M_n = \{(a, b, c) \colon (1, b, 1) \in I\} \cup \{0\}.$$

Let $(1, b, 1) \in I$ and let $(x, y, z), (u, v, w) \in M_n$. By Proposition 11(8), the second component of $(x, y, z) \otimes (1, b, 1)$ is of the form ubv for some strings u and v, whereas the second component of $((x, y, z) \otimes (1, b, 1)) \otimes (u, v, w)$ is of the form $u'bv'$ for some strings u' and v' from the usual way of writing the product. However,

$$(1, u'bv', 1) = (1, u', 1) \otimes (1, b, 1) \otimes (1, v', 1),$$

with either $(1, u', 1)$ or $(1, v', 1)$ omitted if u' or v' is the empty string. But $(1, u'bv', 1) \in I$, since I is an ideal of C_n^0. Consequently, every element of $M_n I M_n$ is of the form (a, b, c) for some $(1, b, 1) \in I$. Conversely, if (a, b, c) is such that $(1, b, 1) \in I$ then, since

$$(a, b, c) = (1, a, 1)^{-1}(1, b, 1)(1, c, 1)^{-1},$$

it is clear that $(a, b, c) \in M_n I M_n$. The fact that

$$M_n I M_n \cap C_n^0 = I$$

now follows easily from the above result.

We now prove that if J is an ideal of M_n then $M_n(J \cap C_n^0)M_n = J$. By definition

$$J \cap C_n^0 = \{(1, b, 1) \colon (1, b, 1) \in J\}.$$

Clearly, $M_n(J \cap C_n^0)M_n \subseteq J$. Now let $(a, b, c) \in J$. Then $(1, b, 1) \in J$ by Proposition 11(6). Thus $(1, b, 1) \in J \cap C_n^0$. It follows by Proposition 11(6) again that $(a, b, c) \in M_n(J \cap C_n^0)M_n$. ■

9.5 Tiling semigroups

The inverse semigroups we discuss in this section have their origins in solid-state physics. Solids can be modelled by tilings, with each tile of the tiling representing a particular atom or molecule. Classically, such tilings were assumed to be periodic, but with the discovery of quasi-crystals (Section 1.3) non-periodic tilings, such as Penrose tilings, have become potential models.

A basic question in solid-state physics is to determine the spectrum of a particle moving in a solid. Mathematically, this involves carrying out calculations on the associated tiling which, for quasi-crystals, is non-periodic. Such calculations are currently too difficult, consequently Bellisard proposed using K-theoretical gap-labelling to obtain qualitative information on the spectrum. This technique involves computing the K_0-group of a C^*-algebra associated with the tiling. Johannes Kellendonk showed that the appropriate C^*-algebra can be obtained in three stages:

(1) An algebraic object is constructed from the tiling; this is an inverse semigroup which we call a tiling semigroup.

(2) A topological groupoid is constructed from the tiling semigroup using the technique described in Section 9.2.

(3) A C^*-algebra is constructed from the topological groupoid in a standard way: see the books by Paterson [302] and Renault [350] for a description of how this is carried out.

The aim of this section is to describe how to associate an inverse semigroup with any tiling: periodic or non-periodic. Remarkably, tiling semigroups are very similar to the free inverse semigroups described in Chapter 6: they are always 0-E-unitary, and the form of the multiplication is essentially the same as the multiplication of Munn trees.

Kellendonk's use of inverse semigroup theory is rather deep, particularly in his theory of the topological equivalence of tilings [170] where a class of prehomomorphisms provides the right notion of morphism.

We begin with a review of some of the basic terms concerning tilings; however, only an intuitive idea of tilings will be needed to understand this section.

Definition and basic properties

The tilings we shall consider live in $\mathbb{R}^n$. A *tile* in $\mathbb{R}^n$ is a connected, bounded subset of $\mathbb{R}^n$ which is the closure of its interior. An *n-dimensional tiling* is an infinite set of tiles which cover $\mathbb{R}^n$ overlapping, at most, at their boundaries. A finite subset of a tiling which is a union of tiles is called a *pattern*; it is important to bear in mind in the subsequent work that a given pattern is located at a specific place in the tiling. If the subset of $\mathbb{R}^n$ covered by a pattern is connected then we say the pattern is *connected*. All patterns will be assumed to be connected.

From now on, we shall assume that a tiling $\mathcal{T}$ in $\mathbb{R}^n$ has been chosen, and make all our definitions relative to this tiling. The first step in defining a semigroup from $\mathcal{T}$ consists in defining an appropriate notion of 'equivalence' on the set of patterns of a tiling. Two patterns P and Q will be termed *equivalent* if there is a translation τ of $\mathbb{R}^n$ such that $\tau(P) = Q$. It is important to note that we do *not* require the translation τ to be a symmetry of the tiling; in general, the tilings of most interest will have no translational symmetries. An equivalence class of patterns is called a *pattern class*. If P is a pattern then $[P]$ will denote the corresponding pattern class. If a pattern consists of a single tile, the corresponding class is called a *tile class*.

A *doubly pointed pattern* is a triple (p_2, P, p_1) consisting of a pattern, P, and two distinguished tiles from the pattern p_1 and p_2. The notion of equivalence between patterns can be extended to doubly pointed patterns in the following way: two doubly pointed patterns (p_2, P, p_1) and (q_2, Q, q_1) are said to be equivalent if there is a translation τ of $\mathbb{R}^n$ such that

$$\tau(P) = Q,\ \tau(p_1) = q_1 \text{ and } \tau(p_2) = q_2.$$

The equivalence class containing (p_2, P, p_1) is denoted by $[p_2, P, p_1]$ and called a *doubly pointed pattern class*. We call p_1 the *in-tile* and p_2 the *out-tile* (for the representative P).

Define $S(\mathcal{T})$ to consist of all doubly pointed pattern classes of the tiling $\mathcal{T}$ together with a new element 0. We shall now define a binary operation on $S(\mathcal{T})$ which will convert it into an inverse semigroup.

Let $[p_2, P, p_1]$ and $[q_2, Q, q_1]$ be two doubly pointed pattern classes. Suppose that there are translations τ_1 and τ_2 of $\mathbb{R}^n$ such that: $\tau_1(P)$ and $\tau_2(Q)$ are patterns and $\tau_1(p_1) = \tau_2(q_2)$; define

$$[p_2, P, p_1][q_2, Q, q_1] = [\tau_1(p_2), \tau_1(P) \cup \tau_2(Q), \tau_2(q_1)],$$

where $\tau_1(P) \cup \tau_2(Q)$ is the pattern obtained by taking the union of the patterns $\tau_1(P)$ and $\tau_2(Q)$. If no such translations can be found then the product is defined to be 0. Also any product involving 0 is defined to be 0.

Theorem 1 *The above binary operation is well-defined and endows $S(\mathcal{T})$ with the structure of an inverse semigroup with zero.*

Proof We begin by showing that the binary operation is well-defined. Let $[p_2, P, p_1]$ and $[q_2, Q, q_1]$ be two doubly pointed pattern classes with a non-zero product. Let τ_1 and τ_2 be a pair of translations such that $\tau_1(P)$ and $\tau_2(Q)$ are patterns and $\tau_1(p_1) = \tau_2(q_2)$. By definition the corresponding product is

$$[\tau_1(p_2), \tau_1(P) \cup \tau_2(Q), \tau_2(q_1)].$$

Now let τ_3 and τ_4 be another pair of translations such that $\tau_3(P)$ and $\tau_4(Q)$ are patterns and $\tau_3(p_1) = \tau_4(q_2)$. By definition the corresponding product is

$$[\tau_3(p_2), \tau_3(P) \cup \tau_4(Q), \tau_4(q_1)].$$

In particular,

$$\tau_1(p_1) = (\tau_2\tau_4^{-1}\tau_3)(p_1).$$

But τ_1 and $\tau_2\tau_4^{-1}\tau_3$ are translations and so $\tau_1 = \tau_2\tau_4^{-1}\tau_3$. Consider the translation $\tau_3\tau_1^{-1}$. Observe that $\tau_3\tau_1^{-1}\tau_2 = \tau_4$. Thus

$$\tau_3\tau_1^{-1}(\tau_2(q_1)) = \tau_4(q_1) \text{ and } \tau_3\tau_1^{-1}(\tau_1(p_2)) = \tau_3(p_2)$$

and

$$(\tau_3\tau_1^{-1})(\tau_1(P) \cup \tau_2(Q)) = \tau_3(P) \cup \tau_4(Q).$$

Hence

$$[\tau_1(p_2), \tau_1(P) \cup \tau_2(Q), \tau_2(q_1)] = [\tau_3(p_2), \tau_3(P) \cup \tau_4(Q), \tau_4(q_1)]$$

and so the product is well-defined.

To prove associativity of the product, let $[p_2, P, p_1]$, $[q_2, Q, q_1]$ and $[r_2, R, r_1]$ be three doubly pointed pattern classes. Assume that the product

$$([p_2, P, p_1][q_2, Q, q_1])[r_2, R, r_1]$$

is nonzero. Let τ_1 and τ_2 be translations such that $\tau_1(P)$ and $\tau_2(Q)$ are patterns and $\tau_1(p_1) = \tau_2(q_2)$. Then

$$[p_2, P, p_1][q_2, Q, q_1] = [\tau_1(p_2), \tau_1(P) \cup \tau_2(Q), \tau_2(q_1)].$$

Now let τ_3 and τ_4 be translations such that $\tau_3(\tau_1(P) \cup \tau_2(Q))$ and $\tau_4(R)$ are patterns and $\tau_3(\tau_2(q_1)) = \tau_4(r_2)$. Then

$$([p_2, P, p_1][q_2, Q, q_1])[r_2, R, r_1] = [\tau_3(\tau_1(p_2)), \tau_3(\tau_1(P) \cup \tau_2(Q)) \cup \tau_4(R), \tau_4(r_1)].$$

By assumption, $\tau_3(\tau_1(P) \cup \tau_2(Q))$ is a pattern so that both $\tau_3\tau_1(P)$ and $\tau_3\tau_2(Q)$ are patterns. Thus the product can be written as

$$[(\tau_3\tau_1)(p_2), (\tau_3\tau_1)(P) \cup (\tau_3\tau_2)(Q) \cup \tau_4(R), \tau_4(r_1)].$$

The proof of associativity is now straightforward.

To show that the semigroup is inverse, we begin by locating its idempotents. Suppose that $[p_2, P, p_1]^2 = [p_2, P, p_1]$. Then by definition there exist translations τ_1 and τ_2 such that $\tau_1(p_1) = \tau_2(p_2)$ and $\tau_1(P)$ and $\tau_2(P)$ are both patterns. Thus

$$[p_2, P, p_1][p_2, P, p_1] = [\tau_1(p_2), \tau_1(P) \cup \tau_2(P), \tau_2(p_1)]$$

which is equal to $[p_2, P, p_1]$ by assumption. Thus from the definition of pattern classes there is a translation τ such that

$$\tau(\tau_1(p_2)) = p_2, \ \tau(\tau_1(P) \cup \tau_2(P)) = P, \text{ and } \tau(\tau_2(p_1)) = p_1.$$

Thus $\tau^{-1} = \tau_1$ and $\tau^{-1} = \tau_2$, so that $\tau_1 = \tau_2$. Hence $p_1 = p_2$. Thus if $[p_2, P, p_1]$ is an idempotent then $p_1 = p_2$; it is easy to check that every element satisfying this condition is an idempotent.

Let $[p, P, p]$ and $[q, Q, q]$ be two idempotents. Suppose that $[p, P, p][q, Q, q]$ is non-zero. Then there are translations τ_1 and τ_2 such that $\tau_1(P)$ and $\tau_2(Q)$ are patterns and $\tau_1(p) = \tau_2(q)$. Thus

$$[p, P, p][q, Q, q] = [\tau_1(p), \tau_1(P) \cup \tau_2(Q), \tau_2(q)],$$

which is an idempotent. This shows that the set of idempotents is closed under multiplication; the proof that idempotents commute is now straightforward.

Finally, it is easy to check that

$$[p_1, P, p_2] = [p_1, P, p_2][p_2, P, p_1][p_1, P, p_2]$$

and

$$[p_1, P, p_2] = [p_2, P, p_1][p_1, P, p_2][p_2, P, p_1].$$

Thus $[p_1, P, p_2]^{-1} = [p_1, P, p_2]$. ■

We call $S(\mathcal{T})$ the *tiling semigroup* of the tiling $\mathcal{T}$.

Theorem 2 *Tiling semigroups are combinatorial, completely semisimple, 0-E-unitary inverse semigroups.*

Proof To show that the tiling semigroup is combinatorial suppose that

$$[p_2, P, p_1] \, \mathcal{H} \, [q_2, Q, q_1].$$

Then by Theorem 1, we have that

$$[p_1, P, p_1] = [q_1, Q, q_1] \text{ and } [p_2, P, p_2] = [q_2, Q, q_2].$$

Thus there are translations τ_1 and τ_2 such that

$$\tau_1(P) = Q \text{ and } \tau_1(p_1) = q_1$$

and

$$\tau_2(P) = Q \text{ and } \tau_2(p_2) = q_2.$$

Hence $\tau_1 = \tau_2$ and so

$$[p_2, P, p_1] = [q_2, Q, q_1].$$

Before proving that the semigroup is completely semisimple we first determine the nature of the $\mathcal{D}$-classes. Suppose that

$$[p_2, P, p_1] \, \mathcal{D} \, [q_2, Q, q_1].$$

Then there is a doubly pointed pattern class $[r_2, R, r_1]$ such that

$$[p_1, P, p_1] = [r_2, R, r_2] \text{ and } [q_1, Q, q_1] = [r_1, R, r_1].$$

Thus there are translations τ_1 and τ_2 such that

$$\tau_1(p_1) = r_2 \text{ and } \tau_1(P) = R$$

and

$$\tau_2(q_1) = r_1 \text{ and } \tau_2(Q) = R.$$

Hence $(\tau_2^{-1}\tau_1)(P) = Q$, and so P and Q are equivalent patterns. Conversely, it is easy now to show that if P and Q are equivalent patterns then

$$[p_2, P, p_1] \, \mathcal{D} \, [q_2, Q, q_1].$$

Thus the non-zero $\mathcal{D}$-classes of the tiling semigroup correspond to the pattern classes. Since each pattern is finite there are only finitely many ways of choosing a pair of distinguished tiles. Thus every $\mathcal{D}$-class is finite. In particular, tiling semigroups are completely semisimple.

The fact that the semigroup is 0-E-unitary will follow readily once we have characterised the natural partial ordering. If

$$[p_2, P, p_1] \leq [q_2, Q, q_1]$$

then

$$[p_2, P, p_1] = [q_2, Q, q_1][p_1, P, p_1].$$

Thus there are translations τ_1 and τ_2 such that $\tau_1(Q)$ and $\tau_2(P)$ are patterns and $\tau_1(q_1) = \tau_2(p_2)$. Hence

$$[p_2, P, p_1] = [\tau_1(q_2), \tau_1(Q) \cup \tau_2(P), \tau_2(p_1)].$$

Thus there is a translation τ such that

$$\tau(p_2) = \tau_1(q_2),\ \tau(P) = \tau_1(Q) \cup \tau_2(P) \text{ and } \tau(p_1) = \tau_2(p_1).$$

But then $\tau = \tau_2$ so that

$$(\tau^{-1}\tau_1)(q_1) = p_1,\ (\tau^{-1}\tau_1)(q_2) = p_2 \text{ and } (\tau^{-1}\tau_1)(Q) \subseteq P.$$

Thus we have found a translation which maps the doubly pointed pattern (q_2, Q, q_1) so that it lies inside (p_2, P, p_1) with distinguished tiles matching. Conversely, it can easily be proved that if such a translation exists then the elements are related via the natural partial ordering. It is now immediate that any element above a non-zero idempotent is an idempotent. ∎

The form taken by the multiplication in tiling semigroups, together with the properties derived above, are strongly reminiscent of the behaviour of free inverse semigroups described in Chapter 6. We saw there that the free inverse semigroup on n generators can be constructed from the Cayley graph of the free group on n generators. In some sense, a Cayley graph of a free group is a tiling, but on a 1-dimensional space; the underlying graph of the Cayley graph of a free group is a tree which can be regarded as a metric space.

1-dimensional tiling semigroups

We shall now analyse in more detail the structure of 1-dimensional tiling semigroups, by which we mean tilings in $\mathbb{R}$. We shall show that they are closely related to the McAlister semigroups of Section 9.4.

A 1-dimensional tiling can be regarded as a bi-infinite string over an aphabet Σ; that is, as a function from $\mathbb{Z}$ to Σ. Pattern classes of such a tiling can be identified with finite, non-empty substrings of the bi-infinite string and doubly-pointed pattern classes with patterns having two letters distinguished. For this reason we shall refer to such classes as *doubly pointed strings.*

Before saying something about tiling semigroups themselves, we first introduce a class of auxillary inverse semigroups K_n where n is the cardinality of Σ, and then show how a tiling gives rise to an ideal in K_n; the tiling semigroup will then be the Rees quotient of K_n by the ideal.

The elements of K_n consist of 0, except when $n = 1$ when the zero is omitted, and the following types of strings:

- $x\breve{a}y$ where $x, y \in \Sigma^*$ and $a \in \Sigma$.
- $u\grave{a}v\acute{b}w$ where $u, v, w \in \Sigma^*$ and $a, b \in \Sigma$.
- $u\acute{a}v\grave{b}w$ where $u, v, w \in \Sigma^*$ and $a, b \in \Sigma$.

The accents are used simply to mark the in- and out-tiles: the acute accent marks the 'in-letter' and the grave accent marks the 'out-letter'. The check is to be interpreted as both an in- and an out-letter.

Let p and q be non-zero patterns in that order. Place p above q so that the in-letter of p is above the out-letter of q. We say that p and q *match* if, ignoring accents, they agree everywhere on their overlap.

We may define a binary operation on K_n as follows: if p and q are non-zero doubly pointed strings which do not match then define $p \otimes q = 0$. If p and q do match, then glue the strings together along their overlap, erasing the in-accent of p and the out-accent of q and carry forward the remaining accents. If a grave and acute accent occur on the same letter in the resulting string then rewrite as a check. The resulting doubly pointed string is defined to be $p \otimes q$. A few examples of this product should clarify this definition.

Examples

1. Let $p = \acute{a}\grave{b}$ and $q = b\grave{a}b\acute{b}$. To compute $p \otimes q$ write $p = \acute{a}\grave{b}$ above $q = b\grave{a}b\acute{b}$ lining up the $\acute{a}$ of p above the $\grave{a}$ of q. The strings match on their overlap and their product is $ba\grave{b}\acute{b}$.

2. Let $p = \grave{a}\acute{b}$ and $q = \grave{b}\acute{a}$. Write $p = \grave{a}\acute{b}$ above $q = \grave{b}\acute{a}$ lining up the $\acute{b}$ of p above the $\grave{b}$ of q. The strings match on their overlap and their product is $\grave{a}b\acute{a}$.

3. Let $p = \grave{a}b\acute{a}$ and $q = \acute{a}b\grave{a}$. Write $p = \grave{a}b\acute{a}$ above $q = \acute{a}b\grave{a}$ lining up the $\acute{a}$ of p above the $\grave{a}$ of q. The strings match on their overlap and their product is $\breve{a}ba$.

The proof of the following is essentially the same as the proof of Theorem 1.

Theorem 3 $(K_n, \otimes)$ *is an inverse semigroup which has a zero if* $n \geq 2$. *The (non-zero) idempotents are the elements of the form* $x\breve{a}y$ *where* $x, y \in \Sigma^*$ *and* $a \in \Sigma$. *The inverse of* $u\grave{a}v\acute{b}w$, *where* $u, v, w \in \Sigma^*$ *and* $a, b \in \Sigma$, *is the element* $u\acute{a}v\grave{b}w$. ■

We call the inverse semigroup K_n the *Kachel-semigroup* on n letters, where *Kachel* is a German word meaning 'tile'. Observe that K_n does not have n generators.

Let us consider in more detail the structure of the semigroup K_1.

Theorem 4 K_1 *is the free inverse monoid on one generator.*

Proof Let $\Sigma = \{a\}$. Then the elements of K_1 consist of all doubly pointed non-empty strings in a. The string $\breve{a}$ is the identity of this semigroup, so that K_1 is a monoid.

Now consider the element $x = \grave{a}\acute{a}$. Then $x^2 = \grave{a}a\acute{a}$, and by induction, we see that for $n \geq 1$ the string x^n has length $n+1$, its first letter carries a grave accent and its last letter carries an acute accent. We can move the grave accent of x^n one letter to the right by multiplying x^n on the left by x^{-1}. Thus if $m \leq n$ then $x^{-m}x^n$ carries the grave accent on the $m+1$th letter from the left and the acute accent on the last letter. Likewise, if $p \leq n$ then $x^n x^{-p}$ carries the grave accent on the first letter and the acute accent on the $p+1$th letter from the right. Thus the string $x^{-m}x^n x^{-p}$ where $m, n, p \in \mathbb{N}$ and $n \geq 1$ and $m, p \leq n$ has length $n+1$, carries the grave accent on the $m+1$th letter from the left, and carries the acute accent on the $p+1$th letter from the right.

We conclude that K_1 is generated as an inverse monoid by $x = \grave{a}\acute{a}$ and that every non-identity element can be written uniquely in the form $x^{-m}x^n x^{-p}$ where $m, n, p \in \mathbb{N}$, $n \geq 1$ and $m, p \leq n$. Hence K_1 is the free inverse monoid on one generator. ■

The semigroups K_n can be embedded in M_n. The idea of the embedding is to encode a doubly pointed string over the alphabet Σ by means of an allowable triple from $\Sigma^* \times \Sigma^\dagger \times \Sigma^*$, and so by means of an element of M_n.

Define a function ϕ: $K_n \to M_n$ as follows:

- $\phi(0) = 0$,
- $\phi(x\breve{a}y) = (x, xay, ay)$,
- $\phi(u\grave{a}v\acute{b}w) = (u, uavbw, bw)$,
- $\phi(u\acute{a}v\grave{b}w) = (uav, uavbw, avbw)$.

Thus the first component is the prefix of the string which precedes the letter carrying the grave accent, the second component is the whole string, and the third component is the suffix which begins at the letter carrying the acute accent.

Theorem 5 *The function ϕ: $K_n \to M_n$ is an injective homomorphism, whose image is the subset of M_n consisting of 0 together with all elements of the form (x, y, z) where $x \neq y$ and $z \neq 1$.*

Proof It is evident from the definition, that the image of ϕ consists of elements of the stated form. It is also clear that ϕ is injective. It remains therefore to prove that ϕ is onto the given set. Let (x, y, z) be such that $x \neq y$ and $z \neq 1$. There are three cases to consider.

Suppose first that (x, y, z) is an idempotent. Then $xz = y$. By assumption, $z \neq 1$ and so $z = au$ for some letter $a \in \Sigma$. From the definition, $\phi(x\breve{a}u) = (x, y, z)$.

Next, suppose that (x, y, z) is not an idempotent and $y = xuz$ where u is non-empty. Since both u and z are non-empty we can write $u = av$ and $z = bw$ for some letters $a, b \in \Sigma$ and strings v and w. From the definition, $\phi(x\grave{a}v\acute{b}w) = (x, y, z)$.

Finally, suppose that (x, y, z) is not an idempotent and $x = pu$, $z = ur$ and $y = pur$, where u is a non-empty string. Then $u = am$ for some letter $a \in \Sigma$ and string m. Observe that if r were empty then x would equal y which would contradict our assumption that $x \neq y$. Thus r is non-empty and so $r = bs$ for some letter $b \in \Sigma$ and string s. From the definition $\phi(p\acute{a}m\grave{b}s) = (x, y, z)$.

It remains to show that ϕ is a homomorphism. The proof is essentially a case-by-case analysis of the various types of products, but the calculations can be made a little easier by using Theorem 3.1.5 and showing that ϕ preserves the product of idempotents, the natural partial order, and the restricted multiplication. Since the details are tedious, the proof will merely be outlined.

Let $e = u\breve{a}v$ and $f = w\check{b}x$ be two idempotents whose product is non-zero. Thus in particular, $a = b$. Four possible ways exist for these two strings to match: w is a suffix of u and x is a prefix of v, w is a suffix of u and v is a prefix of x, u is a suffix of w and x is a prefix of v, and u is a suffix of w and v is a prefix of x. In each case, it can be verified that $\phi(e)\phi(f) = \phi(e \otimes f)$.

The natural partial order in K_n is essentially a special case of the natural partial order described in Theorem 2. Thus if $u, v \in K_n$ then $u \leq v$ if, and only if, $u = pvq$ for some strings $p, q \in \Sigma^*$. To prove that $\phi(u) \leq \phi(v)$ it is enough to show that $(pa, pbq, cq) \leq (a, b, c)$. But it is straightforward to check that $(pa, pbq, cq) = (a, b, c)(p[bc^{-1}], pbq, cq)$ where $(p[bc^{-1}], pbq, cq)$ is an idempotent.

Finally, we consider restricted products. It is easy to check that $(a, b, c) \cdot (u, w, z)$ exists in M_n precisely when $bc^{-1} = u$, $b = w$ and $c = u^{-1}w$. In which case, $(a, b, c) \cdot (u, w, z) = (a, b, z)$. We have to check that if $\exists u \cdot v$ then $\exists \phi(u) \cdot \phi(v)$, and that $\phi(u) \cdot \phi(v) = \phi(u \cdot v)$. The labour in carrying out these calculations is reduced to checking five kinds of restricted products by the fact that $\phi(u^{-1}) = \phi(u)^{-1}$.

We conclude that ϕ is a homomorphism. ∎

We now return to tiling semigroups of tilings over Σ. Let K_n be the Kachel-semigroup over Σ. There is an obvious function $\delta: K_n \to \Sigma^\dagger \cup \{0\}$ which deletes diacritics and sends zero to zero. Let $\mathcal{T}$ be a tiling over Σ and let $F = F(\mathcal{T})$ be the set of all finite substrings of $\mathcal{T}$. This set has the property that if $uxv \in I$ then $x \in I$. It follows that the complement of F in $\Sigma^\dagger \cup \{0\}$ is an ideal. Let $I = I(\mathcal{T})$ be the inverse image under δ of this ideal in K_n. Then I is an ideal of K_n. We may therefore form the Rees quotient semigroup K_n/I. This semigroup can be regarded as the set $\delta^{-1}(F)$ of all doubly pointed strings of the tiling $\mathcal{T}$ together with zero. The product of two non-zero elements of K_n/I

is non-zero precisely when they match *and* their product is again a doubly pointed string. We have therefore proved the following result.

Theorem 6 *The tiling semigroup of a 1-dimensional tiling $\mathcal{T}$ over Σ, where Σ contains n elements, is isomorphic to a Rees quotient semigroup K_n/I where the ideal I is determined by the tiling $\mathcal{T}$. In particular, every 1-dimensional tiling semigroup over an alphabet with n elements divides M_n.* ■

Apart from any intrinsic interest this result might have, it also tells us something about tiling semigroups. It is evident that K_1 is the tiling semigroup (with the zero removed) of the 1-dimensional tiling consisting of one kind of tile repeated along the line. Thus tiling semigroups not only have algebraic properties akin to free inverse semigroups but the free inverse semigroup on one generator is an example of such a semigroup.

9.6 Notes on Chapter 9

Section 9.1

The class of 0-E-unitary semigroups was first defined by Mária Szendrei [404], although she called them *E^*-unitary.* The term 0-E-unitary appears to be due to Meakin and Sapir [251].

The strong compatibility relation and some of its basic properties were tacitly introduced in [170]. The relation β was introduced in [260]. The theory of 0-E-unitary semigroups which are categorical at zero and for which S/β is a Brandt semigroup is described explicitly in [102].

To describe more general kinds of 0-E-unitary semigroups the trick appears to be to replace idempotent pure homomorphisms by idempotent pure prehomomorphisms; this was implicitly recognised by Bulman-Fleming, Fountain and Gould [37] and explicitly by myself [194]. The theory of 0-E-unitary semigroups is a good deal more complex than the theory of E-unitary semigroups.

Section 9.2

The topological groupoid described in this section was first defined in [168], but our description is based on [170]. Paterson's topological groupoid is described in [302]; this is related to but not identical with Kellendonk's topological groupoid.

Section 9.3

Polycyclic monoids were introduced and their basic properties determined by Nivat and Perrot [286]. They were then rediscovered by Cuntz [54] as part of his construction of a class of simple C*-algebras; this work is described in Renault [350]. Polycyclic monoids are very closely related to context-free languages: on the one hand they are the syntactic monoids of the bracketing languages (see [286]), on the other hand they may be used to obtain an algebraic model of pushdown automata [93], [285], [364]. Although polycyclic monoids are congruence-free they have a rich substructure lattice as Meakin and Sapir have shown [251].

The results on rings which contain a copy of the polycyclic P_2 were sketched out by Girard in [95], and formalised in inverse semigroup-theoretic terms in [136]. These results are equivalent to a technique in K-theory called 'halving projections' [426]. The fact that $B(l^2)$ contains a strongly embedded copy of P_2 was used by J.-Y. Girard in his work on linear logic [95]. Recall that Jacobson [151] described those rings which contain a copy of the bicyclic monoid (see Section 3.4).

The results on embeddings of P_2 and P_∞ in the symmetric inverse monoid $I(\mathbb{N})$ were discovered by Girard [95]. They were used, together with the ring results above, to study linear logic. They were also used by Asperti, Danos, Laneve and Regnier in their work on β-reduction in λ-calculus [13], [58], [59]. Peter Hines and I formalised some of the inverse semigroup-theoretic aspects of this work, and Peter subsequently used them as the basis of his thesis [135]; I am grateful to Peter for a number of discussions on this area.

Section 9.4

McAlister semigroups were effectively introduced in [228], but the connection with inverse semigroups generated by orthogonal elements appears to be new and comes from [193]; my approach was motivated by my work on 1-dimensional tiling semigroups.

McAlister's paper [228] is of considerable interest. The central concept is that of an inverse semigroup separated over a subsemigroup. This is a technique for constructing inverse semigroups from semigroups which are themselves not inverse by means of 'fractions'; the procedure is analogous to the construction of the rationals from the integers. In the case of McAlister semigroups, we have already noted that these consist of fractions with one numerator and two denominators drawn from a free semigroup (see the comments following Theorem 9.4.12). The motivation for this work appears to come from Clifford [47] and Eberhart and Selden [64]. The theory was subsequently developed in [245]. McAlister's review of [64] is itself interesting (MR45#5258).

The solution to the word problem for M_n reduces to Gluskin's solution to the word problem for free inverse semigroups on one generator described in Chapter 6. Thus the semigroups M_n are in some ways very natural generalisations of free inverse semigroups on one generator: the normal forms for the elements are linear structures rather than the branching trees used in the solution of the word problem for free inverse semigroups.

The fact that the free inverse semigroup on one generator embeds into the direct product of the bicyclic monoid with itself was first proved by Scheiblich [365].

Section 9.5

The theory of tiling semigroups was originated by Johannes Kellendonk in a series of papers [168], [169], [170]. The discussion of 1-dimensional tiling semigroups is mainly due to myself, but Theorem 5 was discovered by Johannes.

I am grateful to Johannes Kellendonk for many discussions concerning the material of this chapter.

Chapter 10

Category actions and inverse semigroups

In this chapter, we return to the relationship between inverse semigroups and partial symmetries which we first described in Chapter 1. Our aim is to refine the Wagner–Preston representation theorem (Theorem 1.5.1) in such a way that we obtain a new characterisation of inverse semigroups.

Recall that the Wagner–Preston theorem asserts that every inverse semigroup can be faithfully represented by an inverse semigroup of partial bijections on a set. We show here that each inverse semigroup is isomorphic to the inverse semigroup of *all* partial symmetries (belonging to a well-defined class) of some structure; the structures in question are a class of category actions.

The proof of this result is not deep, but it is a little involved. There are three basic steps:

(1) In Sections 10.1 and 10.2, we show that from every category acting on a set satisfying what we call the orbit condition we may construct an inverse semigroup with zero. We describe this inverse semigroup in two ways: firstly, in a coordinate-free manner as an inverse semigroup of partial symmetries of the action; and secondly, in a coordinatised manner, equipped with a multiplication which generalises the form of the multiplication in polycyclic monoids (Section 9.3).

(2) In Section 10.3, we show that from every inverse semigroup with zero we may construct a category action satisfying the orbit condition. It transpires that the category actions arising in this way satisfy some additional properties and are what we term systems. Thus as a result of the first two steps we know how to construct inverse semigroups with zero from

systems and vice-versa. These constructions are shown to be functorial in Section 10.4.

(3) In Section 10.6, we study what happens when we iterate the constructions. One case is straightforward: if we construct a system from an inverse semigroup S, and then construct the corresponding inverse semigroup, it will be isomorphic to S (Theorem 10.6.1(2)). The other case is more complex. If we construct an inverse semigroup from a system, and then construct the corresponding system, it need not be isomorphic to the one we started with. Instead, it is equivalent to it (Theorem 10.6.2(1)); what we mean by an equivalence of systems is described in Section 10.5.

We conclude the chapter by discussing a number of applications of this theory to polycyclic monoids and their generalisations, inverse monoids, 0-bisimple inverse semigroups, and finally 0-E-unitary semigroups.

10.1 Inverse semigroups from category actions

The definition of a category acting on a set is a natural generalisation of a monoid acting on a set. Before making our definition we need some notation. Let C be a category and X a set. If $\mathbf{p}\colon X \to C_o$ is a function then $C * X$ is defined to be the set

$$C * X = \{(a, x) \in C \times X \colon \mathbf{d}(a) = \mathbf{p}(x)\}.$$

We shall write $\exists a \cdot x$ if $(a, x) \in C * X$.

An *action of C on X on the left* is determined by a function $\mathbf{p}\colon X \to C_o$ and a function $C * X \to X$, denoted by $(a, x) \mapsto a \cdot x$, such that the following axioms hold:

(A1) $\exists \mathbf{p}(x) \cdot x$ and $\mathbf{p}(x) \cdot x = x$ for all $x \in X$.

(A2) If $\exists a \cdot x$ then $\mathbf{p}(a \cdot x) = \mathbf{r}(a)$.

(A3) If $\exists ab$ in C and $\exists (ab) \cdot x$ then $\exists b \cdot x$ and $\exists a \cdot (b \cdot x)$ and $(ab) \cdot x = a \cdot (b \cdot x)$.

The axioms (A1) and (A2) generalise the identity axiom for monoid actions, and (A3) generalises the associativity axiom for monoid actions. When the category C is a monoid then C_o consists of just the identity of the monoid and so the function $\mathbf{p}$ can be omitted. Thus category actions generalise monoid actions.

We also say that X is a *left C-system* and write ${}_C X$ or (C, X) to indicate the fact that X is a set on which C acts on the left.

In order to define partial symmetries of category actions we need to single out a class of substructures; these will be the cyclic subsystems defined below.

Let X be a left C-system. For $x \in X$ put

$$C \cdot x = \{a \cdot x \colon a \in C \text{ and } \exists a \cdot x\}.$$

Observe that $x \in C \cdot x$ by axiom (A1). If $X' \subseteq X$ then define

$$C \cdot X' = \bigcup \{C \cdot x \colon x \in X'\}.$$

A subset $Y \subseteq X$ is said to be *C-invariant* or a *left C-subsystem* if $C \cdot Y \subseteq Y$. If Y is C-invariant then it can be regarded as a left C-system in its own right. Subsets of the form $C \cdot x$ are always C-invariant and so form a special class of left C-subsystems called *cyclic*. Alternatively, $C \cdot x$ can be regarded as the *orbit* of x under the action of C.

Let X be a left C-system. We say that it satisfies the *orbit condition* if $C \cdot x \cap C \cdot y$ non-empty implies that $C \cdot x \cap C \cdot y = C \cdot z$ for some $z \in X$. We shall denote any element z as above by $x \wedge y$; it need not, of course, be unique. We denote by $x * y$ and $y * x$ elements of C chosen so that

$$x \wedge y = (x * y) \cdot y = (y * x) \cdot x.$$

We shall almost exclusively be interested in category actions which satisfy the orbit condition.

We now define morphisms between actions. Let $C * X \to X$ and $D * Y \to Y$ be two actions. A *morphism* from ${}_C X$ to ${}_D Y$ is a pair (F, θ) consisting of a functor $F \colon C \to D$ and a function $\theta \colon X \to Y$ satisfying the following two axioms:

(M1) $\mathbf{p}(\theta(x)) = F(\mathbf{p}(x))$ for all $x \in X$.

(M2) If $\exists a \cdot x$ in $C * X$ then $\theta(a \cdot x) = F(a) \cdot \theta(x)$.

If $C = D$ and F is the identity functor then we may replace the pair (F, θ) by θ and under these circumstances we say that θ is a *C-homomorphism*. A bijective C-homomorphism is called a *C-isomorphism*. Some basic properties of C-isomorphisms are listed below.

Lemma 1 *Let X and Y be two left C-systems, let A and B be C-invariant subsets of X, and let $\theta \colon A \to B$ be a C-isomorphism.*

(1) *If $\phi \colon X \to Y$ is a C-isomorphism, then $\phi^{-1} \colon Y \to X$ is a C-isomorphism.*

(2) *If $A' \subseteq A$ is a C-invariant subset of A then $\theta(A')$ is a C-invariant subset of B.*

(3) *If $B' \subseteq B$ is a C-invariant subset of B then $\theta^{-1}(B')$ is a C-invariant subset of A.*

(4) *Let $C \cdot x \subseteq A$ be a cyclic C-invariant subset of A. Then $\theta(C \cdot x)$ is a cyclic C-invariant subset of B equal to $C \cdot \theta(x)$.*

(5) *Let $C \cdot y \subseteq B$ be a cyclic C-invariant subset of B. Then $\theta^{-1}(C \cdot y)$ is a cyclic C-invariant subset of A equal to $C \cdot \theta^{-1}(y)$.*

Proof (1) We check that the axioms (M1) and (M2) hold.

Axiom (M1) holds: let $y \in Y$ and put $x = \phi^{-1}(y)$. Then

$$\mathbf{p}(\phi^{-1}(y)) = \mathbf{p}(x) = \mathbf{p}(\phi(x)) = \mathbf{p}(y),$$

using the fact that ϕ satisfies axiom (M1).

Axiom (M2) holds: let $y \in Y$ and put $x = \phi^{-1}(y)$. Suppose that $\exists a \cdot y$. By definition, $\mathbf{d}(a) = \mathbf{p}(y)$. Since axiom (M1) holds for ϕ^{-1} we have that $\mathbf{p}(y) = \mathbf{p}(x)$. Thus $\exists a \cdot x$, and so $\exists a \cdot \phi^{-1}(y)$. But ϕ satisfies axiom (M2), and so $\phi(a \cdot x) = a \cdot \phi(x)$. Hence $\phi(a \cdot \phi^{-1}(y)) = a \cdot y$, consequently $\phi^{-1}(a \cdot y) = a \cdot \phi^{-1}(y)$, as required.

(2) Let $y \in \theta(A')$, and suppose that $\exists a \cdot y$ where $a \in C$. Put $\theta(x) = y$. By definition, $\mathbf{d}(a) = \mathbf{p}(y)$. But θ is a C-homomorphism so that $\mathbf{p}(y) = \mathbf{p}(\theta(x)) = \mathbf{p}(x)$. Hence $\mathbf{d}(a) = \mathbf{p}(x)$, and so $\exists a \cdot x$. But A' is C-invariant and $x \in A'$. Hence $a \cdot x \in A'$, and so $\theta(a \cdot x) \in \theta(A')$. But $\theta(a \cdot x) = a \cdot \theta(x) = a \cdot y$. Hence $a \cdot y \in \theta(A')$. Thus $\theta(A')$ is a C-invariant subset of B.

(3) Immediate from (1) and (2).

(4) By (2), $\theta(C \cdot x)$ is a C-invariant subset of B. Thus it only remains to show that $\theta(C \cdot x)$ is cyclic. We claim that $\theta(C \cdot x) = C \cdot \theta(x)$. Clearly, $\theta(C \cdot x) \subseteq C \cdot \theta(x)$. Let $a \cdot \theta(x) \in C \cdot \theta(x)$. Then $\mathbf{d}(a) = \mathbf{p}(\theta(x))$. By axiom (M1), $\mathbf{p}(\theta(x)) = \mathbf{p}(x)$. Thus $\exists a \cdot x$. But $a \cdot x \in C \cdot x$ and $\theta(a \cdot x) = a \cdot \theta(x)$, and so $C \cdot \theta(x) \subseteq \theta(C \cdot x)$. Hence $\theta(C \cdot x) = C \cdot \theta(x)$.

(5) Immediate from (1) and (4). ■

Let X be a left C-system. Denote by $I({}_C X)$ the set of all C-isomorphisms between C-invariant subsets of X.

Proposition 2 *$I({}_C X)$ is an inverse subsemigroup of $I(X)$, the symmetric inverse semigroup on X.*

Proof The set $I({}_C X)$ is closed under inverses by Lemma 1. Let $\theta, \phi \in I({}_C X)$. Then $\operatorname{dom}\theta$ and $\operatorname{im}\phi$ are both C-invariant, and so $A = \operatorname{dom}\theta \cap \operatorname{im}\phi$ is C-invariant. By Lemma 1, $\phi^{-1}(A)$ is C-invariant. The functions

$$(\varphi \mid \phi^{-1}(A)) \colon \phi^{-1}(A) \to A \text{ and } (\theta \mid A) \colon A \to \theta(A)$$

are C-isomorphisms and so

$$\theta\phi = (\theta \mid A)(\phi \mid \phi^{-1}(A))$$

is a C-isomorphism. ∎

We now come to our most important definition. Denote by $J(_CX)$ or $J(C, X)$ the subset of $I(_CX)$ consisting of those C-isomorphisms between cyclic C-subsystems of X together with the empty map.

Theorem 3 *Let X be a left C-system. Then $J(_CX)$ is an inverse subsemigroup of $I(_CX)$ if, and only if, $_CX$ satisfies the orbit condition.*

Proof We proved in Proposition 2 that $I(_CX)$ is an inverse semigroup. Suppose that $J(_CX)$ is an inverse subsemigroup of $I(_CX)$. The non-zero idempotents of $J(_CX)$ are the identity functions on the cyclic C-subsystems. The orbit condition now follows from the fact that the product of two idempotents in $J(_CX)$ is the identity function on the intersection of their domains.

Conversely, suppose that $_CX$ satisfies the orbit condition. Let $\theta: C\cdot x \to C\cdot y$ be a C-homomorphism, and let $C \cdot u \subseteq C \cdot x$. From Lemma 1, we have that $\theta(C\cdot u) = C\cdot\theta(u)$. The proof that $J(_CX)$ is an inverse subsemigroup of $I(_CX)$ is now straightforward. ∎

We have succeded in associating an inverse semigroup with zero $J(_CX)$ to every left C-system X satisfying the orbit condition. We shall ultimately prove that every inverse semigroup with zero is isomorphic to such an inverse semigroup.

Inverse semigroups without zero can be constructed when a strengthened form of the orbit condition holds. Let X be a left C-system. Then $_CX$ satisfies the *strong orbit condition* if for all $x, y \in X$ there exists $z \in X$

$$C \cdot x \cap C \cdot y = C \cdot z.$$

Put $J^*(_CX) = J(_CX) \setminus \{0\}$. The proof of the following is straightforward.

Theorem 4 *Let the category C act on the set X on the left and satisfy the strong orbit condition. Then the product of any two non-zero elements of $J(_CX)$ is nonzero. Consequently, $J^*(_CX)$ is an inverse semigroup.* ∎

10.2 A coordinatisation theorem

In this section, we shall obtain an explicit description of the multiplication in $J(_CX)$ by means of a coordinatisation of its elements. The crucial ingredient in this alternative description of the inverse semigroup $J({}_C X)$ is an equivalence relation $\mathcal{R}^*$ which is defined on the set X and is determined by the action of C.

Let X be a left C-system. We define a relation $\mathcal{R}^*$ on X as follows: $(x, y) \in \mathcal{R}^*$ if, and only if, $\mathbf{p}(x) = \mathbf{p}(y)$ and for all $a, b \in C$ such that $a \cdot x$ and $b \cdot x$ are defined, we have that

$$a \cdot x = b \cdot x \Leftrightarrow a \cdot y = b \cdot y;$$

both sides of these two equations exist since from $\mathbf{p}(x) = \mathbf{p}(y)$ we have that $\exists a \cdot x \Leftrightarrow \exists a \cdot y$ and $\exists b \cdot x \Leftrightarrow \exists b \cdot y$.

It is clear that $\mathcal{R}^*$ is an equivalence relation on X, and that if $(x, y) \in \mathcal{R}^*$ and $\exists c \cdot x$ and $\exists c \cdot y$ then $(c \cdot x, c \cdot y) \in \mathcal{R}^*$.

Lemma 1 *Let X be a left C-system and $x, y \in X$. Then the following are equivalent:*

(1) $(x, y) \in \mathcal{R}^*$.

(2) *There exists $\theta \in I({}_C X)$ such that $\theta(x) = y$.*

Proof (1) $\Rightarrow$ (2). Define a function θ: $C \cdot x \to C \cdot y$ by $\theta(a \cdot x) = a \cdot y$. First, θ is well-defined, for if $a \cdot x = a' \cdot x$ then $a \cdot y = a' \cdot y$ since $(x, y) \in \mathcal{R}^*$. Clearly, $\theta(x) = y$. Next, θ is injective, for suppose that $\theta(a \cdot x) = \theta(b \cdot x)$. Then $a \cdot y = b \cdot y$, and so $a \cdot x = b \cdot x$, since $(x, y) \in \mathcal{R}^*$. To see that θ is surjective, let $a \cdot y \in C \cdot y$. Then $\mathbf{d}(a) = \mathbf{p}(y) = \mathbf{p}(x)$ since $(x, y) \in \mathcal{R}^*$. Thus $\exists a \cdot x$ and clearly $\theta(a \cdot x) = a \cdot y$. We finish off by showing that θ is a C-homomorphism.

Axiom (M1) holds: let $x' \in C \cdot x$, where $x' = a \cdot x$. Then

$$\mathbf{p}(\theta(x')) = \mathbf{p}(\theta(a \cdot x)) = \mathbf{p}(a \cdot y) = \mathbf{r}(a).$$

But $\mathbf{r}(a) = \mathbf{p}(a \cdot x) = \mathbf{p}(x')$. Hence $\mathbf{p}(\theta(x')) = \mathbf{p}(x')$.

Axiom (M2) holds: let $x' = a \cdot x$ and $a' \in C$ such that $\exists a' \cdot x'$. Then

$$\begin{aligned} \theta(a' \cdot x') &= \theta(a' \cdot (ax)) \\ &= \theta((a'a) \cdot x) \text{ by (A3)} \\ &= (a'a) \cdot y = a' \cdot (a \cdot y) = a' \cdot \theta(a \cdot x) = a' \cdot \theta(x'). \end{aligned}$$

(2) $\Rightarrow$ (1). Let $\theta \in I({}_C X)$ be such that $\theta(x) = y$. Then $\mathbf{p}(x) = \mathbf{p}(y)$ by axiom (M1). Suppose that $a \cdot x = b \cdot x$. Then $a \cdot x \in \operatorname{dom} \theta$ since $\operatorname{dom} \theta$ is a left C-system. Hence $\theta(a \cdot x) = \theta(b \cdot x)$. Now θ is a C-homomorphism and so $a \cdot \theta(x) = b \cdot \theta(x)$. But $\theta(x) = y$ and so $a \cdot y = b \cdot y$. Conversely, $\theta^{-1}(y) = x$ and $\theta^{-1} \in I({}_C X)$ by Lemma 10.1.1. Thus $a \cdot y = b \cdot y$ implies $a \cdot x = b \cdot x$. Hence $(x, y) \in \mathcal{R}^*$. ∎

Before proving our next result we need some notation. Let X be a left C-system, and let x and y be a pair of elements such that $\mathbf{p}(x) = \mathbf{p}(y)$. Then if $u \in C$ is such that $\exists u \cdot x$ (and so $\exists u \cdot y$) then we write $(u \cdot x, u \cdot y) = u \cdot (x, y)$.

Lemma 2 *Let X be a left C-system satisfying the orbit condition. On the set of ordered pairs $\mathcal{R}^*$ define a relation $\sim$ by*

$$(x, y) \sim (x', y') \Leftrightarrow (x, y) = u \cdot (x', y') \text{ and } (x', y') = v \cdot (x, y)$$

for some $u, v \in C$. Then $\sim$ is an equivalence relation.

Proof This is almost immediate; the only case that requires any comment is reflexivity, and this follows from the fact that if $(x, y) \in \mathcal{R}^*$ then $\mathbf{p}(x) = \mathbf{p}(y)$ and so $(x, y) = \mathbf{p}(x) \cdot (x, y)$ by axiom (A1). ■

Denote by $[x, y]$ the $\sim$-equivalence class containing the pair (x, y). We can now obtain an explicit description of the multiplication in $J({}_C X)$.

Theorem 3 *Let X be a left C-system satisfying the orbit condition. Let S be the set of $\sim$-equivalence classes together with a new symbol 0. Define a product on S as follows:*

$$[x, y] \otimes [w, z] = \begin{cases} [(w * y) \cdot x, (y * w) \cdot z] & \text{if } C \cdot y \cap C \cdot w \neq \emptyset \\ 0 & \text{else} \end{cases}$$

and all other products equal to 0. Then $(S, \otimes)$ is a semigroup isomorphic to $J({}_C X)$. In the semigroup $(S, \otimes)$ we have that

$$[x, y]^{-1} = [y, x], \ [x, y]^{-1} \otimes [x, y] = [y, y], \text{ and } [x, y] \otimes [x, y]^{-1} = [x, x],$$

the idempotents are all the elements of the form $[x, x]$ where $x \in X$ and the natural partial order is given by

$$[x, y] \leq [w, z] \Leftrightarrow (x, y) = u \cdot (w, z) \text{ for some } u \in C.$$

Proof Since the proof is rather long, we split it into four parts.

1. *$(w * y) \cdot x$ is $\mathcal{R}^*$-related to $(y * w) \cdot z$.*

By definition $(w * y) \cdot y = (y * w) \cdot w$, so that $\mathbf{p}((w * y) \cdot y) = \mathbf{p}((y * w) \cdot w)$. Thus $\mathbf{r}(w * y) = \mathbf{r}(y * w)$ by axiom (A2). But $\mathbf{p}((w * y) \cdot x) = \mathbf{r}(w * y)$ and $\mathbf{p}((y * w) \cdot z) = \mathbf{r}(y * w)$ by axiom (A2). Hence

$$\mathbf{p}((w * y) \cdot x) = \mathbf{p}((y * w) \cdot z).$$

Now suppose that

$$a \cdot [(w * y) \cdot x] = b \cdot [(w * y) \cdot x].$$

Then $(a(w*y)) \cdot x = (b(w*y)) \cdot x$ by axiom (A3). Thus $(a(w*y)) \cdot y = (b(w*y)) \cdot y$ since $(x, y) \in \mathcal{R}^*$, and so

$$a \cdot [(w * y) \cdot y] = b \cdot [(w * y) \cdot y]$$

by axiom (A3). By definition $(w * y) \cdot y = (y * w) \cdot w$. Thus

$$a \cdot [(y * w) \cdot w] = b \cdot [(y * w) \cdot w].$$

But $(a(y*w)) \cdot w = (b(y*w)) \cdot w$ by axiom (A3). Hence $(a(y*w)) \cdot z = (b(y*w)) \cdot z$ since $(w, z) \in \mathcal{R}^*$. We thus have

$$a \cdot [(y * w) \cdot z] = b \cdot [(y * w) \cdot z]$$

by axiom (A3). We may similarly show that

$$a \cdot [(y * w) \cdot z] = b \cdot [(y * w) \cdot z]$$

implies

$$a \cdot [(w * y) \cdot x] = b \cdot [(w * y) \cdot x].$$

Hence $(w * y) \cdot x$ is $\mathcal{R}^*$-related to $(y * w) \cdot z$.

2. $\otimes$ *is a well-defined binary operation.*

Let $[x, y] = [x', y']$ and $[w, z] = [w', z']$. We show that

$$[x, y] \otimes [w, z] = [x', y'] \otimes [w', z'].$$

From the definition there are elements $u, v, a, b \in C$ such that

$$(x, y) = u \cdot (x', y') \text{ and } (x', y') = v \cdot (x, y)$$

and

$$(w, z) = a \cdot (w', z') \text{ and } (w', z') = b \cdot (w, z).$$

Now $y = u \cdot y'$ and $y' = v \cdot y$ imply that $C \cdot y = C \cdot y'$. Similarly $C \cdot w = C \cdot w'$. Thus

$$C \cdot y \cap C \cdot w \neq \emptyset \Leftrightarrow C \cdot y' \cap C \cdot w' \neq \emptyset.$$

Hence

$$[x, y] \otimes [w, z] = 0 \Leftrightarrow [x', y'] \otimes [w', z'] = 0.$$

We shall consider the case where $C \cdot y \cap C \cdot w \neq \emptyset$. Let

$$C \cdot y \cap C \cdot w = C \cdot (y \wedge w) \text{ and } C \cdot y' \cap C \cdot w' = C \cdot (y' \wedge w')$$

for some $y \wedge w$ and $y' \wedge w'$. Using the notation introduced earlier, we have that

$$y \wedge w = (y * w) \cdot w = (w * y) \cdot y$$

and

$$y' \wedge w' = (y' * w') \cdot w' = (w' * y') \cdot y'.$$

From the definition

$$[x, y] \otimes [w, z] = [(w * y) \cdot x, (y * w) \cdot z]$$

and

$$[x', y'] \otimes [w', z'] = [(w' * y') \cdot x', (y' * w') \cdot z'].$$

Since $C \cdot (y \wedge w) = C \cdot (y' \wedge w')$ there exist elements $c, d \in C$ such that

$$c \cdot (y \wedge w) = y' \wedge w' \text{ and } d \cdot (y' \wedge w') = y \wedge w.$$

Now $y \wedge w = (y * w) \cdot w$ and so $c \cdot (y \wedge w) = (c(y * w)) \cdot w$. Thus $y' \wedge w' = (c(y * w)) \cdot w$. But $w = aw'$ and so $y' \wedge w' = (c(y * w)a) \cdot w'$. Also $y' \wedge w' = (y' * w') \cdot w'$. Hence

$$(y' * w') \cdot w' = (c(y * w)a) \cdot w'.$$

Now $(w', z') \in \mathcal{R}^*$ and so

$$(y' * w') \cdot z' = (c(y * w)a) \cdot z'.$$

We can similarly show that

$$(w' * y') \cdot x' = (c(w * y)u) \cdot x'.$$

But $a \cdot z' = z$ and $u \cdot x' = x$ and so

$$(y' * w') \cdot z' = c \cdot [(y * w) \cdot z] \text{ and } (w' * y') \cdot x' = c \cdot [(w * y) \cdot x].$$

Hence

$$((w' * y') \cdot x', (y' * w') \cdot z') = c \cdot ((w * y) \cdot x, (y * w) \cdot z).$$

We may similarly prove that

$$d \cdot ((w' * y') \cdot x', (y' * w') \cdot z') = ((w * y) \cdot x, (y * w) \cdot z).$$

Hence

$$[(w * y) \cdot x, (y * w) \cdot z] = [(w' * y') \cdot x', (y' * w') \cdot z'].$$

3. *For each $(x,y) \in \mathcal{R}^*$ define a function $\theta_{(x,y)} : C \cdot y \to C \cdot x$ by $\theta_{(x,y)}(a \cdot y) = a \cdot x$. From the proof of Lemma 1 we have that $\theta_{(x,y)} \in J({}_C X)$. We claim that*

$$\theta_{(x,y)} = \theta_{(x',y')} \Leftrightarrow (x,y) \sim (x',y').$$

Suppose that $\theta_{(x,y)} = \theta_{(x',y')}$. Then

$$C \cdot y = C \cdot y', \; C \cdot x = C \cdot x' \text{ and } (x,y),(x',y') \in \mathcal{R}^*.$$

Thus, in particular, $y = a \cdot y'$ and $y' = b \cdot y$ for some $a, b \in C$. Now

$$\theta_{(x,y)}(y) = \theta_{(x,y)}(\mathbf{p}(y) \cdot y) = x$$

and

$$\theta_{(x',y')}(y) = \theta_{(x',y')}(a \cdot y') = a \cdot x'.$$

But $\theta_{(x,y)}(y) = \theta_{(x',y')}(y)$. Thus $x = a \cdot x'$. Similarly, $x' = b \cdot x$. Thus $(x,y) = a \cdot (x',y')$ and $(x',y') = b \cdot (x,y)$, and so $(x,y) \sim (x',y')$.

Now suppose that $(x,y) \sim (x',y')$. We show that $\theta_{(x,y)} = \theta_{(x',y')}$. By assumption,

$$(x,y) = a \cdot (x',y') \text{ and } (x',y') = b \cdot (x,y)$$

for some $a, b, \in C$. Clearly, $C \cdot x = C \cdot x'$ and $C \cdot y = C \cdot y'$. We show that the functions $\theta_{(x,y)}$ and $\theta_{(x',y')}$ take the same values. Let $d \cdot y = d' \cdot y' \in C \cdot y = C \cdot y'$. Then $\theta_{(x,y)}(d \cdot y) = d \cdot x$ and $\theta_{(x',y')}(d' \cdot y') = d' \cdot x'$. Now $d \cdot y = d' \cdot y'$ implies $d \cdot y = (d'b) \cdot y$. But $(x,y) \in \mathcal{R}^*$ so that $d \cdot x = (d'b) \cdot x$. Hence $d \cdot x = d' \cdot x'$, as required.

4. *Define a function Θ: $S \to J({}_C X)$ as follows: $\Theta(0)$ is just the empty function on X and $\Theta([x,y]) = \theta_{(x,y)}$. Then θ is an isomorphism.*

Our calculations in (3) above show that the definition of $\Theta([x,y])$ is independent of the choice of representative of $[x,y]$. It also follows from our calculations in (3) that Θ is an injective function. To show that Θ is surjective let $\phi \in J({}_C X)$ where ϕ: $C \cdot y \to C \cdot x$. Clearly, $C \cdot \phi(y) = C \cdot x$, and so by Lemma 2, we have that $(y, \phi(y)) \in \mathcal{R}^*$. It is now clear that $\phi = \theta_{(\phi(y),y)}$ and so $\Theta([\phi(y), y]) = \phi$.

To show that Θ is a homomorphism we compute the product $\theta_{(x,y)}\theta_{(w,z)}$. Suppose that $C \cdot y \cap C \cdot w \neq \emptyset$. Then $C \cdot y \cap C \cdot w = C \cdot (y \wedge w)$, and with the notation introduced earlier we have

$$y \wedge w = (y * w) \cdot w = (w * y) \cdot y.$$

Now

$$\theta^{-1}_{(w,z)}(C \cdot (y \wedge w)) = \theta^{-1}_{(w,z)}(C \cdot (y * w) \cdot w) = C \cdot (y * w) \cdot z.$$

Also

$$\theta_{(x,y)}(C \cdot (y \wedge w)) = \theta_{(x,y)}(C \cdot (w * y) \cdot y) = C \cdot (w * y) \cdot x.$$

Thus $\theta_{(x,y)}\theta_{(w,z)}$ has domain $C \cdot (y * w) \cdot z$ and image $C \cdot (w * y) \cdot x$. We now calculate the effect of this composite function:

$$\begin{aligned}
(\theta_{(x,y)}\theta_{(w,z)})((a(y * w)) \cdot z) &= \theta_{(x,y)}(\theta_{(w,z)}((a(y * w)) \cdot z)) \\
&= \theta_{(x,y)}((a(y * w)) \cdot w) \\
&= \theta_{(x,y)}(a \cdot (y \wedge w)) \\
&= \theta_{(x,y)}((a(w * y)) \cdot y) \\
&= (a(w * y)) \cdot x \\
&= \theta_{((w*y)\cdot x,(y*w)\cdot z)}((a(y * w)) \cdot z).
\end{aligned}$$

We have shown that if $C \cdot y \cap C \cdot w \neq \emptyset$ then

$$\theta_{(x,y)}\theta_{(w,z)} = \theta_{((w*y)\cdot x,(y*w)\cdot z)}.$$

On the other hand, if $C \cdot y \cap C \cdot w = \emptyset$ then $\theta_{(x,y)}\theta_{(w,z)}$ is the empty function. It is now immediate that Θ is a homomorphism and so an isomorphism of semigroups. ■

10.3 Category actions from inverse semigroups

In Sections 10.1 and 10.2, we provided two equivalent ways of constructing an inverse semigroup with zero from a left C-system satisfying the orbit condition. In this section, we shall go in the opposite direction and show how to construct a category action from an inverse semigroup with zero. The first step is to construct an appropriate category from an inverse semigroup.

Let S be an inverse semigroup with zero. Put

$$C'(S) = \{(s, e) \in S \times E(S) \colon s^{-1}s \leq e\}.$$

Define $\mathbf{d}(s, e) = (e, e)$, $\mathbf{r}(s, e) = (ss^{-1}, ss^{-1})$ and a partial product

$$(s, e) \cdot (t, f) = \begin{cases} (st, f) \text{ if } e = tt^{-1} \\ \text{undefined else} \end{cases}$$

Put $Z = \{(0, e) \colon e \in E(S)\}$ and $C(S) = C'(S) \setminus Z$.

Proposition 1 *Let S be an inverse semigroup with zero.*

(1) *$(C'(S), \cdot)$ is a right cancellative category.*

(2) *The isomorphisms in $C'(S)$ are the elements of the form $(s, s^{-1}s)$.*

(3) *$C(S)$ is a full subcategory of $C'(S)$.*

Proof (1) Observe that $\exists (s,e) \cdot (t,f)$ precisely when $\mathbf{d}(s,e) = \mathbf{r}(t,f)$. It is now a simple matter to show that $(C'(S), \cdot)$ is a category with identities $\{(e,e) \colon e \in E(S)\}$. We now show that $(C'(S), \cdot)$ is right cancellative. Suppose that

$$(s,e) \cdot (t,f) = (u,i) \cdot (t,f).$$

Then $(st, f) = (ut, f)$ so that $st = ut$. Thus $stt^{-1} = utt^{-1}$. But $e = tt^{-1} = i$ and $s^{-1}s, u^{-1}u \leq e$. Hence $s = u$ and $e = i$, and so $(s,e) = (u,i)$.
(2) Observe first that the products

$$(s, s^{-1}s) \cdot (s^{-1}, ss^{-1}) \text{ and } (s^{-1}, ss^{-1}) \cdot (s, s^{-1}s)$$

are defined, and that

$$(s, s^{-1}s) \cdot (s^{-1}, ss^{-1}) = (ss^{-1}, ss^{-1}) = \mathbf{r}(s, s^{-1}s),$$

and

$$(s^{-1}, ss^{-1}) \cdot (s, s^{-1}s) = (s^{-1}s, s^{-1}s) = \mathbf{d}(s, s^{-1}s).$$

Thus $(s, s^{-1}s)$ is invertible with inverse (s^{-1}, ss^{-1}). Conversely, suppose that (s,e) is an invertible element. Then there exists an element (t,f) such that $(s,e) \cdot (t,f)$ and $(t,f) \cdot (s,e)$ are defined and

$$(s,e) \cdot (t,f) = \mathbf{r}(s,e) \text{ and } (t,f) \cdot (s,e) = \mathbf{d}(s,e).$$

Thus

$$(st, f) = (ss^{-1}, ss^{-1}) \text{ and } (ts, e) = (e, e).$$

Then $e = tt^{-1}$, $f = ss^{-1}$, $st = ss^{-1}$ and $ts = e$. From $st = ss^{-1}$ we obtain $sts = s$. From $ts = e$ we obtain $tst = et = tt^{-1}t = t$. Thus $t = s^{-1}$. Hence $e = s^{-1}s$.
(3) Consider the full subcategory of $C'(S)$ determined by all the identities of $C'(S)_o$ except $(0,0)$. We have that (s,e) belongs to this subcategory if, and only if, $\mathbf{d}(s,e), \mathbf{r}(s,e) \neq (0,0)$. It is easy to check that (s,e) belongs to this subcategory if, and only if, $(s,e) \notin Z$. ■

The category $C(S)$ is one of the key ingredients in our construction. Observe that if $S = \{0\}$ then $C(S)$ is the empty category.

We say that a left C-system X satisfies the *right cancellation condition* if whenever $c \cdot x$ and $d \cdot x$ exist and $c \cdot x = d \cdot x$ then $c = d$. We shall call a pair (C, X) a *system* if the following axioms hold:

(S1) C is a right cancellative category acting on the set X on the left.

(S2) The orbit condition holds.

(S3) The function $\mathbf{p}$: $X \to C_O$ is surjective.

(S4) The right cancellation condition holds.

Let S be an inverse semigroup with zero. Put $X_S = S \setminus \{0\}$. Define $\mathbf{p}$: $X_S \to C(S)_o$ by $\mathbf{p}(x) = (xx^{-1}, xx^{-1})$ and define a function $C(S) * X_S \to X_S$ by $(s,e) \cdot x = sx$ if $\mathbf{d}(s,e) = \mathbf{p}(x)$.

Theorem 2 *For every inverse semigroup with zero S the pair $(C(S), X_S)$ is a system.*

Proof Axiom (S1) holds: observe first that the function $C(S) * X_S \to X_S$ is well-defined. For suppose $\exists (s,e) \cdot x$ where $s, e, x \neq 0$ and $(s,e) \cdot x = sx = 0$. Then $sxx^{-1} = 0$. But $e = xx^{-1}$ and $s^{-1}s \leq e$, and so $s = 0$, which contradicts our choice of s. Thus if $\exists (s,e) \cdot x$ then $(s,e) \cdot x \in X_S$. Next we show that X_S is a left $C(S)$-system by checking that the axioms (A1), (A2) and (A3) hold.

Axiom (A1) holds: by definition, $\mathbf{p}(x) = (xx^{-1}, xx^{-1})$ and $\mathbf{d}(\mathbf{p}(x)) = (xx^{-1}, xx^{-1})$. Thus $\exists \mathbf{p}(x) \cdot x$. By definition $\mathbf{p}(x) \cdot x = xx^{-1}x = x$.

Axiom (A2) holds: suppose that $\exists (s,e) \cdot x$. Then by definition $(s,e) \cdot x = sx$. Now

$$\mathbf{p}(sx) = ((sx)(sx)^{-1}, (sx)(sx)^{-1}) = (sxx^{-1}s^{-1}, sxx^{-1}s^{-1}).$$

But $\mathbf{d}(s,e) = \mathbf{p}(x)$ and so $e = xx^{-1}$, also $s^{-1}s \leq e$ and so $ses^{-1} = ss^{-1}$. Thus

$$\mathbf{p}(sx) = (ss^{-1}, ss^{-1}) = \mathbf{r}(s,e).$$

Axiom (A3) holds: suppose that $\exists (s,e)(t,f)$ and $\exists ((s,e)(t,f)) \cdot x$. Then

$$(s,e)(t,f) = (st, f) \text{ and } ((s,e)(t,f)) \cdot x = stx.$$

From the definitions

$$e = tt^{-1}, s^{-1}s \leq e, t^{-1}t \leq f \text{ and } f = xx^{-1}.$$

$\exists (t,f) \cdot x$ since $f = xx^{-1}$, and by definition $(t,f) \cdot x = tx$. Now $\mathbf{d}(s,e) = (e,e)$ and

$$\begin{aligned} \mathbf{p}(tx) &= ((tx)(tx)^{-1}, (tx)(tx)^{-1}) = (txx^{-1}t^{-1}, txx^{-1}t^{-1}) \\ &= (tft^{-1}, tft^{-1}) = (tt^{-1}, tt^{-1}) = (e,e). \end{aligned}$$

Thus $\exists (s,e) \cdot (tx)$ and $(s,e) \cdot (tx) = stx$. By Proposition 1, $C(S)$ is a right cancellative category.

Axiom (S2) holds: we begin by describing the cyclic $C(S)$-subsystems of X_S. By definition,

$$\begin{aligned} C(S) \cdot x &= \{(s,e) \cdot x \colon (s,e) \in C(S),\ \exists (s,e) \cdot x\} \\ &= \{sx \colon s^{-1}s \leq e = xx^{-1} \text{ where } s, e, x \neq 0\}. \end{aligned}$$

Clearly, $C(S) \cdot x \subseteq Sx \backslash \{0\}$. Let $sx \in Sx\backslash\{0\}$ and let $e = xx^{-1}$. Then

$$sx = (sxx^{-1})x = (se)x.$$

Consider the ordered pair (se, e). Clearly, $(se)^{-1}se \leq e$. Thus $(se, e) \in C(S)$. Furthermore, $(se, e) \cdot x = sx$. Thus $C(S) \cdot x = Sx\backslash\{0\}$. Suppose now that

$$C(S) \cdot x \cap C(S) \cdot y \neq \emptyset.$$

Then there exists $a \in Sx \cap Sy$ with $a \neq 0$. Now $a = sx = ty$ for some $s, t \in S$, and $ax^{-1}x = a$ and $ay^{-1}y = a$. Thus $ax^{-1}xy^{-1}y = a$. It follows that $x^{-1}xy^{-1}y \neq 0$. Clearly, $Sx \cap Sy = Sx^{-1}xy^{-1}y$, and so

$$C(S) \cdot x \cap C(S) \cdot y = C(S) \cdot (x^{-1}xy^{-1}y).$$

Axiom (S3) holds: if (e, e) is any identity of $C(S)$ then $e \in S \setminus \{0\}$ and $\mathbf{p}(e) = (e, e)$.

Axiom (S4) holds: suppose that

$$(s, e) \cdot x = (t, f) \cdot x$$

Then $sx = tx$, and $e = xx^{-1} = f$. Thus $sxx^{-1} = txx^{-1}$ and so $se = te$. It follows that $s = t$ since $s^{-1}s \leq e$ and $t^{-1}t \leq f = e$. Hence $(s, e) = (t, f)$. ■

The description of the cyclic $C(S)$-systems contained in the proof of the above theorem makes the proof of the following result immediate.

Theorem 3 *Let S be an inverse semigroup without zero. Then $(C(S), X_S)$ is a system satisfying the strong orbit condition.* ■

Systems are much easier to handle than arbitrary category actions, as the following result indicates.

Lemma 4 *Let (C, X) be a system.*

(1) *For all $x, y \in X$ we have that*

$$(x, y) \in \mathcal{R}^* \Leftrightarrow \mathbf{p}(x) = \mathbf{p}(y).$$

(2) *In the construction of Theorem 10.2.3, we have that*

$$(x, y) \sim (x', y') \Leftrightarrow (x, y) = u \cdot (x', y')$$

where u is an isomorphism in C.

Proof (1) ($\Rightarrow$) Immediate from the definition of $\mathcal{R}^*$.
($\Leftarrow$) Suppose that $\mathbf{p}(x) = \mathbf{p}(y)$, and $a \cdot x = b \cdot x$. Then by the cancellation condition $a = b$. Thus $a \cdot y = b \cdot y$. The converse is similar. Hence $(x, y) \in \mathcal{R}^*$.
(2)($\Leftarrow$) Immediate.
($\Rightarrow$) Suppose that $(x, y) \sim (x', y')$. Then from the definition there are elements u and v of C such that

$$x = u \cdot x', \; y = u \cdot y', \; x' = v \cdot x \text{ and } y' = v \cdot y.$$

Now $x = u \cdot x' = (uv) \cdot x$. Thus by the cancellation condition $\mathbf{r}(x) = uv$. Similarly, $\mathbf{r}(x') = vu$. Thus u and v are mutually inverse isomorphisms. ∎

10.4 Functors between systems and semigroups

We have seen how to construct inverse semigroups from systems and systems from inverse semigroups. We now place the results of the previous sections in their proper categorical setting by taking account of the morphisms between systems, to be defined below, and the appropriate morphisms between inverse semigroups with zero.

Let (C, X) and (D, Y) be systems and let (F, θ) be a morphism from (C, X) to (D, Y). We say that (F, θ) is a *system morphism* if the axiom (M3) below also holds:

(M3) For all $x, y \in X$, if $C \cdot x \cap C \cdot y = \emptyset$ then $D \cdot \theta(x) \cap D \cdot \theta(y) = \emptyset$; whereas if there exists $z \in X$ such that $C \cdot x \cap C \cdot y = C \cdot z$ then $D \cdot \theta(x) \cap D \cdot \theta(y) = D \cdot \theta(z)$.

Denote the category of systems and system morphisms by **Sys**. The category **Inv** consists of inverse semigroups with zero and those semigroup homomorphisms θ: $S \to T$ such that $\theta^{-1}(0) = \{0\}$.

Define a function $\mathbf{J}$ from the category **Sys** to the category **Inv** as follows: if (C, X) is a system then $\mathbf{J}(C, X)$ is the inverse semigroup constructed in Theorem 10.2.3. If (F, θ): $(C, X) \to (D, Y)$ is a morphism of systems then $\mathbf{J}(F, \theta)$: $\mathbf{J}(C, X) \to \mathbf{J}(D, Y)$ is defined by $\mathbf{J}(F, \theta)([x, y]) = [\theta(x), \theta(y)]$ and $\mathbf{J}(F, \theta)(0) = 0$.

Theorem 1 *With the above notation* $\mathbf{J}$: **Sys** $\to$ **Inv** *is a functor.*

Proof We begin by checking that $\mathbf{J}(F,\theta)$ is a well-defined function. Let $[x,y] \in \mathbf{J}(C,X)$. Then $(x,y) \in \mathcal{R}^*$. But by Lemma 10.3.4(1), this is equivalent to $\mathbf{p}(x) = \mathbf{p}(y)$. But $F(\mathbf{p}(x)) = F(\mathbf{p}(y))$ implies that $\mathbf{p}(\theta(x)) = \mathbf{p}(\theta(y))$ by axiom (M1). Thus by Lemma 10.3.4(1), we have that $(\theta(x),\theta(y)) \in \mathcal{R}^*$. It follows that $[\theta(x),\theta(y)] \in \mathbf{J}(D,Y)$. Now suppose that $[x,y] = [x',y']$. Then $(x,y) = u \cdot (x',y')$ for some isomorphism u in C by Lemma 10.3.4(2). Clearly, $(\theta(x),\theta(y)) = \theta(u) \cdot (\theta x',\theta y')$, and $\theta(u)$ is an isomorphism in D. Thus by Lemma 10.3.4(2), we have that $[\theta(x),\theta(y)] = [\theta(x'),\theta(y')]$.

We now show that $\mathbf{J}(F,\theta)$ is a homomorphism in **Inv**. Let $[x,y],[w,z] \in \mathbf{J}(C,X)$. Suppose first that the product of $[x,y]$ and $[w,z]$ is zero. Then by definition of the product, we have that $C \cdot y \cap C \cdot w$ is empty. But (F,θ) is a system morphism and so $D \cdot \theta(y) \cap D \cdot \theta(w)$ is also empty. It follows that $[\theta(x),\theta(y)] \otimes [\theta(w),\theta(z)]$ is zero. Suppose now that $[x,y] \otimes [w,z]$ is non-zero. Then by assumption $C \cdot y \cap C \cdot w = C \cdot (y \wedge w)$, and so

$$[x,y] \otimes [w,z] = [(w * y) \cdot x, (y * w) \cdot z].$$

Again (F,θ) is a system morphism and so $D \cdot \theta(y) \cap D \cdot \theta(w) = D \cdot \theta(y \wedge w)$. Also $F(w * y) \cdot \theta(y) = \theta(y \wedge w)$ and $F(y * w) \cdot \theta(w) = \theta(y \wedge w)$. It follows that we can take $\theta(w) * \theta(y)$ to be $F(w * y)$ and $\theta(y) * \theta(w)$ to be $F(y * w)$. It is now straightforward to show that $\mathbf{J}(F,\theta)$ is a homomorphism, and that $\mathbf{J}$ is a functor. ∎

Define a function $\mathbf{C}$ from the category **Inv** to the category **Sys** as follows: if S is an inverse semigroup then $\mathbf{C}(S) = (C(S), X_S)$, as defined in Theorem 10.3.2. If θ: $S \to T$ is a homomorphism in **Inv** then $\mathbf{C}(\theta)$: $(C(S),X_S) \to (C(T),X_T)$ is defined to be $\mathbf{C}(\theta) = (F_\theta,\theta)$ where F_θ: $C(S) \to C(T)$ is defined by $F_\theta(s,e) = (\theta(s),\theta(e))$ and θ: $X_S \to X_T$ is the restriction of θ to $S \setminus \{0\}$.

Theorem 2 *With the above notation* $\mathbf{C}$: **Inv** $\to$ **Sys** *is a functor.*

Proof We show that (F_θ,θ) is a system morphism.

Axiom (M1) holds: let $x \in X_S$. Then $\mathbf{p}(\theta(x)) = (\theta(x)\theta(x)^{-1},\theta(x)\theta(x)^{-1})$. Whereas

$$F_\theta(\mathbf{p}(x)) = F_\theta(xx^{-1},xx^{-1}) = (\theta(xx^{-1}),\theta(xx^{-1})).$$

Hence $\mathbf{p}(\theta(x)) = F_\theta(\mathbf{p}(x))$.

Axiom (M2) holds: suppose $\exists(s,e) \cdot x$ in $(C(S),X_S)$. By definition $(s,e) \cdot x = sx$ and so $\theta((s,e) \cdot x) = \theta(sx)$. On the other hand, $F_\theta(s,e) = (\theta(s),\theta(e))$, so that $F_\theta(s,e) \cdot \theta(x) = \theta(s)\theta(x)$.

Axiom (M3) holds: suppose that $C(S) \cdot x \cap C(S) \cdot y$ is empty. Then

$$C(S) \cdot (x^{-1}xy^{-1}y) = (Sx^{-1}xy^{-1}y) \setminus \{0\}$$

by the proof of Theorem 10.3.2. Hence $x^{-1}xy^{-1}y = 0$. It follows that

$$\theta(x)^{-1}\theta(x)\theta(y)^{-1}\theta(y) = 0.$$

Thus $C(T) \cdot \theta(x) \cap C(T) \cdot \theta(y)$ is also empty. A similar argument shows that

$$C(S) \cdot x \cap C(S) \cdot y = C(S) \cdot z$$

implies

$$C(T) \cdot \theta(x) \cap C(T) \cdot \theta(y) = C(T) \cdot \theta(z).$$

It is now easy to check that **C** is a functor. ∎

10.5 Equivalent systems

In this section, we introduce a special class of system morphisms called equivalences; these are best viewed as generalisations of isomorphisms. We shall prove that if there is an equivalence between two systems then their associated inverse semigroups are isomorphic.

In order to define equivalences, we need to describe some extra structures which the categories **Sys** and **Inv** possess. We begin by defining transformations between system morphisms; these arise since system morphisms are essentially functors and so we can consider natural transformations between them.

Let (F, θ) and (G, φ) be system morphisms from (C, X) to (D, Y). A *transformation* τ from (F, θ) to (G, φ) is defined by the following two axioms:

(T1) τ: $F \to G$ is a natural transformation.

(T2) $\varphi(x) = \tau_{\mathbf{p}(x)} \cdot \theta(x)$ for all $x \in X$.

The right-hand side of axiom (T2) makes sense because $\tau_e \in \hom(F(e), G(e))$ for each identity e in C. Thus

$$\mathbf{d}(\tau_{\mathbf{p}(x)}) = F(\mathbf{p}(x)) = \mathbf{p}(\theta(x)),$$

by axiom (M1), and so $\tau_{\mathbf{p}(x)} \cdot \theta(x)$ is defined.

The *identity transformation* from (F, θ) to (F, θ) is the natural transformation 1_F from F to itself; condition (T2) holds automatically. The transformation τ is said to be an *isomorphism* if τ: $F \to G$ is a natural isomorphism. In this case τ^{-1} denotes the transformation from G to F defined by $\tau_e^{-1} = (\tau_e)^{-1}$ for each identity e in C.

Lemma 1 *Let (C, X) and (D, Y) be systems.*

(1) *The transformations between the system morphisms from (C, X) to (D, Y) form a category which is a preorder.*

(2) *If τ: $(F, \theta) \to (G, \varphi)$ and σ: $(G, \varphi) \to (F, \theta)$ are transformations then τ is an isomorphism and $\sigma = \tau^{-1}$.*

Proof (1) Let μ be a transformation from (F, θ) to (G, φ) and let ν be a transformation from (G, φ) to (H, ψ). By axiom (T1), μ is a natural transformation from F to G and ν is a natural transformation from G to H. Thus $\nu\mu$ is a natural transformation from F to H. For each $x \in X$ we have by axiom (T2) that

$$\varphi(x) = \mu_{\mathbf{p}(x)} \cdot \theta(x)$$

and

$$\psi(x) = \nu_{\mathbf{p}(x)} \cdot \varphi(x).$$

Thus

$$\psi(x) = (\nu\mu)_{\mathbf{p}(x)} \cdot \theta(x).$$

Hence $\nu\mu$ is a transformation. It is easy to check that if τ is a transformation from (F, θ) to (G, ϕ) then $\tau 1_F = \tau = 1_G \tau$. It is now evident that transformations form a category. To prove that this category is a preorder, suppose that μ and ν are both transformations from (F, θ) to (G, ϕ). Then by axiom (T2)

$$\mu_{\mathbf{p}(x)} \cdot \theta(x) = \nu_{\mathbf{p}(x)} \cdot \theta(x)$$

for all $x \in X$. Thus by the right cancellation condition, $\mu_{\mathbf{p}(x)} = \nu_{\mathbf{p}(x)}$. But $\mathbf{p}$: $X \to C_o$ is a surjection, and so $\mu_e = \nu_e$ for every $e \in C_o$. Thus $\mu = \nu$.
(2) Immediate from (1). ■

The category **Inv** also has some extra structure, which arises from the natural partial order on inverse semigroups. Let θ, φ: $S \to T$ be two homomorphisms in **Inv**. We write $\varphi \leq \theta$ if $\varphi(s) \leq \theta(s)$ for all $s \in S$. Thus the set of all homomorphisms from S to T is a partially ordered set. The link between transformations in **Sys** and the order relation between homomorphisms in **Inv** is provided by the following result.

Lemma 2 (1) *Let $(F, \theta), (G, \varphi)$: $(C, X) \to (D, Y)$ be morphisms of systems. If τ: $(F, \theta) \to (G, \varphi)$ is a transformation then $\mathbf{J}(F, \theta) \leq \mathbf{J}(G, \varphi)$ in* **Inv**.

(2) *Let θ, φ: $S \to T$ be homomorphisms in* **Inv** *such that $\varphi \leq \theta$. Then there is a transformation τ: $\mathbf{C}(\varphi) \to \mathbf{C}(\theta)$.*

Proof (1) By Theorem 10.4.1,

$$\mathbf{J}(F,\theta)([x,y]) = [\theta(x),\theta(y)] \text{ and } \mathbf{J}(G,\varphi)([x,y]) = [\varphi(x),\varphi(y)].$$

By axiom (T2), we have that

$$\varphi(x) = \tau_{\mathbf{p}(x)} \cdot \theta(x) \text{ and } \varphi(y) = \tau_{\mathbf{p}(y)} \cdot \theta(y).$$

But $\mathbf{p}(x) = \mathbf{p}(y)$. Thus by Theorem 10.2.3, we have that

$$[\varphi(x),\varphi(y)] \le [\theta(x),\theta(y)].$$

Hence $\mathbf{J}(F,\theta) \le \mathbf{J}(G,\phi)$.
(2) Recall that $\mathbf{C}(\theta) = (F_\theta,\theta)$ and $\mathbf{C}(\varphi) = (F_\varphi,\varphi)$ where

$$F_\theta(s,e) = (\theta(s),\theta(e)) \text{ and } \theta\colon X_S \to X_T$$

and

$$F_\varphi(s,e) = (\varphi(s),\varphi(e)) \text{ and } \varphi\colon X_S \to X_T.$$

For each $(e,e) \in C(S)_o$ put $\tau_{(e,e)} = (\varphi(e),\theta(e))$. Then $\tau_{(e,e)} \in C(S)$ since $\varphi(e) \le \theta(e)$ by assumption. Clearly,

$$\mathbf{d}(\tau_{(e,e)}) = F_\theta(e,e) \text{ and } \mathbf{r}(\tau_{(e,e)}) = F_\varphi(e,e).$$

Thus

$$\tau_{(e,e)} \in \hom(F_\theta(e,e),F_\varphi(e,e)).$$

We show that τ is a transformation of system morphisms.

Axiom (T1) holds: let $(e,e),(f,f) \in C(S)_o$ and $(s,e) \in \hom((e,e),(f,f))$. Then

$$F_\varphi(s,e)\tau_{(e,e)} = (\varphi(s),\varphi(e))(\varphi(e),\theta(e)) = (\varphi(s)\varphi(e),\varphi(e))$$

which is equal to $(\varphi(s),\theta(e))$ since $s^{-1}s \le e$. Whereas

$$\tau_{(f,f)}F_\theta(s,e) = (\varphi(f),\theta(f))(\theta(s),\theta(e)) = (\varphi(f)\theta(s),\theta(e))$$

which is equal to $(\varphi(s),\theta(e))$, since $\varphi(s) \le \theta(s)$ and $\varphi(s) = \varphi(ss^{-1})\theta(s) = \varphi(f)\theta(s)$.

Axiom (T2) holds: let $x \in X_S$. Then

$$\mathbf{p}(x) = (xx^{-1},xx^{-1}) \text{ and } \tau_{\mathbf{p}(x)} = (\varphi(xx^{-1}),\theta(xx^{-1})).$$

Now

$$\tau_{\mathbf{p}(x)} \cdot \theta(x) = (\varphi(xx^{-1}),\theta(xx^{-1})) \cdot \theta(x) = \varphi(xx^{-1})\theta(x)$$

which is equal to $\varphi(x)$ since $\varphi(x) \le \theta(x)$. ■

We now single out a special class of system morphisms. In the definition below, we use the following notation: if (C, X) is a system then $(1_C, 1_X)$ is the identity system morphism at (C, X), where 1_C is the identity functor on C and 1_X is the identity function on X.

Let (C, X) and (D, Y) be systems. A system morphism (F, θ) from (C, X) to (D, Y) is said to be an *equivalence (of systems)* if there is a system morphism (G, φ) from (D, Y) to (C, X) and isomorphisms (i.e. transformations)

$$\sigma\colon (1_D, 1_Y) \to (FG, \theta\varphi) \text{ and } \tau\colon (1_C, 1_X) \to (GF, \varphi\theta).$$

The key properties of equivalences are described in our next result.

Theorem 3 (1) *If (C, X) is a system then $(1_C, 1_X)$ is an equivalence.*

(2) *The composition of equivalences is an equivalence.*

(3) *Let $(F, \theta)\colon (C, X) \to (D, Y)$ be an equivalence. Then $\mathbf{J}(F, \theta)\colon \mathbf{J}(C, X) \to \mathbf{J}(D, Y)$ is an isomorphism.*

Proof The proofs of (1) and (2) are straightforward.
(3) By definition there is a system morphism (G, φ) from (D, Y) to (C, X) and isomorphisms σ and τ such that:

$$\sigma\colon (1_D, 1_Y) \to (FG, \theta\varphi) \text{ and } \tau\colon (1_C, 1_X) \to (GF, \varphi\theta).$$

Since $\mathbf{J}$ is a functor we have that

$$\mathbf{J}(F, \theta)\mathbf{J}(G, \phi) = \mathbf{J}(FG, \theta\varphi) \text{ and } \mathbf{J}(G, \phi)\mathbf{J}(F, \theta) = \mathbf{J}(GF, \varphi\theta).$$

Now $\mathbf{J}(1_D, 1_Y)$ is the identity homomorphism on $\mathbf{J}(D, Y)$ and $\mathbf{J}(1_C, 1_X)$ is the identity homomorphism on $\mathbf{J}(C, X)$. Thus by Lemma 2, $\mathbf{J}(FG, \theta\varphi)$ is an identity homomorphism, as is $\mathbf{J}(GF, \varphi\theta)$. It is now immediate that $\mathbf{J}(F, \theta)$ is an isomorphism. ■

Two systems are said to be *equivalent* if there is an equivalence of systems between them. The above result implies that equivalent systems give rise to isomorphic inverse semigroups. An explicit description of equivalences is contained in the following result.

Proposition 4 *A system morphism (F, θ) from (C, X) to (D, Y) is an equivalence of systems if, and only if, the following axioms hold:*

(ES1) *F is an equivalence of categories.*

(ES2) *For each $y \in Y$ there exists an isomorphism $u \in D$ and an element $x \in X$ such that $y = u \cdot \theta(x)$.*

(ES3) *If* $y_1 = a \cdot y_2$ *in* ${}_DY$ *and* $\theta(x_1) = y_1$ *and* $\theta(x_2) = y_2$ *then there exists* $a' \in C$ *such that* $x_1 = a' \cdot x_2$.

Proof Let (F, θ) be an equivalence of systems from (C, X) to (D, Y).

Axiom (ES1) holds: there is a system morphism (G, φ) from (D, Y) to (C, X) and isomorphisms σ and τ such that

$$\sigma\colon (1_D, 1_Y) \to (FG, \theta\varphi) \text{ and } \tau\colon (1_C, 1_X) \to (GF, \varphi\theta).$$

Thus

$$\sigma\colon 1_D \to FG \text{ and } \tau\colon 1_C \to GF$$

are natural isomorphisms. In particular, F is an equivalence of categories.

Axiom (ES2) holds: let $y \in Y$. Since σ is an isomorphism from $(1_D, 1_Y)$ to $(FG, \theta\phi)$, we have that $\theta(\varphi(y)) = \sigma_{\mathbf{p}(y)} \cdot y$ by axiom (T2). Put $x = \varphi(y)$ and $u = (\sigma_{\mathbf{p}(y)})^{-1}$. Then $y = u \cdot \theta(x)$, as required.

Axiom (ES3) holds: let $y_1 = a \cdot y_2$ in ${}_DY$ and let $\theta(x_1) = y_1$ and $\theta(x_2) = y_2$. Since there is an isomorphism τ from $(1_C, 1_X)$ to $(GF, \phi\theta)$, we have that $x = \tau_{\mathbf{p}(x)} \cdot (\varphi\theta)(x)$ for all $x \in X$ by axiom (T2). Thus

$$x_1 = \tau_{\mathbf{p}(x_1)} \cdot (\varphi\theta)(x_1) \text{ and } x_2 = \tau_{\mathbf{p}(x_2)} \cdot (\varphi\theta)(x_2).$$

Now

$$x_1 = \tau_{\mathbf{p}(x_1)} \cdot (\varphi\theta)(x_1) = \tau_{\mathbf{p}(x_1)} \cdot \varphi(\theta(x_1)).$$

However $\theta(x_1) = y_1$ and so

$$x_1 = \tau_{\mathbf{p}(x_1)} \cdot \varphi(y_1).$$

Also $\varphi(y_1) = G(a) \cdot \varphi(y_2)$. Thus

$$x_1 = \tau_{\mathbf{p}(x_1)} \cdot (G(a) \cdot \varphi(y_2)).$$

Since $\varphi(y_2) = \varphi(\theta(x_2))$, we have that

$$x_1 = \tau_{\mathbf{p}(x_1)} \cdot (G(a) \cdot \varphi(\theta(x_2))).$$

But

$$(\varphi\theta)(x_2) = \tau^{-1}_{\mathbf{p}(x_2)} \cdot x_2.$$

Put $a' = \tau_{\mathbf{p}(x_1)} G(a) \tau^{-1}_{\mathbf{p}(x_2)}$. Then $x_1 = a' \cdot x_2$.

To prove the converse, let $(F, \theta)\colon (C, X) \to (D, Y)$ be a system morphism satisfying axioms (ES1), (ES2) and (ES3). We shall prove that it is an equivalence of systems. For each identity e in D there exists, by axiom (ES1), an identity $G(e)$ in C and isomorphism $\sigma_e \in \hom(e, F(G(e)))$. In the usual way (see [210]), this information may be used to construct a functor $G\colon D \to C$.

Let $y \in Y$. By axiom (ES2), there is an isomorphism u in D and element x of X, such that $y = u \cdot \theta(x)$. Now $\sigma_{\mathbf{p}(y)} \cdot y$ is defined and so

$$\sigma_{\mathbf{p}(y)} \cdot y = (\sigma_{\mathbf{p}(y)} u) \cdot \theta(x).$$

Also

$$\sigma_{\mathbf{p}(y)} u \in \hom(F(\mathbf{p}(x)), F(G(\mathbf{p}(y))))$$

and so since F is an equivalence of categories there exists a unique isomorphism u' in C such that $u' \in \hom(\mathbf{p}(x), G(\mathbf{p}(y)))$ and $F(u') = \sigma_{\mathbf{p}(y)} u$. Now

$$\sigma_{\mathbf{p}(y)} \cdot y = F(u') \cdot \theta(x) = \theta(u' \cdot x)$$

by axiom (M2). Thus $y = \sigma^{-1}_{\mathbf{p}(y)} \cdot \theta(u' \cdot x)$. Define $\varphi(y) = u' \cdot x$. It is straightforward to check that (G, φ) is an equivalence of systems.

By Theorem 3(2), $(FG, \theta\varphi)$ is an equivalence from (D, Y) to itself. We have seen that there is an isomorphism $\sigma_e \colon e \to F(G(e))$ for each identity e in D. Together these isomorphisms are the components of a natural isomorphism σ from 1_D to FG. It is also easy to check that axiom (T2) holds. Thus σ is an isomorphism from $(1_D, 1_Y)$ to $(FG, \theta\varphi)$.

We shall now define an isomorphism τ from $(1_C, 1_X)$ to $(GF, \varphi\theta)$. We begin by constructing a natural isomorphism from 1_C to GF. Let f be an identity in C. Then by the above $\sigma_{F(f)}$ is an isomorphism from $F(f)$ to $F(G(F(f)))$. Since F is an equivalence of categories, there exists a unique isomorphism τ_f from f to $G(F(f))$ such that $F(\tau_f) = \sigma_{F(f)}$. It is now easy to check that the τ_f are the components of a natural isomorphism from 1_C to GF. We finish off by showing that τ satisfies axiom (T2). Let x_1 be an arbitrary element of X. Put $y = \theta(x_1)$. With the notation as before, there is an isomorphism u in D and element x in X such that $\theta(x_1) = y = u \cdot \theta(x)$. As above, there exists a unique isomorphism u' from $\mathbf{p}(x)$ to $G(\mathbf{p}(y))$ such that $F(u') = \sigma_{\mathbf{p}(y)} u$. Now, $y = \theta(x_1) = u \cdot \theta(x)$, and so by axiom (ES3), there is a unique morphism v from $\mathbf{p}(x)$ to $\mathbf{p}(x_1)$ such that $F(v) = u$ and $x_1 = v \cdot x$. In fact, v is an isomorphism since F is an equivalence. Now,

$$F(u')u^{-1} = F(\tau_{\mathbf{p}(x_1)}),$$

and $u^{-1} = F(v)^{-1}$. The category product $u'v^{-1}$ is defined and so

$$F(u'v^{-1}) = F(\tau_{\mathbf{p}(x_1)}).$$

Both $u'v^{-1}$ and $\tau_{\mathbf{p}(x_1)}$ are morphisms from $\mathbf{p}(x_1)$ to $G(F(\mathbf{p}(x_1)))$, and so $\tau_{\mathbf{p}(x_1)} = u'v^{-1}$ since F is an equivalence. By definition $\varphi(y) = u' \cdot x$ and so

$$\varphi(y) = \varphi(\theta(x_1)) = u' \cdot x = (\tau_{\mathbf{p}(x_1)} v) \cdot x,$$

which is equal to $\tau_{\mathbf{p}(x_1)} \cdot x_1$, as required. ■

10.6 Composing the functors

In this section, we shall prove that every inverse semigroup can be constructed from a system. More generally, the functors $\mathbf{J}$: $\mathbf{Sys} \to \mathbf{Inv}$ and $\mathbf{C}$: $\mathbf{Inv} \to \mathbf{Sys}$ constructed in Section 10.4 will be composed, and we shall compare S with $(\mathbf{JC})(S)$, and (C, X) with $(\mathbf{CJ})(C, X)$.

We start by comparing S with $(\mathbf{JC})(S)$. Let S be any inverse semigroup with zero. Then S acts on itself on the left by means of multiplication. Let $\check{S}$ be the inverse semigroup of all (left) S-isomorphisms between *principal* left ideals of S. Define Θ: $S \to \check{S}$ by $\Theta(s)$: $Ss^{-1}s \to Sss^{-1}$ and $\Theta(s) = as^{-1}$ for each $a \in Ss^{-1}s$. Then Θ is an isomorphism by the dual of Theorem 1.5.4.

We may now prove that every inverse semigroup with zero is isomorphic to an inverse semigroup arising from a category acting on a set satisfying the orbit condition.

Theorem 1 *Let S be an inverse semigroup with zero. Then S and $(\mathbf{JC})(S)$ are isomorphic.*

Proof We prove first, that the $C(S)$-isomorphisms from $C(S) \cdot x$ to $C(S) \cdot y$ induce, and are induced by, S-isomorphisms from Sx to Sy. Let

$$\theta\colon\ C(S) \cdot x \to C(S) \cdot y$$

be a $C(S)$-isomorphism. By the proof of Theorem 10.3.2, we have that $C(S) \cdot x = Sx \setminus \{0\}$ and $C(S) \cdot y = Sy \setminus \{0\}$. Extend θ to a function from Sx to Sy by defining $\theta(0) = 0$. Clearly, θ: $Sx \to Sy$ is a bijection. We show that θ is an S-isomorphism. As a first step, we prove for all $s \in S$ that $sx \,\mathcal{R}\, \theta(sx)$. We consider the cases where sx is non-zero and zero separately. Suppose that sx is non-zero. Then $sx \in C(S) \cdot x$. By axiom (M1), $\mathbf{p}(\theta(sx)) = \mathbf{p}(sx)$. Then $\theta(sx)\theta(sx)^{-1} = (sx)(sx)^{-1}$ and so $sx \,\mathcal{R}\, \theta(sx)$. If $sx = 0$ then $\theta(sx) = 0$ by definition and $sx \,\mathcal{R}\, \theta(sx)$ is immediate.

We can now show that θ: $Sx \to Sy$ is an S-homomorphism. We need to show for all $t \in S$ and $sx \in Sx$ that $\theta(t(sx)) = t\theta(sx)$. Suppose that $t(sx) = 0$ but $t, sx \neq 0$. From the result above $\theta(sx) \,\mathcal{R}\, sx$ and so since $\mathcal{R}$ is a left congruence, $t\theta(sx) \,\mathcal{R}\, tsx = 0$. Thus $t\theta(sx) = 0$ and so $\theta(t(sx)) = t\theta(sx)$. Thus we may suppose that $t(sx) \neq 0$. Put $e = (sx)(sx)^{-1}$. Then $t(sx) = t(e(sx))$. Consider the ordered pair (te, e). Observe that $e \neq 0$, for if $e = 0$ then $sx = esx = 0$ which is a contradiction. Also, $te \neq 0$ since $te = 0$ implies that $tesx = tsx = 0$ which is a contradiction. Thus $te, e \neq 0$. Also

$$(te)^{-1}te = et^{-1}te \leq e$$

It follows that $(te, e) \in C(S)$. Now

$$\mathbf{d}(te, e) = (e, e) = \mathbf{p}(sx).$$

Thus $\exists(te, e) \cdot (sx)$ and

$$(te, e) \cdot (sx) = tesx = tsx.$$

Since θ is a $C(S)$-homomorphism, we have that

$$\theta((te, e) \cdot (sx)) = (te, e) \cdot \theta(sx).$$

Hence $\theta(tsx) = te\theta(sx)$. Now

$$e = (sx)(sx)^{-1} \,\mathcal{R}\, \theta(sx)$$

by the result above. Thus $te\theta(sx) = t\theta(sx)$. It follows that $\theta(t(sx)) = t\theta(sx)$.

Conversely, let θ: $Sx \to Sy$ be an S-isomorphism. We can define a $C(S)$-isomorphism θ' from $C(S) \cdot x$ to $C(S) \cdot y$ by $\theta'((s, e) \cdot x) = \theta(sx)$. Suppose $(t, f) \cdot x = (s, e) \cdot x$. Then by Theorem 10.3.2, $(t, f) = (s, e)$ and so θ' is well-defined. It is easy to check that $\theta'((t, f) \cdot ((s, e) \cdot x)) = (t, f) \cdot \theta'((s, e) \cdot x)$.

We can now prove the theorem. Define a function ι: $S \to J({}_{C(S)}X_S)$ by

$$\iota(s) = \rho_{s^{-1}}\colon C(S) \cdot (s^{-1}s) \to C(S) \cdot (ss^{-1})$$

if s is non-zero, and $\iota(0) = 0$. By our result above, and the fact that S and $\check{S}$ are isomorphic, it is clear that $\iota(s) \in J({}_{C(S)}X_S)$, and ι is an isomorphism. ■

The above theorem also yields the result that every inverse semigroup without zero is isomorphic to an inverse semigroup arising from a category acting on a set satisfying the strong orbit condition.

We now compare (C, X) with $(\mathbf{CJ})(C, X)$; it is here that the notion of equivalence comes into play.

Theorem 2 *Let (C, X) be a system.*

(1) *For each function $\mathbf{q}$: $C_o \to X$, such that $\mathbf{p}(\mathbf{q}(e)) = e$ for each $e \in C_o$, there exists an equivalence of systems $(F_\mathbf{q}, \theta_\mathbf{q})$: $(C, X) \to (\mathbf{CJ})(C, X)$.*

(2) *Let $\mathbf{q}, \mathbf{q}'$: $C_o \to X$ be functions from C_o to X such that*

$$\mathbf{p}(\mathbf{q}(e)) = e = \mathbf{p}(\mathbf{q}'(e))$$

for each $e \in C_o$. Then there is an isomorphic transformation from $(F_\mathbf{q}, \theta_\mathbf{q})$ to $(F_{\mathbf{q}'}, \theta_{\mathbf{q}'})$.

Proof (1) Define $(F, \theta) = (F_\mathbf{q}, \theta_\mathbf{q})$ as follows: F: $C \to C(J({}_CX))$ is the function defined by

$$F(s) = ([\mathbf{q}(\mathbf{r}(s)), s \cdot \mathbf{q}(\mathbf{d}(s))], [\mathbf{q}(\mathbf{d}(s)), \mathbf{q}(\mathbf{d}(s))])$$

and θ: $X \to X_{J({}_C X)}$ is the function defined by

$$\theta(x) = [\mathbf{q}(\mathbf{p}(x)), x].$$

In what follows we shall use the following notation:

$$e_s = \mathbf{q}(\mathbf{d}(s)) \text{ and } f_s = \mathbf{q}(\mathbf{r}(s)).$$

The fact that θ and F are well-defined functions is straightforward to check from the definitions. The proof of the theorem consists of a series of verifications. We begin by showing that F is a functor. It is straightforward to check that F maps identities to identities. Now suppose that $\exists st$ in C. Then by definition

$$F(s) = ([f_s, s \cdot e_s], [e_s, e_s]) \text{ and } F(t) = ([f_t, t \cdot e_t], [e_t, e_t]).$$

Now

$$\mathbf{d}(F(s)) = ([e_s, e_s], [e_s, e_s]) \text{ and } \mathbf{r}(F(t)) = ([f_t, f_t], [f_t, f_t]).$$

But by assumption $\mathbf{d}(s) = \mathbf{r}(t)$. Thus $e_s = f_t$ and so $\mathbf{d}(F(s)) = \mathbf{r}(F(t))$. We now compute $F(s)F(t)$. By definition

$$F(s)F(t) = ([f_s, s \cdot e_s] \otimes [f_t, t \cdot e_t], [e_t, e_t]).$$

Now

$$C \cdot (s \cdot e_s) \cap C \cdot f_t = C \cdot (s \cdot e_s).$$

Thus

$$e_s * (s \cdot e_s) = \mathbf{r}(s) \text{ and } (s \cdot e_s) * f_t = s.$$

Hence

$$F(s)F(t) = ([f_s, (s \cdot t) \cdot e_t], [e_t, e_t]),$$

which is equal to $F(st)$. Thus F is a functor.

To show that (F, θ) is a morphism of systems, we have to check that three axioms hold. Both axioms (M1) and (M2) are straightforward to check. We show that axiom (M3) holds. From the proof of Theorem 10.3.2,

$$C(J({}_C X)) \cdot \theta(w) = J({}_C X) \otimes [w, w] \setminus \{0\} = \{[d, u \cdot w] \in J({}_C X) \setminus \{0\}: u \in C\}.$$

If

$$C(J({}_C X)) \cdot \theta(x) \cap C(J({}_C X)) \cdot \theta(y)$$

is non-empty, then we can find elements in the intersection such that $[a, u \cdot x] = [b, v \cdot y]$. But then $u \cdot x = p \cdot (v \cdot y)$ for some $p \in C$, and so $C \cdot x \cap C \cdot y$ is non-empty. Now suppose that $C \cdot x \cap C \cdot y = C \cdot z$. It is straightforward to check that

$$C(J(_C X)) \cdot \theta(x) \cap C(J(_C X)) \cdot \theta(y) = C(J(_C X)) \cdot \theta(z).$$

To show that (F, θ) is an equivalence of systems, we have to check that three axioms hold.

Axiom (ES1) holds: we have to show that F is full, faithful and dense.

The functor F is full: let e and f be identities in C and let $(\mathbf{s}, \mathbf{e})$ be a morphism in $C(J(_C X))$ such that

$$\mathbf{d}(\mathbf{s}, \mathbf{e}) = F(e) \text{ and } \mathbf{r}(\mathbf{s}, \mathbf{e}) = F(f).$$

Let $(\mathbf{s}, \mathbf{e}) = ([x, y], [z, z])$. Then from $\mathbf{s}^{-1}\mathbf{s} \leq \mathbf{e}$ we obtain $y = u \cdot z$ for some $u \in C$. From $\mathbf{d}(\mathbf{s}, \mathbf{e}) = (\mathbf{e}, \mathbf{e})$ we obtain $\mathbf{e} = [\mathbf{q}(e), \mathbf{q}(e)]$, and so $[z, z] = [\mathbf{q}(e), \mathbf{q}(e)]$, from which we have that $z = a \cdot \mathbf{q}(e)$ for some isomorphism a. From

$$([x, x], [x, x]) = \mathbf{r}(s, e) = ([\mathbf{q}(f), \mathbf{q}(f)], [\mathbf{q}(f), \mathbf{q}(f)]),$$

we obtain $x = b \cdot \mathbf{q}(f)$ for some isomorphism b. Thus

$$(\mathbf{s}, \mathbf{e}) = ([b \cdot \mathbf{q}(f), (ua) \cdot \mathbf{q}(e)], [a \cdot \mathbf{q}(e), a \cdot \mathbf{q}(e)]).$$

Now

$$\mathbf{d}(b^{-1}ua) = \mathbf{d}(a) = \mathbf{p}(\mathbf{q}(e)) = e,$$

and

$$\mathbf{r}(b^{-1}ua) = \mathbf{r}(b^{-1}) = \mathbf{d}(b) = \mathbf{p}(\mathbf{q}(f)) = f,$$

and $F(b^{-1}ua) = (\mathbf{s}, \mathbf{e})$.

The functor F is faithful: suppose that $F(s) = F(t)$ and

$$\mathbf{r}(s) = \mathbf{r}(t) = f \text{ and } \mathbf{d}(s) = \mathbf{d}(t) = e.$$

We show that $s = t$. By assumption,

$$[\mathbf{q}(\mathbf{r}(s)), s \cdot \mathbf{q}(\mathbf{d}(s))] = [\mathbf{q}(\mathbf{r}(t)), t \cdot \mathbf{q}(\mathbf{d}(t))].$$

Thus from the definition of $\sim$-equivalence we have that

$$\mathbf{q}(f) = u \cdot \mathbf{q}(f) \text{ and } s \cdot \mathbf{q}(e) = u \cdot (t \cdot \mathbf{q}(e))$$

for some isomorphism u in C. By the cancellation condition, $u = f$ and so $s = t$.

The functor F is dense: let $([x, x], [x, x])$ be any identity in $C(J(_C X))$. Consider the ordered pair

$$([x, \mathbf{q}(\mathbf{p}(x))], [\mathbf{q}(\mathbf{p}(x)), \mathbf{q}(\mathbf{p}(x))]).$$

It is easy to check that it is a well-defined isomorphism in $C(J(_CX))$, and that

$$\mathbf{r}([x,\mathbf{q}(\mathbf{p}(x))],[\mathbf{q}(\mathbf{p}(x)),\mathbf{q}(\mathbf{p}(x))]) = ([x,x],[x,x]),$$

and

$$\mathbf{d}([x,\mathbf{q}(\mathbf{p}(x))],[\mathbf{q}(\mathbf{p}(x)),\mathbf{q}(\mathbf{p}(x))]) = F(\mathbf{p}(x)).$$

Axiom (ES2) holds: let $[x,y]$ be any element of $X_{J(_CX)}$. Then the ordered pair

$$\mathbf{a} = ([x,\mathbf{q}(\mathbf{p}(y))],[\mathbf{q}(\mathbf{p}(y)),\mathbf{q}(\mathbf{p}(y))]),$$

is a well-defined isomorphism in $C(J(_CX))$ and

$$[x,y] = \mathbf{a}\cdot\theta(y).$$

Axiom (ES3) holds: let

$$\theta(x) = (\mathbf{s},\mathbf{e})\cdot\theta(w).$$

By assumption,

$$\mathbf{d}(\mathbf{s},\mathbf{e}) = \mathbf{p}([\mathbf{q}(\mathbf{p}(w)),w])$$

and so $\mathbf{e} = [\mathbf{q}(\mathbf{p}(w)),\mathbf{q}(\mathbf{p}(w))]$. Let $\mathbf{s} = [a,b]$. Then $[b,b] \leq \mathbf{e}$ and so $b = r\cdot\mathbf{q}(\mathbf{p}(w))$ for some $r \in C$. Now

$$[\mathbf{q}(\mathbf{p}(x)),x] = (\mathbf{s},\mathbf{e})\cdot[\mathbf{q}(\mathbf{p}(w)),w]$$

and so

$$[\mathbf{q}(\mathbf{p}(x)),x] = \mathbf{s}\otimes[\mathbf{q}(\mathbf{p}(w)),w].$$

Now

$$\mathbf{s}\otimes[\mathbf{q}(\mathbf{p}(w)),w] = [a,b]\otimes[\mathbf{q}(\mathbf{p}(w)),w]$$

which is equal to

$$[(\mathbf{q}(\mathbf{p}(w)) * b)\cdot a,(b * \mathbf{q}(\mathbf{p}(w)))\cdot w].$$

But

$$C\cdot b \cap C\cdot\mathbf{q}(\mathbf{p}(w)) = C\cdot b,$$

and so

$$b * \mathbf{q}(\mathbf{p}(w)) = r \text{ and } \mathbf{q}(\mathbf{p}(w)) * b = \mathbf{p}(b).$$

Thus

$$[\mathbf{q}(\mathbf{p}(x)),x] = [\mathbf{p}(b)\cdot a, r\cdot w].$$

Therefore, for some isomorphism u in C we have that $x = (ur)\cdot w$. Now we compute $F(ur)$. By definition,

$$F(ur) = ([\mathbf{q}(\mathbf{r}(ur)),(ur)\cdot\mathbf{q}(\mathbf{d}(ur))],[\mathbf{q}(\mathbf{d}(r)),\mathbf{q}(\mathbf{d}(r))]).$$

But

$$\mathbf{p}(x) = \mathbf{r}(u) \text{ and } \mathbf{d}(r) = \mathbf{p}(w),$$

and

$$b = r \cdot \mathbf{q}(\mathbf{p}(w)).$$

Thus

$$F(ur) = ([\mathbf{q}(\mathbf{p}(x)), u \cdot b], \mathbf{e}).$$

But $\mathbf{q}(\mathbf{p}(x)) = u \cdot a$. Hence

$$F(ur) = ([a, b], \mathbf{e}) = (\mathbf{s}, \mathbf{e}).$$

(2) Define $\tau : F_{\mathbf{q}} \to F_{\mathbf{q}'}$ by

$$\tau_e = ([\mathbf{q}'(e), \mathbf{q}(e)], [\mathbf{q}(e), \mathbf{q}(e)])$$

for each $e \in C_o$. It is easy to check that τ_e is a well-defined isomorphism in $\mathbf{C}(\mathbf{J}(C, X))$ and that $\tau_e \in \hom(F_{\mathbf{q}}(e), F_{\mathbf{q}'}(e))$.

Axiom (T1) holds: let $s \in \hom(e, f)$. Straightforward, if somewhat unwieldy, calculations show that

$$F_{\mathbf{q}'}(s)\tau_e = ([\mathbf{q}'(f), s \cdot \mathbf{q}(e)], [\mathbf{q}(e), \mathbf{q}(e)]) = \tau_f F_{\mathbf{q}}(s).$$

Axiom (T2) holds: for each $x \in X$ we have that $\tau_{\mathbf{p}(x)} \cdot \theta_{\mathbf{q}}(x) = \theta_{\mathbf{q}'}(x)$. ■

10.7 Special cases

In this section, we shall examine some applications of the preceding theory to some special cases.

Right cancellative categories

Let C be a category. Then C *acts on itself on the left* as follows: define $\mathbf{p}$: $C \to C_o$ by $\mathbf{p}(x) = \mathbf{r}(x)$, and define $s \cdot x = sx$ if $\mathbf{d}(s) = \mathbf{p}(x)$, the usual category product in C.

Proposition 1 *Let C be a right cancellative category considered as a left C-system.*

(1) *C satisfies the orbit condition if, and only if, any two morphisms $s, t \in C$ such that $us = vt$ for some $u, v \in C$ have a pushout in C.*

(2) *If C satisfies the condition in (1), then it is a system.*

Proof (1) Let C be a right cancellative category satisfying the orbit condition. Let $s, t \in C$ such that $us = vt$ for some $u, v \in C$. We show that s and t have a pushout. Since $C \cdot s \cap C \cdot t$ is non-empty we have that $C \cdot s \cap C \cdot t = C \cdot p$ for some $p \in C$. Let $p = as = bt$ for some $a, b \in C$. Now let h and k be any elements of C such that $hs = kt$. Then $hs = kt \in C \cdot p$ and so $hs = kt = cp$ for some $c \in C$. But then $hs = cp = cas$ and so $h = ca$ by right cancellativity. Similarly, $kt = cp = cbt$ and so $k = cb$. The element c is unique by right cancellativity. Consequently, (a, b) is the pushout of (s, t).

Conversely, suppose that C is right cancellative and that C has pushouts of pairs of morphisms which can be completed to a commutative square. Suppose that $C \cdot s \cap C \cdot t$ is non-empty. Then s and t can be completed to a commutative square and so, by assumption, have a pushout. Let $p = as = bt$ where (a, b) is the pushout of (s, t). We claim that $C \cdot s \cap C \cdot t = C \cdot p$. Let $z \in C \cdot s \cap C \cdot t$. Then $z = us = vt$ for some $u, v \in C$. But then by the property of pushouts there exists $c \in C$ such that $u = ca$ and $v = cb$. Thus $z = us = cas = cp$. Hence $z \in C \cdot p$. Conversely, let $z \in C \cdot p$. Then $z = cp$ for some $c \in C$, and so $z = cas = cbt$. Hence $z \in C \cdot s \cap C \cdot t$.
(2) By definition axiom (S1) holds, and axiom (S2) holds by assumption. Axiom (S3) holds from the definition of **p** and axiom (S4) holds since the action is just the category product and the category is right cancellative. ■

We now give some simple examples of this construction. Consider the monoid $(\mathbb{N}, +)$ as a one-object category acting on itself on the left. Clearly, $(\mathbb{N}, +)$ is right cancellative. Observe that

$$(\mathbb{N} + m) \cap (\mathbb{N} + n) = \mathbb{N} + \max\{m, n\}.$$

Thus ${}_{\mathbb{N}}\mathbb{N}$ satisfies the strong orbit condition. Hence $J^*({}_{\mathbb{N}}\mathbb{N})$ is an inverse semigroup by Theorem 10.1.4. Using the notation of Section 3.4 we can write

$$\max\{m, n\} = (n \dot{-} m) + m = (m \dot{-} n) + n.$$

Clearly the only isomorphism in $\mathbb{N}$ is 0. Thus the underlying set of $J^*({}_{\mathbb{N}}\mathbb{N})$ is $\mathbb{N} \times \mathbb{N}$. The multiplication is given by

$$(a, b) \otimes (c, d) = ((c \dot{-} b) + a, (b \dot{-} c) + d).$$

Thus $J^*({}_{\mathbb{N}}\mathbb{N})$ is just the *bicyclic monoid.*

Our second example generalises our first; the natural numbers under addition form a free monoid on one generator. Let X be any non-empty set with at least two elements. The free monoid X^* on X is a cancellative monoid. In Lemma 9.3.2, we saw that if $X^*u \cap X^*v$ is non-empty then

$$X^*u \cap X^*v = X^*z$$

where $z = (vu^{-1})u = (uv^{-1})v$. Thus the free monoid acting on itself on the left satisfies the orbit condition. Clearly, the only isomorphism in X^* is 1 and so the underlying set of $J({}_{X^*}X^*)$ is just $(X^* \times X^*) \cup \{0\}$. The product in $J({}_{X^*}X^*)$ is given by

$$(u,v) \otimes (x,y) = \begin{cases} ((xv^{-1})u, (vx^{-1})y) & \text{if } X^*v \cap X^*x \neq \emptyset \\ 0 & \text{else.} \end{cases}$$

This is just the *polycyclic monoid on* $|X|$*-generators.*

The bicyclic monoid and the polycyclic monoids are special cases of the following more general construction.

Let $\mathcal{G}$ be a digraph consisting of a set of arrows G and a set of vertices G_o together with source function α and target function ω. If $x, y \in G$ then x and y are said to be *composable* if $\alpha(x) = \omega(y)$. A sequence of arrows $(x_1, \ldots, x_n)$ is called a *composable sequence* if (x_i, x_{i+1}) is composable for $i = 1, \ldots, n-1$. The *free category* G^* *generated by* G is defined to be the set of all composable sequences together with the set $\{1_e : e \in G_o\}$. We define $\mathbf{d}(1_e) = 1_e = \mathbf{r}(1_e)$ and

$$\mathbf{d}(x_1, \ldots, x_n) = 1_{\alpha(x_n)} \text{ and } \mathbf{r}(x_1, \ldots, x_n) = 1_{\omega(x_1)}.$$

A partial multiplication is defined in G^* as follows: if

$$\mathbf{d}(x_1, \ldots, x_m) = \mathbf{r}(y_1, \ldots, y_n)$$

then

$$(x_1, \ldots, x_m)(y_1, \ldots, y_n) = (x_1, \ldots, x_m, y_1, \ldots, y_n),$$

and the elements $\{1_e : e \in G_o\}$ act as identities. In this way, G^* is a category. It is easy to see that G^* is a right cancellative category (in fact, it is cancellative).

We let G^* act on itself on the left. To show that the orbit condition holds, we first define a new partial binary operation on G^* as follows: let $u, v \in G^*$ with $\mathbf{d}(u) = \mathbf{d}(v)$. Define

$$uv^{-1} = \begin{cases} h & \text{if } u = hv \\ \mathbf{r}(v) & \text{else.} \end{cases}$$

Suppose that

$$G^*(x_1, \ldots, x_m) \cap G^*(y_1, \ldots, y_n)$$

is non-empty. Then either $(x_1, \ldots, x_m)$ is a suffix of $(y_1, \ldots, y_n)$, in which case

$$G^*(x_1, \ldots, x_m) \cap G^*(y_1, \ldots, y_n) = G^*(y_1, \ldots, y_n)$$

or $(y_1, \ldots, y_n)$ is a suffix of $(x_1, \ldots, x_m)$ in which case

$$G^*(x_1, \ldots, x_m) \cap G^*(y_1, \ldots, y_n) = G^*(x_1, \ldots, x_m).$$

Observe that in both cases $\alpha(x_m) = \alpha(y_n)$. Now suppose that the intersection

$$G^* 1_e \cap G^*(x_1, \ldots, x_n)$$

is non-empty. The set $G^* 1_e$ consists of all strings $(y_1, \ldots, y_m)$ with

$$\mathbf{d}(y_1, \ldots, y_m) = 1_e.$$

But $\mathbf{d}(x_1, \ldots, x_n) = 1_e$. Hence

$$G^* 1_e \cap G^*(x_1, \ldots, x_n) = G^*(x_1, \ldots, x_n).$$

It is now straightforward to check for all $u, v \in G^*$ that if $G^* u \cap G^* v$ is non-empty then

$$G^* u \cap G^* v = G^*(vu^{-1})u = G^*(uv^{-1})v.$$

Thus $P(\mathcal{G}) = J({}_{G^*}G^*)$ is an inverse semigroup. It is clear that the only isomorphisms in G^* are the identities. Thus the underlying set of $P(\mathcal{G})$ is just $\{(x, y) \in G^* \times G^* \colon \mathbf{r}(x) = \mathbf{r}(y)\} \cup \{0\}$. The multiplication is given by

$$(u, v) \otimes (x, y) = \begin{cases} ((xv^{-1})u, (vx^{-1})y) & \text{if } G^* v \cap G^* x \neq \emptyset \\ 0 & \text{else.} \end{cases}$$

The inverse semigroups $P(\mathcal{G})$ are called *generalised polycyclic semigroups*; they occur naturally in the study of Cuntz–Krieger C^*-algebras. The polycyclic monoids are special cases: they arise as the semigroups $P(\mathcal{G})$ when the graph $\mathcal{G}$ is a bouquet of circles.

Cyclic systems and inverse monoids

In this section, we shall describe how our theory simplifies when the inverse semigroup has an identity. A system ${}_C X$ is said to be *cyclic* if $X = C \cdot x_0$ for some $x_0 \in X$.

Lemma 2 *If S is an inverse monoid then $(C(S), X_S)$ is a cyclic $C(S)$-system.*

Proof We claim that $X_S = C(S) \cdot 1$. Let $x \in X_S$. Then $x \in S \backslash \{0\}$. Now $(x, 1) \in C(S)$ since $x^{-1}x \leq 1$. Also,

$$\mathbf{d}(x, 1) = (1, 1) \text{ and } \mathbf{p}(1) = (1, 1),$$

so that $\exists (x, 1) \cdot 1$. But $(x, 1) \cdot 1 = x1 = x$. ■

The converse of the above result is proved below.

Proposition 3 *Let $_C X$ be a cyclic system such that $X = C \cdot x_0$. Put $1 = \mathbf{p}(x_0)$.*

(1) *$J(_C X)$ is a monoid.*

(2) *For every $e \in C_o$, $\hom(1, e) \neq \emptyset$.*

(3) *For all $s, t \in C$, $us = vt$ for some $u, v \in C$ implies that s and t have a pushout.*

(4) *Put $C' = \{a \in C \colon \mathbf{d}(a) = 1\}$. Then C acts on C' on the left by restricting the action by left multiplication of C on itself, (C, C') is a system, and $_C C'$ is equivalent to $_C X$.*

Proof (1) We show that $[x_0, x_0]$ is the identity of $J(_C X)$. Let $[x, y]$ be any non-zero element of $J(_C X)$. Then

$$[x, y] \otimes [x_0, x_0] = [(x_0 * y) \cdot x, (y * x_0) \cdot x_0].$$

Now $C \cdot y \cap C \cdot x_0 = C \cdot y$ and so $x_0 * y = \mathbf{p}(y)$ and $(y * x_0) \cdot x_0 = y$. Hence we obtain $[x, y] \otimes [x_0, x_0] = [x, y]$. We may similarly prove that $[x_0, x_0] \otimes [x, y] = [x, y]$.
(2) Let $e \in C_o$. Since $\mathbf{p}$ is surjective there exists $x \in X$ such that $\mathbf{p}(x) = e$. Now $x \in C \cdot x_0$ and so $x = a \cdot x_0$ for some $a \in C$. But then

$$\mathbf{d}(a) = \mathbf{p}(x_0) = 1 \text{ and } e = \mathbf{p}(x) = \mathbf{p}(a \cdot x_0) = \mathbf{r}(a).$$

Thus $a \in \hom(1, e)$.
(3) By assumption $us = vt$ for some $u, v \in C$. Thus $\mathbf{d}(s) = \mathbf{d}(t)$. Let $z \in X$ such that $\mathbf{p}(z) = \mathbf{d}(s) = \mathbf{d}(t)$. Thus $\exists s \cdot z$ and $\exists t \cdot z$. Put $x = s \cdot z$ and $y = t \cdot z$. Since $z \in C \cdot x_0$ then $z = a \cdot x_0$ for some $a \in C$. Thus $x = (sa) \cdot x_0$ and $y = (ta) \cdot x_0$ and so $u \cdot x = v \cdot y$. Hence $C \cdot x \cap C \cdot y$ is non-empty. By assumption $C \cdot x \cap C \cdot y = C \cdot w$ for some w. Let $w = b \cdot x = c \cdot y$ for some $b, c \in C$. Now

$$w = b \cdot x = (bsa) \cdot x_0 \text{ and } w = c \cdot y = (cta) \cdot x_0.$$

Thus $(bsa) \cdot x_0 = (cta) \cdot x_0$. By the right cancellation condition we have that $bsa = cta$. By right cancellation in C, we have that $bs = ct$.

We now show that (b, c) is a pushout of (s, t). Suppose that $hs = gt$ for some $g, h \in C$. Then $(hs) \cdot z = (gt) \cdot z$, so that $h \cdot (s \cdot z) = g \cdot (t \cdot z)$. But $s \cdot z = x$ and $t \cdot z = y$ and so $h \cdot x = g \cdot y$. Now

$$h \cdot x = g \cdot y \in C \cdot x \cap C \cdot y = C \cdot w$$

and so $h \cdot x = g \cdot y = d \cdot w$ for some $d \in C$. Now $h \cdot x = (hsa) \cdot x_0$ and $d \cdot w = (dbsa) \cdot x_0$. Thus $(hsa) \cdot x_0 = (dbsa) \cdot x_0$. By the right cancellation condition $hsa = dbsa$, and by right cancellation $h = db$. Also $g \cdot y = (gta) \cdot x_0$ and $d \cdot w = (dcta) \cdot x_0$. Thus $(gta) \cdot x_0 = (dcta) \cdot x_0$. By the right cancellation condition $gta = dcta$, and by right cancellation $g = dc$. Thus $(h, g) = d(b, c)$. The element d is unique by right cancellation.

(4) We begin by showing that ${}_C C'$ is a system.

Axiom (S1) holds: the action of C on C' is the restriction of the action of C on itself. The function $\mathbf{p}: C' \to C_o$ is defined by $\mathbf{p}(x) = \mathbf{r}(x)$. To show that C' is C-invariant, let $x \in C'$ and $a \in C$ be such that $\exists ax$. Then $\mathbf{d}(ax) = \mathbf{d}(x) = 1$ so that $ax \in C'$.

Axiom (S2) holds: in order to show that the orbit condition holds, we shall define a function $\theta: X \to C'$ as follows: let $x \in X$. Then $x \in C \cdot x_0$ so that $x = a \cdot x_0$ for some $a \in C$. But a is uniquely determined by the right cancellation condition. Thus we may define θ by the condition $x = \theta(x) \cdot x_0$. It is straightforward to check that θ is a bijection. We show first that

$$C \cdot x \cap C \cdot y = \emptyset \Leftrightarrow C\theta(x) \cap C\theta(y) = \emptyset.$$

If $C\theta(x) \cap C\theta(y)$ is non-empty, then $a\theta(x) = b\theta(y)$ for some $a, b \in C$. Thus

$$a \cdot x = (a\theta(x)) \cdot x_0 = (b\theta(y)) \cdot x_0 = b \cdot y \in C \cdot x \cap C \cdot y.$$

Hence $C \cdot x \cap C \cdot y$ is non-empty. Conversely, suppose that $C \cdot x \cap C \cdot y$ is non-empty. Then $a \cdot x = b \cdot y$ for some $a, b \in C$. Thus $(a\theta(x)) \cdot x_0 = (b\theta(y)) \cdot x_0$, and so $a\theta(x) = b\theta(y)$. Hence $C\theta(x) \cap C\theta(y)$ is non-empty. We now show that

$$C \cdot x \cap C \cdot y = C \cdot z \Leftrightarrow C\theta(x) \cap C\theta(y) = C \cdot \theta(z).$$

Suppose that $C \cdot x \cap C \cdot y = C \cdot z$. Let $w \in C\theta(x) \cap C\theta(y)$. Then $w = a\theta(x) = b\theta(y)$ for some $a, b \in C$. Thus

$$(a\theta(x)) \cdot x_0 = (b\theta(y)) \cdot x_0 = w \cdot x_0$$

and so

$$w \cdot x_0 = a \cdot x = b \cdot y.$$

Hence, by assumption, $w \cdot x_0 = c \cdot z$ for some $c \in C$. Thus $w \cdot x_0 = (c\theta(z)) \cdot x_0$. Hence $w = c\theta(z)$ and so $w \in C\theta(x) \cap C\theta(y)$ and

$$C\theta(x) \cap C\theta(y) \subseteq C\theta(z).$$

Now suppose that $c\theta(z) \in C\theta(z)$. Then $c\theta(z) \cdot x_0 = c \cdot z \in C \cdot x \cap C \cdot y$. Thus $c \cdot z = d \cdot x$ for some $d \in C$, and so $(c\theta(z)) \cdot x_0 = (d\theta(x)) \cdot x_0$. Hence $c\theta(z) = d\theta(x)$ and so $c\theta(z) \in C\theta(x)$. Similarly, we can show that $c\theta(z) \in C\theta(y)$. Hence

$$C\theta(x) \cap C\theta(y) = C\theta(z).$$

The converse is proved similarly. We can now prove that the orbit condition holds. Suppose that $Ca \cap Cb$ is non-empty, where $a, b \in C'$. By the result above $C \cdot \theta^{-1}(a) \cap C \cdot \theta^{-1}(b)$ is non-empty. Thus $C \cdot \theta^{-1}(a) \cap C \cdot \theta^{-1}(b) = C \cdot z$ for some $z \in X$. Hence $Ca \cap Cb = C\theta(z)$.

Axiom (S3) holds: for if $e \in C_o$ then $\hom(1, e)$ is non-empty, so that there is an element $c \in C'$ such that $\mathbf{d}(c) = 1$ and $\mathbf{r}(c) = e$. Thus $c \in C'$ and $\mathbf{p}(c) = e$.

Axiom (S4) holds: since C is right cancellative.

We now prove that ${}_CC'$ is equivalent to ${}_CX$. Let $I\colon C \to C$ be the identity functor, and let θ be as defined in the proof of axiom (S2). We show that (I, θ) is an equivalence of systems; we begin by showing that it is a morphism of systems.

Axiom (M1) holds: $\mathbf{p}(\theta(x)) = \mathbf{r}(\theta(x)) = \mathbf{p}(x)$, since $x = \theta(x) \cdot x_0$.

Axiom (M2) holds: let $a \in C$ and $x \in X$ be such that $a \cdot x$ exists. Now by definition $a \cdot x = (a\theta(x)) \cdot x_0$. Hence $\theta(a \cdot x) = a\theta(x)$.

Axiom (M3) holds: we have already shown this in the proof of axiom (S2) above.

It remains to show that (I, θ) satisfies axioms (ES1), (ES2) and (ES3) of Proposition 10.5.4. Of these, axiom (ES1) is immediate; axiom (ES2) holds because θ is a bijection; and the proof of axiom (ES3) is straightforward. ■

The above result shows that inverse monoids are determined by categories C satisfying the following conditions:

(L1) There is an identity $1 \in C$, called a *weak initial object*, such that $\hom(1, e)$ is non-empty for every identity e.

(L2) C is right cancellative.

(L3) If $s, t \in C$ are such that $us = vt$ for some $u, v \in C$, then s and t have a pushout.

The inverse monoid is constructed by taking $X = \{a \in C\colon \mathbf{d}(a) = 1\}$ and letting C act on X on the left by multiplication.

Monoid systems and 0-bisimple semigroups

We saw in Section 10.3 that we could construct a system from every inverse semigroup. In certain cases, it is possible to construct simpler equivalent systems. A good example of this is provided by 0-bisimple inverse semigroups. We begin by constructing a system from a 0-bisimple inverse semigroup which is different from the usual one.

Proposition 4 *Let S be a 0-bisimple inverse semigroup, and let e be any non-zero idempotent in S. Put $X_e = R_e$ and $C_e = R_e \cap eSe$. Then C_e is a right cancellative monoid, $C_e \times X_e \to X_e$ defined by $(a, x) \mapsto ax$ is a monoid action and (C_e, X_e) is a system when* $\mathbf{p}$: $X_e \to C_e$ *is defined by* $\mathbf{p}(x) = e$.

Proof We begin by showing that C_e is a right cancellative monoid. First note that C_e contains the idempotent e and so is non-empty. Let $a, b \in C_e$. Then in particular $b \mathcal{R} e$ and $a \in R_e \cap eSe$. Since $\mathcal{R}$ is a left congruence $ab \mathcal{R} ae$. But $ae = a$ and $a \mathcal{R} e$ and so $ab \mathcal{R} e$. Thus $ab \in C_e$, and so C_e is closed under multiplication. It is immediate that C_e is a monoid with identity e. Now suppose that $ac = bc$ where $a, b, c \in C_e$. Then $acc^{-1} = bcc^{-1}$. But $cc^{-1} = e$, $ae = a$ and $be = b$. Hence $a = b$, and so C_e is a right cancellative monoid.

We now show that $C_e \times X_e \to X_e$ defined by $(a, x) \mapsto ax$ is a well-defined monoid action. Let $a \in C_e$ and $x \in X_e$. Then $x \mathcal{R} e$ and so $ax \mathcal{R} ae$, since $\mathcal{R}$ is a left congruence. But $ae = a$ and $a \mathcal{R} e$. Thus $ax \mathcal{R} e$ and so $ax \in X_e$. Clearly, e fixes every element of X_e and the associativity of the action follows from the associativity of the multiplication. Observe that if $ax = bx$ where $a, b \in C_e$ and $x \in X_e$ then $axx^{-1} = bxx^{-1}$ and so $a = b$.

To prove that (C_e, X_e) is a system it only remains to check that the orbit condition holds. To do this we first prove that $C_e \cdot x = Sx \cap R_e$. for $x \in X_e$. Let $y \in Sx \cap R_e$. Then $y \mathcal{R} e$ and $y = sx$ for some $s \in S$. Thus $y = (yx^{-1})x$. We show that $yx^{-1} \in R_e \cap eSe$. Since $ey = y$ and $ex = x$ it follows that $yx^{-1} \in eSe$. We now calculate $yx^{-1}(yx^{-1})^{-1}$. This is just $yx^{-1}xy^{-1}$. But $y = sx$ and so

$$yx^{-1}xy^{-1} = sxx^{-1}xx^{-1}s^{-1} = sxx^{-1}s^{-1} = yy^{-1} = e.$$

Hence $y \in (R_e \cap eSe)x$. Conversely, let $y \in (R_e \cap eSe)x$. Then $y = ax$ where $a \in R_e \cap eSe$. Clearly $y \in Sx$. To show that $y \in R_e$ we calculate yy^{-1} thus

$$yy^{-1} = (ax)(ax)^{-1} = axx^{-1}a^{-1} = aea^{-1} = aa^{-1} = e.$$

Hence $y \in Sx \cap R_e$.

We can now set about showing that the orbit condition holds. It is easy to check that $Sx \cap Sy = Sx^{-1}xy^{-1}y$. Thus

$$C_e \cdot x \cap C_e \cdot y = S(x^{-1}xy^{-1}y) \cap R_e.$$

The right-hand side is empty precisely when $x^{-1}xy^{-1}y = 0$; to see this, suppose first that $x^{-1}xy^{-1}y$ is non-zero. Then there exists $z \in L_{x^{-1}xy^{-1}y} \cap R_e$, since S is 0-bisimple; in which case

$$C_e \cdot x \cap C_e \cdot y = Sz \cap R_e = C_e \cdot z,$$

and is non-empty. Conversely, suppose that $w \in Sx^{-1}xy^{-1}y \cap R_e$. Then $w(x^{-1}xy^{-1}y) = w$ and $w \mathcal{R} e$. Since e is non-zero, w must be non-zero. But then $x^{-1}xy^{-1}y$ must be non-zero. ■

The relationship between the above system and the standard system constructed according to Section 10.3 is described below.

Theorem 5 *Let S be a 0-bisimple inverse semigroup, and e a non-zero idempotent. Then (C_e, X_e) is equivalent to $(C(S), X_S)$.*

Proof Define (F, θ) as follows: F: $C_e \to C(S)$ is defined by $F(a) = (a, e)$, and $\theta(x) = x$. The function F is well-defined since $a \in R_e \cap eSe$ and so $aa^{-1} = e$ and $a^{-1}a \leq e$. Clearly, $\mathbf{r}(a, e) = (e, e)$ and so $F(C_e) = \text{end}((e, e))$. We show first that (F, θ) is a morphism of systems by checking that axioms (M1), (M2) and (M3) hold.

Axiom (M1) holds:

$$\mathbf{p}(\theta(a)) = \mathbf{p}(a, e) = (aa^{-1}, aa^{-1}) = (e, e)$$

and

$$F(\mathbf{p}(a)) = F(aa^{-1}) = F(e) = (e, e).$$

Axiom (M2) holds: $\theta(ax) = ax$, $F(a) = (a, e)$, $\theta(x) = x$ and $F(a) \cdot \theta(x) = ax$.

Axiom (M3) holds: suppose that $C_e \cdot x \cap C_e \cdot y$ is empty. Then by Proposition 1, $x^{-1}xy^{-1}y = 0$ and so $Sx \cap Sy = 0$. But

$$C(S) \cdot x = Sx \setminus \{0\} \text{ and } C(S) \cdot y = Sy \setminus \{0\}$$

by the proof of Theorem 10.3.2. Thus $C(S) \cdot x \cap C(S) \cdot y$ is empty. Now suppose that $C_e \cdot x \cap C_e \cdot y = C_e \cdot z$. Then by Proposition 1, the idempotent $x^{-1}xy^{-1}y$ is non-zero and $C_e \cdot z = Sz \cap R_e$ where $z \in L_{x^{-1}xy^{-1}y} \cap R_e$. But $Sx \cap Sy = Sx^{-1}xy^{-1}y$ and $z \mathcal{L} x^{-1}xy^{-1}y$ implies that $Sx \cap Sy = Sz$. It follows that $C(S) \cdot x \cap C(S) \cdot y = C(S) \cdot z$.

We now show that (F, θ) is an equivalence of systems by showing that axioms (ES1), (ES2) and (ES3) of Proposition 10.5.4 hold.

Axiom (ES1) holds: let (f, f) be any identity of $C(S)$. Then since S is 0-bisimple there exists $a \in S$ such that $e = a^{-1}a$ and $aa^{-1} = f$. Now, $(a, e) \in C(S)$, and $\mathbf{d}(a, e) = (e, e)$ and $\mathbf{r}(a, e) = (f, f)$. Thus (a, e) is an isomorphism.

Axiom (ES2) holds: Let $x \in X_S$, so that $x \in S \setminus \{0\}$. Since S is 0-bisimple there exists $y \in S$ such that $y^{-1}y = x^{-1}x$ and $yy^{-1} = e$. Thus $y \in R_e = X_e$. Put $s = xy^{-1}$. Then $s^{-1}s = e$. Thus (s, e) is an isomorphism in $C(S)$. Also,

$\exists(s,e)\cdot y$ since $\mathbf{d}(s,e)=(e,e)=\mathbf{p}(y)$, and $(s,e)\cdot y = sy = xy^{-1}y = x$. But $y=\theta(y)$. Thus $x=(s,e)\cdot\theta(y)$.

Axiom (ES3) holds: suppose that $\theta(x)=(s,e)\cdot\theta(y)$. Then $x,y\in X_e = R_e$ and $x=sy$ where $s^{-1}s\leq e$. Also $\mathbf{r}(s,e)=\mathbf{p}(\theta(x))$ and so $ss^{-1}=xx^{-1}$. Thus $e=xx^{-1}=ss^{-1}$ and so $s\in R_e$. Also $s^{-1}s\leq e$ and so $se=s$. Hence $s\in R_e\cap eSe$. ∎

Systems in which the acting category is in fact a monoid we shall term *monoid systems*. We can now prove that 0-bisimple inverse semigroups can be precisely described by monoid systems.

Theorem 6 (1) *$J({}_CX)$ is 0-bisimple if, and only if, for all $x,y\in X$ there exist isomorphisms $u,v\in C$ such that $\mathbf{p}(u\cdot x)=\mathbf{p}(v\cdot y)$.*

(2) *If C is a monoid and (C,X) a system then $J({}_CX)$ is 0-bisimple.*

(3) *$J({}_CX)$ is 0-bisimple if, and only if, ${}_CX$ is equivalent to a system of the form ${}_{C'}X'$ where C' is a monoid.*

Proof (1) Suppose that $J({}_CX)$ is 0-bisimple and let $x,y\in X$. Then $[x,x]$ and $[y,y]$ are non-zero idempotents of $J({}_CX)$. Since $J({}_CX)$ is 0-bisimple there is an element $[a,b]$ of $J({}_CX)$ such that

$$[b,b]=[y,y] \text{ and } [a,a]=[x,x].$$

Thus $b=v\cdot y$ and $a=u\cdot x$ for some u and v isomorphisms. Since $[a,b]\in J({}_CX)$, we have that $\mathbf{p}(a)=\mathbf{p}(b)$ and so $\mathbf{p}(v\cdot y)=\mathbf{p}(u\cdot x)$.

Conversely, let $[x,x]$ and $[y,y]$ be two non-zero idempotents in $J({}_CX)$. By assumption, there are isomorphisms u and v such that $\mathbf{p}(u\cdot x)=\mathbf{p}(v\cdot y)$. Hence $[u\cdot x, v\cdot y]$ is a well-defined element of $J({}_CX)$. But

$$[v\cdot y, v\cdot y]=[y,y] \text{ and } [u\cdot x, u\cdot x]=[x,x].$$

Thus $J({}_CX)$ is 0-bisimple.
(2) The function $\mathbf{p}$ maps X to $\{1\}$ the set containing the identity of the monoid. Thus for all $x,y\in X$ we have that $\mathbf{p}(x)=\mathbf{p}(y)$. Thus (1) is trivially satisfied, and so $J({}_CX)$ is 0-bisimple.
(3) Suppose that $J({}_CX)$ is 0-bisimple. Let e be any identity of C. Put $C'=\text{end}(e)$. Let f be any identity. Since $\mathbf{p}$ is surjective there are elements x and y in X such that $\mathbf{p}(x)=f$ and $\mathbf{p}(y)=e$. By (1), there exist isomorphisms u and v in C such that $u\cdot x=v\cdot y$. But

$$\mathbf{d}(u)=\mathbf{p}(x)=f,\ \mathbf{r}(u)=\mathbf{r}(v) \text{ and } \mathbf{d}(v)=\mathbf{p}(y)=e.$$

Thus $v^{-1}u \in \hom(f,e)$ is a well-defined isomorphism and so C' is a dense subcategory of C. Put $X' = \{x \in X\colon \mathbf{p}(x) = e\}$. It is straightforward to check that ${}_{C'}X'$ is a system equivalent to ${}_CX$.

Conversely, suppose that (C, X) is equivalent to (C', X'), a monoid system. Then $J({}_CX)$ is isomorphic to $J({}_{C'}X')$ by Theorem 10.5.3. By (2) the latter is 0-bisimple. ■

The above two theorems tell us that monoid systems completely characterise 0-bisimple inverse semigroups. It is worth making explicit the axioms which such systems satisfy. Let C be a monoid with identity 1 and let X be a set. Then

(MS1) There is a monoid action $C \times X \to X$.

(MS2) C is right cancellative.

(MS3) If $a \cdot x = b \cdot x$ then $a = b$.

(MS4) $C \cdot x \cap C \cdot y \neq \emptyset$ implies $C \cdot x \cap C \cdot y = C \cdot z$ for some $z \in X$.

By applying some of our previous results some special cases can be deduced rather easily. Bisimple inverse semigroups are obtained when we replace axiom (MS4) by the following axiom:

(MS5) $C \cdot x \cap C \cdot y = C \cdot z$ for all $x, y \in X$ and for some $z \in X$.

Bisimple inverse monoids are completely determined by right cancellative monoids in which the intersection of any two principal left ideals is again a principal left ideal.

0-E-unitary inverse semigroups

E-unitary and 0-E-unitary inverse semigroups have played an important rôle throughout this book so it is natural to enquire how they may be characterised in terms of category actions.

A system (C, X) satisfies the *left cancellation condition* if

$$a \cdot x = a \cdot y \Rightarrow x = y$$

for all $a \in C$ and $x, y \in X$.

Proposition 7 *If S is 0-E-unitary then $(C(S), X_S)$ satisfies the left cancellation condition.*

Proof Let $(s, e) \cdot x = (s, e) \cdot y$. Then

$$sx = sy, \; e = xx^{-1} = yy^{-1} \text{ and } s^{-1}s \leq e.$$

Put $w = s^{-1}sx = s^{-1}sy$. Then $w \leq x, y$ and so $w^{-1}w \leq x^{-1}y$. Suppose that $w^{-1}w = 0$. Then $x^{-1}s^{-1}sx = 0$ and so $sx = 0$. But this is a contradiction. Thus $w^{-1}w$ is a non-zero idempotent. Hence $x^{-1}y$ is an idempotent. But $x^{-1}x\mathcal{R}x^{-1}y$ and so $x^{-1}y = x^{-1}x$. Thus $x = y$. ∎

The desired characterisation of 0-E-unitary semigroups now follows.

Theorem 8 *Let (C, X) be a system. Then $J(_CX)$ is 0-E-unitary if, and only if, (C, X) satisfies the left cancellation condition.*

Proof Suppose that the left cancellation condition holds and that $[x, x] \leq [y, z]$ in $J(_CX)$. Then $(x, x) = a \cdot (y, z)$ for some $a \in C$. Hence $x = a \cdot y = a \cdot z$. Thus $y = z$ by the left cancellation condition, and so $[y, z]$ is an idempotent.

Conversely, suppose that $J(_CX)$ is 0-E-unitary. Let $x = a \cdot y = a \cdot z$. Then $\mathbf{d}(a) = \mathbf{p}(y)$ and $\mathbf{d}(a) = \mathbf{p}(z)$. Thus $[y, z]$ is an element of $J(_CX)$. Now $[x, x] \leq [y, z]$, and $[x, x]$ is a non-zero idempotent, so, by assumption, $[y, z]$ is an idempotent. Hence $y = z$, and consequently the left cancellation condition holds. ∎

10.8 Notes on Chapter 10

Sections 10.1–10.6

This chapter is based on my paper [191].

The basic idea which lies behind the construction of inverse semigroups from category actions was first described by Clifford [47]. He showed that every bisimple inverse monoid could be described in terms of a right cancellative monoid in which the set of principal left ideals was closed under finite intersections.

This result was subsequently generalised to bisimple inverse semigroups by Reilly [339], but to accomplish this, right cancellative monoids were replaced by what Reilly termed *RP*-systems. These systems were viewed as partial semigroups satisfying certain cancellation conditions. Later, McAlister [230] observed that 0-bisimple inverse semigroups could be described in terms of generalised *RP*-systems.

This work was developed in two important ways. Firstly, McAlister showed [234] that semigroups with zero in which any two principal left ideals are either disjoint or intersect in a principal left ideal can be used to construct 0-bisimple inverse monoids. The lack of any cancellation condition is overcome by the use of the $\mathcal{R}^*$-relation. Secondly, Leech [197] directly generalised Clifford's result to arbitrary inverse monoids: he showed that inverse monoids could be described by means of right cancellative categories with pushouts of all pairs of morphisms with a common domain and having a weak initial object.

In this chapter, we have essentially provided a joint generalisation of the work of both McAlister and Leech; in this way, we obtain a description of all inverse semigroups.

There are two key observations underpinning this work. The first concerns RP-systems. Their description as partial semigroups does not generalise, but even a superficial study of them reveals that they are nothing other than a special class of monoid actions. In view of Leech's work, this suggests that arbitrary inverse semigroups will arise from category rather than monoid actions. Our second observation clarified the nature of these actions, and arose from reading a paper of Girard [97] on linear logic. In the course of this paper, a semigroup was introduced which, to my surprise, was inverse, and whose multiplication was similar to the multiplication defined in [230]. I call this semigroup the clause semigroup. It is defined in terms of the Unification Algorithm. It is described in [191]. Studying this semigroup led to the correct definition of category actions needed for our generalisation. Such actions are discussed in [211].

Section 10.7

The multiplication in polycyclic monoids is of course the prototype for the multiplication defined in Section 10.2.

The theory of inverse monoids and cyclic systems is essentially due to Leech [197]. The only difference is axiom (L3); Leech has the following axiom instead:

(L3)* Any two elements $s, t \in C$ with $\mathbf{d}(s) = \mathbf{d}(t)$ have a pushout.

The reason for this is that we take inverse monoids with zero as our basic class of semigroups, whereas Leech takes just the class of inverse monoids: the zero is not a distinguished element. Leech's original construction of inverse monoids from categories satisfying axioms (L1), (L2) and (L3)* is in fact a special case of a general construction in category theory: the construction of categories of partial functions from categories of functions. However, Leech discovered two important facts: firstly, that Clifford's original construction of bisimple inverse monoids [47] could be interpreted categorically; and secondly, that every inverse monoid could be constructed from a suitable category.

The theory of monoid systems is equivalent to the classical theory of 0-bisimple inverse semigroups due to Clifford [47], Reilly [339] and McAlister [230]. The most general of these papers [230] introduces generalised RP-systems which are easily shown (see [191]) to be equivalent to monoid systems.

'When you finish reading this book, bind a stone to it, and cast it into the midst of the Euphrates . . . ' Jeremiah: Ch. 51–v. 63.

Bibliography

[1] **J. Adámek, H. Herrlich and G. Strecker**, *Abstract and concrete categories*, John Wiley and Sons, New York, 1990.

[2] **C. Albert, P. Molino**, *Pseudogroupes de Lie transitifs I. Structures principales*, Hermann, Paris, 1984.

[3] **C. Albert, P. Molino**, *Pseudogroupes de Lie transitifs II. Théorèmes d'intégrabilité*, Hermann, Paris, 1987.

[4] **D. Allouch**, Extensions de demi-groupes inverses, *Semigroup Forum* **16** *(1978), 111–116.*

[5] **J. Almeida, M. Delgado**, Sur certains systèmes d'equations avec contraintes dans un groupe libre, *Preprint, Universidade do Porto, 1997.*

[6] **E. Artin**, *Geometric algebra*, Wiley Interscience, New York, 1988.

[7] **M. Aschbacher**, *Finite group theory*, Cambridge University Press, Cambridge, 1988.

[8] **C. J. Ash**, Uniform labelled semilattices, *Journal of the Australian Mathematical Society (Series A)* **28** *(1979), 385–397.*

[9] **C. J. Ash**, The $\mathcal{J}$-classes of an inverse semigroup, *Journal of the Australian Mathematical Society (Series A)* **28** *(1979), 427–432.*

[10] **C. J. Ash**, Finite semigroups with commuting idempotents, *Journal of the Australian Mathematical Society (Series A)* **43** *(1987), 81–90.*

[11] **C. J. Ash**, Inevitable graphs: a proof of the type II conjecture and some related decision problems, *International Journal of Algebra and Computation* **1** *(1991), 127–146.*

[12] **C. J. Ash, T. Hall and J.-E. Pin**, On the varieties of languages associated to some varieties of finite monoids with commuting idempotents, *Information and Computation* **86** *(1990), 32–42.*

[13] **A. Asperti, V. Danos, C. Laneve, and L. Regnier**, Paths in the lambda-calculus, in *Proceedings of the Ninth annual Symposium on Logic in Computer Science*, IEEE Computer Society Press, Paris, 1994.

[14] **K. Auinger**, The congruence lattice of a strict regular semigroup, *Journal of Pure and Applied Algebra* **81** *(1992), 219–245.*

[15] **K. Auinger**, The congruence lattice of a combinatorial strict inverse semigroup, *Proceedings of the Edinburgh Mathematical Society* **37** *(1993), 25–37.*

[16] **K. Auinger**, Free strict inverse semigroups, *Journal of Algebra* **157** *(1993), 26–42.*

[17] **K. Auinger**, Residual finiteness of free products of combinatorial strict inverse semigroups, *Proceedings of the Royal Society of Edinburgh* **124A** *(1994), 137–147.*

[18] **B. B. Baird**, Inverse semigroups of homeomorphisms between open sets, *Journal of the Australian Mathematical Society (Series A)* **24** *(1977), 92–102.*

[19] **B. A. Barnes**, Representation of the l^1-algebra of an inverse semigroup, *Transactions of the American Mathematical Society* **218** *(1976), 361–396.*

[20] **M. Barr, C. Wells**, *Category theory for computing science*, Prentice Hall, London, 1990.

[21] **H. Behnke et al**, *Fundamentals of mathematics Volume II Geometry*, MIT Press, Cambridge Massachsetts, 1983.

[22] **P. Bennett**, Amalgamated free products of inverse semigroups, *Journal of Algebra* **198** *(1997), 499–537.*

[23] **P. Bennett**, On the structure of inverse semigroup amalgams, *International Journal of Algebra and Computation* **7** *(1997), 577–604.*

[24] **J. Berstel**, *Transductions and context-free languages*, B. G. Teubner, Stuttgart, 1979.

[25] **B. Billhardt**, On inverse semigroups the closure of whose set of idempotents is a Clifford semigroup, *Semigroup Forum* **44** *(1992), 320–331.*

[26] **B. Billhardt**, On a semigroup theoretic generalization of the Kalužnin Krasner theorem and normal extensions of inverse semigroups, *Semigroup Forum* **44** *(1992), 364–372.*

[27] **B. Billhardt**, On a wreath product embedding and idempotent pure congruences on inverse semigroups, *Semigroup Forum* **45** *(1992), 45–54.*

[28] **B. Billhardt**, On a wreath product embedding for regular semigroups, *Semigroup Forum* **46** *(1993), 62–72.*

[29] **B. Billhardt, I. Szittyai**, On embeddability of idempotent separating extensions of inverse semigroups, *To appear Semigroup Forum.*

[30] **J.-C. Birget, S. W. Margolis, and J. C. Meakin**, The word problem for inverse monoids presented by one idempotent relator, *Theoretical Computer Science* **123** *(1994), 273–289.*

[31] **J.-C. Birget, S. Margolis, J. C. Meakin and P. Weil**, PSPACE-complete problems for subgroups of free groups and inverse finite automata, *Preprint, University of Nebraska–Lincoln, 1998.*

[32] **J.-C. Birget, J. Rhodes**, Group theory via global semigroup theory, *Journal of Algebra* **120** *(1989), 284–300.*

[33] **W. M. Boothby**, *An introduction to differentiable manifolds and Riemannian geometry*, 2nd edition, Academic Press, Orlando, 1986.

[34] **H. Brandt**, Über eine Verallgemeinerung des Gruppenbegriffes, *Mathematische Annalen* **96** *(1926), 360–366.*

[35] **D. A. Bredikhin**, Inverse semigroups of local automorphisms of universal algebras, *Sibirskiĭ Matematicheskiĭ Zhurnal* **17** *(1976), 499–507* (Russian).

[36] **J. Brzozowski, R. Knast**, The dot-depth hierarchy of star-free events is infinite, *Journal of Computer and System Sciences* **16** *(1978), 37–55.*

[37] **S. Bulman-Fleming, J. Fountain and V. Gould**, Strongly E^*-unitary inverse semigroups, *Preprint, University of York, UK, 1997.*

[38] **X. Caicedo, A. M. Sette**, Logics and pseudogroups, in *Contemporary Mathematics* **69**, American Mathematical Society, Providence, Rhode Island, 1988, 101–115.

[39] **P. J. Cameron**, *Oligomorphic permutation groups*, Cambridge University Press, Cambridge, 1990.

[40] **P. J. Cameron, H. Deza**, On permutation geometries, *Journal of the London Mathematical Society (2)* **20** *(1979), 373–386.*

[41] **S. Y. Chen, S. C. Hsieh**, Factorisable inverse semigroups, *Semigroup Forum* **8** *(1974), 283–297.*

[42] **A. Cherubini, J. Meakin and B. Piochi**, Amalgams of free inverse semigroups, *Semigroup Forum* **54** *(1997), 199–220.*

[43] **A. Cherubini, M. Petrich**, The inverse hull of right cancellative semigroups, *Journal of Algebra* **111** *(1987), 74–113.*

[44] **D. M. Clark, P. H. Krauss**, Relatively homogeneous structures, in *Logic Colloquium 76* (eds R. Gandy, M. Hyland), North-Holland, Amsterdam, 1977, 255–285.

[45] **A. H. Clifford**, Semigroups admitting relative inverses, *Annals of Mathematics* **42** *(1941), 1037–1049.*

[46] **A. H. Clifford**, Matrix representations of completely simple semigroups, *American Journal of Mathematics* **64** *(1942), 327–342.*

[47] **A. H. Clifford**, A class of d-simple semigroups, *American Journal of Mathematics* **75** *(1953), 547–556.*

[48] **P. M. Cohn**, *Universal algebra*, D. Reidel Publishing Company, Dordrecht, 1981.

[49] **J. B. Conway, J. Duncan and A. L. T. Paterson**, Monogenic inverse semigroups and their C^*-algebras, *Proceedings of the Royal Society of Edinburgh* **98A** *(1984), 13–24.*

[50] **A. Coudron**, Sur les extensions de demi-groupes réciproques, *Bulletin de la Société Royale de Liège* **37** *(1968), 409–419.*

[51] **D. F. Cowan**, A class of varieties of inverse semigroups, *Journal of Algebra* **141** *(1991), 115–142.*

[52] **D. F. Cowan**, Inverse monoids of dot-depth two, *International Journal of Algebra and Computation* **3** *(1993), 411–424.*

[53] **D. F. Cowan, N. R. Reilly**, Partial cross-sections of symmetric inverse semigroups, *International Journal of Algebra and Computation* **5** *(1995), 259–287.*

[54] **J. Cuntz**, Simple C^*-algebras generated by isometries, *Communications in Mathematical Physics* **57** *(1977), 173–185.*

[55] **J. Cuntz, W. Krieger**, A class of C^*-algebras and topological Markov chains, *Inventiones Mathematicae* **56** *(1980), 251–268.*

[56] **J. Cuntz**, A class of C^*-algrebras and topological Markov chains and the Ext-functor for C^*-algebras, *Inventiones Mathematicae* **63** *(1981), 25–40.*

[57] **H. D'Alarcao**, Idempotent separating extensions of inverse semigroups, *Journal of the Australian Mathematical Society* **9** *(1969), 211–217.*

[58] **V. Danos, L. Regnier**, Local and asynchronous beta-reduction, in *Proceedings of the Eighth Annual Symposium on Logic in Computer Science*, Montreal, IEEE Computer Society Press, 1993.

[59] **V. Danos, L. Regnier**, Proof-nets and the Hilbert space, in *Advances in linear logic*, (eds J.-Y. Girard, Y. Lafont and L. Regnier), Cambridge University Press, 1995, 307–328.

[60] **P. Dedecker**, Quelques aspects de la théorie des structures locales, *Bulletin de la Société Mathématique de Belgique* **V** *(1952), 26–43.*

[61] **O. I. Domanov**, Semigroups of all partial automorphisms of universal algebras, *Izvestiya Vysshikh Uchebnykh Zavedeniĭ Matematika* **11** *(1971), 52–58* (Russian).

[62] **P. Dubreil**, Contribution à la théorie des demigroupes, *Mémoires de l' Académie des Sciences de l'Institut de France (2)* **63** *(1941), 1–52.*

[63] **H.-D. Ebbinghaus, J. Flum and W. Thomas**, *Mathematical logic*, 2nd edition, Springer-Verlag, New York, 1994.

[64] **C. Eberhart, J. Selden**, One-parameter inverse semigroups, *Transactions of the American Mathematical Society* **168** *(1972), 53–66.*

[65] **C. Ehresmann**, Structures locales, *Annali di matematica pura ed applicata* **36** *(1954), 133–142.*

[66] **C. Ehresmann**, Gattungen von lokalen Strukturen, *Jahresbericht der Deutschen Mathematiker-Vereinigung* **60** *(1957), 49–77.*

[67] **C. Ehresmann**, Catégories inductives et pseudogroupes, *Annales de l'Institut Fourier, Grenoble* **10** *(1960), 307–336.*

[68] **C. Ehresmann**, Groupoïdes sous-inductifs, *Annales de l'Institut Fourier, Grenoble* **13** *(1963), 1–60.*

[69] **C. Ehresmann**, Catégories ordonnées, holonomie et cohomologie, *Annales de l'Institut Fourier, Grenoble* **14** *(1964), 205–268.*

[70] **C. Ehresmann**, *Catégories et structures*, Dunod, Paris, 1965.

[71] **C. Ehresmann**, *Oeuvres complètes et commentées*, (ed. A. C. Ehresmann) Supplements to *Cahiers de Topologie et Géométrie Différentielle*, Amiens, 1980–83.

[72] **S. Eilenberg**, Foundations of fiber bundles, *Lecture Notes, University of Chicago, Summer 1957.*

[73] **S. Eilenberg**, *Automata, languages and machines*, Volume A, Academic Press, New York, 1974.

[74] **D. B. A. Epstein et al**, *Word processing in groups*, Jones and Bartlett, Boston, 1992.

[75] **C. Faith**, *Algebra I Rings, modules and categories*, Springer-Verlag, Berlin, 1981.

[76] **R. P. Feynman**, *Feynman lectures in computation*, Addison-Wesley, Reading, Massachusetts, 1996.

[77] **K. Fichtner**, Zur Symmetriebeschreibung von OD-Kristallstrukturen durch Brandtsche und Ehresmannsche Gruppoide, *Beiträge zur Algebra und Geometrie* **6** *(1977), 71–99.*

[78] **K. Fichtner**, Über geometrische Äquivalenz von Symmetrie-Gruppoiden, in *Algebraische Modelle, Kategorien und Gruppoide*, (ed. H.-J. Hoehnke) Akademie-Verlag, Berlin, 1979, 85–90.

[79] **K. Fichtner**, On groupoids in crystallography, *Match no.* **9** *(1980), 21–40.*

[80] **K. Fichtner**, Non-space-group symmetry in crystallography, *Computers and Mathematics with Applications* **12B** *(1986), 751–762.*

[81] **D. G. FitzGerald**, Inverse semigroups of bicongruences on algebras, particularly semilattices, in *Lattices, semigroups and universal algebra* (eds J. Almeida et al), Plenum, New York, 1990, 59–66.

[82] **D. G. FitzGerald**, Topics in dual symmetric inverse semigroups, in *Proceedings of the Monash Conference on Semigroup Theory* (eds T. E. Hall et al), World Scientific, Singapore, 1991, 62–67.

[83] **D. G. FitzGerald**, Normal bands and their inverse semigroups of bicongruences, *Journal of Algebra* **185** *(1996), 502–526.*

[84] **J. B. Fountain, J.-E. Pin and P. Weil**, Covers for monoids, *Preprint, University of York, UK, 1998.*

[85] **R. Fraïssé**, *Course of mathematical logic Volume I*, D. Reidel Publishing Company, Dordrecht, 1973.

[86] **R. Fraïssé**, *Course of mathematical logic Volume II*, D. Reidel Publishing Company, Dordrecht, 1974.

[87] **R. Fraïssé**, *Theory of relations*, North-Holland, Amsterdam, 1986.

[88] **R. Fricke, A. Ostrowski eds**, *Felix Klein gesammelte mathematische Abhandlungen*, Erster Band, Verlag von Julius Springer, Berlin, 1921.

[89] **R. Fritzsche, H.-J. Hoehnke**, Heinrich Brandt 1886-1986, Martin-Luther-Universität Halle-Wittenberg Wissenschaftliche Beiträge 1986.

[90] **D. Gaboriau, G. Levitt, and F. Paulin**, Pseudogroups of isometries of $\mathbb{R}$ and Rips' theorem on free actions on $\mathbb{R}$-trees, *Israel Journal of Mathematics* **87** *(1994), 403–428.*

[91] **G. Gelbrich**, Self-similar tilings and expanding homomorphisms of groups, *Archiv der Mathematik* **65** *(1995), 481–491.*

[92] **S. M. Gersten**, Intersection of finitely generated subgroups of free groups and resolutions of graphs, *Inventiones Mathematicae* **71** *(1983), 567–591.*

[93] **R. H. Gilman**, Formal languages and infinite groups, *DIMACS Series Discrete Mathematics and Theoretical Computer Science American Mathematical Society* **25** *(1996), 27–51.*

[94] **J.-Y. Girard**, Towards a geometry of interaction, *Contemporary Mathematics* **92** *(1989), 69–108.*

[95] **J.-Y. Girard**, Geometry of interaction 1: interpretation of system F, in *Logic Colloquium '88* (eds R. Ferro et al), North-Holland, 1989, 221–260.

[96] **J.-Y. Girard**, Geometry of interaction 2: deadlock-free algorithms, *Proceedings of COLOG-88 (Berlin)* (eds P. Martin-Löf, G. Mints), Springer Lecture Notes in Computer Science 417, 1989, 76–93.

[97] **J.-Y. Girard**, Geometry of interaction III: the general case, *Advances in linear logic* (eds J.-Y. Girard, Y. Lafont and L. Regnier), Cambridge University Press, Cambridge, 1993.

[98] **L. M. Gluskin**, Elementary generalized groups, *Matematicheskiĭ Sbornik* **41** *(1957), 23–36* (Russian).

[99] **S. M. Goberstein**, Partial automorphisms of inverse semigroups, in *Proceedings of the 1984 Marquette conference on semigroups* (eds K. Byleen, P. Jones and F. Pastijn), 29–43.

[100] **S. M. Goberstein**, On the structure of partial automorphism semigroups, in *Lattices, semigroups and universal algebra* (eds J. Almeida et al), Plenum, New York, 1990, 71–79.

[101] **S. Gołąb**, Über den Begriff der "Pseudogruppe von Transformationen", *Mathematische Annalen* **116** *(1939), 768–780.*

[102] **G. M. S. Gomes, J. M. Howie**, A P-theorem for inverse semigroups with zero, *Portugaliae Mathematica* **53** *(1996), 257–278.*

[103] **G. M. S. Gomes, M. B. Szendrei**, Idempotent pure extensions by inverse semigroups via quivers, *Journal of Pure and Applied Algebra* **127** *(1998), 15–38.*

[104] **M. Grandis**, Concrete representations for inverse and distributive exact categories, *Rendiconti Accademia Nazionale delle Scienze detta dei XL Memorie di Matematica* **VIII** *(1984), 99–120.*

[105] **M. Grandis**, Cohesive categories and manifolds, *Annali di Matematica pura ed applicata,* **CLVII** *(1990), 199–244.*

[106] **J. W. Gray**, A theory of pseudogroups with applications to contact structures, *Technical Report No. 65 Applied Mathematics and Statistics Laboratory Stanford University, California, 1957.*

[107] **D. G. Green**, Extensions of a semilattice by an inverse semigroup, *Bulletin of the Australian Mathematical Society* **9** *(1973), 21–31.*

[108] **D. G. Green**, The lattice of congruences on an inverse semigroup, *Pacific Journal of Mathematics* **57** *(1975), 141–152.*

[109] **P. A. Grillet**, *Semigroups*, Marcel Dekker, New York, 1995.

[110] **M. Hacque**, Prolongements d'isomorphismes et théorie de Galois, *Journal of Algebra* **113** *(1988), 136–200.*

[111] **D. E. Haile**, The Brauer monoid of a field, *Journal of Algebra* **81** *(1983), 521–539.*

[112] **D. E. Haile, R. G. Larson and M. E. Sweedler**, A new invariant for $\mathbb{C}$ over $\mathbb{R}$, *American Journal of Mathematics* **CV** *(1983), 689–814.*

[113] **M. Hall**, Coset representations in free groups, *Transactions of the American Mathematical Society* **67** *(1949), 421 432.*

[114] **M. Hall**, A topology for free groups and related groups, *Annals of Mathematics* **52** *(1950), 127–139.*

[115] **T. E. Hall**, Free products with amalgamation of inverse semigroups, *Journal of Algebra* **34** *(1975), 375–385.*

[116] **T. E. Hall**, Inverse and regular semigroups and amalgamation: a brief survey, *Proceedings of a Symposium on regular semigroups, Northern Illinois University, DeKalb*, 1979, 49–79.

[117] **T. E. Hall**, Biprefix codes, inverse semigroups and syntactic monoids of injective automata, *Theoretical Computer Science* **32** *(1984), 201–213.*

[118] **P. R. Halmos**, *Finite-dimensional vector spaces*, Springer-Verlag, New York, 1974.

[119] **R. Hancock, I. Raeburn**, The C^*-algebras of some inverse semigroups, *Bulletin of the Australian Mathematical Society* **42** *(1990), 335–348*

[120] **M. Hasse, L. Michler**, *Theorie der Kategorien*, VEB Deutscher Verlag der Wissenschaften, Berlin, 1966.

[121] **S. Haataja, S. W. Margolis, and J. Meakin**, Bass–Serre theory for groupoids and the structure of full regular semigroup amalgams, *Journal of Algebra* **183** *(1996), 38–54.*

[122] **T. Hawkins**, The *Erlanger Programm* of Felix Klein: reflections on its place in the history of mathematics, *Historica Mathematica* **11** *(1984), 442–470.*

[123] **G. Hector, U. Hirsch**, *Introduction to the geometry of foliations Part A*, 2nd edition, Vieweg & Sohn, Braunschweig, 1986.

[124] **G. Hector, U. Hirsch**, *Introduction to the geometry of foliations Part B*, 2nd edition, Vieweg & Sohn, Braunschweig, 1987.

[125] **K. Henckell, S. W. Margolis, J.-E. Pin and J. Rhodes**, Ash's type II theorem, profinite topology and Malcev products part I, *International Journal of Algebra and Computation* **1** *(1991), 411–436.*

[126] **K. Henckell, J. Rhodes**, The theorem of Knast, the $PG = BG$ and type-II conjectures, in *Monoids and semigroups with applications* (ed. J. Rhodes), World Scientific, Singapore, 1991, 453–463.

[127] **T. Herbst, R. M. Thomas**, Group presentations, formal languages and characterizations of one-counter groups, *Theoretical Computer Science* **112** *(1993), 187–213.*

[128] **B. Herwig**, Extending partial isomorphisms on finite structures, *Combinatorica* **15** *(1995), 365–371.*

[129] **B. Herwig**, Extending partial isomorphisms for the small index property of many ω-categorical structures, *Preprint, Albert-Ludwigs-Universität Freibourg, 1997.*

[130] **B. Herwig, D. Lascar**, Extending partial automorphisms and the profinite topology on the free groups, *Preprint, Albert-Ludwigs-Universität Freibourg and University Paris 7, 1997.*

[131] **J. B. Hickey**, A type of product for uniform semilattices, *Semigroup Forum* **6** *(1973), 198–215.*

[132] **J. B. Hickey**, A family of bisimple inverse semigroups, *Semigroup Forum* **11** *(1975), 30–48.*

[133] **Ph. J. Higgins**, *Categories and groupoids*, Van Nostrand Reinhold, New York, 1971.

[134] **G. Higman, B. H. Neumann and H. Neumann**, Embedding theorems for groups, *Journal of the London Mathematical Society* **24** *(1949), 247–254.*

[135] **P. Hines**, *The algebra of self-similarity and its applications*, PhD Thesis, University of Wales, Bangor, 1998.

[136] **P. Hines, M. V. Lawson**, An application of polycyclic monoids to rings, *Semigroup Forum* **56** *(1998), 146–149.*

[137] **W. Hodges**, What is a structure theory?, *Bulletin of the London Mathematical Society* **19** *(1987), 209–237.*

[138] **W. Hodges, I. Hodkinson, D. Lascar and S. Shelah**, The small index property for ω-stable ω-categorical structures and for the random graph, *Journal of the London Mathematical Society (2)* **48** *(1993), 204–218*

[139] **H.-J. Hoehnke**, On certain classes of categories and monoids constructed from abstract Mal'cev clones I, in *Proceedings of the V Universal Algebra Symposium* (eds K. Hałkowska, B. Stawski), World Scientific, Singapore, 1989, 149–176.

[140] **H.-J. Hoehnke**, On certain classes of categories and monoids constructed from abstract Mal'cev clones II, in *Potsdamer Forschungen der Pädagogischen Hochschule 'Karl Liebknecht'* (eds K. Denecke, H.-J. Vogel), Heft 62, Potsdam, 1989, 30–64.

[141] **K. H. Hofmann, M. W. Mislove**, All compact Hausdorff lambda models are degenerate, *Fundamenta Informatica* **22** *(1995), 23–52.*

[142] **K. H. Hofmann, M. W. Mislove**, Principles underlying the degeneracy of topological models of the untyped lambda calculus, in *Semigroup theory and its applications* (eds K. H. Hofmann, M. W. Mislove), Cambridge University Press, Cambridge, 1996, 123–155.

[143] **A. Horn, N. Kimura**, The category of semilattices, *Algebra Universalis* **1** *(1971), 26–38.*

[144] **C. H. Houghton**, Embedding inverse semigroups in wreath products, *Glasgow Mathematics Journal* **17** *(1976), 77–82.*

[145] **J. M. Howie**, The maximum idempotent-separating congruence on an inverse semigroup, *Proceedings of the Edinburgh Mathematical Society (2)* **14** *(1964), 71–79.*

[146] **J. M. Howie**, *Introduction to semigroup theory*, Academic Press, London, 1976.

[147] **J. M. Howie**, A congruence-free inverse semigroup associated with a pair of infinite cardinals, *Journal of the Australian Mathematical Society (Series A)* **31** *(1981), 337–342.*

[148] **J. M. Howie**, *Fundamentals of semigroup theory*, Clarendon Press, Oxford, 1995.

[149] **J. M. Howie, M. P. O. Marques-Smith**, Inverse semigroups generated by nilpotent transformations, *Proceedings of the Royal Society of Edinburgh* **99A** *(1984), 153–162.*

[150] **E. Hrushovski**, Extending partial isomorphisms of graphs, *Combinatorica* **12** *(1992), 411–416.*

[151] **N. Jacobson**, Some remarks on one-sided inverses, *Proceedings of the American Mathematical Society* **1** *(1950), 352–355.*

[152] **S. M. Jekel**, Pseudogroups and homology, *Lecture Notes, Northeastern University Topology Seminar, 1985.*

[153] **P. T. Johnstone**, *Stone spaces*, Cambridge University Press, Cambridge, 1986.

[154] **P. R. Jones**, The lattice of inverse subsemigroups of a reduced inverse semigroup, *Glasgow Mathematics Journal* **17** *(1976), 161–172.*

[155] **P. R. Jones**, On congruence lattices of regular semigroups, *Journal of Algebra* **82** *(1983), 18–39.*

[156] **P. R. Jones**, Free products of inverse semigroups, *Transactions of the American Mathematical Society* **282** *(1984), 293–318.*

[157] **P. R. Jones**, Lattice isomorphisms of free products of inverse semigroups, *Journal of Algebra* **89** *(1984), 280–290.*

[158] **P. R. Jones**, Inverse semigroups and their lattices of inverse subsemigroups, in *Lattices, semigroups and universal algebra* (eds J. Almeida et al), Plenum Press, New York, 1990, 115–127.

[159] **P. R. Jones**, Free products of inverse semigroups: a survey, with examples, *Proceedings of the Monash Conference on Semigroup Theory* (eds T. E. Hall, P. R. Jones and J. C. Meakin) World Scientific, Singapore, 1991, 140–156.

[160] **P. R. Jones, P. G. Trotter**, The Howson property for free inverse semigroups, *Simon Stevin* **63** *(1989), 277–284.*

[161] **P. R. Jones, S. W. Margolis, J. Meakin and J. B. Stephen**, Free products of inverse semigroups II, *Glasgow Mathematical Journal* **33** *(1991), 373–387.*

[162] **G. Joubert**, Contribution à l'étude des catégories ordonnées, applications aux structures feuilletées, *Cahiers de Topologies et Géométrie Différentielle* **VIII** *(1966), 1–117.*

[163] **J. Kadourek**, On varieties of combinatorial inverse semigroups I, *Semigroup Forum* **43** *(1991), 305–330.*

[164] **J. Kadourek**, On varieties of combinatorial inverse semigroups II, *Semigroup Forum* **44** *(1992), 53–78.*

[165] **J. Karnofsky, J. Rhodes**, Decidability of complexity one-half for finite semigroups, *Semigroup Forum* **24** *(1982), 55–66.*

[166] **J. Kastl**, Inverse categories, in *Algebraische Modelle, Kategorien und Gruppoide* (ed. H.-J. Hoehnke), Akademie Verlag, Berlin, 1979, 51–60.

[167] **M. Keenan, G. Lallement**, On certain codes admitting inverse semigroups as syntactic monoids, *Semigroup Forum* **8** *(1974), 312–331.*

[168] **J. Kellendonk**, Noncommutative geometry of tilings and gap labelling, *Reviews in Mathematical Physics* **7** *(1995), 1133–1180.*

[169] **J. Kellendonk**, The local structure of tilings and their integer group of coinvariants, *Communications in Mathematical Physics* **187** *(1997), 115–157.*

[170] **J. Kellendonk**, Topological equivalence of tilings, *Journal of Mathematical Physics* **38** *(1997), 1823–1842.*

[171] **B. P. Kochin**, The structure of inverse ideally-simple ω-semigroups, *Vestnik Leningradskogo Universiteta* **23** *(1968), 41–50* (Russian).

[172] **A. Kock**, The algebraic theory of moving frames, *Cahiers de Topologie et Géométrie Différentielle* **XXIII** *(1982), 347–362.*

[173] **A. Kock, I. Moerdijk**, Local groupoids and étendues, *Talk at 54 PSSL, University of Sussex, 1994.*

[174] **A. Kock, I. Moerdijk**, Presentations of étendues, *Cahiers de Topologie et Géométrie Différentielle Catégoriques* **32** *(1991), 145–164.*

[175] **A. Kumjian**, On localizations and simple C^*-algebras, *Pacific Journal of Mathematics* **112** *(1984), 141–192.*

[176] **G. Lallement**, *Semigroups and combinatorial applications*, John Wiley and Sons, New York, 1979.

[177] **G. Lallement, E. Milito**, Recognizable languages and finite semilattices of groups, *Semigroup Forum* **11** *(1975), 181–184.*

[178] **S. Lang**, *Algebra*, 3rd edition, Addison-Wesley, Reading Massachusetts, 1993.

[179] **H. Lausch**, Cohomology of inverse semigroups, *Journal of Algebra* **35** *(1975), 273–303.*

[180] **M. V. Lawson**, The geometric theory of inverse semigroups I: E-unitary inverse semigroups, *Journal of Pure and Applied Algebra* **67** *(1990), 151–177.*

[181] **M. V. Lawson**, Semigroups and ordered categories I: the reduced case, *Journal of Algebra* **141** *(1991), 422–462.*

[182] **M. V. Lawson**, The geometric theory of inverse semigroups II: E-unitary covers of inverse semigroups, *Journal of Pure and Applied Algebra* **83** *(1992), 121–139.*

[183] **M. V. Lawson**, Congruences on ordered groupoids, *Semigroup Forum* **47** *(1993), 150–167.*

[184] **M. V. Lawson**, A note on a paper of Joubert, *Semigroup Forum* **47** *(1993), 389–392.*

[185] **M. V. Lawson**, An equivalence theorem for inverse semigroups, *Semigroup Forum* **47** *(1993), 7–14.*

[186] **M. V. Lawson**, Coverings and embeddings of inverse semigroups, *Proceedings of the Edinburgh Mathematical Society* **36** *(1993), 399–419.*

[187] **M. V. Lawson**, Extending partial isomorphisms and McAlister's covering theorem, *Semigroup Forum* **48** *(1994), 18–27.*

[188] **M. V. Lawson**, Almost factorisable inverse semigroups, *Glasgow Mathematical Journal* **36** *(1994), 97–111.*

[189] **M. V. Lawson**, A class of actions of inverse semigroups, *Journal of Algebra* **179** *(1996), 570–598.*

[190] **M. V. Lawson**, Enlargements of regular semigroups, *Proceedings of the Edinburgh Mathematical Society* **39** *(1996), 425–460*

[191] **M. V. Lawson**, Constructing inverse semigroups from category actions, *To appear Journal of Pure and Applied Algebra.*

[192] **M. V. Lawson**, The Möbius inverse monoid, *Journal of Algebra* **200** *(1998), 428–438.*

[193] **M. V. Lawson**, McAlister semigroups, *Journal of Algebra* **202** *(1998), 276–294.*

[194] **M. V. Lawson**, The structure of 0-E-unitary inverse semigroups I: the monoid case, *Accepted for publication by the Proceedings of the Edinburgh Mathematical Society.*

[195] **F. W. Lawvere**, Metric spaces, generalized logic and closed categories, *Rend. Sem. Matematico è Fisico Univ. di Milano* **43** *(1974), 135–166.*

[196] **F. W. Lawvere**, Qualitative distinctions between some toposes of generalized graphs, *Contemporary Mathematics* **92** *(1989), 261–299.*

[197] **J. Leech**, Constructing inverse monoids from small categories, *Semigroup Forum* **36** *(1987), 89–116.*

[198] **J. Leech**, Inverse monoids with a natural semilattice ordering, *Proceedings of the London Mathematical Society (3)* **70** *(1995), 146–182.*

[199] **G. Levitt**, La dynamique des pseudogroupes de rotations, *Inventiones Mathematicae* **113** *(1993), 633–670.*

[200] **A. E. Liber**, On symmetric generalised groups, *Matematicheskiĭ Sbornik* **33** *(1953), 531–544* (Russian).

[201] **A. E. Liber**, On the theory of generalised groups, *Doklady Akademiĭ Nauk SSSR* **97** *(1954), 25–28* (Russian).

[202] **A. L. Libih**, On the theory of inverse semigroups of local automorphisms, in *Theory of semigroups and its applications*, Saratov University Press, Saratov, **3** (1974), 46–59 (Russian).

[203] **L. M. Lima**, Nilpotent local automorphisms of an independence algebra, *Proceedings of the Royal Society of Edinburgh Section* **124A** *(1994), 423–436.*

[204] **S. L. Lipscomb**, *Symmetric inverse semigroups*, American Mathematical Society, Providence, Rhode Island, 1996.

[205] **E. S. Ljapin**, Canonical form of elements of an associative system given by defining relations, *Leningradskiĭ Gosudarstvennyiĭ Pedagogicheskiĭ Institut Uchenye Zapiski* **89** *(1953), 45–54.* (Russian)

[206] **M. Loganathan**, Cohomology of inverse semigroups, *Journal of Algebra* **70** *(1981), 375–393.*

[207] **A. L. Mackay**, Post-classical crystallography, *Journal de Physique, Colloque C3, supplément au No. 7* **47** *(1986), 153–163.*

[208] **A. L. Mackay**, Generalised crystallography, *Computers and Mathematics with Applications* **12B** *(1986), 21–37.*

[209] **K. Mackenzie**, *Lie groupoids and lie algebroids in differential geometry*, Cambridge University Press, Cambridge, 1987.

[210] **S. MacLane**, *Categories for the working mathematician*, Springer-Verlag, Berlin, 1971.

[211] **S. MacLane, I. Moerdijk**, *Sheaves in geometry and logic*, Springer-Verlag, Berlin, 1992.

[212] **W. Magnus, A. Karrass, and D. Solitar**, *Combinatorial group theory*, 2nd revised edition, Dover Publications, Inc., New York, 1976.

[213] **S. W. Margolis**, Consequences of Ash's proof of the Rhodes Type II Conjecture, in *Proceedings of the Monash Conference on Semigroup Theory* (eds T. E. Hall, P. R. Jones and J. C. Meakin), World-Scientific, Singapore, 1991, 180–205.

[214] **S. W. Margolis**, Kernels and expansions: an historical and technical perspective, in *Monoids and semigroups with applications* (ed. J Rhodes), World Scientific, Singapore, 1991, 3–30

[215] **S. W. Margolis, J. C. Meakin**, E-unitary inverse monoids and the Cayley graph of a group presentation, *Journal of Pure and Applied Algebra* **58** *(1989), 45–76.*

[216] **S. W. Margolis, J. C. Meakin**, Inverse monoids and rational Schreier subsets of the free group, *Semigroup Forum* **44** *(1992), 137–148.*

[217] **S. W. Margolis, J. C. Meakin**, Inverse monoids, trees and context-free languages, *Transactions of the American Mathematical Society* **335** *(1993), 259–276.*

[218] **S. W. Margolis, J. C. Meakin**, Free inverse monoids and graph immersions, *International Journal of Algebra and Computation* **3** *(1993), 79–99.*

[219] **S. W. Margolis, J. Meakin and M. Sapir**, Algorithmic problems in groups, semigroups and inverse semigroups, in *Semigroups, formal languages and groups* (ed. J. Fountain), Kluwer Academic Publishers, 1995, 147–214.

[220] **S. W. Margolis, J. Meakin, and J. B. Stephen**, Some decision problems for inverse monoid presentations, in *Semigroups and their applications* (eds S. M. Goberstein, P. M. Higgins), Reidel, Dordrecht, 1987, 99–110.

[221] **S. W. Margolis, J. C. Meakin and J. B. Stephen**, Free objects in certain varieties of inverse semigroups, *Canadian Journal of Mathematics* **42** *(1990), 1084–1097.*

[222] **S. W. Margolis, J.-E. Pin**, Languages and inverse semigroups, in *Proceedings, 11th ICALP, Lecture Notes in Computer Science* **172**, 1984, 337–345.

[223] **S. W. Margolis, J.-E. Pin**, Inverse semigroups and extensions of groups by semilattices, *Journal of Algebra* **110** *(1987), 277–297.*

[224] **S. W. Margolis, J.-E. Pin**, Expansions, free inverse semigroups, and Schützenberger product, *Journal of Algebra* **110** *(1987), 298–305.*

[225] **S. W. Margolis, J.-E. Pin**, Inverse semigroups and varieties of finite semigroups, *Journal of Algebra* **110** *(1987), 306–323.*

[226] **S. W. Margolis, J.-E. Pin**, New results on the conjecture of Rhodes and on the topological conjecture, *Journal of Pure and Applied Algebra* **80** *(1992), 305–313.*

[227] **D. B. McAlister**, A homomorphism theorem for semigroups, *Journal of the London Mathematical Society* **43** *(1968), 355–366.*

[228] **D. B. McAlister**, Inverse semigroups separated over a subsemigroup, *Transactions of the American Mathematical Society* **182** *(1973), 85–117.*

[229] **D. B. McAlister**, On 0-simple inverse semigroups, *Semigroup Forum* **8** *(1974), 347–360.*

[230] **D. B. McAlister**, 0-bisimple inverse semigroups, *Proceedings of the London Mathematical Society (3)* **28** *(1974), 193–221.*

[231] **D. B. McAlister**, Groups, semilattices and inverse semigroups, *Transactions of the American Mathematical Society* **192** *(1974), 227–244.*

[232] **D. B. McAlister**, Groups, semilattices and inverse semigroups II, *Transactions of the American Mathematical Society* **196** *(1974), 251–270.*

[233] **D. B. McAlister**, $\vee$-prehomomorphisms on inverse semigroups, *Pacific Journal of Mathematics* **67** *(1976), 215–231.*

[234] **D. B. McAlister**, One-to-one partial right translations of a right cancellative semigroup, *Journal of Algebra* **43** *(1976), 231–251.*

[235] **D. B. McAlister**, Some covering and embedding theorems for inverse semigroups, *Journal of the Australian Mathematical Society (Series A)* **22** *(1976), 188–211.*

[236] **D. B. McAlister**, Inverse semigroups generated by a pair of subgroups, *Proceedings of the Royal Society of Edinburgh* **77A** *(1977), 9–22.*

[237] **D. B. McAlister**, E-unitary inverse semigroups over semilattices, *Glasgow Mathematical Journal* **19** *(1978), 1–12.*

[238] **D. B. McAlister**, Amenably ordered inverse semigroups, *Journal of Algebra* **65** *(1980), 118–146.*

[239] **D. B. McAlister**, Embedding inverse semigroups in coset semigroups, *Semigroup Forum* **20** *(1980), 255–267.*

[240] **D. B. McAlister**, Regular semigroups, fundamental semigroups and groups, *Journal of the Australian Mathematical Society (Series A)* **29** *(1980), 475–503.*

[241] **D. B. McAlister**, A random ramble through inverse semigroups, in *Semigroups* (eds T. E. Hall et al), Academic Press, 1980, 1–20.

[242] **D. B. McAlister**, Rees matrix covers for regular semigroups, *Journal of Algebra* **89** *(1984), 264–279.*

[243] **D. B. McAlister**, Some covering theorems for locally inverse semigroups, *Journal of the Australian Mathematical Society* **39** *(1985), 63–74.*

[244] **D. B. McAlister**, Quasi-ideal embeddings and Rees matrix covers for regular semigroups, *Journal of Algebra* **152** *(1992), 166–183.*

[245] **D. B. McAlister, R. McFadden**, Zig-zag representations and inverse semigroups, *Journal of Algebra* **32** *(1974), 178–206.*

[246] **D. B. McAlister, N. R. Reilly**, E-unitary covers for inverse semigroups, *Pacific Journal of Mathematics* **68** *(1977), 161–174.*

[247] **R. McFadden, L. O'Carroll**, F-inverse semigroups, *Proceedings of the London Mathematical Society (3)* **22** *(1971), 652–666.*

[248] **J. Meakin**, Coextensions of inverse semigroups, *Journal of Algebra* **45** *(1977), 315–333.*

[249] **J. Meakin**, The partially ordered set of $\mathcal{J}$-classes of a semigroup, *Journal of the London Mathematical Society (2)* **21** *(1980), 244–256.*

[250] **J. Meakin**, An invitation to inverse semigroup theory, in *Ordered structures and algebra of computer languages* (eds K. P. Shum, P. C. Yuen), World Scientific, Singapore, 1993, 91–115.

[251] **J. Meakin, M. Sapir**, Congruences on free monoids and submonoids of polycyclic monoids, *Journal of the Australian Mathematical Society (Series A)* **54** *(1993), 236–254.*

[252] **J. Meakin, M. Sapir**, The word problem in the variety of inverse semigroups with abelian covers, *Journal of the London Mathematical Society (2)* **53** *(1996), 79–98.*

[253] **L. Michler**, Bemerkungen über induktive Gruppoide, *Mathematische Nachrichten* **43** *(1970), 47–64.*

[254] **L. Michler, J. Schreckenberger**, Über die induktiven Gruppoide der partiellen Automorphismen von Algebren, *Mathematische Nachrichten* **72** *(1976), 219–230.*

[255] **L. Michler, J. Schreckenberger**, Über die Vollständigkeit algebraischer induktiver Gruppoide, in *Algebraische Modelle, Kategorien und Gruppoide*, Akademie-Verlag, Berlin, 1979, 61–83.

[256] **P. Molino**, *Riemannian foliations*, Birkhäuser, Boston, 1988.

[257] **W. D. Munn**, Matrix representations of semigroups, *Proceedings of the Cambridge Mathematical Society* **53** *(1957), 5–12.*

[258] **W. D. Munn**, The characters of the symmetric inverse semigroup, *Proceedings of the Cambridge Philosophical Society* **53** *(1957), 13–18.*

[259] **W. D. Munn**, A class of irreducible matrix representations of an arbitrary inverse semigroup, *Proceedings of the Glasgow Mathematical Association* **5** *(1961), 41–48.*

[260] **W. D. Munn**, Brandt congruences on inverse semigroups, *Proceedings of the London Mathematical Society* **14** *(1964), 154–164.*

[261] **W. D. Munn**, A certain sublattice of the lattice of congruences on a regular semigroup, *Proceedings of the Cambridge Philosophical Society* **60** *(1964), 385–391.*

[262] **W. D. Munn**, The lattice of congruences on a bisimple ω-semigroup, *Proceedings of the Royal Society of Edinburgh* **67A** *(1966), 175–184.*

[263] **W. D. Munn**, Uniform semilattices and bisimple inverse semigroups, *Quarterly Journal of Mathematics Oxford (2)* **17** *(1966), 151–159.*

[264] **W. D. Munn**, The idempotent separating congruences on a regular 0-bisimple semigroup, *Proceedings of the Edinburgh Mathematical Society (2)* **15** *(1967), 233–240.*

[265] **W. D. Munn**, Regular ω-semigroups, *Glasgow Mathematical Journal* **9** *(1968), 46–66.*

[266] **W. D. Munn**, Fundamental inverse semigroups, *Quarterly Journal of Mathematics Oxford (2)* **21** *(1970), 157–170.*

[267] **W. D. Munn**, 0-bisimple inverse semigroups, *Journal of Algebra* **15** *(1970), 570–588.*

[268] **W. D. Munn**, On simple inverse semigroups, *Semigroup Forum* **1** *(1970), 63–74.*

[269] **W. D. Munn**, Free inverse semigroups, *Semigroup Forum* **5** *(1973), 262–269.*

[270] **W. D. Munn**, Free inverse semigroups, *Proceedings of the London Mathematical Society (3)* **29** *(1974), 385–404.*

[271] **W. D. Munn**, Congruence-free inverse semigroups, *Quarterly Journal of Mathematics Oxford (2)* **25** *(1974), 463–484.*

[272] **W. D. Munn**, A note on congruence-free inverse semigroups, *Quarterly Journal of Mathematics Oxford (2)* **26** *(1975), 385–387.*

[273] **W. D. Munn**, A note on E-unitary inverse semigroups, *Bulletin of the London Mathematical Society* **8** *(1976), 71–76.*

[274] **W. D. Munn**, Semiunitary representations of inverse semigroups, *Journal of the London Mathematical Society (2)* **18** *(1978), 75–80.*

[275] **W. D. Munn**, Ideals of free inverse semigroups, *Journal of the Australian Mathematical Society* **30** *(1980), 157–167.*

[276] **W. D. Munn**, An embedding theorem for free inverse semigroups, *Glasgow Mathematical Journal* **22** *(1981), 217–227.*

[277] **W. D. Munn**, Congruence-free regular semigroups, *Proceedings of the Edinburgh Mathematical Society* **28** *(1985), 113–119.*

[278] **W. D. Munn, R. Penrose**, A note on inverse semigroups, *Proceedings of the Cambridge Philosophical Society* **51** *(1955), 396–399.*

[279] **W. D. Munn, N. R. Reilly**, Congruences on a bisimple ω-semigroup, *Proceedings of the Glasgow Mathematical Association* **7** *(1966), 184–192.*

[280] **K. S. S. Nambooripad**, *Structure of regular semigroups I*, Memoirs of the American Mathematical Society **224** (1979).

[281] **K. S. S. Nambooripad**, The natural partial order on a regular semigroup, *Proceedings of the Edinburgh Mathematical Society* **23** *(1980), 249–260.*

[282] **K. S. S. Nambooripad, F. J. Pastijn**, Amalgamation of regular semigroups, *Houston Journal of Mathematics* **15** *(1989), 249–254.*

[283] **K. S. S. Nambooripad, R. Veeramony**, Subdirect products of regular semigroups, *Semigroup Forum* **28** *(1983), 265–307.*

[284] **T. Nishimori**, Some remarks in a qualitative theory of similarity pseudogroups, *Hokkaido Mathematical Journal* **24** *(1995), 161–177.*

[285] **M. Nivat**, Sur les automates a memoir pile, in *Proceedings of the International Computing symposium, Bonn 1970*, (ed. W. Itzfeld), North-Holland, 1970, 655–663.

[286] **M. Nivat, J.-F. Perrot**, Une Généralisation du monoïde bicyclique, *Comptes Rendus de l'Académie des Sciences de Paris* **271** *(1970), 824–827.*

[287] **L. O'Carroll**, A class of congruences on a posemigroup, *Semigroup Forum* **3** *(1971), 173–179.*

[288] **L. O'Carroll**, A note on free inverse semigroups, *Proceedings of the Edinburgh Mathematical Society (2)* **19** *(1974), 17–23.*

[289] **L. O'Carroll**, Quasi-reduced inverse semigroups, *Journal of the London Mathematical Society* **9** *(1974), 142–150.*

[290] **L. O'Carroll**, Reduced inverse and partially ordered semigroups, *Journal of the London Mathematical Society (2)* **9** *(1974), 293–301.*

[291] **L. O'Carroll**, Inverse semigroups as extensions of semilattices, *Glasgow Mathematics Journal* **16** *(1975), 12–21.*

[292] **L. O'Carroll**, Idempotent determined congruences on inverse semigroups, *Semigroup Forum* **12** *(1976), 233–244.*

[293] **L. O'Carroll**, Embedding theorems for proper inverse semigroups, *Journal of Algebra* **42** *(1976), 26–40.*

[294] **L. O'Carroll**, Strongly E-reflexive inverse semigroups, *Proceedings of the Edinburgh Mathematical Society (2)* **20** *(1977), 339–354.*

[295] **L. O'Carroll**, Strongly E-reflexive inverse semigroups II, *Proceedings of the Edinburgh Mathematical Society (2)* **21** *(1978), 1–10.*

[296] **L. O'Carroll**, A note on strongly E-reflexive inverse semigroups, *Proceedings of the American Mathematical Society* **79** *(1980), 352–354.*

[297] **L. O'Carroll**, On some embedding theorems for inverse semigroups, *Proceedings of the American Mathematical Society* **83** *(1981), 8–10.*

[298] **L. O'Carroll**, Reduced inverse semigroups, *Semigroup Forum* **8** *(1984), 270–276.*

[299] **P. J. Olver**, *Applications of Lie groups to differential equations*, Springer-Verlag, New York, 1986.

[300] **F. Pastijn**, L'enveloppe conjugée d'un demi-groupe inverse à metacentre idempotent, *Canadian Journal of Mathematics* **34** *(1982), 900–909.*

[301] **F. Pastijn**, Inverse semigroups, groupoids and amalgamation, in *Semigroups* (eds K. P. Shum, Y. Q. Guo, M. Ito, and Y. Fong), Springer-Verlag, New York, 1998.

[302] **A. L. T. Paterson**, *Groupoids, inverse semigroups, and their C^*-algebras*, In preparation, Birkhäuser, 1998.

[303] **G. Pazderski**, Induktive Isomorphie von Gruppen deren Kommutatorgruppe Primzahlordnung besitzt, *Archiv der Mathematik* **41** *(1983), 410–418.*

[304] **J. P. Pécuchet**, Automates boustrophedon, semi-groupe de Birget et monoïde inversif libre, *R.A.I.R.O Informatique théorique* **19** *(1985), 71–100.*

[305] **D. Pedoe**, *Geometry*, Dover Publications, Inc., New York, 1988.

[306] **P. Perkins**, Bases for equational theories of semigroups, *Journal of Algebra* **11** *(1969), 298–314.*

[307] **J. F. Perrot**, Codes de Brandt, in *Théorie des codes, actes de la 7ème Ecole de Printemps d'informatique Théorique* (ed. D. Perrin), Jougne, 1979, 177–183.

[308] **M. Petrich**, Congruences on inverse semigroups, *Journal of Algebra* **55** *(1978), 231–256.*

[309] **M. Petrich**, Some categorical equivalences for E-unitary inverse semigroups, *Transactions of the American Mathematical Society* **259** *(1980), 493–503.*

[310] **M. Petrich**, The conjugate hull of an inverse semigroup, *Glasgow Mathematical Journal* **21** *(1980), 103–124.*

[311] **M. Petrich**, Extensions normales de demi-groupes inverses, *Fundamenta Mathematicae* **112** *(1981), 187–203.*

[312] **M. Petrich**, *Inverse semigroups*, John Wiley and Sons, New York, 1984.

[313] **M. Petrich, N. R. Reilly**, A representation of E-unitary inverse semigroups, *Quarterly Journal of Mathematics Oxford (2)* **30** *(1979), 339–350.*

[314] **M. Petrich, N. R. Reilly**, E-unitary covers and varieties of inverse semigroups, *Acta Scientiarum Mathematicarum (Szeged)* **46** *(1983), 59–72.*

[315] **J.-E. Pin**, *Varieties of formal languages*, North Oxford Academic, London, 1986.

[316] **J.-E. Pin**, On the languages accepted by finite reversible automata, in *14th ICALP* Lecture Notes in Computer Science 267, Springer-Verlag, Berlin, 1987, 237–249.

[317] **J.-E. Pin**, A topological approach to a conjecture of Rhodes, *Bulletin of the Australian Mathematical Society* **38** *(1988), 421–431.*

[318] **J.-E. Pin**, On a conjecture of Rhodes, *Semigroup Forum* **39** *(1989), 1–15.*

[319] **J.-E. Pin**, Topologies for the free monoid, *Journal of Algebra* **137** *(1991), 297–337.*

[320] **J.-E. Pin**, On reversible automata, in *Proceedings of the first LATIN Conference, Saõ-Paulo*, Lecture Notes in Computer Science 583, Springer-Verlag, 1992, 401–416.

[321] **J.-E. Pin**, Polynomial closure of group languages and open sets of the Hall topology, *Theoretical Computer Science* **169** *(1996), 185–200.*

[322] **J.-E. Pin, C. Reutenauer**, A conjecture on the Hall topology for the free group, *Bulletin of the London Mathematical Society* **23** *(1991), 356–362.*

[323] **J. F. Plante**, Measure preserving pseudogroups and a theorem of Sacksteder, *Annales de l'Institut de Fourier, Grenoble* **25** *(1975), 237–249.*

[324] **J. F. Pommaret**, *Systems of partial differential equations and Lie pseudogroups*, Gordon & Breach, New York, 1978.

[325] **G. B. Preston**, Inverse semi-groups, *Journal of the London Mathematical Society* **29** *(1954), 396–403.*

[326] **G. B. Preston**, Inverse semi-groups with minimal right ideals, *Journal of the London Mathematical Society* **29** *(1954), 404–411.*

[327] **G. B. Preston**, Representations of inverse semi-groups, *Journal of the London Mathematical Society* **29** *(1954), 411–419.*

[328] **G. B. Preston**, The structure of normal inverse semigroups, *Proceedings of the Glasgow Mathematical Association* **3** *(1956), 1–9.*

[329] **G. B. Preston**, Congruences on Brandt semigroups, *Mathematische Annalen* **139** *(1959), 91–94.*

[330] **G. B. Preston**, Free inverse semigroups, *Journal of the Australian Mathematical Society* **16** *(1973), 443–453.*

[331] **G. B. Preston**, Monogenic inverse semigroups, *Journal of the Australian Mathematical Society (Series A)* **40** *(1986), 321–342.*

[332] **G. B. Preston**, The semidirect product of an inverse semigroup and a group, *Bulletin of the Australian Mathematical Society* **33** *(1986), 261–272.*

[333] **P. Prusinkiewicz, A. Lindenmayer**, *The algorithmic beauty of plants*, Springer-Verlag, New York, 1990.

[334] **A. R. Rajan**, A structure theorem for combinatorial pseudo-inverse semigroups, in *Colloquia Mathematica Societatis János Bolyai* **39** *Semigroups, Szeged (Hungary), 1981, 325–346.*

[335] **D. Rees**, On the group of a set of partial transformations, *Journal of the London Mathematical Society* **22** *(1947) 281–284.*

[336] **D. Rees**, On the ideal structure of a semi-group satisfying a cancellation law, *Quarterly Journal of Mathematics (Oxford)* **19** *(1948), 101–108.*

[337] **N. R. Reilly**, Embedding inverse semigroups in bisimple inverse semigroups, *Quarterly Journal of Mathematics (Oxford) (2)* **16** *(1965), 183–187.*

[338] **N. R. Reilly**, Bisimple ω-semigroups, *Proceedings of the Glasgow Mathematical Association* **7** *(1966), 160–167.*

[339] **N. R. Reilly**, Bisimple inverse semigroups, *Transactions of the American Mathematical Society* **132** *(1968), 101–114.*

[340] **N. R. Reilly**, Congruences on bisimple inverse semigroups in terms of RP-systems, *Proceedings of the London Mathematical Society (3)* **23** *(1971), 99–127.*

[341] **N. R. Reilly**, Free generators in free inverse semigroups, *Bulletin of the Australian Mathematical Society* **7** *(1972), 407–424.*

[342] **N. R. Reilly**, Semigroups of order preserving partial transformations of a totally ordered set, *Journal of Algebra* **22** *(1972), 409–427.*

[343] **N. R. Reilly**, Free inverse semigroups, in *Colloquia Mathematica Societatis János Bolyai* **20** *Algebraic theory of semigroups, Szeged (Hungary), 1976, 479–508.*

[344] **N. R. Reilly**, Enlarging the Munn representation of inverse semigroups, *Journal of the Australian Mathematical Society (Series A)* **23** *(1977), 28–41.*

[345] **N. R. Reilly, A. H. Clifford**, Bisimple inverse semigroups as semigroups of ordered triples, *Canadian Journal of Mathematics* **20** *(1968), 25–39.*

[346] **N. R. Reilly, W. D. Munn**, E-unitary congruences on inverse semigroups, *Glasgow Mathematical Journal* **17** *(1976), 57–75.*

[347] **N. R. Reilly, H. E. Scheiblich**, Congruences on regular semigroups, *Pacific Journal of Mathematics* **23** *(1967), 349–360.*

[348] **N. R. Reilly, P. G. Trotter**, Properties of relatively free inverse semigroups, *Transactions of the American Mathematical Society* **294** *(1986), 243–262.*

[349] **B. L. Reinhart**, *Differential geometry of foliations*, Springer-Verlag, Berlin, 1983.

[350] **J. Renault**, *A groupoid approach to C^*-algebras*, Lecture Notes in Mathematics 793, Springer-Verlag, Berlin, 1980.

[351] **C. Reutenauer**, Une topologie du monoïde libre, *Semigroup Forum* **18** *(1979), 33–49.*

[352] **C. Reutenauer**, Sur mon article 'Une topologie du monoïde libre', *Semigroup Forum* **22** *(1981), 93–95.*

[353] **M. A. Reynolds**, A new construction of free inverse semigroups, *Semigroup Forum* **30** *(1984), 291–296.*

[354] **J. Rhodes**, Survey of global semigroup theory, in *Lattices, semigroups and universal algebra*, (eds J. Almeida et al), Plenum Press, New York, 1990.

[355] **L. Ribes, P. A. Zalesskiĭ**, On the profinite topology on a free group, *Bulletin of the London Mathematical Society* **25** *(1993), 37–43.*

[356] **F. Rimlinger**, R-trees and normalization of pseudogroups, *Experimental Mathematics* **1** *(1992), 95–114.*

[357] **W. Rinow**, Über die Vervollständigung induktiver Gruppoide, *Mathematische Nachrichten* **25** *(1963), 199–222.*

[358] **W. Rinow**, Vervollständigung geordneter Kategorien, *Mathematische Nachrichten* **33** *(1966), 129–175.*

[359] **E. P. Robinson, G. Rosolini**, Categories of partial maps, *Information and Computation* **79** *(1988), 95–130.*

[360] **J. Roe**, *Elementary geometry*, Oxford University Press, Oxford, 1993.

[361] **J. S. Rose**, *A course on group theory*, Cambridge University Press, Cambridge, 1978.

[362] **R. Sacksteder**, Foliations and pseudogroups, *American Journal of Mathematics* **87** *(1965), 79–102.*

[363] **T. Saîto**, Proper ordered inverse semigroups, *Pacific Journal of Mathematics* **15** *(1965), 649–666.*

[364] **J. Sakarovitch**, *Syntaxe des langages de Chomsky: Essai sur le déterminisme*, Thése de Doctorat d'Etat, Université Paris VII, 1979.

[365] **H. E. Scheiblich**, A characterisation of free elementary inverse semigroups, *Semigroup Forum* **2** *(1971), 76–79.*

[366] **H. E. Scheiblich**, Free inverse semigroups, *Semigroup Forum* **4** *(1972), 351–359.*

[367] **H. E. Scheiblich**, Concerning congruences on symmetric inverse semigroups, *Czechoslovak Mathematical Journal* **23** *(1973), 1–10.*

[368] **H. E. Scheiblich**, Free inverse semigroups, *Proceedings of the American Mathematical Society* **38** *(1973), 1–7.*

[369] **H. E. Scheiblich**, Kernels of inverse semigroup homomorphisms, *Journal of the Australian Mathematical Society* **18** *(1974), 289–292.*

[370] **B. M. Schein**, A system of axioms for semigroups embeddable in groups, *Soviet Mathematics Doklady* **1** *(1961), 1180–1183.*

[371] **B. M. Schein**, Representations of generalized groups, *Izvestiya Vysshikh Uchebnykh Zavedeniĭ Matematika* **3** *(1962), 164–176* (Russian).

[372] **B. M. Schein**, On the theory of generalized groups, *Doklady Akademiĭ Nauk SSSR* **153** *(1963), 296–299* (Russian).

[373] **B. M. Schein**, On the theory of generalized groups and generalized heaps, in *The theory of semigroups and its applications I*, University of Saratov, Saratov, 1965, 286–324 (Russian).

[374] **B. M. Schein**, Semigroups of strong subsets, *Volzhskiĭ Matematicheskiĭ Sbornik* **4** *(1966), 180–186* (Russian).

[375] **B. M. Schein**, A remark concerning congruences on (0-)bisimple inverse semigroups, *Semigroup Forum* **3** *(1971), 80–83.*

[376] **B. M. Schein**, Completions, translational hulls, and ideal extensions of inverse semigroups, *Czechoslovak Journal of Mathematics* **23** *(1973), 575–610.*

[377] **B. M. Schein**, Injectives in certain classes of semigroups, *Semigroup Forum* **9** *(1974), 159–171.*

[378] **B. M. Schein**, A new proof of the McAlister P-Theorem, *Semigroup Forum* **10** *(1975), 185–188.*

[379] **B. M. Schein**, Free inverse semigroups are not finitely presentable, *Acta Mathematica Academiae Scientiarum Hungaricae* **26** *(1975), 41–52.*

[380] **B. M. Schein**, Injective monars over inverse semigroups, in *Colloquia Mathematica Societatis János Bolyai* **20**, *Algebraic theory of semigroups, Szeged (Hungary), 1976, 519–544.*

[381] **B. M. Schein**, On the theory of inverse semigroups and generalized grouds, *American Mathematical Society Translations (2)* **113** *(1979), 89–122.*

[382] **B. M. Schein**, V. V. Vagner 1908-1981, *Semigroup Forum* **23** *(1981), 189–200.*

[383] **B. M. Schein**, Representation of inverse semigroups by local automorphisms and multi-automorphisms of groups and rings, *Semigroup Forum* **32** *(1985), 55–60.*

[384] **B. M. Schein**, Prehistory of the theory of inverse semigroups, *Proceedings of the 1986 Semigroup Conference Baton Rouge, Louisiana State University*, (1986), 72–76.

[385] **B. M. Schein**, Multigroups, *Journal of Algebra* **111** *(1987), 114–132.*

[386] **B. M. Schein**, Embedding semigroups in inverse semigroups, *American Mathematical Society Translations* **139** *(1988), 93–116.*

[387] **B. M. Schein**, Cosets in groups and semigroups, in *Semigroups with applications*, (eds J. M. Howie, W. D. Munn, and H. J. Weinert), World-Scientific, Singapore, 1992, 205–221.

[388] **B. M. Schein**, Subsemigroups of inverse semigroups, *Le Matematiche* **LI** *(1996), 205–227.*

[389] **E. T. Schmidt**, Universale Algebren mit gegebenen Automorphismengruppen und Unteralgebrenverbänden, *Acta Scientiarum Mathematicarum (Szeged)* **24** *(1963), 251–254.*

[390] **B. Schultz–Heinecke**, Zur Vollständigkeit der induktiven Gruppoide der partiellen Automorphismen von Algebren, *Periodica Mathematica Hungarica* **20** *(1989), 115–145.*

[391] **B. Schweizer, A. Sklar**, Function systems, *Mathematische Annalen* **172** *(1967), 1–16.*

[392] **M. Senechal**, *Quasicrystals and geometry*, Cambridge University Press, Cambridge, 1996.

[393] **K. Shoji**, Completions and injective hulls of E-reflexive semigroups, *Semigroup Forum* **36** *(1987), 55–68.*

[394] **P. V. Silva**, Rational languages and inverse monoid presentations, *International Journal of Algebra and Computation* **2** *(1992), 187–207.*

[395] **P. V. Silva**, Normal-convex embeddings of inverse semigroups, *Glasgow Mathematical Journal* **35** *(1993), 115–121.*

[396] **J. R. Stallings**, Topology of finite graphs, *Inventiones Mathematicae* **71** *(1983), 551–565.*

[397] **J. B. Stephen**, Presentations of inverse monoids, *Journal of Pure and Applied Algebra* **63** *(1990), 81–112.*

[398] **J. B. Stephen**, The word problem for inverse monoids and related questions, in *Monoids and semigroups with applications* (ed. J. Rhodes), World Scientific, Singapore, 1991.

[399] **I. Stewart, M. Golubitsky**, *Fearful symmetry*, Penguin Books, London, 1993.

[400] **H. Straubing, P. Weil**, On a conjecture concerning dot-depth two languages, *Theoretical Computer Science* **104** *(1992), 161–183.*

[401] **R. Strecker**, Induktive Isomorphie unendlicher abelscher Gruppen, *Monatsberichte der Deutschen Akademie der Wissenschaften zu Berlin* **13** *(1971), 81–86.*

[402] **M. Suzuki**, *Group theory I*, Springer-Verlag, Berlin, 1982.

[403] **L. Szabó**, Characterization of some related semigroups of universal algebras, *Acta Scientiarum Mathematicarum (Szeged)* **37** *(1975), 143–147.*

[404] **M. B. Szendrei**, A generalization of McAlister's P-theorem for E-unitary regular semigroups, *Acta Scientiarum Mathematicarum (Szeged) 57 (1987), 229–249.*

[405] **M. B. Szendrei**, E-unitary regular semigroups, *Proceedings of the Royal Society of Edinburgh* **106A** *(1987), 89–102.*

[406] **M. B. Szendrei**, A note on Birget-Rhodes expansion of groups, *Journal of Pure and Applied Algebra* **58** *(1989), 93–99.*

[407] **M. B. Szendrei**, Regular semigroups and semidirect products, in *Semigroups, automata and languages* (eds J. Almeida, G. M. S. Gomes and P. V. Silva), World Scientific, Singapore, 1994, 233–246.

[408] **S. Talwar**, Morita equivalence for semigroups, *Journal of the Australian Mathematical Society (Series A)* **59** *(1995), 81–111.*

[409] **B. R. Tennison**, *Sheaf theory*, Cambridge University Press, Cambridge, 1975.

[410] **R. Thom**, On the origin and stability of symmetries, in *Symmetrie*, (ed. R. Wille), Springer-Verlag, Berlin, 1988, 73–78.

[411] **R. Thom**, *Structural stability and morphogenesis*, Addison-Wesley, Redwood City, 1989.

[412] **B. Tilson**, Categories as algebra: an essential ingredient in the theory of monoids, *Journal of Pure and Applied Algebra* **48** *(1987), 83–198.*

[413] **B. Tilson**, Monoids kernels, semidirect products and their adjoint relationship, in *Monoids and semigroups with applications* (ed. J. Rhodes), World Scientific, Singapore, 1991, 31–54.

[414] **P. G. Trotter**, Projective inverse semigroups, *Semigroup Forum* **26** *(1983), 167–190.*

[415] **P. G. Trotter**, Relatively free inverse semigroups, *Quarterly Journal of Mathematics Oxford (2)* **37** *(1986), 357–374.*

[416] **T. Uesugi**, A note on the topological groupoid associated with a pseudogroup of transformations, *Bull. of Kyushu Inst. of Technology No. 12 (1965), 35–40.*

[417] **T. Uesugi**, On the relations between the groupoid and the pseudogroup of transformations, *Bull. Kyushu Inst. of Technology No. 12 (1965), 41–55.*

[418] **O. Veblen, J. H. C. Whitehead**, *The foundations of differential geometry*, Cambridge Tract No. 24, Cambridge University Press, Cambridge, 1932.

[419] **V. V. Wagner**, Generalized groups, *Doklady Akademiĭ Nauk SSSR* **84** *(1952), 1119–1122* (Russian).

[420] **V. V. Wagner**, The theory of generalised heaps and generalised groups, *Matematicheskiĭ Sbornik (N.S.)* **32** *(1953), 545–632* (Russian).

[421] **V. V. Wagner**, Generalised heaps and generalised groups with a transitive relation of compatibility, *Uchenye Zapiski Saratov Gos. Univ. Ser. Mekh-Mat.* **70** *(1961), 25–39* (Russian).

[422] **V. V. Wagner**, Algebraic topics of the general theory of partial connections in fiber bundles, *Izvestiya Vysshikh Uchebnykh Zavedeniĭ Matematika* issue **11** *(1968), 26–32* (Russian).

[423] **V. V. Wagner**, On the theory of antigroups, *Izvestiya Vysshikh Uchebnykh Zavedeniĭ Matematika* issue **4** *(1971), 3–15* (Russian).

[424] **V. V. Wagner**, t-simple representations of antigroups, *Izvestiya Vysshikh Uchebnykh Zavedeniĭ Matematika* issue **9** *(1971), 18–29* (Russian).

[425] **V. V. Wagner**, On the theory of involuted semigroups, *Izvestiya Vysshikh Uchebnykh Zavedeniĭ Matematika* issue **10** *(1971), 24–35* (Russian).

[426] **N. E. Wegge-Olsen**, *K-Theory and C^*-algebras*, Oxford University Press, Oxford, 1994.

[427] **P. Weil**, *Inverse monoids and the dot-depth hierarchy*, Ph.D Thesis, University of Lincoln-Nebraska, 1988.

[428] **P. Weil**, Inverse monoids of dot-depth two, *Theoretical Computer Science* **66** *(1989), 233–245.*

[429] **P. Weil**, Some results on the dot-depth hierarchy, *Semigroup Forum* **46** *(1993), 352–370.*

[430] **A. Weinstein**, Affine Poisson structures, *International Journal of Mathematics* **1** *(1990), 343–360.*

[431] **A. Weinstein**, Groupoids: unifying internal and external symmetry, *Notices of the American Mathematical Society* **43** *(1996), 744–752.*

[432] **H. Weyl**, *Symmetry*, Princeton University Press, 1969.

[433] **J. H. C. Whitehead**, *Mathematical works of J. H. C. Whitehead*, (ed. I. M. James), Pergamon Press, Oxford, 1962.

[434] **R. Wilkinson**, A description of E-unitary inverse semigroups, *Proceedings of the Royal Society of Edinburgh* **95A** *(1983), 239–242.*

[435] **H. Wussing**, *The genesis of the abstract group concept*, MIT Press, Cambridge Massachusetts, 1984.

[436] **I. M. Yaglom**, *Felix Klein and Sophus Lie*, Birkhäuser, Boston, 1988.

[437] **A. Yamamura**, HNN extensions of inverse semigroups and applications, *International Journal of Algebra and Computation* **7** *(1997), 605–624.*

[438] **G. I. Zhitomirskiĭ**, On imprimitivity relations of generalized groups of transformations, *Matematicheskiĭ Sbornik (N.S.)* **73** *(1967), 500–512* (Russian).

[439] **G. I. Zhitomirskiĭ**, Generalized heaps and generalized groups with modular lattice of congruence relations, *Izvestiya Vysshikh Uchebnykh Zavedeniĭ Matematika issue* **6** *(1972), 26–35* (Russian).

[440] **G. I. Zhitomirskiĭ**, Inverse semigroups and fiberings, in *Semigroups—algebraic theory and applications*, (eds C. Bonzini, A. Cherubini and C. Tibiletti), World Scientific, Singapore, 1993, 311–321.

[441] **G. I. Zhitomirskiĭ**, Restrictive semigroups and sheaves, in *Semigroups, automata and languages* (eds J. Almeida, G. Gomes and P. Silva), World Scientific, Singapore, 1996, 289–299.

Index

action
  by category, 332
  by endomorphisms, 148
  by group, 198
  by monoid, 37
  by semigroup, 37
algebraic lattice, 44
alphabet, 50
atlas, 10, 41, 43
  chart in, 10, 252
  compatible with pseudogroup, 13
  complete, 14, 252
  partial, 10
  transition function of, 11
automaton, 51
  flower, 53
  folding operation on, 54
  group, 52
  injective, 52
  inverse, 52
  language recognised by, 52
  linear inverse, 53
  prolongation of, 52
  pushdown, 328
  reversible, 72
  transition monoid of, 52
  two-way, 72

Bass–Serre theory, 196
bi-congruence, 124, 132
bi-equivalence, 123, 132
Billhardt
  congruence, 151, 202, 230
  split congruence, 156, 165
  transversal, 151
Birget–Rhodes expansion, 230

C^*-algebra, 42, 119, 132, 318
  Cuntz, 328
  Cuntz–Krieger, 361
Cantor set, 17, 285
categorical at zero, 275
category, 78
  action, 332
  comma, 252
  derived, 167
  endomorphism monoid in, 78
  functor between, 79
  hom-set in, 78
  homomorphism in, 78
  identity in, 78
  inverse, 6, 64, 70, 73
  morphism in, 78
  of partial functions, 40
  star in, 79
  symmetric monoidal, 297
  weak initial object in, 364
category action, 332
  C-homomorphism, 333
  C-isomorphism, 333
  cyclic, 333
  morphism of, 333
  orbit condition on, 333
  strong orbit condition on, 335
  system, 342

with right cancellation, 342
centraliser of the idempotents, 139
chain of groups, 147, 170
chart, 10, 252
closed subset (order), 34
closure of a subset (order), 34
compatibility relation, 12, 24, 43, 65
left, 24
right, 24
strong, 274, 327
compatible subset, 26
congruence, 58
Billhardt, 151, 230
class, 58
generated by a relation, 59
idempotent pure, 65, 137, 230, 277
idpt.-separating, 90, 104, 137, 139
Kernel of, 134
lattice, 59, 73
left, 58
max. idpt.-separating, 140
minimum group, 63, 73
normal, 134
ordered, 126
pair, 135
perfect, 59
permutable, 124
preserves maximal group homomorphic images, 228
primitive, 276
proper, 58
right, 58
split Billhardt, 156, 165
syntactic, 61
trace of, 134
0-restricted, 276
congruence extension theorem, 241
conjugation, 18
converse relation, 60, 122
coset, 34, 43, 121
cover
E-unitary, 58, 72, 245, 246, 271
E-unitary over a group, 67
F-inverse, 226, 230, 232, 248
of an inverse semigroup, 31
crystallography, 42
post-classical, 15
quasi-crystal, 15
cycle-free n-partitioned graph, 55

Dedekind infinite set, 97
derived covering, 231
di-functional relation, 122
diffeomorphism, 3
differential manifold, 14
digraph, 76
division, 31
division thm for inv. semigrps, 203

Ehr.–Schein–Nam. thm, 114
elementary transition, 60
enlargement, 271
almost factorisable, 248
choice function for, 241
congruence extension theorem, 241
of inverse semigroup, 224
of ordered groupoid, 234
Erlanger Programm, 2, 15, 40

factorisable embedding, 246
strict, 66
factorisable envelope, 247, 248
first isomorphism theorem
semigroups, 59
special ordered functors, 128
foliation, 41
topological, 43, 231
fractal, 42
Cantor set, 17, 285
Sierpiński gasket, 17
frame, 45

free group, 51, 175
    Cayley graph of, 187, 323
    closed rational subset of, 71
    incomparable strings in, 178
    left-system in, 55
    prefix order in, 178
    profinite topology on, 53
    proper prefix in, 178
    rational subgroup of, 70
    recognisable subgroup of, 70
    reduced string in, 51
free product
    of inverse semigroups, 195
    with amalgamation, 195
functor, 82
    covering, 79
    inductive, 108
    ordered, 108
    special ordered, 125
    star bijective, 79
    star injective, 79
    star surjective, 79
    transportable, 250

geometry of interaction, ix, 297
Girard monoid, 297
graph, 187
    connected, 187
    orientation of, 187
    tree, 187
Green's relations, 82
    preservation of, 90
Grothendieck construction, 231
group
    countable abelian, 49
    extension, 137
    Galois, 48
    Möbius, 217
    of germs, 64
    residually finite, 52
    split extension, 153
group coset semigroup, 36, 43, 122
groupoid, 8, 40, 78
    adjoining a zero to, 95
    associated, 79, 84
    connected, 78, 96, 105
    connected component of, 78
    coverings of, 70
    factorisable ordered, 243
    inductive, 108, 195
    Kellendonk's topological, 284
    of germs, 64
    ordered, 108
    topological, 42, 64, 73, 282, 327

$\mathcal{H}$-coextension, 132
HNN-property, 49
homomorphism
    factorisation theorem, 228
    full, 204
    idpt.-separating, 57
    involution, 30
    join-preserving, 31
    kernel of, 58
    $\mathcal{L}$-bijective, 200
    $\mathcal{L}$-injective, 201
    meet-preserving, 31
    natural, 59
    of monoids, 31
    of semigroups with zero, 31

ideal, 20
    extension, 60
    generator of, 20
    left, 20
    principal, 20
    principal left, 20
    principal right, 20
    retract, 43
    right, 20
idempotent, 6
    primitive, 93
inverse monoid, 361
    bicyclic, 98, 103, 105, 160, 162, 328, 329, 359

bisimple, 40
dual symmetric, 124
F-, 202, 218
factorisable, 56, 72, 122, 204
free monogenic, 324, 329
Galois, 48
Möbius, 217, 221
polycyclic, 286, 293, 328, 360
symmetric, 6

inverse semigroup, 6
almost factorisable, 203, 205, 231
antigroup, 168
arrow diagram of, 76
bisimple, 85
Brandt, 40, 86, 94, 277
Bruck–Reilly, 170
Clifford, 40, 139, 145, 168
cohomology, 167, 231, 272
combinatorial, 83, 104, 171, 186, 313
complete, 27
completely semisimple, 92, 103, 171, 186, 313, 321
congruence-free, 290
E^*-unitary, 327
E-reflexive, 43, 86, 87, 104, 222
E-unitary, viii, 31, 40, 57, 65, 73, 170, 171, 186, 199, 259
F-, 222, 225, 231
free, 171
free monogenic, 193
fundamental, 140, 142, 168
generalised polycyclic, 361
group coset, 36, 43, 122
HNN-extension of, 196
infinitely distributive, 28
Kachel, 324
McAlister, 304, 328
meet complete, 27, 202
monogenic, 174
Munn, 141, 180
ω-, 40, 100, 160, 163, 164, 169
P-, 213
presentation, 174
primitive, 94, 95
Reilly, 164
restricted product in, 76
self-injective, 43
strongly E-reflexive, 272
tiling, 321, 329
topologically complete, 143
with zero, 20
0-bisimple, 85, 104, 364
0-E-unitary, 273, 369

inverse subsemigroup, 7
full, 19
generated by a set, 19
lattice, 73
normal, 134
self-conjugate, 134

K-theoretic gap-labelling, 318
Kachel-semigroup, 324
Kaloujnine–Krasner theorem, 151
kernel-normal system, 167

L-semigroup, 269
Lallement's lemma, 30
λ-calculus, 103, 328
language, 51, 168
context-free, 328
correct bracketing, 72, 103, 328
star-free, 72

left
C-system, 332
cyclic C-system, 333
cyclic S-subsystem, 38
S-homomorphism, 37
S-isomorphism, 38
S-subsystem, 38
S-system, 37

linear logic, 19, 43, 297, 328, 370
local homeomorphism, 14

local structure, 14, 41, 73, 249, 251
local submonoid, 83

matrix ring, 293
maximum enlargement thm, 256
McAlister covering theorem, viii, 57, 73, 245
McAlister semigroup, 304, 309
McAlister triple, 212, 239, 260
McAlister–O'Carroll triple, 265
model theory, 48
  elementary equivalence, 69
  finitely isomorphic structures, 69
  homogeneous structure, 48
  small index property, 49
Möbius transformation, 216
monoid, 6
  action of, 37
  cancellative, 68
  free, 51
  free with involution, 51
  Girard, 297
  homomorphism, 31
  left reversible, 68
  presentation, 62
  syntactic, 61
monus, 98
Munn representation theorem, 141, 142
Munn tree, 189

natural partial order, 8, 23, 43
  homomorphism reflects, 31
normal extension, 138
  equivalent, 138
  idempotent pure, 138, 270
  idempotent-separating, 138
  of a semilattice, 138
  problem, 138, 167
  triple, 138
normal-convex embedding, 230

order automorphism, 198
order isomorphism, 30
order-preserving function, 30
ordered congruence
  extension theorem, 241
  identity-separating, 240
ordered functor, 108
  covering, 108
  first isomorphism theorem for, 128
  inductive, 108
  ordered embedding, 111
  pseudoproduct preserving, 115
  special, 125
  star bijective, 108
  star injective, 108, 115
  star surjective, 108
ordered groupoid, viii, 7, 45, 108, 130
  corestriction in, 108
  enlargement of, 234
  inductive, 108, 131
  ordered congruence on, 126
  ordered subgroupoid of, 111
  pseudoproduct in, 112, 130
  restriction in, 108
ordered representation, 262
Ore embedding theorem, 68, 69
orthogonality relation, 26

P-semigroup, 195, 213, 259
P-theorem, 215, 231
  coordinate-free, 239
  for free inv. semigrps, 185
partial automorphism, 17, 44
partial bijection, 4
  compatible, 12
  extending, 49, 209
  inverse of, 4
partial equivalence relation, 35
partial function, 4
  composition of, 4

corestriction of, 9
domain of, 4
empty, 4
full product of, 8
image of, 4
partial identity, 4
restricted product of, 8
restriction of, 9
partial isometry, 49, 116, 132
special, 120
partial right translation, 38
permissible subset, 31
poset, 22
directed, 275
greatest lower bound in, 23
lower bound in, 23
order ideal in, 22
principal order ideal in, 22
prehomomorphism, 80
dual, 80
idempotent pure, 91, 115, 327
$\mathcal{L}$-bijective, 91, 122
$\mathcal{L}$-injective, 91
$\mathcal{L}$-surjective, 91
$\mathcal{R}$-bijective, 91
$\mathcal{R}$-injective, 91
$\mathcal{R}$-surjective, 91
presheaf of groups, 144, 168
projection, 116
halving, 328
pseudogroup, 3, 7, 11, 37, 40, 63, 73
complete, 12, 249
Lie, 3, 40

quasi-crystal, 15, 318

$\mathcal{R}^*$-relation, 336, 370
Ramsey's theorem, 71
Rees quotient semigroup, 60
regular semigroup, 6
representation
by partial bijections, 31
effective, 35
faithful, 31
transitive, 35
restricted product, 76
river
Euphrates, 371
Thames, v

Scheiblich normal form, 177
self-similarity, 15, 292, 294
semidirect product
classical, 148
λ-, 150, 169
λ-wreath, 151
of a poset by a group, 236
of a semilattice by a group, 199, 230
semigroup, 6
action of, 37
adjoining a zero to, 20
adjoining an identity to, 20
amalgam, 195
completely 0-simple, 94
congruence-free, 89
finitely generated, 61
finitely presented, 61
free, 51
free with involution, 172
of matrix units, 86
of right S-isomorphisms, 38
partially ordered, 23
presentation, 61
quotient, 59
Rees quotient, 60
simple, 88
$*$-, 116
subgroup in, 83
with commuting idempotents, 70
with involution, 18
0-simple, 88

semigroupoid, 107
semilattice
  meet, 23
  of idempotents, 23
  of the free inverse semigroup, 182
  uniform, 170
sheaf, 168
Sierpiński gasket, 17
solid-state physics, 318
strict factorisable embedding, 66
string, 50
  cancellation operation, 287
  concatenation of, 50
  factor of, 50
  prefix of, 50
  suffix of, 50
strong amalgamation property, 195
strong semilattice of groups, 145
structure mapping, 131
symmetric monoidal category, 297
symmetry, 15, 16
  global, 17
  partial, 17, 292, 331
system, 342
  cyclic, 361
  description of inv. semigrps, 353
  equivalent, 350
  generalised RP-, 369
  monoid, 367
  morphism, 345
  RP-, 169, 369
  with left cancellation, 368
system morphism, 345
  transformation between, 347

tiling, 318
  Penrose, 318
  self-similar, 170
  semigroup, 321, 329
Todd–Coxeter algorithm, 56
topological conjecture, 71
topological space
  basis of, 53, 143
  coarser topology on, 53
  discrete, 53
  hausdorff, 168
  product of, 282
  subbasis of, 282
  subspace of, 282
  T_0, 143
  T_1, 168
  T_2, 168
translational hull, 43
type II conjecture, 50, 54, 70

unitary subset, 64
  left, 64
  right, 64

variety
  of languages, 71
  of semigroups, 71

Wagner–Preston theorem, viii, 36, 331
  for ordered groupoids, 115
  refinement of, 39
Witt property, 49
Witt's lemma, 49
word, 50
word problem, 61

0-direct union of semigroups, 96